DNA-BASED MARKERS IN PLANTS

Advances in Cellular and Molecular Biology of Plants

VOLUME 6

The titles published in this series are listed at the end of this volume.

DNA-Based Markers in Plants

2nd Edition

Edited by

Ronald L. Phillips
Department of Agronomy and Plant Genetics,
University of Minnesota,
St. Paul, Minnesota, U.S.A.

and

Indra K. Vasil
Laboratory of Plant Cell and Molecular Biology,
University of Florida,
Gainesville, Florida, U.S.A.

KLUWER ACADEMIC PUBLISHERS
DORDRECHT / BOSTON / LONDON

Library of Congress Cataloging-in-Publication data

DNA-based markers in plants / edited by Ronald L. Phillips and Indra K. Vasil.-- 2nd ed.
   p. cm. -- (Advances in cellular and molecular biology of plants ; v. 6)
   ISBN 0-7923-6865-7 (hb : alk. paper)
   1. Plant genome mapping. 2. Genetic polymorphisms. 3. Plant molecular genetics. I.
Phillips, Ronald L. II. Vasil, I. K. III. Series.

QK981.45 .D53 2001
572.8'62--dc21
                                                                                    2001029075
ISBN 0-7923-6865-7 (2nd edition)
ISBN 0-7923-2714-4 (1st edition)

Published by Kluwer Academic Publishers,
P.O. Box 17, 3300 AA Dordrecht, The Netherlands.

Sold and distributed in North, Central and South America
by Kluwer Academic Publishers,
101 Philip Drive, Norwell, MA 02061, U.S.A.

In all other countries, sold and distributed
by Kluwer Academic Publishers,
P.O. Box 322, 3300 AH Dordrecht, The Netherlands.

Printed on acid-free paper

First published 1994
Reprinted   1998
Second edition 2001

Printed in the Netherlands.

Contents

---

* *Reprinted without change from the first edition.*

General preface

The double helix architecture of DNA was elucidated in 1953. Twenty years later, in 1973, the discovery of restriction enzymes helped to create recombinant DNA molecules in vitro. The implications of these powerful and novel methods of molecular biology, and their potential in the genetic manipulation and improvement of microbes, plants and animals, became increasingly evident, and led to the birth of modern biotechnology. The first transgenic plants in which a bacterial gene had been stably integrated were produced in 1983, and by 1993 transgenic plants had been produced in all major crop species, including the cereals and the legumes. These remarkable achievements have resulted in the production of crops that are resistant to potent but environmentally safe herbicides, or to viral pathogens and insect pests. In other instances genes have been introduced that delay fruit ripening, or increase starch content, or cause male sterility. Most of these manipulations are based on the introduction of a single gene — generally of bacterial origin — that regulates an important monogenic trait, into the crop of choice. Many of the engineered crops are now under field trials and are expected to be commercially produced within the next few years.

The early successes in plant biotechnology led to the realization that further molecular improvement of plants will require a thorough understanding of the molecular basis of plant development, and the identification and characterization of genes that regulate agronomically important multigenic traits. During the past ten years there has been a resurgence of molecular and related cellular studies in plants, including the molecular mapping of plant genomes. A great deal of interesting and useful information has been generated about the molecular basis of important plant processes. This series of volumes is intended to chronicle the most important advances in the cellular and molecular biology of plants, and to stimulate further interest and research in the plant sciences. The success and usefulness of these volumes depends on the timeliness of the subjects discussed, and the authoritative and insightful accounts provided by distinguished and internationally respected contributing authors. In this, I have been greatly aided by the advice of members of our Editorial Advisory Board and the editors of individual volumes, to whom I owe a debt of gratitude. I also thank Dr Ad. C. Plaizier of Kluwer Academic Publishers in helping me to launch this series, and his competent and helpful staff in the preparation of the volumes for publication. The various volumes already in press and in preparation have provided me with the opportunity to know and work with many colleagues, and have helped me to improve my own understanding and appreciation of plant molecular biology.

Indra K. Vasil

Preface

As with numerous other technological advances, many insights have emerged from application of the new tools of molecular biology. The concept of DNA-based markers has revolutionized our ability to follow chromosome segments, including minute regions, and has led to new opportunities such as map-based cloning and directed plant breeding. Species with little genetic information available in the past now have hundreds of genetic markers. In some cases, the map from one species can be transferred almost directly to another species, such as from tomato to potato. Additional genomic structure information, such as homoeologous chromosome identification, is forthcoming for several species. The suggestion of ancient polyploidy has been made even for maize, the most genetically studied plant species. The ancestral relationships of species and the pedigree relatedness of lines have been identified in hundreds of situations.

This volume records the recently developed linkage maps of many important crop species. This recording, as impressive as it appears, will perhaps serve as a historical log of the development of the first molecular genetic marker maps for several species. The initial chapters provide aspects of the theory, technology, and applications of DNA-based markers. This book is intended to provide a solid base upon which students and researchers can build their understanding of genetic linkage and the applications of that knowledge. It also provides a statistical appreciation of DNA-based markers for the genetic dissection of qualitative and quantitative traits. Some of the newest ideas and concepts are expressed here for the first time. The locations of QTLs (Quantitative Trait Loci) for the important agronomic and quality traits, as well as those that differentiate species, represent extremely useful knowledge. Genomic markers for regions with major effects for such traits as disease resistance already are being used for gene cloning and breeding purposes.

We recognize that genetic maps are a dynamic entity and will continue to expand and evolve over the years. At the same time, genetic map-based information is rigorous by its nature, and provides a set of basic tenets that should not change with future advancements. Thus, an understanding of the numerous basic tenets of mapping carefully described in these pages will establish a framework for future enhancement of the theory and applications of this important and useful technology.

Clearly, our ability to genetically manipulate more than just a few plant species has increased several orders of magnitude with the advent of DNA-based markers. The future will be exciting.

We gratefully acknowledge the diligent efforts of each of the contributors to this volume for providing a clear statement of a difficult and still evolving subject. They represent pioneers in the field. Their interest and cooperation in making this volume a reality reflects a sincere and shared belief in the broad application of DNA-markers in plants.

Ronald L. Phillips
Indra K. Vasil

1. Some concepts and new methods for molecular * mapping in plants

BENJAMIN BURR

Biology Department, Brookhaven National Laboratory, Upton, NY 11973, U.S.A.
<burr@bnl.gov>

Contents

1.   Introduction

The use of molecular markers is based on naturally occurring polymorphism whose magnitude has only recently been appreciated. Understanding the nature of naturally occurring polymorphism is a basis for designing strategies to exploit it for applied genetic practices. The purpose of this chapter is to express a personal view on a few key concepts and to draw attention to new technology.

One of the unheralded revolutions in biological thinking in the 1960's was the realization of the enormous amount of polymorphism present in natural populations. Although population geneticists should have realized this sooner from the variety of genetically inherited blood types that were known, most believed that there was little variation among wild-type alleles. The finding of Hubby and Lewontin (1966) that at least 30% of the loci they could assay in wild *Drosophila pseudoobscura* populations encoded electrophoretically polymorphic proteins changed forever the view that there was little polymorphism in natural populations, and seriously brought into question the notion that much of the change in allele frequencies was the result of natural selection. Consequently, the idea that there was widespread polymorphism in natural populations was soon confirmed for many species and provided geneticists with new tools for genetic analysis. Previously, our ability to map genes depended on synthesizing stocks multiply marked with mutations, which frequently reduced their viability. Now, we can make use of differences in wild-type alleles that potentially distinguish any two strains for linkage analysis.

*Reprinted without change from the first edition.

1

R.L. Phillips and I.K. Vasil (eds.), DNA-Based Markers in Plants, 1 - 8.
© 2001 *Kluwer Academic Publishers. Printed in the Netherlands.*

2.   The nature of naturally occurring polymorphism

It is important to understand something of the nature of naturally occurring genetic variation because it bears on the detection methods we employ. One of the most extensive studies was conducted by Kreitman (1983) who sequenced eleven wild-type *Alcohol dehydrogenase (Adh)* alleles of *Drosophila melanogaster*. He detected 43 basepair changes and ten length polymorphisms, all less than 40 basepairs. Only one of these changes led to a change in a charged amino acid in the encoded protein. One of the surprising findings from this study was that polymorphisms in introns, untranslated, and flanking regions (1.9%) were slightly less frequent than in coding regions (2.5%). In maize, most spontaneous mutations are the result of some form of insertion or deletion. For example, Wessler and Varagona (1985) found that 12 out of 17 spontaneous *waxy* mutations were associated with detectable (greater than 50 basepairs) deletions or insertions. On the other hand, when wild-type alleles are compared, small changes predominate. Ralston, English and Dooner (1988) sequenced three alleles of maize *bronzel*. In a region of over 3200 basepairs they detected 115 (77%) basepair changes, 28 (19%) deletions/insertions less than 20 basepairs, and only 7 (4%) major deletions or insertions. As is the case with *Drosophila Adh*, the frequency of polymorphism (1.5%) within the coding region was the same as non-translated and flanking regions. Of the 21 differences that distinguished the coding sequence of the two wild-type alleles, nine led to amino acid substitutions, but only three involved charged amino acids. To summarize, the lessons from a variety of studies indicate that basepair changes are more frequent than large rearrangements, that heterogeneity is not restricted by coding regions, and, not surprisingly, polymorphisms at the DNA level are much more frequent than are charge changes in proteins.

The level of variation in a species affects our ability to perform linkage analysis. The degree of polymorphism can be assessed by sequencing or making restriction maps of random wild-type alleles. Often this information is generated in the characterization of mutant alleles at a gene locus. The polymorphism present in natural populations differs markedly in different species (Evola et al. 1986). Differences at the DNA level vary from 1 to 2 basepairs per 1000 in humans to more than 40 per 1000 in maize. Even among plant species, these differences are remarkable. Shattuck-Eidens et al. (1990) sequenced six to eight RFLP alleles at each of four loci in maize and found an average of 1.2% variation. The spectrum of types of changes was similar to that observed at the *bronzel* locus. By contrast six alleles of three RFLP loci in melon (*Cucumus melo*) revealed only single base substitutions in each of two alleles at one locus for a level of variation of 0.02%, or about 50 times less than that observed for maize.

A commonly held view is that maize is exceptionally polymorphic because of the transposable element activity associated with the species. In addition to creating rearrangements, transposable elements can also generate small sequence changes. Schwarz-Sommer et al. (1985) pointed out that when transposable elements are excised from a locus, they leave behind a 'footprint', most commonly a variant of the short sequence duplicated upon insertion, as a mechanism for generating diversity. When elements were inserted in exons, the revertant can sometimes have altered activity.

However, Aquadro (1992) reminds us that among three closely related species, *Drosophila melanogaster* has eight times more copies of transposable elements than do *D. simulans* or *D. pseudoobscura*, but that the level of diversity in *D. melanogaster* is one-third to one-fifth the level of polymorphism in the other two species.

According to Nei (1983), if mutations are neutral, heterozygosity is a function of the effective species population size and the mutation rate. In fact, these factors tend to overestimate heterozygosity so Nei concluded that selection must be limiting heterozygosity to some extent. Helentjaris et al. (1985) compared the level of polymorphism as judged by RFLP among a number of plant species and noted that species that were self-pollinating showed much less variation than those that used out-crossing. This observation is consistent with Nei's hypothesis because self-pollination would be expected to reduce the effective population size. Hannah (pers. comm.) has noted out-crossing plants, because they are more likely to be heterozygous, can create new alleles by recombination and thus elevate the level of variation beyond that created by the mutation rate.

There are some regions of the genome that are significantly more polymorphic than single copy sequences. Assaying these regions is a way of increasing the level of detectable polymorphism between individuals. Tandem repeats of 15 to 60 basepair long sequences (mini-satellites or variable number tandem repeats, VNTRs) were originally discovered adjacent to unique sequence genes and found to account for an exceptionally high rate of RFLP (Jeffreys 1987; Nakamura et al. 1987). In fact, VNTRs form the basis for current methods of DNA typing in humans. While it may be difficult to identify new VNTRs, much simpler sequence repeats of one to four basepairs (microsatellites) that also exhibit a high degree of polymorphism (Webber and May 1989), are much easier to discover. Polymorphism of tandem repeats is thought to be generated by unequal crossing over or by slippage during DNA polymerization. Whatever the mechanism, the use of short tandem repeats has recently been a minor revolution in human and mammalian genetics in that it has provided ample polymorphism where it had been difficult to detect with more conventional molecular markers (Dietrich et al. 1992; Serikawa et al. 1992; Weissenbach et al. 1992). The tandem repeats are present throughout the genome; although, in humans, VNTRs may be clustered in the vicinity of telomeres, and microsatellites appear to be sparse in terminal regions (Weissenbach et al. 1992). Thus an oligonucleotide probe for one of these repeats will identify clones from other regions of the genome containing the repeat. Unique sequences that flank the tandem repeats can be used as highly polymorphic probes or for making PCR primers.

3.   PCR-based mapping methods

Microsatellites are especially attractive for a number of reasons in addition to the fact that they are frequently highly polymorphic. First of all, the assay is PCR based. It is sufficient to merely separate the amplification products by electrophoresis to observe the results. This considerably reduces the time required to obtain a result compared with methods that are based upon Southern blotting. Secondly, the use of radioisotopes can be avoided because the size polymorphism between alleles is frequently large enough to be seen in agarose gels. However, it is possible to think of radionucleotide-based assays for microsatellites that could be automated and completely avoid the gel electrophoresis step. Methods that are based upon DNA sequence have another advantage over probe based strategies. The maintenance and distribution of probes has proven to be time consuming and frequently error prone. The only step in dissemination of sequence-based methods is to publish the DNA sequence along with the mapping results so that individual laboratories can synthesize their own oligonucleotides. Despite the high degree of polymorphism, microsatellite alleles seem to be surprisingly stable. Lander and colleagues (Dietrich et al. 1992) estimate a mutation rate of 4.5 x 10^{-5} in mice. In our own preliminary work we have observed that allelic differences are stable in recombinant inbreds that have been selfed eight times beyond the F2. The major disadvantage of microsatellites is the cost of establish polymorphic primer sites and the investment in synthesizing the oligonucleotides. Once these costs are incurred, however, the long term use of this method will pay off in reduced man power and material costs.

While microsatellites are well established for human and mammalian genetics, there are also two recent surveys of their practicality in plants (Akkaya et al. 1992; Morgante and Olivieri 1993). Both of these studies demonstrated the segregation of microsatellites as co-dominant markers and showed that this type of polymorphism is prevalent in soybean, a species in which it has been difficult to find polymorphic markers in the past. There are well established methods of finding microsatellites in phage libraries by screen with oligonucleotide probes. But a quicker, if limited, approach is to examine sequence databanks for their presence. The major result, noted by Morgante and Olivieri (1993), is that while AC repeats are the most frequent class in humans and rats (Beckmann and Weber 1992), it is the AT repeat that is the most prevalent in higher plants.

It is interesting to observe that while the use of tandem repeats dominates animal work, another method based on random priming is used almost exclusively in plant molecular mapping. This technique was developed in two laboratories (Williams et al. 1990; Welch and McClelland 1990) and is frequently referred to with the acronym RAPD. It depends on the observation that single short oligonucleotide primers can frequently recognize similar sequences that are opposed to each other at distances close enough for the intervening sequence to be amplified in the PCR. Entire maps have been made using these primers exclusively (Reiter et al. 1992; Chaparro et al. 1992). One of the most intense uses of RAPDs is for finding new markers that are tightly linked with a specific locus. Because of the availability of random oligonucleotides and the relatively easy assay to look for linkage, it is a simple matter to screen many loci rather rapidly. The application of an ingenious idea makes this process even simpler. Bulk segregant analysis (Michelmore et al. 1991) uses two groups of individuals, optimally each group is homozygous for alternate alleles that govern the trait. If members of a segregating population are pooled on the basis of their phenotype, then the groups can be expected to be heterozygous at all unlinked loci but markedly skewed in the direction of one or the other parental allele for a linked locus. Giovannoni et al. (1991) demonstrated the use of this technique to target a region after individuals had been pooled on the basis of their RFLP phenotype. As powerful as it is, there are two properties of RAPD analysis that limit its application. Priming polymorphism appears to be based on mismatches with target sequences so that alleles are either present or absent. Since there is no guarantee that the dominant allele will be present in a second population, it is not always possible to use a mapped RAPD locus with new parents or to map a RAPD, that has been found to be linked with an interesting locus, in a new population. One way around this problem is to convert RAPD bands to RFLPs by cloning the amplification product. This has worked in species with small genomes like tomato and rice, but may prove very difficult in plants with larger genomes and more repeated DNA because there is no preference for amplifying single copy sequences. An alternative approach is to sequence the ends of a RAPD band so that stable primers can be synthesized to amplify this band preferentially in several genotypes (Paran and Michelmore 1993).

Two final PCR-based methods deserve mention. Allele-specific PCR is based on the choice of primers so that one allele or set of alleles will not amplify and another allele, the mutant one is interested in following for instance, will. Shattuck-Eidens et al. (1991) have demonstrated the technique for a *wx* allele of maize and have shown that non-isotopic hybridization techniques are sufficient to distinguish the mutant from the wild-type allele in dot blots of the amplification reactions, thus avoiding a gel electrophoresis step. It has been known for some time that the conformation of single stranded DNA during gel electrophoresis is dependent on DNA sequence and can be altered by single nucleotide substitutions to produce altered mobility. Single strand conformation polymorphism can be demonstrated for short (200 to 400 nucleotide) PCR products to demonstrate polymorphism (if it exists) when other methods have failed (Orita et al. 1989).

4.   Synteny

A recent and important observation in the use of plant molecular markers is the finding that many distantly related species have co-linear maps for portions of their genomes. This was first demonstrated for members of the Solanaceae (Bonierbale et al. 1988; Tanksley et al. 1988) and has been extended to a demonstration of extensive co-linearity between maize and sorghum (Hulbert et al. 1990; Whitkus et al. 1992) and more remarkably between maize and rice (Ahn 1993). There are indications that these relations might be conserved in even more distantly related plants because the duplicated sucrose synthase loci are on the same linkage group in both maize and wheat (Marana et al. 1988). The obvious result of this work is that, once careful interspecific comparisons have been made, predictions for gene locations mapped in one species can be made in the second. Since both sorghum and rice have smaller genomes than maize, it may be feasible to use one of these specifies to clone a gene sought in maize. In the short run, regions that are sparsely mapped in one species might be filled by markers from the molecular map of the second species.

5.   Acknowledgement

Work was performed at Brookhaven National Laboratory under the auspices of the U.S. Department of Energy Office of Basic Energy Sciences.

6.   References

Aquadro, C.F. (1992) Why is the genome variable? Insights from Drosophila. Trends Genet. 8: 355-362.

Ahn, S.N. (1993) Comparative mapping of the rice and maize genomes and mapping quality genes of rice using restriction fragment length polymorphism (RFLP) markers. Ph.D. thesis, Cornell University.

Akkaya, M.S., Bhagwat, A.A. and Cregan, P.B. (1992) Length polymorphisms of simple sequence repeat DNA in soybean. Genetics 32: 1131-1139.

Beckmann, J.S. and Weber, J.L. (1992) Survey of human and rat microsatellites. Genomics 12: 627-631.

Bonierbale, M.W., Plaisted, R.L. and Tanksley, S.D. (1988) RFLP maps based on a common set of clones revealed modes of chromosomal evolution in potato and tomato. Genetics 120: 1095-1103.

Chaparro, J., Wilcos, P., Grattapaglia, D., O'Malley, D., McCord, S., Sederoff, R., McIntyre, L.M. and Whetten, R.(1992) Genetic mapping of pine using RAPD markers. In: Advances in Gene Technology: Feeding the World in the 21st Century, Miami Winter Symposium, Miami, FL.

Dietrich, W., Katz, H., Lincoln, S.E., Shin, H.-S., Friedman, J., Dracopoli, N.C. and Lander, E.S. (1992) A genetic map of the mouse suitable for typing intraspecific crosses. Genetics 131: 423-447.

Evola, S.V., Burr, F.A. and Burr, B. (1986) The suitability of restriction fragment length polymorphisms as genetic markers in maize. Theor. Appl. Genet. 71: 765-771.

Giovannoni, J.J., Wing, R.A., Ganal, M.W. and Tanksley, S.D. (1991) Isolation of molecular markers from specific chromosomal intervals using DNA pools from existing mapping populations. Nucl. Acids Res. 19: 6553-6558.

Helentjaris, T., King, G., Slocum, M., Siederstrang, C. and Wegman, S. (1985) Restriction fragment polymorphisms as probes for plant diversity and their development as tools for applied plant breeding. Plant Mol. Biol. 5: 109-118.

Hubby, J.L. and Lewontin, R.C. (1966) A molecular approach to the study of genic heterozygosity in natural populations. I. The number of alleles at different loci in Drosophila pseudoobscura. Genetics 54: 577-594.

Hulbert, Sl.H., Richter, T.E., Axtell, J.D. and Bennetzen, J.L. (1990) Genetic mapping and characterization of sorghum and related crops by means of maize DNA probes. Proc. Natl. Acad. Sci. U.S.A. 87: 4251-4255.

Jeffreys, A.J. (1987) Highly variable minisatellites and DNA fingerprints. Biochem. Soc. Trans. 15: 309-317.

Kreitman, M. (1983) Nucleotide polymorphism at the alcohol dehydrogenase locus of Drosophila melanogaster. Nature 304: 412-417.

Marana, C., Garcia-Olmedo, F. and Carbonero, P. (1988) Linked sucrose synthase genes in group-7 chromosomes in hexaploid wheat (Triticum aestivum L.). Gene 63: 253-260.

Michelmore, R.W., Paran, I. and Kesseli, R.V. (1991) Identification of markers linked to disease resistance genes by bulked segregant analysis: a rapid method to detect markers in specific genomic regions using segregating populations. Proc. Natl. Acad. Sci. U.S.A. 88: 9828-9832.

Morgante, M. and Olivieri, A.M. (1993) PCR-amplified microsatellites as markers in plant genetics. Plant J. 3: 175-182.

Nakamura, Y., Leppert, M., O'Connell, P., Wolf, R., Holm, T., Culver, M., Martin, C., Fugimoto, E., Hoff, M., Kumlin, E. and White, R. (1987) Variable number of tandem repeat (VNTR) markers for human gene mapping. Science 235: 1616-1662.

Nei, M. (1983) Genetic polymorphism and the role of mutation in evolution. In: M. Nei and R.K. Koehn (eds.), Evolution of Genes and Proteins, pp. 165-190. Sinauer Assoc., Sunderland, MA.

Orita, M., Susuki, Y., Sekiya, T. and Hayashi, K. (1989) Rapid and sensitive detection of point mutations and DNA polymorphisms using the polymerase chain reaction. Genomics 5: 875-879.

Paran, I. and Michelmore, R.W. (1993) Sequence characterized amplified regions (SCARs) as a codominant genetic marker in Lactuca spp. Theor. Appl. Genet. 85: 985-993.

Ralston, E.J., English, J.J. and Dooner, H.K. (1988) Sequence of three bronze alleles of maize and correlation with the genetic fine structure. Genetics 199: 185-197.

Reiter, R.S., Williams, J., Feldman, K.A., Rafalski, J.A., Tingey, S.V. and Skolnick, P.A. (1992) Global and local genome mapping in Arabidopsis thaliana by using recombinant inbred lines and random amplification polymorphic DNAs. Proc. Natl. Acad. Sci. U.S.A. 89: 1477-1481.

Schwarz-Sommer, Z., Gierl, A., Cuypers, H., Peterson, P.A. and Saedler, H. (1985) Plant transposable elements generate the sequence diversity needed in evolution. EMBO J. 4: 591-597.

Serikawa, T., Kuramoto, T., Hilbert, P., Mori, M., Yamada, J., Dubay, C.J., Lindpainter, K., Ganten, D., Guenet, J.-L., Lathrop, G.M. and Beckmann, J.S. (1992) Rat gene mapping using PCR-analyzed microsatellites. Genetics 131: 701-721.

Shattuck-Eidens, D.M., Bell, R.N., Neuhausen, S.L. and Helentjaris, T. (1990) DNA sequence variation within maize and melon: observations from polymerase chain reaction amplification and direct sequencing. Genetics 126: 207-217.

Shattuck-Eidens, D.M., Bell, R.N., Mitchell, J.T. and McWorter, V.C. (1991) Rapid detection of maize DNA sequence variation. GATA 8: 240-245.

Tanksley, S.D., Bernatzky, R., Lapitan, N.L. and Prince, J.P. (1988) Conservation of gene repertoire but not gene order in pepper and tomato. Proc. Natl. Acad. Sci. U.S.A. 85: 6419-6423.

Weber, L. and May, P.E. (1998) Abundant class of human DNA polymorphisms which can be typed using the polymerase chain reaction. Am. J. Hum.Genet. 44: 388-396.

Weissenback, J., Gyapay, G., Dib, C., Vignal, A., Morissette, J., Millasseau, Pl, Vaysseix, G. and Lathrop, M. (1992) A second-generation linkage map of the human genome. Nature 359: 794-801.

Welsh, J. and McClelland, M. (1990) Fingerprinting genomes using PCR with arbitrary primers.

Nucl. Acids Res. 18: 7213-7218.

Wessler, S.R. and Varagona, M.J. (1985) Molecular basis of mutations at the waxy locus of maize: Correlation with the fine structure genetic map. Proc. Natl. Acad. Sci. U.S.A. 82: 4177-4181.

Whitkus, R., Doebley, J. and Lee, M. (1992) Comparative genome mapping of sorghum and maize. Genetics 132: 1119-1130.

Williams, J.G.K., Kubelik, A.R., Livak, K.J., Rafalski, J.A. and Tingey, S.V. (1990) DNA polymorphisms amplified by arbitrary primers are useful as genetic markers. Nucl. Acids Res. 18: 6531-6535.

2. PCR-based marker systems

ROBERT REITER

Monsanto, Agricultural Technology, Molecular Breeding Group, Ankeny, IA 50021, U.S.A.
<robert.s.reiter@monsanto.com>

Contents

1. Introduction

There has been a continual evolution in methods used to determine the genetic makeup of individuals. The complete genetic makeup or genotype of an individual is represented by its entire DNA sequence; however, until entire genomes can be sequenced with little effort, geneticists will continue to rely on genetic markers as a means of characterising the genotypic variation of individuals.

Mendel first used phenotype-based or morphological markers to study inheritance in pea plants. Later work in Drosophila, again using phenotypic markers, demonstrated that some of these genetic markers were inherited in a non-random fashion establishing the theory of genetic linkage. These first genetic markers possessed some key features which continue to be important even for today's more direct methods of assaying DNA variation. These markers were easily scored, they exhibited complete penetrance, and they were qualitative in nature. Unfortunately, phenotype-based markers also suffered from a serious drawback. Despite the cataloguing of as many as several hundred genetic markers in some organisms, when any two strains were examined very few of the marker loci exhibited phenotypic variation. This occurred because many of the naturally-occurring or induced allelic variants studied, had strong and undesirable morphological phenotypes. Consequently, they were rare alleles in the population. In order to circumvent this problem significant effort was dedicated to the creation of genetic stocks composed of mutant alleles at several

R.L. Phillips and I.K. Vasil (eds.), DNA-Based Markers in Plants, 9 - 29.

genetic marker loci. These stocks have served geneticists well; however, both a practical and biological limit to the number of mutant alleles an organism can tolerate exists.

A more evolutionarily "neutral" set of genetic markers was needed. The first solution was a direct examination of protein variation. Biochemical methods had been developed to study the properties of proteins and these were applied to examining protein variability. Allelic variation was discovered at the protein level which appeared to have little or no detrimental phenotypic effect on the organism. These enzyme variants or allozymes (Stuber and Goodman 1983) appeared to possess the key feature lacking with phenotypic markers, namely allelic variants could readily be found among strains. Although allozymes were used successfully in some species, notably Drosophila and corn, the available number of polymorphic allozyme markers in many other species was low.

Molecular biology ushered in a new era with techniques which directly assayed DNA. Although DNA sequencing techniques had been developed, they were laborious and not amenable to rapid genotype determination. A significant breakthrough in genetic analysis came when the first genetic map using restriction fragment length polymorphisms (RFLPs) (Botstein *et al.* 1980) was constructed. Although more costly, and technically more demanding than allozymes, the abundance of polymorphic marker loci made RFLPs the technology of choice for detailed genetic analysis.

Since the first study of quantitative variation with genetic markers by Sax (Sax 1923), there has been great interest in the genetic dissection of quantitative variation. In plants, the first attempts at performing genome-wide analysis of quantitative variation used allozymes (Tanskley *et al.* 1982, Edwards *et al.* 1987). It was recognized however that the available allozyme marker loci provided only limited coverage of the genome. However, with the development of RFLPs, a dramatic proliferation in the construction and use of marker-based genetic maps for analysing quantitative variation occurred. RFLP methods were streamlined and impressive throughput capabilities were obtained by many laboratories, yet RFLPs remain technically complex, laborious and difficult to automate. More seriously, within some species, especially self-pollinated plants, only low levels of polymorphism were observed. Thus, despite the popularity of RFLPs the search continued for improved genetic marker systems.

2.   The polymerase chain reaction

The *in vitro* amplification of DNA by the polymerase chain reaction (PCR) (Saiki *et al.* 1985) has proven to be a revolutionary technique in molecular biology. PCR facilitates the *in vitro* amplification of DNA by using two oligonucleotide primers complementary to opposing DNA strands. These oligonucleotides act as sequence-dependent priming sites for simultaneous primer extension on the opposing DNA strands. By appropriate selection of priming sites and reaction conditions the primer extension reactions will result in the synthesis of new DNA template for the opposing extension reaction. By combining a thermal-stable DNA polymerase with repeated cycling through reaction conditions favoring DNA template denaturation, followed by template-primer annealing and finally primer extension, exponential DNA amplification occurs (Fig. 1).

Beginning with as little as one molecule of template DNA, microgram quantities of a specific DNA fragment are produced in a couple of hours.

PCR is rapid, inexpensive and technically simple. PCR has proven also to be extremely flexible. Because of these features a multitude of new genotyping methods based upon PCR are being used for genetic analysis in many organisms, including plants. These PCR-based genotyping methods can be divided into two categories: 1) those which are sequence-arbitrary methods, and 2) those which require *a priori* sequence information.

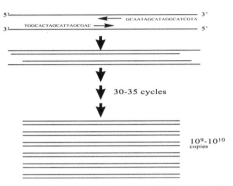

Figure 1. The polymerase chain reaction. Exponential DNA synthesis occurs using a thermal-stable DNA polymerase and repeated cycles of template denaturing, and oligonucleotide-directed annealing and extension

3. Sequence-arbitrary methods

In its traditional form, PCR is performed using two sequence-dependent oligonucle-otide primers. Sequence information is therefore necessary in order to design primers which would facilitate successful amplification of a specific DNA fragment. This initially limited the utility of PCR as a method for large-scale genetic mapping. First, the availability of large amounts of sequence information is limited in many species and obtaining additional sequence information can be cost-prohibitive. Second, based upon available sequence, the degree of polymorphism revealed by PCR amplification and fragment size separation is low. These problems were especially evident in many plant species. What was needed was a DNA amplification-based method which obviated the need for *a priori* sequence information, yet revealed DNA polymorphism. It is therefore not surprising that the development and use of sequence-arbitrary methods has occurred primarily with plant species.

3.1. Random amplified polymorphic DNAs (RAPDs)

Two groups (Williams *et al.* 1990, Welsh and McClelland 1990) independently discovered that reproducible and heritable DNA fragments could be amplified from genomic DNA using only single, sequence-arbitrary primers. The first of these methods was termed random amplified polymorphic DNAs (RAPDs) (Williams *et al.* 1990).

RAPDs are the result of amplification using a single, short, sequence-arbitrary design oligonucleotide (10mer typical) (Fig. 2). Amplification is allowed to proceed by incorporating a low annealing temperature (35-37°C typical) in the thermal-cycling profile. The resulting amplification products are usually size separated on agarose gels (Fig. 2). A DNA fragment pattern of low complexity is observed following ethidium bromide staining. When the same primer is used with DNA from different species, unique fragment patterns are observed. More importantly, when DNA from different strains from the same species are used, fragment patterns are nearly similar with the exception that an occasional amplified fragment is present in one strain and absent in another. It was determined that these fragment polymorphisms were heritable and constituted a new class of genetic markers (Williams *et al.* 1990). The frequency at which these polymorphisms are observed is dependent upon the species studied, ranging from 0.2-1 polymorphism per primer in select angiosperms (Tingey *et al.* 1992). It was also determined that an entire genetic map could be constructed using a set of arbitrary primers (Reiter *et al.* 1991, Carlson *et al.* 1991).

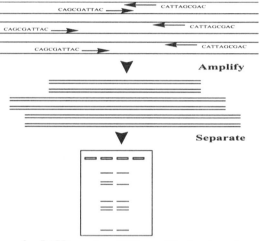

Figure 2. The steps of a RAPD reaction. DNA amplification is performed using a single 10mer oligonucleotide. The resulting amplification products are size separated on an agarose gel and visualized

RAPDs have proven to be extremely popular with plant scientists. A recent bibliographic search identified over 2,000 publications in which RAPD markers have been used. This popularity has been fueled by the simplicity of the method and the low capital investment required. No more than a thermal cycler and inexpensive horizontal gel electrophoresis equipment is needed. Also, low cost sets of random 10mer oligonucleotides may be purchased. In addition, these oligonucleotides are universal in that they may be used with any species of interest.

Most RAPDs are amplification polymorphisms which are the result of either a nucleotide base change at one of the priming sites or an insertion/deletion event within the amplified region (Williams *et al.* 1990, Parks *et al.* 1991). Therefore, depending upon the sequence composition of the DNA template in an individual, amplification of a given fragment will either be successful or unsuccessful. DNA amplification from individuals heterozygous at RAPD marker loci, generally are indistinguishable from the positive parent. Thus RAPD markers are typically dominant markers, although occasional co-dominant RAPD marker loci are observed as fragment length polymorphisms. Dominant marker systems are less informative than co-dominant markers for genetic mapping in some types of segregating populations (Allard, 1956). A more serious drawback are concerns with fragment allelism.

When several strains are amplified using the same 10mer primer, similar DNA fragment patterns are observed. It is however not known whether all fragments of the same size class (the resolution of fragment size being dependent on the method of size separation) are in fact composed of the same DNA sequence and hence are allelic. Fragments from related strains are likely to be allelic; however, as genetic distance increases the probability of non-allelism increases. It is expected that this same relationship would hold when no fragments are observed i.e. closely related strains being scored as minus for a given RAPD fragment are more likely to contain identical DNA sequence at the target site which fails to support DNA amplification than unrelated individuals. Fragment allelism is an important concern when RAPDs are used for population genetic studies, varietal fingerprinting and some forms of marker-assisted selection.

Finally, concerns regarding RAPD pattern reproducibility exist. Any sequence-arbitrary amplification method will be considerably less robust than conventional PCR for the following reasons: 1) multiple amplicons are present competing for available enzyme and substrate; and 2) low-stringency thermal-cycling permits mismatch annealing between primer and template. To overcome these problems it is important to create optimal and consistent amplification conditions. In addition, the quantity and quality of reaction components (DNA, magnesium, enzyme etc) should be carefully controlled (Tingey *et al.* 1992)

3.2. Arbitrarily primed polymerase chain reaction (AP-PCR)

Arbitrarily primed polymerase chain reaction (AP-PCR) (Welsh and McClelland 1990), like RAPDs, is based upon the use of a single, sequence-arbitrary design primer in a DNA amplification reaction. Unlike RAPDs, the oligonucleotide length is typical for conventional PCR and primer concentrations are 10-fold higher (Welsh and McClelland 1990). The thermal-cycling profile begins with one or two cycles incorporating a low annealing temperature (40-48°C typical) followed by 30 cycles using a high stringency annealing temperature and radiolabeled dCTP to label the newly synthesized fragments (Welsh and McClelland 1990, Welsh *et al* 1991). The DNA fragments are size separated on polyacrylamide and visualized via radiography. The moderately complex fragment pattern is a result of the increased sensitivity afforded by both radiolabeling of amplification products and the use of polyacrylamide gels for size separation.

As with RAPDs, AP-PCR generated fragments are present in some strains and absent in other strains of the same species and these plus/minus DNA amplification-based polymorphisms can be used as genetic markers (Welsh *et al.* 1991). Using mouse strains, an average of one polymorphism per primer was found (Welsh *et al.* 1991). As with RAPDs, AP-PCR polymorphisms are likely the result of either sequence divergence at one of the priming sites or insertion/deletion within the amplification region.

Because AP-PCR is also based upon DNA amplification polymorphisms, concerns over allelism exist. The probability of fragment allelism is increased for both RAPDs and AP-PCR if higher resolution size separation methods are used. AP-PCR reproducibility has received less attention than RAPDs. However, it is unknown whether AP-PCR reactions are in fact more robust or whether because of limited use reproducibility has not become a more central issue.

3.3. DNA amplification fingerprinting (DAF)

A third DNA amplification-based method for genotyping which uses a single, short arbitrary primer is termed DNA amplification fingerprinting (DAF) (Caetano-Anolles *et al.* 1991). DAF uses a single, short oligonucleotide of 5-8 nucleotides at high concentration (3-30mM) in combination with either low or high stringency annealing temperature, high resolution fragment detection (silver staining) and high resolution fragment separation (polyacrylamide) (Caetano-Anolles *et al.* 1991). The resulting DNA fragment pattern is highly complex.

DAF shares those features common to AP-PCR and RAPDs; namely it results in plus/minus heritable amplification polymorphisms, a preponderance of dominant marker loci, and unknown allelism between fragments of equivalent molecular weight. DAF has not been used in a large number of labs and therefore the issue of reproducibility has not been well addressed. As with all sequence-arbitrary DNA amplification methods, consistent results can only be obtained by careful attention to optimization and standardization of protocols.

DAF patterns contain many more bands than AP-PCR or RAPD patterns. Because more DNA fragments are seen, the likelihood is increased of observing polymorphism between strains. This enhanced ability, versus RAPDs, to detect polymorphism was confirmed in a comparison of *Chrysanthemum* sports and bacterial strains (Caetano-Anolles 1996).

3.4. Other sequence arbitrary methods

Several variations of the previously described methods have been developed. Using various strategies, each of these methods attempts to enhance the ability to detect polymorphism by either increasing the number of observed amplification products or increasing the ratio of polymorphic amplification products to non-polymorphic products.

Either by using more than one arbitrary primer (Callahan *et al.* 1993, Micheli *et al.* 1993) or by using a degenerate primer in the amplification reaction (Caetano-Anolles 1994) increasingly complex RAPD and DAF amplification patterns have been observed.

With degenerate primers, pattern complexity was increased only when using primers containing conventional base substitutions. This is equivalent to using multiple primers at lower primer concentration. Interestingly, inosine substitutions generally simplified DAF patterns (Caetano-Anolles 1994).

One approach to increasing the proportion of polymorphisms being revealed involves pre-digestion of the template DNA with restriction endonucleases (Caetano-Anolles *et al.* 1993). This method has been termed template endonuclease cleavage multiple arbitrary amplicon profiling (tecMAAP). Despite the fact that endonuclease digestion should result in the destruction of potential amplicons, it appears that new amplicons containing priming sites with probably higher levels of primer-template mismatch, are amplified. Pre-digestion of template DNA may also be creating new endogenous sets of arbitrary primers from highly repeated DNA regions which by themselves or in concert with the arbitrary primer enable amplification of DNA products. Why these new amplicons show higher levels of polymorphism is not known (Caetano-Anolles 1994).

Another approach to enhancing the level of informativeness from DAF reactions is by using primers which contain both a 5' mini-hairpin sequence and a short 3' arbitrary sequence either alone (Caetano-Anolles and Gresshoff 1994) or in a two step amplification procedure called arbitrary signatures from amplification profiles (ASAP) (Caetano-Anolles and Gresshoff 1996). Using ASAP, a DAF amplification is first performed and then an aliquot of the resulting amplification reaction is used as template in a second amplification primed with either a mini-hairpin based primer or alternatively a 5'-anchored simple sequence repeat (SSR) primer (Caetano-Anolles and Gresshoff 1996). It is not understood as to why these strategies increase the probability of detecting polymorphism, however they appear to be sampling regions of the genome which are less conserved.

Other methods have been tested which are not sequence arbitrary *per se*, but require only limited sequence information. Most of these methods attempt to exploit the ubiquitous and highly polymorphic nature of single sequence repeats SSRs. One such approach used various simple sequence repeat patterns to promote inter-SSR amplification (Gupta *et al.* 1994). Using this strategy, Gupta *et al.* (1994) found that tetra-nucleotide repeat primers provided informative, moderately complex patterns. Tri-nucleotide repeats were less informative and di-nucleotide repeat primers resulted in smeared patterns. To prevent smearing and to take advantage of the abundance of di-nucleotide repeats, Zietkiewicz *et al.*(1994) generated complex and highly informative fingerprints using 3' and 5' anchored SSR primers. This was accomplished by arbitrarily extending the primer sequence by 2 to 4 bases outside of the repeat motif, in order to anchor the primer location during inter-SSR amplification. High resolution acrylamide gels and radiolabeling of the amplification products resulted in highly complex patterns (Zietkiewicz *et al.* 1994).

Two other methods have been described that combine aspects of RAPDs with SSRs. The first of these methods termed random amplified microsatellite polymorphisms (RAMPs) (Wu *et al.* 1994) is an extension of the anchored SSR method described by Zietkiewicz *et al.*(1994). By combining a radiolabeled, anchored SSR primer with a 10mer RAPD primer, informative fingerprint patterns are observed (Wu *et al.* 1994).

Because the anchored repeat primer is labeled, only those amplification products containing the target repeat motif are visualized (Wu *et al*. 1994).

Finally, in an effort to derive greater information from RAPD patterns, the strategy of hybridizing SSR repeat primers to RAPD amplification patterns has been described (Cifarelli *et al*. 1995, Richardson *et al*. 1995). The method has been called either random amplified hybridization microsatellites (RAHM) (Cifarelli *et al*. 1995) or random amplified microsatellite polymorphisms (RAMPO) (Richardson *et al*. 1995). In most instances the bands observed following hybridization with the SSR probe were amplification products not observed using only ethidium bromide staining (Richardson *et al*. 1995) and thus may uncover additional polymorphism.

3.5. AFLPs

The AFLP technique (Vos *et al*. 1995) is a sequence-arbitrary, amplification-based method which is a significant departure from the previously described methods. In the AFLP technique, genomic DNA is first digested with one or more restriction endonucleases. Next an adapter of known sequence is ligated to the digested genomic DNA. DNA amplification is carried out using primers with sequence specificity for the adapter. The primer(s) also contains one or more bases at their 3' ends which provide amplification selectivity by limiting the number of perfect sequence matches between the primer and the pool of available adapter/DNA templates. The resulting amplification products are typically observed by radiolabeling of one of the primers, followed by fragment separation on acrylamide gels (Vos *et al*. 1995) (Fig. 3). Fluorescence-based labeling in conjunction with automated sequence detection equipment is also frequently used (Applied Biosystems, Foster City, CA). In complex genetic organisms, like plants, two amplification steps are performed. The first is a pre-amplification using primers which may or may not contain selective nucleotides. Diluted product from the pre-amplification is used as template for the second amplification along with primers with selective nucleotides (Vos *et al*. 1995) (Fig. 3).

Typically, observed AFLP patterns are moderately complex; however, both pattern complexity and concomitantly the number of observed polymorphisms is dependent upon both primer and restriction endonuclease selection. This flexibility is an attractive feature of the AFLP method. This flexibility allows the researcher to adjust reaction conditions in order to reveal either more or less DNA fragments and concomitantly more or less polymorphic bands. AFLP polymorphisms are generally of the plus/minus variety common to the other sequence arbitrary methods. Using fragment signal intensity, software has been designed which facilitates the scoring of many AFLP polymorphisms as co-dominant marker loci (KeyGene N.V., Wageningen, The Netherlands). Heterozygous and homozygous positive signals are differentiated quantitatively.

The AFLP technique has proven to be quite popular. A multitude of high density genetic maps have been constructed in numerous plant species using AFLPs (Mendendez *et al* 1997, Vuylsteke *et al*. 1997). In addition, a number of QTL mapping (Nandi *et al* 1997, Powell *et al* 1997) and fingerprinting (Hill *et al* 1996, Powell *et al*, 1996) studies have been carried out using AFLPs. As with the other sequence arbitrary

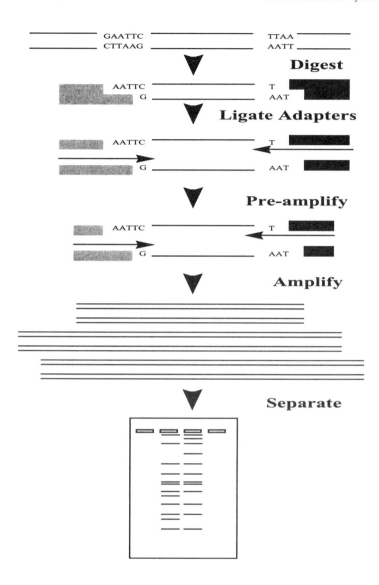

Figure 3. Steps in AFLP analysis. Genomic DNA is restriction digested, adapters are ligated and one or two rounds of selective amplification are carried out. Final amplification products are size separated on acrylamide gels and visualized.

methods, the issue of allelism must be considered. Observed fragments are presumed to be allelic when the progeny of experimental populations only have one parental source for an observed allele. Fragments of equal size may not be allelic when the source of those alleles is unknown.

Although reproducibility is of concern with AFLPs, the AFLP technique appears to be somewhat more robust than many of the previously described methods. This improvement is likely due to PCR amplification using long primers which are perfect matches to adapter-genomic DNA templates. Amplification conditions are used which facilitate the amplification of perfect matches between template and primer and reduce the likelihood of mis-match amplification.

Regardless, careful control of both reagents and methodology can result in highly consistent results.

3.6. Selective amplification of microsatellite polymorphic loci (SAMPL)

SAMPL is a hybrid method which attempts to exploit features of both SSR sequence based methods and the AFLP method (Morgante and Vogel 1996). Genomic DNA is first restriction endonuclease digested using one or more endonucleases and adapters ligated as in the AFLP method. Amplification is then performed using a single oligonucleotide primer composed of an SSR motif with additional anchoring bases and an oligonucleotide primer corresponding to the adapter sequence. The amplification products are typically radiolabeled and size separated on high resolution polyacrylamide gels (Morgante and Vogel 1996). The method is especially effective at revealing polymorphism when the SSR motif is a perfect compound SSR (Morgante and Vogel 1996). The method can be varied somewhat by first performing a pre-amplification step using adapter-specific primers, followed by selective amplification which incorporates an SSR primer (Morgante and Vogel 1996). While the method has not been used extensively, it is another of the possible tools to help in the identification and exploitation of DNA polymorphism.

3.7. Summary

Most polymorphisms revealed by sequence-arbitrary marker systems are DNA amplification based, with the exception being polymorphisms revealed by the AFLP technique. Typically, polymorphisms are revealed when a DNA fragment is amplified in one strain and fails to amplify in another. As a result the maximum allele diversity observable is two (presence or absence). This limitation is not important in experimental populations derived from two fully inbred parents where a maximum of two alleles at a locus can occur; however this is a confounding limitation in the comparison of entries where more than two alleles may occur. In these instances allelism is assumed where in fact it may not exist. In these instances fragments are assumed to be identical by descent when in fact often they are only identical in state.

The popularity of sequence-arbitrary methods is primarily driven by their absence or near-absence of development cost and their ability to readily detect polymorphism. These attributes make sequence-arbitrary methods well suited for the saturation of specific genetic locations and has been especially valuable in map-based cloning experiments.

Both for map-based cloning as well as for marker-assisted selection it is frequently useful to clone and/or sequence an observed polymorphic fragment in order to create a sequence-dependent marker. A polymorphism may be detected using the cloned

DNA as an RFLP probe, or the fragment sequence may be queried for the presence of polymorphism via PCR and conventional gel electrophoresis (see section 4.2.2). In instances where polymorphism is not initially detected, other methods (see sequence-dependent methods below) may reveal polymorphism. It should be noted, that a lack of detectable polymorphism may occur if the original DNA polymorphism was resident in the sequence complementary to the arbitrary primer. These polymorphisms will not be represented in the population of primer-containing amplification products. Also polymorphisms revealed by sequence-arbitrary methods frequently reside in repetitive DNA sequences and this may limit conversion to sequence-dependent markers.

4.   Sequence-dependent methods

Despite their considerable development cost, sequence-dependent, PCR-based marker systems are now being exploited in many plant species. Sequence-dependent markers exhibit key features which make them especially attractive, including allelism, co-dominance, and assay robustness. These features make sequence-dependent markers especially attractive for marker-assisted plant breeding.

4.1   Simple sequence repeats (SSRs)

Although known by many names and acronyms, including simple tandem repeats (STRs), microsatellites, and simple sequence repeats (SSRs), SSRs have received considerable attention and are probably the current marker system of choice for marker-based genetic analysis and marker-assisted plant breeding (Tautz 1989, Akkaya *et al.* 1992, Chin *et al.* 1996).

SSRs are ubiquitous sets of tandemly repeated DNA motifs. The repeat regions are generally composed of di- tri- tetra- and sometimes greater perfectly repeated nucleotide sequence (Tautz and Renz 1984). Compound repeats composed of two or more repeat motifs are also frequently found (Morgante and Vogel 1996). The length of a given repeat sequence varies greatly, with different alleles varying in the number of units of the repeat motif. For example, di-nucleotide SSRs have alleles which either differ by two base pairs or multiples of two base pairs. The variability in the number of repeat units is typically the basis of observed polymorphism.

The ability to detect polymorphism is improved considerably over the previous sequence-dependent method of choice RFLPs (Powell *et al.* 1996). Even in those self-pollinated species where the level of polymorphism detected by RFLPs is prohibitively low, acceptable levels of polymorphism are observed with SSRs (Akkaya *et al.* 1992, Powell *et al.* 1996). The high degree of observed polymorphism appears to be the result of increased rates of sequence mutation affecting the number of repeat motifs present at an SSR locus (Edwards *et al.* 1992), with the observed variation likely due to replication slippage (Edwards *et al.* 1992) or unequal crossing over. The rate of mutation is dependent upon the SSR locus examined (Edwards *et al.* 1992) Because of mutation, two alleles of equal length at a locus, though likely identical by descent may in rare instances only be identical in state.

Unlike sequence-arbitrary methods, each SSR locus must be cloned and sequenced before a useful marker can be generated. Small fragment genomic libraries, preferably enriched for SSR-repeat motifs, are screened for clones containing the SSR sequence using an oligonucleotide probe complementary to the repeat motif. In order to obtain single-copy DNA sequence flanking the SSR marker, each positive clone is sequenced. Oligonucleotides primers complementary to unique DNA sequence flanking both sides of the repeat are synthesized and used for PCR amplification of the SSR (Fig 4). Amplification products may be size separated using either high percentage agarose, Metaphor (FMC, Rockland, ME), polyacrylamide gels or direct mass measurement (see MALDI-TOF MS below). Complete resolution of all allele sizes is only accomplished on either polyacrylamide or via direct mass measurement; however, sufficient polymorphism for many experiments is frequently revealed using lower resolution agarose gel systems. PCR amplification products can be labeled via either incorporation or end-labeling of either a radio- or fluorescent-labeled deoxynucleotide. Fluorescent-labeled amplification products can be analyzed using several fluorescent detection systems (Applied Biosystems, Foster City, CA; LiCor, Lincoln, NE).

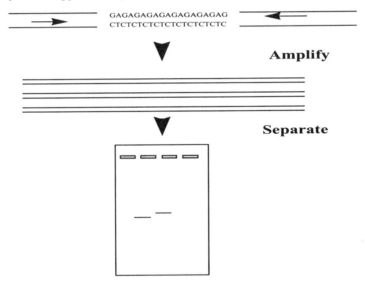

Figure 4. Steps in the amplification of simple sequence repeats. Oligonucleotides complementary to DNA sequence flanking a simple sequence repeat facilitate amplification of the simple sequence repeat region. Resulting amplification products are size separated and visualized.

SSR markers are generally co-dominant, such that all three genotypic classes are observed in F2 populations of species which exhibit disomic inheritance. The number of alleles at SSR loci vary dramatically, but are generally high. Maughan *et al.* (1995) observed from five to fifteen alleles when diverse soybean germplasm was assayed at five SSR loci. With good primer design, most SSR markers in corn and soybean amplify alleles from only a single locus. Occasionally, allele(s) at a second duplicated locus are co-amplified. The frequency of duplicate loci being revealed is dependent upon both the degree of genome duplication and sequence similarity for each species in question.

SSR markers developed for one species are generally of limited utility in other more distantly related species. For example, corn SSR markers will only occasionally generate amplification products in rice and few soybean SSR markers will amplify products in other legumes (Cregan 1997). Amplification is usually successful however in closely related species like *Teosinte* (corn) and *Glycine soja* (soybean). It is also not uncommon to observe null alleles within the same species. For example, in corn the observed frequency of null alleles is 0-3%.

Because SSR amplification is a sequence-dependent amplification reaction, the procedures are very robust. In order to ensure consistent amplification, good primer design is essential. In comparison to previously described sequence-arbitrary methods, a disadvantage of SSR markers is the reduced amount of information generated from each amplification reaction and each gel lane. Limited pre- or post- amplification multiplexing can be accomplished with SSR markers to alleviate this problem, although pre-amplification multiplexing can be inconsistent. Through careful experimentation however, sets of primer pairs can be identified which exhibit consistent co-amplification.

4.2. Single nucleotide polymorphisms (SNPs)

The vast majority of polymorphisms which exist in DNA sequence are single base pair differences. Recently, much effort has been focused on the exploitation of single nucleotide polymorphisms (SNPs) (Marshall 1997, Nikiforov *et al.* 1994) and sequence-dependent methods for detecting SNPs.

SNPs are of interest because of their abundance and presence in single-copy DNA sequence. They may be found in both transcribed and non-transcribed regions and in some instances are the direct cause for observed phenotypic variation. In humans, SNPs have been found at a frequency greater than 1 per 1000 base pairs (Wang *et al.* 1996). Several methods have been developed to assay SNPs and some of these are described below.

4.2.1. *Allele-specific amplification*

Allele-specific amplification (D'Ovidio and Anderson 1994,, Mohler and Jahoor 1996, Williams *et al.* 1996, Xu and Hall 1994) takes advantage of the sensitivity to PCR amplification inhibition which occurs when mismatch exists between a PCR primer and template DNA. This inhibition is especially pronounced by mismatch at one or more of the three 3' bases. If an SNP is known, then the polymorphic base pair may be queried by designing a primer so that its 3' end overlaps the query base. When a primer encounters template DNA which consists of perfect match sequence then the amplification is facilitated. If on the other hand the template DNA contains a mismatch at the query base then amplification is inhibited. Typically, paired reactions are performed with one reaction containing the perfect match primer to the reference allele and the second reaction containing perfect match primer to the mutant allele. Heterozygous template DNA would result in positive amplification using either primer pair. Marker assays based upon allele-specific amplification can be difficult to develop and use because of the reliance on a single base mismatch resulting in differential PCR amplification.

4.2.2. *Sequence characterized amplified regions (SCARs)*

Sequence characterized amplified regions (SCARs) (Paran and Michelmore, 1993) are an example of a class of sequence-dependent marker derived from sequence-arbitrary marker loci. By cloning and sequencing RAPD amplification products, Paran and Michelmore (1993) were able to design sequence-dependent primers and amplify PCR products unique for each marker locus. The ability to observe polymorphism was dependent upon the SCAR. Some SCARs were functionally dominant markers whereas some SCARs were co-dominant (Paran and Michelmore, 1993).

4.2.3. *Cleaved amplified polymorphic sequence (CAPS)*

Instances where direct gel electrophoresis of PCR products does not reveal polymorphism may necessitate the use of more elaborate detection methods. Cleaved amplified polymorphic sequences (CAPS) are PCR amplification products which are subsequently cleaved using restriction endonucleases and size separated in order to reveal polymorphism (Konieczny and Ausubel 1993).

4.2.4. *Single-strand conformation polymorphism*

Single-strand conformation polymorphism (SSCP) takes advantage of differences in the three-dimensional conformation of single stranded DNA, with the conformation being dependent upon the DNA sequence (Orita *et al.* 1989,). In SSCP analysis, PCR amplification products are first denatured and then electrophoresed under non-denaturing conditions. Single base differences can result in changes in three-dimensional conformation and result in differences in the rate of migration during electrophoresis (Orita *et al.* 1989). Modification of electrophoresis conditions can be used to enhance the ability to detect polymorphism; however not all polymorphisms can be detected using this method. SSCP analysis is especially useful as a way of scanning a DNA sequence for potential polymorphism (To *et al* 1993).

4.2.5. *Denaturing gradient gel electrophoresis*

Denaturing gradient gel electrophoresis (DGGE) is a sequence scanning method which differentiates PCR amplification products based upon differences in their melting profiles under denaturing conditions (Fischer and Lerman 1979). By using either a chemical denaturant, usually either urea or formamide, or heat denaturing during electrophoresis, differences are revealed in the rate of migration of DNA molecules. These migration differences are the result of differences in DNA melting which occur upon exposure of the DNA to a linearly increasing denaturing gradient. DNA melting occurs in melting groups of 50-300bp. Each groups stability is dependent upon the sequence composition and the melting profile can be simulated using a computer algorithm (Eng and Vijg 1997). Single base differences in DNA sequence can lead to detectable differences using DGGE (Myers *et al*, 1988) The method has been shown to be an extremely effective method of scanning for SNPs (Fischer and Lerman 1979). A disadvantage is that not all possible SNPs can be detected using DGGE and that specialized electrophoresis equipment is required.

4.2.6. *Genetic bit analysis*

Numerous primer-directed nucleotide incorporation assays have been described (Syvanen *et al*. 1990, Kuppuswami *et al*. (1991). Perhaps one of the most useful is Genetic Bit Analysis (GBA) (Nikiforov *et al*. 1994). GBA is a method specifically developed for high throughput detection of known SNPs. It differs from some of the previously mentioned methods in that a known bimorphic base is queried.

First, PCR amplification is performed using primers for a region of DNA containing a query base. Single-stranded DNA is next generated by exonuclease digestion of the resulting PCR product. One of the strands of DNA is protected from exonuclease digestion by the introduction, during oligonucleotide synthesis, of several phosphorothioate groups at the primer 5' end. The resulting single-stranded DNA is hybridized with an oligonucleotide which immediately flanks the query base at its 3' end. The hybridization takes place in an ELISA plate onto which the oligonucleotide has been immobilized. Next either a biotin- or fluorescein-labeled dideoxynucleoside is used to extend the 3' end of the oligonucleotide at the query base. The success of the single base extension is dependent upon the base sequence of the hybridized, single-stranded DNA template. Detection of labeled dideoxynucleoside is achieved using either horseradish peroxidase, antibiotin or alkaline phosphatase, antifluorescein conjugates in combination with enzyme-specific, colorimetric determination using a plate reader (Nikiforov *et al*. 1994) (Fig. 5). Enzyme reactions are performed sequentially with each enzyme detection system. A positive signal from both enzyme detection systems indicate a heterozygous sample. The assay provides an easily automated plate readout which can be easily interpreted (Nikiforov *et al*. 1994).

4.2.7. *Oligonucleotide chip-based hybridization*

Potentially one of the most exiting new technology platforms for genetic analysis are oligonucleotide chips (Chee *et al*. 1996, Yershov *et al*. 1996). Silicon chips onto which thousands of oligonucleotides are arrayed have been used to sequence and scan DNA for mutation (Chee *et al*. 1996), and simutaneously diagnose several hundred SNPs (Wang *et al* 1996). The most discussed technology, pioneered by Affymetrix (Santa Clara, CA), involves the use of photolithography to simultaneously synthesiz thousands of oligonucleotides on a chip. Using these oligonucleotide chips, thousands of SNPs may be queried simultaneously by allele-specific hybridization with PCR amplified DNA from each polymorphic region. Hybridization is conveniently detected via

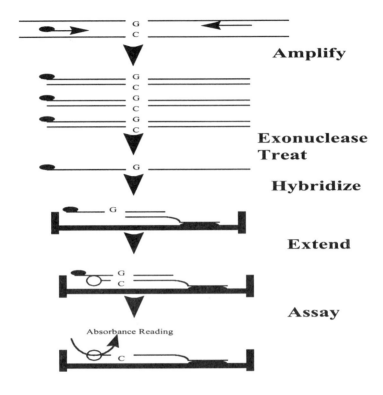

Figure 5. Steps in Genetic Bit Analysis. Target site DNA is amplified using one endonuclease-protected and one unprotected oligonucleotide. The amplification product is made single-stranded by exonuclease treatment and hybridized to a previously immobilized third oligonucleotide. Single base extension and detection is carried out using a labeled dideoxynucleotide.

fluorescence-labeling of the target DNA (Chee *et al.* 1996). The fluorescence signal of perfect match between PCR product and immobilized oligonucleotide is compared with several instances of mismatch. This facilitates accurate determination of perfect match versus single base mismatch signal differences.

Notable issues in the use of oligonuleotide chips are their high cost, and current limits on re-use. The cost of custom chips is currently beyond the means of most laboratories. As mass-produced chips, designed for the detection of specific sequence polymorphisms, are introduced, costs should decrease; however these chips will be limited to pre-designed query positions. The efficiency of oligonucleotide synthesis using photolithography is low in comparison with conventional cartridge-based synthesis systems. The reduced efficiency negatively impacts the ability to re-use chips for sequential analyses. Notwithstanding these limits, oligonucleotide chip-based hybridization offers highly parallel and therefore very high throughput diagnostic analysis.

4.2.8. *Matrix assisted laser desorption ionization, time of flight mass spectrometry (MALDI-TOF MS)*

Another exciting new technology for DNA analysis is matrix assisted laser desorption ionization, time of flight mass spectrometry (MALDI-TOF MS) (Haff and Smirnov 1997; Monforte and Becker 1997). Recent breakthroughs in the ability to analyze larger DNA molecules using MALDI-TOF MS have (Monforte and Becker 1997) opened up opportunities in the exploitation of this method for genetic analysis.

The primary advantage which MALDI-TOF MS offers is extremely rapid and accurate fragment size determination. Individual fragment separation times take milliseconds; however, for accurate determination ionization from multiple laser pulses are measured (Monforte and Becker 1997). As a result a single sample is processed in a few seconds or less. Current technical limits restrict accurate size determination to fragments 100 bases or less (Monforte and Becker 1997). Also, if the mass of individual fragments is varied, multiplexing of samples is possible (Haff and Smirnov 1997). MALDI-TOF MS is not limited to the analysis of SNPs and can also be used for the size determination of SSR alleles (Monforte and Becker 1997). Two noteworthy limitations with MALDI-TOF MS are the high cost of instrumentation and the lack of parallel sample processing for those with high throughput requirements. The lack of parallel processing can be significantly offset by the dramatic sample processing speed available using MALDI-TOF MS and determined leveraging of sample multiplexing.

5. Future directions

It can be anticipated that marker systems in plants will continue to follow the developments in mammalian systems. The next generation map in humans is going to be an SNP-based genetic map and therefore SNP-based plant maps are sure to follow. Improvements in sequencing throughput and large-scale EST and genomic sequencing programs in plant species will provide the sequence information necessary for SNP marker development.

Another trend likely to continue for the near term will be the reliance on PCR as key method in most genotyping methods. This will be coupled with significant miniaturization of the PCR process itself using one of several microfabrication techniques (Burns *et al* 1996). Miniaturization promises to significantly reduce cost and increase throughput. It will be interesting to observe the impact of currently expensive, yet promising technologies like oligonucleotide chips, capillary array electrophoresis, and MALDI-TOF MS. Each of these new technologies promises to create unparalleled opportunities in the genetic analysis of plant species and greatly facilitate further understanding of plant genome organization.

6.   Acknowledgements

The author wishes to thank Todd Krone and Peter (Jeff) Maughan for critical reading of the manuscript.

7.   References

Akkaya, M.S., Bhagwat, A.A. Cregan, P.B. (1992) Length polymorphisms of simple sequence repeat DNA in soybean. Genetics 132:1131-1139.

Allard, R.W. (1956) Formulas and tables to facilitate the calculation of recombination value in heredity. Hilgardia 24:235-278.

Botstein, D., White, R.L., Skolnick, M, and Davis R.W.. (1980) Construction of a genetic linkage map in man using restriction fragment length polymorphisms. Am. J. Hum. Genet. 32:314-331

Caetano-Anolles, G., Bassam, B.J., and Gresshoff, P.M. (1991) DNA amplification fingerprinting using very short arbitrary oligonucleotide primers. Bio/Technology 9:292-305.

Caetano-Anolles, G., and Gresshoff, P.M. (1994) DNA amplification fingerprinting using arbitrary mini-hairpin oligonucleotide primers. Bio/Technology 12:619-623.

Caetano-Anolles, G., and Gresshoff, P.M. (1996) Generation of sequence signatures from DNA amplification fingerprints with mini-hairpin and microsatellite primers. Biotechniques 20:1044-1056.

Caetano-Anolles, G., Bassam, B.J., and Gresshoff, P.M. (1993) Enhanced detection of poly-morphic DNA by multiple arbitrary amplicon profiling of endonuclease digested DNA: identification of markers tightly linked to the supernodulation locus of soybean. Mol. Gen. Genet. 241:57-64.

Caetano-Anolles, G. (1994) MAAP: a versatile and universal tool for genome analysis. Plant Mol. Biol. 25:1011-1026.

Caetano-Anolles, G. (1996) Scanning of nucleic acids by in vitro amplification: New develop-ments and applications. Nature Biotechnolgy 14:1668-1674.

Callahan, L.M., Weaver, K.R., Caetano-Anolles, G., Bassam, B.J., and Gresshoff, P.M. (1993) DNA fingerprinting of turfgrasses. Int. Turfgrass Soc. Res. J. 7:761-767.

Carlson, J.E., Tulsieram, L.K., Glaubitz, J.C., Luk, V., Kauffeldt, C., and Rutledge, R. (1991) Segregation of random amplified DNA markers in F1 progeny of conifers. Theor. Appl. Genet. 83:194-200.

Cifarelli, R.A., Gallitelli, M., and Cellini, F. (1995) Random amplified hybridization microsatel-lites (RAHM): isolation of a new class of microsatellite-containing DNA clones. Nucleic Acids Res. 23:3802-3803.

Cregan, P. (1997) Soybean SSR marker development - application in legumes. Final Program and Abstracts Guide, Plant and Animal Genome V, Abs. W74, p 45.

D'Ovidio, R., and Anderson, O.P. (1994) PCR analysis to distinguish between alleles of a member of a multigene family correlated with wheat bread-making quality. Theor. Appl. Genet. 88:759-763.

Edwards, A., Hammond, H.A., Jin, L., Caskey, C.T., and Chakraborty, R. (1992) Genetic variation at five trimeric and tetrameric tandem repeat loci in four human population groups.

Genomics 12:241-253.

Eng, C. and Vijg, J. (1997) Genetic testing: the problems and the promise. Nature Biotechnology 15:422-426.

Fischer, S.G. and Lerman, L.S. (1979) Length-independent separation of DNA restriction fragments in two-dimensional gel electrophoresis. Cell 16:191-200.

Guta, M., Chyi, Y.-S., Romero-Severson, J., and Owen, J.L. (1994) Amplification of DNA markers from evolutionarily diverse genomes using single primers of simple-sequence repeats. Theor. Appl. Genet. 89:998-1006.

Haff, L.A. and Smirnov, I.P. (1997) Multiplex genotyping of PCR products with MassTag-labeled primers. Nucleic Acids Res. 25:3749-3750.

Hill, M., Witsenboer, H., Zabeau, M., Vos, P., Kesseli, R., and Michelmore, R. (1996) PCR-based fingerprinting using AFLPs as a tool for studying genetic relationships in Lactuca spp. Theor. Appl. Genet. 93:1202-1210.

Konieczny, A., and Ausubel, F.M. (1993) A procedure for mapping Arabidopsis mutations using co-dominant ecotype-specific PCR-based markers. The Plant J. 4:403-410.

Kuppuswami, M.N., Hoffmann, J.W., Kasper, C.K., Spitzer, S.G., Groce, S.L., and Bajaj, S.P. (1991) Single nucleotide primer extension to detect genetic diseases: Experimental application to hemophilia B (factor IX) and cystic fibrosis genes. Proc. Natl. Acad. Sci. USA 88:1143-1147.

Marshall, E. (1997) The hunting of the SNP. Science 278:2047.

Maughan, P.J, Saghai Maroof, M.A., and Buss, G.R. (1995) Microsatellite and amplified sequence length polymorphisms in cultivated and wild soybean. Genome 38:715-723.

Monforte, J.A. and Becker, C.H. (1997) High-throughput DNA analysis by time-of-flight mass spectrometry. Nature Medicine 3:360-362.

Myers, R.M., Sheffield, V.C., and Cox, D.R. (1988) Detection of single base changes in DNA: ribonuclease cleavage and denaturing gradient gel electrophoresis In: K.E. Davies (ed.), Genome Analysis: A Practical Approach. pp. 95-139. IRL Press, Oxford.

Nikiforov, T.T., Rendle, R.B., Goelet, P., Rogers, Y-H., Kotewicz, M.L., Anderson, S., Trainor, G.L., and Knapp, M.R. (1994) Genetic Bit Analysis: a solid phase method for typing single nucleotide polymorphisms. Nucleic Acids Res. 22:4167-4175.

Orita, M., Iwahana, H., Kanazawa, H., Hayashi, K. and Sekiya T. (1989) Detection of polymorphisms of human DNA by gel electrophoresis as single-strand conformation polymorphisms. Proc. Natl. Acad. Sci. U.S.A. 86:2766-2770.

Menendez, C.M., Hall, A.E., and Gepts, P. (1997) A genetic-linkage map of cowpea (Vigna-Unguiculata) developed from a cross between 2 inbred, domesticated lines. Theor. Appl. Genet. 95:1210-1217.

Mohler, V., and Jahoor, A. (1996) Allele-specific amplification of polymorphic sites for the detection of powdery mildew resistance loci in cereals. Theor. Appl. Genet. 93:1078-1082.

Morgante, M. and Vogel, J.M. (1996) Compound microsatellite primers for the detection of genetic polymorphisms. World Patent Organization WO96/17082.

Nandi, S., Subudhi, P.K., Senadhira, D., Manigbas, N.L., Senmandi, S., and Huang, N. (1997) Mapping QTLs for submergence tolerance in rice by AFLP analysis and selective genotyping. Mol. Gen. Genet. 255:1-8.

Parks, C., Chang, L.-S., and Shenk, T. (1991) A polymerase chain reaction mediated by a single primer: cloning of genomic sequences adjacent to a serotonin receptor protein coding region.

Nucleic Acids Res. 19:7155-7160.

Powell, W., Morgante, M., Andre, C., Hanafey, M., Vogel, J., Tingey, S., and Rafalski, A. (1996) The comparison of RFLP, RAPD, AFLP and SSR (microsatellite) markers for germplasm analysis. Molecular Breeding 2:225-238.

Powell, W., Thomas, W.T.B., Baird, E., Lawrence, P., Booth, A., Harrower, B., McNicol, J.W., and Waugh, R. (1997) Analysis of quantitative traits in barley by the use of amplfied fragment length polymorphisms. Heredity 79:48-59.

Reiter, R.S., Williams, J.G.K., Feldmann, K.A., Rafalski, J.A., Tingey, S.V. and Scolnik, P.A. (1992) Global and local genome mapping in Arabidopsis thaliana by using recombinant inbred lines and random amplified polymorphic DNAs. Proc. Natl. Acad. Sci. USA 89:1477-1481.

Richardson, T., Cato, S., Ramser, J., Kahl, G., and Weising, K. (1995) Hybridization of microsatellites to RAPD: a new source of polymorphic markers. Nucleic Acids Res. 23:3798-3799.

Saiki, R.K., Scharf, S. Faloona, F., Mullis, K.B., Horn, G.T., Erlich, H.A. and Arnheim, N. (1985) Enzymatic amplification of ß-globin genomic sequences and restriction site analysis for diagnosis of sickle cell anemia. Science 230:1350-1354.

Sax, K. (1923) The association of size differences with seed coat pattern and pigmentation in Phaseolus vulgarus. Genetics 8:552:560.

Stuber, C.W. and Goodman M.M. (1983) Allozyme genotypes for popular and historically important inbred lines of corn, Zea mays L. USDA Agric. Res. Results, Southern Ser., No. 16.

Syvanen, A.-C., Aalto-Setala, K., Harju, L., Kontula, K., and Soderlund, H. (1990) A primer-guided nucleotide incorporation assay in the genotyping of apolipoprotein E. Genomics 8:684-692.

Tingey, S.V., Rafalski, J.A., and Williams, J.G.K. (1992) Genetic analysis with RAPD markers. In: Proceedings of the Symposium, Applications of RAPD Technology to Plant Breeding. Crop Sci. Soc./Amer. Soc. Hort. Sci./Amer. Genet. Assoc. p 3-8.

To, K.-Y., Liu, C.-I., Liu, S.-T., and Chang, Y.-S. (1993) Detection of point mutations in the chloroplast genome by single-stranded conformation polymorphism analysis. Plant J. 3:183-186.

Vos, P., Hogers, R., Bleeker, M., Reijans, M., van de Lee, T., Fornes, M., Frijters, A., Pot, J., Peleman, J., Kuiper, M., and Zabeau, M. (1995) AFLP: a new technique for DNA fingerprinting. Nucleic Acids Res. 23:4407-4414.

Vuylsteke, M., Anonise, R., Bastiaans, E., Senior M., Stuber, C., and Kuiper, M. (1997) A high density AFLP Linkage map of Zea mays L. Final Program and Abstracts Guide, Plant and Animal Genome V, Abs. P206, p 104.

Wang, D., Sapolsky, R., Spencer, J., Rioux, J., Kruglyak, L., Hubbell, E., Ghandour, G., Hawkins, T., Hudson, T., Lipshutz, R., and Lander, E. (1996) Toward a third generation genetic map of the human genome based on bi-allelic polymorphisms. Am. J. Human Genet. 59 supplement p. A3.

Welsh, J. and McClelland, M. (1990) Fingerprinting genomes using PCR with arbitrary primers. Nucleic Acids Res. 18:7213-7218.

Welsh, J., Petersen, C. and McClelland, M. (1991) Polymorphisms generated by arbitrarily primed PCR in the mouse: application to strain identification and genetic mapping. Nucleic Acids Res. 19:303-306.

Williams, J.G.K., Kubelik, A.R., Livak, K.J., Rafalski, J.A. and Tingey S.V. (1990) DNA polymorphisms amplified by arbitrary primers are useful as genetic markers. Nucleic Acids Res. 18:6531-6535.

Williams, K.J., Fisher, J.M., and Langridge, P. (1996) Development of a PCR-based allele-specific assay from an RFLP probe linked to resistance to cereal cyst nematode in wheat. Genome 39:798-801.

Wu, K.-S., Jones, R., Danneberger, L., and P.A. Scolnik (1994) Detection of microsatellite polymorphisms without cloning Nucleic Acids Res. 22:3257-3258.

Xu, L. and Hall, B.G. (1994) SASA: a simplified, reliable method for allele-specific amplification of polymorphic sites. BioTechniques 6:44-45.

Zietkiewicz, E., Rafalski, A., and D. Labuda (1994) Genome fingerprinting by simple sequence repeat (SSR)-anchored polymerase chain reaction amplification. Genomics 20:176-183.

3. Constructing a plant genetic linkage map with DNA markers

NEVIN DALE YOUNG

Department of Plant Pathology, 495 Borlaug Hall, University of Minnesota,
St. Paul, Minnesota 55108 <neviny@tc.umn.ed>

Contents

1. Overview

Scientists are constructing genetic linkage maps composed of DNA markers for a wide range of plant species (O'Brien, 1993). Several types of DNA markers have been widely used (Fig. 1): restriction fragment length polymorphisms (RFLPs) (Botstein et al. 1980), random amplified polymorphic DNAs (RAPDs) (Williams et al. 1990), simple sequence repeats (SSRs or microsatellites) (Litt and Luty 1989) and amplified fragment length polymorphisms (AFLPs) (Vos et al. 1995). In the future, maps built from single nucleotide polymorphisms (SNPs) in combination with DNA chip technology are likely (Wang et al. 1998). All types of DNA markers detect sequence polymorphisms and monitor the segregation of a DNA sequence among progeny of a genetic cross in order to construct a linkage map. While the theory of linkage mapping is the same for DNA markers as in classical genetic mapping, special considerations must be kept in mind. This is primarily a result of the fact that potentially unlimited numbers of DNA markers can be analyzed in a single mapping population. Backcross and F2 populations are suitable for DNA-based mapping, but recombinant inbred (Burr and Burr, 1991) and doubled haploid lines (Huen et al. 1991) provide permanent mapping resources. These types of populations are also better suited for analysis of quantitative traits.

R.L. Phillips and I.K. Vasil (eds.), DNA-Based Markers in Plants, 31 - 47.

RFLP
Restriction Fragment Length
Polymorphism

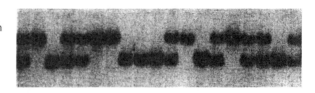

RAPD
Random Amplified
Polymorphic DNA

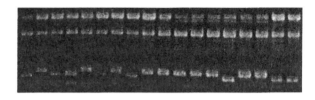

SSR
Simple Sequence Repeats
or "Microsatellites"

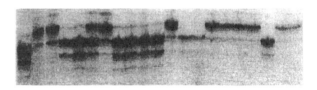

AFLP
Amplified Fragment Length
Polymorphism

Fig. 1.   Different Types of DNA Genetic Markers. Four of the typical DNA markers systems (RFLP, RAPD, SSR, and AFLP) used in plant genetic mapping. Details in text.

The number of DNA markers on published linkage maps ranges from a few hundred to several thousand (Tanksley et al. 1993; Keim et al. 1997; Qi et al. 1998), so the resolution of DNA marker maps can be extremely high. However, competing maps have been constructed in some species without links between maps and additional effort is required to join the information (Beavis and Grant, 1991; Qi et al. 1996). DNA-based maps can be related to existing cytogenetic maps through the use of aneuploid or substitution lines (Helentjaris et al. 1986; Sharp et al. 1989; Young et al. 1987; Rooney et al. 1994). Recently, the focus has shifted to relationships between genetic and physical maps (Schmidt, et al. 1997; Zhang and Wing, 1997).

Applications of DNA markers to plant breeding and genetics have been described in previous reviews (Soller and Beckmann, 1983; Tanksley et al. 1989; Lee, 1995). Moreover, details about recent innovations in DNA sequencing and DNA chip technology

— and their impact on map construction — are beyond the scope of this chapter. Here, practical strategies for constructing genetic linkage maps using DNA markers will be described. Because of the breadth of this area, only an introduction to the concepts and techniques can realistically be covered.

2.   Constructing a linkage map with DNA markers

2.1. The mapping population

One of the most critical decisions in constructing a linkage map with DNA markers is the mapping population. In making this decision, several factors must be kept in mind, the most important of which is the goal of the mapping project. Is the goal simply to generate a framework map to provide a set of mapped loci for the future, or instead, to identify and orient DNA markers near a target gene for eventual map-based cloning? Perhaps the goal is mapping quantitative trait loci (QTL), or the monitoring of several disease resistance loci in the process of pyramiding them into a single background. Whichever goal is the motivating factor behind mapping, it will have a critical influence on which parents are chosen for crossing, the size of the population, how the cross is advanced, and which generations are used for DNA and phenotypic analysis.

2.2. DNA polymorphisms among parents

Sufficient DNA sequence polymorphisms between parents must be present. This cannot be overemphasized, for in the absence of DNA polymorphism, segregation analysis and linkage mapping are impossible. Naturally outcrossing species, such as maize, tend to have high levels of DNA polymorphisms and virtually any cross that does not involve related individuals will provide sufficient polymorphism for mapping (Helentjaris, 1987). However, levels of DNA sequence variation are generally lower in naturally inbreeding species and finding suitable DNA polymorphisms may be more challenging (Miller and Tanksley, 1990). Sometimes mapping of inbreeding species requires that parents be as distantly related as possible, which can often be inferred from geographical, morphological, or isozyme diversity. In some cases, suitable wide crosses may already be available because a frequent goal in plant breeding in the past has been the introduction of desirable characters from wild relatives into cultivars. Moreover, SSR markers tend to exhibit high levels of polymorphism, even in narrow crosses (Rongwen et al. 1995), providing the possibility of constructing maps in crosses between closely related parents.

2.3. Choice of segregating population

Once suitable parents have been chosen, the type of genetic population to use for linkage mapping must be considered. Several different kinds of genetic populations are suitable. The simplest are F2 populations derived from F1 hybrids and backcross populations. For most plant species, populations such as these are easy to construct, although sterility in the F1 hybrid may limit some combinations of parents, particularly in wide crosses.

The major drawback to F2 and backcross populations is that they are ephemeral, that is, seed derived from selfing these individuals will not breed true. This limitation can be overcome to a limited extent by cuttings, tissue culture or bulking F3 plants to provide a constant supply of plant material for DNA isolation. Nevertheless, it is difficult or impossible to measure characters as part of quantitative trait locus (QTL) mapping in several locations or over several years with F2 or backcross populations. For these reasons, permanent resources for genetic mapping are essential.

The best solution to this dilemma is the use of inbred populations that provide a permanent mapping resource. Recombinant inbred (RI) lines derived from individual F2 plants are an excellent strategy (Burr et al. 1988; Burr and Burr, 1991). RI lines are created by single seed descent from sibling F2 plants through at least five or more generations. This process leads to lines that each contain a different combination of linkage blocks from the original parents. The differing linkage blocks in each RI line provide a basis for linkage analysis. However, several generations of breeding are required to generate a set of RIs, so this process can be quite time-consuming. Moreover, some regions of the genome tend to stay heterozygous longer than expected from theory (Burr and Burr, 1991) and obligate outcrossing species are much more difficult to map with RIs because of the difficulty in selfing plants.

Nevertheless, in cases where it is feasible, seed from RI lines is predominantly homogeneous and abundant, so the seed can be sent to any lab interested in adding markers to an existing linkage map previously constructed with the RI lines. Moreover, RI lines can be grown in replicated trials, several locations, and over several years — making them ideal for QTL mapping. Similar types of inbred populations, such as doubled haploids, can also be used for linkage mapping with many of the same advantages of RI lines (Huen et al. 1991), while recurrent intermated populations have been used for genome-wide high resolution mapping (Liu et al. 1996). Still another mapping population of great value is the backcross inbred population (Bernacchi et al, 1998), in which germplasm development and QTL mapping proceed simultaneously.

2.4. Population size

Once an appropriate mapping population has been chosen, the appropriate population size must be determined. Since the resolution of a map and the ability to determine marker order is largely dependent on population size, this is a critical decision. Clearly, population size may be technically limited by how many seeds are available or by the number of DNA samples that can reasonably be prepared. Whenever possible, the larger the mapping population the better. Populations less than 50 individuals generally provide too little mapping resolution to be useful. Moreover, if the goal is high resolution mapping in specific genomic regions or mapping QTLs of minor effect, much larger populations will be required. Based on Monte Carlo simulations, Bevis concluded that populations smaller than 200 individuals would rarely be successful in find most QTLs, and in many cases, populations larger than 500 are required (Bevis, 1994). For example, Messeguer et al. (1991) examined over 1000 F2 plants to construct a high resolution map around the *Mi* gene of tomato, Stuber et al. (1987) analyzed over 1800 maize F2s to find QTLs controlling as little as 1% of the variation in yield components, and Alpert and Tanskley analyzed more than 3,400 individuals to obtain a detailed map around a fruit weight locus (Alpert and Tanskley 1996).

2.5. DNA extractions

No matter what type of population or DNA marker one plans to use, DNA must first be isolated from the plants in the mapping population. Fortunately, plants can be grown in a variety of environments and in different locations and still provide starting material for DNA isolation. This is in contrast to phenotypic markers, such as morphological or disease resistance traits, whose expression tend to be highly dependent upon growth conditions.

Several methods for DNA extraction have been developed, beginning with those aimed at RFLP technology (Dellaporta et al. 1983; Murray and Thompson, 1984; Tai and Tanskley, 1990). More recently, researchers have moved to polymerase chain reaction (PCR)-based markers, which all require smaller amounts of starting material and simpler extraction technologies (Berthomieu and Meyer 1991; Edwards et al. 1991; Lamalay et al. 1990; Lange et al. 1998; Luo et al. 1992; Thomson and Henry 1995; Wang et al. 1993). With these methods, the goals are simplicity, speed, and a small amount of starting material. Simplicity and speed are absolutely essential for processing large numbers of individuals — an obvious necessity when large populations of several hundred, or even thousands, of individuals need to be examined. Small amounts of starting material are advantageous if larger quantities are hard to obtain, such as seeds, seedlings, or physically small plants like *Arabidopsis*.

DNA used for genetic mapping does not need to be highly purified. As long as an extraction provides DNA in sufficient quantity and quality for restriction enzyme digestion or as a template for PCR, the method is probably satisfactory. Further efforts to purify DNA take time and cut down on the number of samples that can be processed. In general, limits to genetic mapping are more often due to small numbers of individuals in a mapping population (or difficulties with associated phenotypic scoring) than to DNA purity.

Still, one must guard against the most troublesome problems of DNA marker analysis. In the case of RFLPs, the major artifact is partial digestion of DNA. Since methods to extract DNA are streamlined, the DNA used in RFLP analysis can be quite impure. Sometimes this leads to partial digestion, which invariably leads to the appearance of extra bands upon autoradiography. It is very frustrating trying to map a "polymorphic band" that turns out to be only a partial digest in one parent. Complete digests of plant genomic DNA have a distinctive appearance upon gel electrophoresis, including a smear of DNA fragments throughout the appropriate size range for the restriction enzyme used, as well as the presence of reproducible DNA bands derived from the chloroplast. Moreover, partial digests lead to bands on autoradiographs that are generally fainter and higher in molecular weight than authentic restriction fragments.

Problems with RFLP mapping can also arise if too little DNA is used. Because RFLPs generally represent single copy sequences, the amount of any one target sequence in a genomic DNA sample can be vanishingly small. If too little DNA is loaded onto the gel for blotting, it may be impossible to see a signal after hybridization and autoradiography. Clearly, this will be related to genome size, and organisms with smaller genomes may require less DNA per sample than species with very large genomes. In practice, at least two micrograms, and potentially as much as ten micrograms or more of DNA should be used for RFLP analysis.

Problems arising from artificatual bands and a lack of reproducibility are also problems with PCR-based markers, especially RAPDs. Because of the enormous amplification associated with PCR, as well as the tenuous association between the decamer primers used in RAPD analysis and template genomic DNA, variations in the set of amplified DNA molecules observed with a single primer are not uncommon. Variables as simple as differences in template and primer concentration can lead to the appearance or disappearance of DNA products. Moreover, PCR reactions in the absence of a plant DNA template can sometimes lead to products, possibly due to the synthesis of "primer dimers" or even minute contamination of foreign DNA template. Because of these artifactual DNA products, special care must be taken to optimize and standardize PCR reactions based on RAPDs. For these reasons, only the most prominent and dependable bands in a RAPD reaction should typically be used for mapping.

Care must also be taken with AFLP markers. AFLP technology involves several steps: restriction digestion, PCR, DNA ligation (Vos et al. 1995). Problems with the DNA at any of these steps can lead to poor gel resolution or unreliable bands that could represent artifacts. Still, when carried out with care, AFLP technology has the potential to produce dozens of mappable bands from a single reaction. In any case, SSR markers are generally the most reliable and highly reproducible. Indeed, SSRs are now widely recognized as the foundation for many framework linkage maps (Akkaya et al. 1995; Bell and Ecker, 1994). Unfortunately, SSR markers are also the most difficult type of marker to generate in the first place.

3.   Relationships among genetic maps

3.1. Relationship between DNA marker and cytogenetic maps

The most common method to relate DNA marker maps to specific chromosomes is the use of aneuploids, such as monosomics (Helentjaris et al. 1986; Rooney et al. 1994), trisomics (Young et al. 1987), and substitution lines (Sharp et al. 1989). In species where aneuploid lines for each chromosome are available, nucleic acid hybridization with a mapped DNA clone indicates its chromosome location by observing the loss of a band (in the case of nullisomics) or a change in the relative signal on an autoradiogram (McCouch et al. 1988). This type of analysis may require "within-lane" standards (such as a second DNA clone of previously determined chromosome location), so that subtle changes in the relative intensity of a band can be compared between lanes.

Using substitution lines to associate mapped DNA markers to specific chromosomes is similar in concept to aneuploid mapping. In cereal species where this approach is most common, lines with known chromosomes or chromosome arms substituted with homoeologous segments from alien species have been developed. Probing a DNA clone onto a blot containing restriction digested DNA from a complete set of substitution lines easily identifies the chromosome location of that clone (Sharp et al. 1989). This is because the substitution line corresponding to the location of a clone shows a different restriction fragment pattern compared to the other substitution lines.

3.2. Relationship between genetic and physical maps

Eventually, distances between DNA markers need to be described not only by recombinational frequency, but by actual physical distance. Soon this kind of information will be abundantly clear in *Arabidopsis* and rice through complete physical mapping and eventual genome sequencing (Schmidt, et al. 1997; Zhang and Wing, 1997). Even in other more complex plant genomes, positional cloning projects based on yeast artificial chromosome (YAC) and bacterial artificial chromosome (BAC) libraries are beginning to shed light on genetic to physical relationships. Fine structure mapping of the same genome region using both recombinational and physical techniques is the best method to compare different types of maps directly. One general observation has been that the relationship between genetic and physical distance varies dramatically according to location on a chromosome (Ganal et al. 1989). In more recent studies, large genomic contigs have provided estimates for the ratio between kilobase pairs (kbp) and centimorgans (cM). In one study in *Arabidopsis,* this ratio was estimated at 160 kbp/cM averaged over 1,440 kbp genomic segment near the top of chromosome V (Thorlby et al. 1997). In tomato, a study of a 610 kbp region found that the ratio changed abruptly from 105-140 kbp/cM to less than 24 kbp/cM (Gorman et al. 1996). Indeed, in the *bronze* locus of maize, the level of recombination has been shown to be more than 100 times greater than the genome as a whole (Dooner and Marinez-Ferez, 1997).

3.3. Parallel mapping in the same species

In the most important plant species there are often multiple efforts to construct DNA-based genome maps. This has led to the unfortunate situation of having several maps for the same species with little or no information correlating one map to another. Of course this makes it difficult to relate the reported location of a gene on one map to its location on another map. It also means that the maps are less saturated, and therefore less powerful, than they could be.

Even where there is no proprietary barrier to relating maps to one another, there are often practical and theoretical problems. The most obvious is that markers polymorphic in one mapping population may not show variation in a second population. The first genetic maps were based on mapping populations optimized for DNA polymorphisms, often including parents from distinct, but cross-compatible species. As researchers move to more narrow crosses, previously excellent genetic markers will be useless for lack of polymorphism. When this happens it will be difficult to relate genetic map location between populations, except by cloning sequences that flank the original marker (a substantial amount of effort) or by testing adjacent DNA markers in hopes that they show more sequence variation.

A similar problem may be observed when one attempts to relate RAPD markers among different crosses. While there are often several bands observed in the analysis of each RAPD primer, only one of the bands may be polymorphic between two individuals (Williams et al. 1990). If an identical RAPD primer is analyzed in a second population, there is no guarantee that the same band (locus) will be the one that segregates. While any bands that do segregate in the second population will be suitable as markers, it is unlikely that they represent the same locus as the original marker. Similar situations can arise with RFLPs if they correspond to a sequence with multiple loci. Finally, there can be theoretical problems in relating linkage order data from one map to another, since each map is based on a different set of segregating individuals. However, the use of appropriate computer algorithms can potentially overcome this problem (Qui et al. 1996; Stam 1993).

Simple sequence repeat markers have played a critical role in merging disparate linkage maps (Akkaya et al. 1995; Bell and Ecker, 1994). Because they are nearly always single locus markers, even in complex genomes like the grasses and soybean, SSRs define specific locations in a genome unambiguously. This makes them suitable to tie multiple maps together. Moreover, being PCR-based, the information necessary to map SSR loci can be shared among labs simply by sharing primer sequence data.

3.4. Parallel mapping in related taxa

One of the most powerful aspects of genetic mapping with DNA markers, particularly RFLPs, is the fact that markers mapped in one genus or species can often be used to construct parallel maps in related, but genetically incompatible, taxa. For this reason, a new mapping project can often build on previous mapping work in related organisms. Examples include a potato map constructed with tomato markers (Bonierbale et al. 1988; Gebhardt et al. 1991; Tanksley et al. 1993), sorghum maps constructed with maize markers (Hulbert et al. 1990; Pereira et al. 1994), a turnip map constructed with markers from cabbage (McGrath and Quiros, 1991), and a mungbean map constructed with markers from both soybean and common bean (Menancio-Hautea et al. 1993).

Not only does a pre-existing map provide a set of previously tested DNA markers, it also gives an indication of linkage groups and marker order. In the case of tomato and potato, only five paracentric inversions involving complete chromosome arms differentiate the two maps (Bonierbale et al. 1988; Gebhardt et al. 1991; Tanksley et al. 1992). Similar conservation of linkage order was observed between sorghum and maize (Hulbert et al. 1990; Pereira et al. 1994) and indeed, among most of the grasses (Bennetzen and Freeling, 1993) as well as among legumes (Boutin et al. 1995). In cases like these, markers can be added to a new map in an optimum manner, either by focusing on markers evenly distributed throughout the genome, or by targeting specific regions of interest (Concibido et al. 1996). In some cases, though, DNA clones may hybridize in multiple taxa, yet show little conservation in linkage group or order.

Even though the tomato and potato maps are nearly homosequential (syntenic) in marker order, both differ significantly from the linkage map of pepper, despite the fact that all were constructed with the same RFLP markers (Prince et al. 1993).

4.   Targeting specific genomic regions

In most cases, genome mapping is directed toward a comprehensive genetic map covering all chromosomes evenly. This is essential for effective marker-assisted breeding, QTL mapping, and chromosome characterization. However, there are special situations in which specific regions of the genome hold special interest. One example is where the primary goal of a research project is map-based cloning. In this case, markers that are very close to a target gene and suitable as starting points for chromosome walking are needed, so the goal is to generate a high density linkage map around that gene as quickly as possible. While the construction of a complete genome map by conventional means eventually leads to a high density map throughout the genome, special strategies for rapidly targeting specific regions have also been developed.

The first strategy for targeting specific regions was based on near isogenic lines (NILs). Over the years, breeders have utilized recurrent backcross selection to introduce traits of interest from wild relatives into cultivated lines. This process led to the development of pairs of NILs; one, the recurrent parent and the other, a new line resembling the recurrent parent throughout most of its genome except for the region surrounding the selected gene(s). This introgressed region, derived from the donor parent and often highly polymorphic at the DNA sequence level, provides a target for rapidly identifying clones located near the gene of interest (Young et al. 1988; Martin et al. 1991; Paran et al. 1991; Muehlbauer et al. 1991). NILs make it easy to determine the location of a marker relative to the target gene. This is in contrast to typical genetic mapping where it would be necessary to test every clone with a complete mapping population to determine whether it mapped near the gene of interest.

Another, more general strategy makes it possible to target specific genomic regions without the need for developing specialized genotypes, generally known as bulked segregant analysis (Michelmore et al. 1991; Giovanonni et al. 1992). The strategy is to select individuals from a segregating population that are homozygous for a trait of interest and pool their DNA. In the pooled DNA sample, the only genomic region that will be homozygous will be the region encompassing the genomic region of interest, which can then be used as a target for screening DNA markers rapidly. This means that any trait that can be scored in an F2, backcross, or RI population can now be rapidly targeted with DNA markers (Zhang et al. 1994). Used in conjunction with AFLP markers, it is possible to identify large numbers of DNA markers in a region of interest in a short time.

Moreover, pooled DNA samples can also be generated based on homozygosity for a DNA marker (as opposed to a phenotypic trait). In this way, any genomic region of interest that has been previously mapped in terms of DNA markers can be rapidly targeted with new markers. This may be especially useful in trying to fill in gaps on a genetic map. All that is required is a pooled DNA sample selected on the basis of DNA markers flanking the genomic region of interest (Giovanonni et al. 1992).

5.   Computer software for genetic mapping

Advances in computer technology have been essential for progress in DNA marker maps. While the theory behind linkage mapping with DNA markers is identical to mapping with classical genetic markers, the complexity of the problem has increased dramatically. Linkage order is still based on maximum likelihood, in other words, the order of markers that yields the shortest distance and requires the fewest multiple crossovers between adjacent markers. Likewise, genetic distance between markers is measured in centimorgans, which is based on the frequency of genetic crossing-over (accounting for the likelihood of, and interference among, multiple crossovers). For these reasons, the concepts and theories previously developed for classical genetic mapping can still be applied to mapping with DNA markers. The difference between classical and DNA-based mapping lies in the number of markers that are mapped in a single population. With DNA-based genetic maps, this number can easily reach into the thousands, and so there is a close connection between progress in DNA markers and advances in computer technology.

In the simplest situations, all that is required to construct a linkage map from DNA marker data are statistics software packages capable of running Chi-squared contingency table analysis. This statistical test determines two-point linkage between markers, which can then form a basis for constructing linkage groups. Unfortunately, as the number of markers begins to grow, this approach becomes increasingly unsuited for comparing possible orders and choosing the best. Still, in research situations where computer power is limiting and where linkage analysis is based on relatively few markers, this strategy is perfectly suitable.

For most mapping projects the most widely-used genetic mapping software is Mapmaker (Lander et al. 1987). Mapmaker is based on the concept of the LOD score, the "log of the odds-ratio" (Morton, 1955). A LOD score indicates the log (10) of the ratio between the odds of one hypothesis (for example, linkage between two loci) versus an alternative hypothesis (no linkage in this example). Through the use of the LOD score, data from different populations can be pooled — one reason that the program has gained so much popularity among human and animal geneticists where population sizes can be limiting. Yet even in plants, Mapmaker has become a virtual standard for constructing genetic linkage maps, as indicated by its widespread use.

Mapmaker's popularity for genetic analysis is based on the ease with which it performs multipoint analysis of many linked loci. Most plant genetic linkage maps have at least one hundred markers, and sometimes one thousand markers or more. Therefore, fast and simple multipoint analysis is absolutely essential to sort out the many different possible marker orders. Mapmaker has several routines that simplify multipoint analysis, including an algorithm that quickly gathers markers into likely linkage groups and another for guessing the best possible order. Once a plausible order has been established, another algorithm compares the strength of evidence for that order compared to possible alternatives in a routine called "ripple". The power of this routine is that it enables the user to confirm the best order in a way that increases only arithmetically with increasing number of loci (as opposed to factorially, if all possible orders must be compared).

Just building a linkage map of DNA markers is generally just a first step in genetic marker analysis. As noted earlier, it is often essential to relate one's map to those derived from other mapping populations. The computer program JoinMap is specifically suited for this application (Stam, 1993). Often one wishes to apply information about a linkage map to QTL analysis. Indeed, Mapmaker has even been modified to carry out quantitative trait locus (QTL) analysis using mathematical models and an interface very much like the original program (Lander and Botstein, 1989). Other programs like QTL Cartographer (Basten et al. 1998) provide very much the same type of analysis. Sometimes, linkage mapping information is intended for marker-assisted breeding. A program like Map Manager (Manly and Cudmore, 1998) helps to keep track of marker data in a population of interest, while Hypergene helps to display graphical genotypes (Young and Tanksley, 1989). The program qGENE seeks to bring all of these important DNA marker tools together into a single package (Nelson, 1997).

These computer programs demonstrate the close connection that is evolving between genetic analysis and computer technology. Indeed, the U.S. Department of Agriculture has established "Genome Mapping Database Projects" for several of the most important crop species, including maize, wheat, soybean, and pine. These databases, which are easily accessed through the world wide web (http://probe.nalusda.gov:8000/index.html) incorporate some of the routines mentioned above, but focus primarily on collecting information from separate mapping groups (and associated phenotype data) into a single repository. Since the databases are graphically-based and use "click-and-point" routines, they are easy to learn and use. Increasingly, these databases will be updated to become more relational in design, as well as to interact with DNA sequence databases and "data-mining" projects.

6.   Perspectives on genetic mapping and DNA markers

In the brief period since DNA marker technology was first applied to plants, there has been an explosion in the development and application of genetic linkage maps. Using these new DNA-based maps, researchers have constructed maps in species where only poorly populated classical maps existed before (Bonierbale et al. 1988; Grattapaglia and Sederoff 1994; Gebhardt et al. 1991; Landry et al. 1987; Menancio-Hautea et al. 1993), located genes for both qualitative and quantitative characters (Concibido et al. 1997; Lin et al. 1995; Mansur et al. 1993), often in great detail, and provided a basis for positional cloning (Tanksley et al. 1995). Despite this incredible progress, DNA marker technology still has a long way to go before its full potential is realized.

With current procedures, the number of plant samples and DNA markers that can reasonably be processed limits the widespread application of mapping technology. Even the most efficient DNA extraction techniques handle only a few hundred samples each day, and once samples have been isolated, significant investments in time and effort are still required to obtain genotypic information. Given the fact that a typical breeding project might include several thousand, or even tens of thousands of individuals, and since information is needed as quickly as possible to make breeding decisions, the current technical limitations are significant. These limitations also constrain the application of DNA marker technology in QTL mapping to genetic factors with relatively major effects. Finally, DNA marker technology is still so technically complex that it is practically impossible for it to be applied where it is needed most — in less-developed countries.

However, better types of DNA genetic markers are on the horizon. The most important are SNPs that can be assayed through DNA chip technology. Already, researchers working with the human genome have constructed a map of more than 2000 SNP markers, many assayed by chip technology (Wang et al. 1998). It may also be possible to create the equivalent of "radiation-hybrids" for plants. This is an extremely powerful resource that has routinely been used in animal genome mapping for decades. Research have recently shown that it is possible to create stable lines of oats that contain a small segment of maize genome (Ananiev et al. 1997). If enough oat lines carrying overlapping maize segments can be generated, extremely fast and efficient high resolution mapping may be achievable.

Even as DNA marker technology advances, parallel achievements are essential in complementary technologies. As the number of markers, genetic resolution, and amount of mapping information grows, so does the need for better computer algorithms and databases. Finally, genetic linkage maps, even those based on DNA markers, are still limited by the range of sexual crosses that can be made. To make the most of genes uncovered through genetic mapping, improvements in making wide crosses, somatic hybrids, and plant transformation will be essential. In the future, better DNA markers, along with advances in these complementary technologies, will enable linkage mapping, one of the oldest genetic techniques, to become one the of most powerful.

7.   Acknowledgements

This paper is published as a contribution of the series of the Minnesota Agricultural Experiment Station on research conducted under Project 015, supported by G.A.R. funds.

8.   References

Akkaya, M.S., Shoemaker, R.C., Specht, J.E., Bhagwat, A.A. and Cregan, P.B. (1995) Integration of simple sequence repeat DNA markers into a soybean linkage map. Crop Sci. 35: 1439-1445.

Alpert, K.B. and Tanksley, S.D. (1996) High-resolution mapping and isolation of a yeast artificial chromosome contig containing fw2.2 - a major fruit weight quantitative trait locus in tomato. Proc. Natl. Acad. Sci. U.S.A. 93: 15503-15507.

Ananiev, E.V., Riera-Lizarazu, O., Rines, H.W. and Phillips, R.L. (1997) Oat-maize chromosome addition lines - a new system for mapping the maize genome. Proc. Natl. Acad. Sci. U.S.A. 94: 3524-3529.

Basten, C.J., Weir, W.S., and Zeng, Z.-B. (1998) QTL cartographer: a suite of programs for mapping quantitative trait loci. Plant and Animal Genome VI (San Diego, CA, January 1998). p. 76.

Beavis, W.D. and Grant, D. (1991) A linkage map based on information from four F2 populations of maize (Zea mays L.). Theor. Appl. Genet. 82: 636-644.

Bell, C.J. and Ecker, J.R. (1994) Assignment of 30 microsatellite loci to the linkage map of Arabidopsis. Genomics 19:137-144.

Bennetzen, J.L. and Freeling, M. (1993) Grasses as a single genetic system: genome composition, collinearity and compatibility. Trends Genet. 9:259-261.

Bernacchi, D., Beck-Bunn, T., Eshed, Y., Lopez, J., Petiard, V., Uhlig, J., Zamir, D. and Tanksley, S.D. (1998) Advanced backcross QTL analysis in tomato. I. Identification of QTLs for traits of agronomic importance from Lycopersicon hirsutum. Theo. Appl. Genet. 97: 381-397.

Berthomieu, P. and Meyer, C. (1991) Direct amplfication of plant genomic DNA from leaf and root pieces using PCR. Plant Molec. Biol. 17: 555-557.

Bevis, W.D. (1994) The power and deceit of QTL experiments: lessons from comparative QTL studies. Proceedings 49th Annual Corn and Sorghum Industry Research Conference, Washington, D.C. American Seed Trade Association.

Bonierbale, M.W., Plaisted, R.L. and Tanksley, S.D. (1988) RFLP maps based on common sets of clones reveal modes of chromosome evolution in potato and tomato. Genetics 120: 1095-1103.

Botstein, D., White, R.L., Skolnick, M. and Davis, R.W. (1980) Construction of a genetic linkage map in man using restriction fragment length polymorphisms. Amer. J. Hum. Genet. 32: 314-331.

Boutin, S., Young, N.D., Olson, T., Yu, Z.H., Shoemaker. R.C. and Vallejos, C.E. (1995) Genome conservation among three legume genera detected with DNA markers. Genome 38: 928-937.

Burr, B. and Burr, F.A. (1991) Recombinant inbreds for molecular mapping in maize. Trends Genet. 7: 55-60.

Burr, B., Burr, F.A., Thompson, K.H., Albertson, M.C. and Stuber, C.W. (1988) Gene mapping with recombinant inbreds in maize. Genetics 118: 519-526.

Concibido, V.C., Lange, D.A., Denny, R.L., Orf, J.H., and Young, N.D. (1997) Genome mapping of soybean cyst nematode resistance loci in 'Peking', PI 90763, and PI 88788 using DNA markers. Crop Sci. 37:258-264.

Concibido, V.C., Young, N.D., Lange, D.A., Denny, R.L., Danesh, D., and Orf, J.H. (1996) Targeted comparative genome analysis and qualitative mapping of a major partial-resistance gene to the soybean cyst nematode. Theor. Appl. Genet. 93: 234-241

Dellaporta, S.L., Wood, J. and Hicks, J.B. (1983) A plant DNA minipreparation: version II. Plant. Molec. Biol. Rep. 1: 19-21.

Dooner, H.K. and Martinez-Ferez, I,M (1997) Recombination occurs uniformly within the bronze gene, a meiotic recombination hotspot in the maize genome Plant Cell. 9:1633-1646.

Edwards, K., Johnstone, C., and Thompson, C. (1991) A simple and rapid method for the preparation of plant genomic DNA for PCR analysis. Nucl. Acids Res. 19: 1349.

Ganal, M.W., Young, N.D. and Tanksley S.D. (1989) Pulsed field gel electrophoresis and physical mapping of large DNA fragments in the Tm-2a region of chromosome 9 in tomato. Mol. Gen. Genet. 215: 395-400.

Gebhardt, C., Ritter, E., Barone, T., Debener, T., Walkemeier, B., Schachtschabel, U., Kaufman, H., Thompson, R.D., Bonierbale, M.W., Ganal, M.W., Tanksley, S.D. and Salamini, F. (1991) RFLP maps of potato and their alignment with the homoeologous tomato genome. Theor. Appl. Genet. 83: 49-57.

Giovanonni, J., Wing, R. and Tanksley, S.D. (1992) Isolation of molecular markers from specific chromosomal intervals using DNA pools from existing mapping populations. Nucl. Acid Res. 19: 6553-6558.

Gorman, S.W., Banasiak, D., Fairley, C. and McCormick, S. (1996) A 610 kb YAC clone harbors 7 cM of tomato (Lycopersicon esculentum) DNA that includes the male sterile 14 gene and a hotspot for recombination. Molec. Gen. Genet. 251:52-59.

Grattapaglia, D. and Sederoff, R. (1994) Genetic linkage maps of eucalyptus grandis and Eucalyptus urophylla using a pseudo-testcross - mapping strategy and RAPD markers. Genetics. 137:1121-1137.

Helentjaris, T. (1987) A genetic linkage map for maize based on RFLPs. Trends Genet 3: 217-221.

Helentjaris, T., Weber, D.F. and Wright, S. (1986) Use of monosomics to map cloned DNA fragments in maize. Proc. Natl. Acad. Sci. U.S.A. 83: 6035-6039.

Huen, M., Kennedy, A.E., Anderson, J.A., Lapitan, N.L.V., Sorrells, M.E. and Tanksley, S.D. (1991) Construction of a restriction fragment length polymorphism map for barley (Hordeum vulgare). Genome 34: 437-447.

Hulbert, S.H., Richter, T.E., Axtell, J.D. and Bennetzen, J.L. (1990) Genetic mapping and characterization of sorghum and related crops by means of maize DNA probes. Proc Natl Acad Sci U.S.A. 87: 4251-4255.

Keim, P., Schupp, J.M., Travis, S.E., Clayton, K., Zhu, T., Shi, L.A., Ferreira, A. and Webb, D.M. (1997) A high-density soybean genetic map based on AFLP markers. Crop Sci. 37: 537-543

Lamalay, J.C., Tejwani, R. and Rufener, G.K.I. (1990) Isolation and analysis of genomic DNA from single seeds. Crop Sci. 30: 1079-1084.

Lander, E.S. and Botstein, D. (1989) Mapping Mendelian factors underlying quantitative traits using RFLP linkage maps. Genetics 121:185-199.

Lander, E.S., Green, P., Abrahamson, J., Barlow, A., Daly, M.J., Lincoln, S.E. and Newburg, L. (1987) MAPMAKER; an interactive computer program for constructing genetic linkage maps of experimental and natural populations. Genomics 1: 174.

Landry, B.S., Kesseli, R.V., Farrara, B. and Michelmore, R.W. (1987) A genetic map of lettuce (Lactuca sativa L.) with restriction fragment length polymorphism, isozyme, disease resistance and morphological markers. Genetics 116: 331-337.

Lange, D.A., Pe– uela, S., Denny, R.L., Mudge, J., Orf, J.H. and Young, N.D. (1998) A plant DNA isolation protocol suitable for polymerase chain reaction marker-assisted selection. Crop Science 38: 217-220

Lee, M. (1995) DNA markers and plant breeding programs. Adv. Agron. 55: 265-344.

Lin, Y.-R., Schertz, K.F. and Paterson, A.H. (1995) Comparative analysis of QTLs affecting plant height and maturity across the Poaceae, in reference to an interspecific sorghum population. Genetics 141: 391-411.

Litt, M. and Luty, J.A. (1989) A hypervariable microsatellite revealed by in vitro amplification of a dinucleotide repeat within the cardiac muscle actin gene. Am. J. Hum. Genet. 44: 397-401.

Liu, S.-C., Kowalski, S.P., Lan, T.-H., Feldmann, K.A. and Paterson, A.H. (1996) Genome-wide high-resolution mapping by recurrent intermating using Arabidopsis thaliana as a model. Genetics 142: 247-258.

Luo, G., Hepburn, A.G. and Widholm, J.M. (1992) Preparation of plant DNA for PCR analysis: a fast, general and reliable procedure. Plant Mol. Biol. Rep. 10: 319-323.

Manley, K.P. and Cudmore, R. H. (1998) Map manager XP. Plant and Animal Genome VI (San Diego, California, January, 1998). p. 75.

Mansur, L.M., Lark, K.G., Kross, H. and Oliveira, A. (1993) Interval mapping of quantitative trait loci for reproductive, morphological, and seed traits of soybean (Glycine max L.). Theor. Appl. Genet. 86: 907-913

Martin, G.B., Williams, J.G.K. and Tanksley, S.D. (1991) Rapid identification of markers linked to a Pseudomonas resistance gene in tomato using random primers and near-isogenic lines. Proc. Natl. Acad. Sci. U.S.A. 88: 2336-2340.

McCouch, S.R., Kochert, G., Yu, Z.Y., Khush, G.S., Coffman, W.R. and Tanksley, S.D. (1988) Molecular mapping of rice chromosomes. Theor. Appl. Genet. 76: 815-829.

McGrath, J.M. and Quiros, C.F. (1991) Inheritance of isozyme and RFLP markers in Brassica campestris and comparison with B. oleracea. Theor. Appl. Genet. 82: 668-673.

Menancio-Hautea, D., Kumar, L., Danesh, D. and Young, N.D. (1993) RFLP linkage map for mungbean (Vigna radiata (L.) Wilczek) in: Genetic Maps, 1992. S. J. O'Brien, ed. Cold Spring Harbor Press, Cold Spring Harbor, NY, pp. 6.259-6.260.

Messeguer, R., Ganal, M., de Vicente, M.C., Young, N.D., Bolkan, H. and Tanksley, S.D. (1991) High resolution RFLP map around the root knot nematode resistance gene (Mi) in tomato. Theor. Appl. Genet. 82: 529-536.

Michelmore, R.W., Paran, I. and Kesseli, R.V. (1991) Identification of markers linked to disease-resistance genes by bulked segregation analysis: A rapid method to detect markers in specific genome regions by using segregating populations. Proc. Natl. Acad. Sci. U.S.A. 88: 9828-9832.

Miller, J.C. and Tanksley, S.D. (1990). RFLP analysis of phylogenetic relationships and genetic variation in the genus Lycopersicon. Theor. Appl. Genet. 80: 437-448.

Morton, N.E. (1955) Sequential tests for the detection of linkage. Amer. J. Hum. Genet. 7: 277-318.

Muehlbauer, G.J., Staswick, P.E., Specht, J.E., Graef, G.L., Shoemaker, R.C. and Keim, P. (1991) RFLP mapping using near-isogenic lines in the soybean [Glycine max (L.) Merr.]. Theor. Appl. Genet. 81: 189-198.

Murray, M. and Thompson, W.F. (1984) Rapid isolation of high molecular weight plant DNA. Nucl. Acid. Res. 8: 4321-4325.

Nelson, J.C. (1997) QGENE - Software for marker-based genomic analysis and breeding. Molec. Breed. 3: 239-245.

O'Brien, S.J. (1993) Genetic Maps. Locus Maps of Complex Genomes. 6th Ed. Cold Spring Harbor Laboratory, Cold Spring Harbor, N.Y.

Paran, I., Kesseli, R. and Michelmore, R. (1991) Identification of restriction fragment length polymorphism and random amplified polymorphic DNA markers linked to downy mildew resistance genes in tomato, using near-isogenic lines. Genome 1021-1027:.

Pereira, M.G., Lee, M., Bramel-Cox, P., Woodman, W., Doebley, J. and Whitkus, R. (1994)

Construction of an RFLP map in sorghum and comparative mapping in maize. Genome 37: 236-243.

Prince, J.P., Pochard, E. and Tanksley, S.D. (1993) Construction of a molecular linkage map of pepper and a comparison of synteny with tomato. Genome 36: 404-417.

Qi, X., Stam, P. and Lindhout, P. (1996) Comparison and integration of four barley genetic maps. Genome 39: 379-394

Qi, X., Stam, P. and Lindhout, P. (1998) Use of locus-specific AFLP markers to construct a high-density molecular map in barley. Theor. Appl. Genet. 96: 376-384

Rongwen, J., Akkaya, M.S., Bhagwat, A.A., Lavi, U. and Cregan, P.B. (1995) The use of micro-satellite DNA markers for soybean genotype identification. Theor. Appl. Genet. 90: 43-48

Rooney, W., Jellen, E., Phillips, R., Rines, H., and Kianian, S. (1994) Identification of homoeologous chromosome in hexaploid oat (A-byzantina cv kanota) using monosomics and RFLP analysis. Theor. Appl. Genet. 89: 329-335

Schmidt, R., Love, K., West, J., Lenehan, Z., and Dean C. (1996) Description of 31 YAC contigs spanning the majority of Arabidopsis thaliana chromosome 5. Plant Journ. 11:563-572.

Sharp, P.J., Chao, S., Desai, S. and Gale, M.D. (1989) The isolation, characterization and application in the Triticeae of a set of wheat RFLP probes identifying each homoeologous chromosome arm. Theor. Appl. Genet. 78: 342-348.

Soller, M. and Beckmann, J.S. (1983) Genetic polymorphism in varietal identification and genetic improvement. Theor. Appl. Genet. 67: 837-843.

Song, K.M., Suzuki, J.Y., Slocum, M.K., Williams, P.H. and Osborn, T.C. (1991) A linkage map of Brassica rapa (syn. campestris) based on restriction fragment length polymorphism loci. Theor. Appl. Genet. 82: 296-304.

Stam, P. (1993) Construction of integrated genetic linkage maps by means of anew computer package, JoinMap. Plant Journ. 3: 739-744

Stuber, C.W., Edwards, M.D. and Wendel, J.F. (1987) Molecular marker-facilitated investigations of quantitative trait loci in maize. II. Factors influencing yield and its component traits. Crop Sci. 27: 639-648.

Tai, T.H. and Tanksley, S.D. (1990) A rapid and inexpensive method for isolation of total DNA from dehydrated plant tissue. Plant Mol. Biol. Rep. 8: 297-303.

Tanksley, S.D., Ganal, M.W., Prince, J.P., deVincente, M.C., Bonierbale, M.W., Broun, P., Fulton, T.M., Giovanonni, J.J., Grandillo, S., Martin, G.B., Messeguer, R., Miller, J.C., Miller, L., Paterson, A.H., Pineda, O., Roder, M., Wing, R.A., Wu, W. and Young, N.D. (1992) High density molecular linkage maps of the tomato and potato genomes: biological inferences and practical applications. Genetics 132: 1141-1160.

Tanksley, S.D., Ganal, M.W. and Martin, G.B. (1995) Chromosome landing: a paradigm for map-based gene cloning in plants with large genomes. Trends Genet. 11: 63-68

Tanksley, S.D., Young, N.D., Paterson, A.H. and Bonierbale, M.W. (1989) RFLP mapping in plant breeding: New tools for an old science. Bio/Technology 7: 257-264.

Thomson, D. and Henry, R. (1995) Single-step protocol for preparation of plant tissue for analysis by PCR. BioTechniques 19: 394-400

Thorlby, G.L., Shlumukov, L., Vizir, I.Y., Yang, C.Y., Mulligan, B.J. and Wilson, Z.A. (1997) Fine-scale molecular genetic (RFLP) and physical mapping of a 8.9 cM region on the top arm of Arabidopsis chromosome 5 encompassing the male sterility gene, ms1. Plant Journ. 12:471-479.

Vos, P., Hogers, R., Bleeker, M., Reijans, M., van de Lee, T., Hornes, M., Frijters, A., Pot, J., Peleman, J., Kuiper, M. and Zabeau, M. (1995) AFLP: a new techinque for DNA fingerprinting. Nucl. Acids Res. 23: 4407-4414

Wang, D.G., Fan, J.-B., Siao, C.J., Berno, A., Young, P., Sapolsky, R., Ghandour, G., Perkins, N., Winchester, E., Spencer, J., Kruglyak, L., Stein, L., Hsie, L., Topaloglou, T., Hubbell, E., Robinson, E., Mittmann, M., Morris, M.S., Shen, N., Kilburn, D., Rioux, J., Nusbaum, C., Rozen, S., Hudson, T.J., Lipshutz, R., Chee, M. and Lander, E.S. (1998) Large-scale identification, mapping, and genotyping of single-nucleotide polymorphisms in the human genome. Science 280: 1077-1082

Wang, H., Qi, M. and Cutler, A.J. (1993) A simple method of preparing plant samples for PCR. Nucl. Acids Res. 21: 4153-4154

Williams, J., Kubelik, A., Livak, K., Rafalski, J. and Tingey, S. (1990) DNA polymorphisms amplified by arbitrary primers are useful as genetic markers. Nucl Acid Res 18: 6531-6535.

Young, N.D., Miller, J. and Tanksley, S.D. (1987) Rapid chromosomal assignment of multiple genomic clones in tomato using primary trisomics. Nucl Acid Res 15: 9339-9348.

Young, N.D. and Tanksley, S.D. (1989) Restriction fragment length polymorphism maps and the concept of graphical genotypes. Theor Appl Genet 77: 95-101.

Young, N.D., Zamir, D., Ganal, M.W. and Tanksley, S.D. (1988) Use of isogenic lines and simultaneous probing to identify DNA markers tightly linked to the Tm-2a gene in tomato. Genetics 120: 579-585.

Zhang, H.B. and Wing, R.A. (1997) Physical mapping of the rice genome with BACS. Plant Molec. Biol. 35:115-127.

Zhang, Q.F., Shen, B.Z., Dai, X.K., Mei, M.H., Saghai-Maroof, M.A. and Li, Z.B. (1994) Using bulked extremes and recessive class to map genes for photoperiod-sensitive genic male sterility in rice. Proc. Natl. Acad. Sci. USA 91: 8675-8679.

4. Use of DNA Markers in introgression and isolation of genes associated with durable resistance to rice blast

D.-H. CHEN[1], R. J. NELSON[2], G.-L WANG[3],
D. J. MACKILL[4], and P. C. RONALD[1]

[1] Department of Plant Pathology, University of California, Davis, 1 Shield Ave, Davis; CA 95616
[2] Centro Internacional de la Papa, Lima 12, Peru
[3] The Institute of Molecular Agrobiology, The National University of Singapore,
1 Research Link, NUS, Singapore, 117604
[4] USDA-ARS, Department of Agronomy and Range Science, University of California,
1 Shields Ave, Davis, CA 95616. - <pcronald@ucdavis.edu>

Contents

1. Introduction

Rice blast caused by *Pyricularia grisea* Sacc. (= *P. oryzae* Cav., teleomorph *Magnaporthe grisea* Barr.), is one of the most widespread and destructive diseases of rice. Incorporation of blast resistance genes into elite rice cultivars has been a priority in rice breeding for decades in virtually all rice growing countries. Although resistance to blast is often short-lived, some cultivars are considered to possess durable resistance (Johnson 1981). Durable resistance is thought to be associated with partial resistance that is in many cases under oligo- or polygenic control (Higashi and Kushibuchi 1978; Higashi and Saito 1985; Wang et al. 1994; Parlevliet 1988). For example, the rice cultivar Moroberekan displays durable resistance to blast in upland conditions (Bidaux 1978; Ahn 1994; Fomba and Taylor 1994). The use of molecular markers has facilitated studies of the genetic basis of this durable blast resistance (Wang et al. 1994; Chen et al. 1997; Inukai and Aya 1997; Inukai et al. 1997).

Molecular marker technology allows us to dissect the complexity of resistance in durable resistant cultivars (Chen et al. 1999; Wang et al. 1994; Naqvi and Chattoo 1996), and facilitate introgression of genes associated with durable resistance. At UC Davis and the International Rice Research Institute (IRRI), we are dissecting the complexity of

R.L. Phillips and I.K. Vasil (eds.), DNA-Based Markers in Plants, 49 - 57.

resistance in the durably resistant cultivar, Moroberekan, by 1) developing recombinant inbred lines for mapping major and quantitative trait loci (QTLs); 2) characterizing the resistance spectrum of targeted genes by phenotypic analysis; and 3) constructing high-resolution maps for positional cloning of genes.

2.   Mapping genes for blast resistance using a recombinant inbred (RI) population

2.1. Identifying and Mapping Resistance Genes in Moroberekan

Moroberekan is a *Japonica* traditional cultivar from West Africa. It showed durable resistance under blast conducive upland conditions in Africa (Fomba and Taylor 1994), and complete resistance or high levels of partial resistance to blast in the majority of trials conducted under International Rice Blast Nursery (IRBN) in many countries (Ahn 1994). Moroberekan was completely resistant to over 300 Philippine isolates of *P. grisea* in greenhouse inoculation tests and was used as a blast resistant donor in many breeding programs. CO39, a highly susceptible *Indica* cultivar, was susceptible to all races except those from lineage 1 of *P. grisea* in the Philippines (Chen 1993; Chen et al. 1995; Zeigler et al. 1995). By single seed descent, a RI population consisting of 281 F_7 stable lines was developed from this cross, which was used for RFLP mapping of genes resistant to rice blast (Wang et al. 1994).

Each of the RI lines was subjected to RFLP analysis at 127 loci, and was also subjected to phenotypic analysis in a greenhouse and in the field tests. Five isolates, collected from different regions of the Philippines, were used in a greenhouse monocyclic test for qualitative resistance analysis. Isolate PO6-6 was used to assess the partial resistance (lesion number, lesion size and diseased leaf area) in a polycyclic test of 131 lines, which were susceptible to the tested five isolates. To determine the associations between DNA markers and phenotypic effects, regression analysis and MapMaker QTL were used to identify the DNA markers associated with the complete resistance to the five tested isolates, and with partial resistance to the isolate PO6-6.

Two dominant loci conferring qualitative resistance to all five isolates were identified to be derived from Moroberekan. One locus, tentatively designated *Pi5(t)*, was mapped to chromosome 4, near DNA marker *RG788*. Another major gene, tentatively designated *Pi7(t)*, was bracketed by markers *RG103A* and *RG16* in chromosome 11. Ten putative QTLs derived from Moroberekan were mapped in eight chromosomes to be associated with the parameters (lesion number, lesion size and diseased leaf area) of partial resistance.

The five isolates used for identifying *Pi5(t)* and *Pi7(t)* were thought to be genetically diverse due to their different origins and different virulent spectra. DNA analysis later showed that they represented only three of more than ten genetic lineages identified in the Philippines (Chen 1993; Chen et al. 1995). Because of the under-representation of diversity, it was hypothesized that more resistance genes could be identified from Moroberekan if diverse isolates were used for phenotypic analysis.

By inoculating the same RI population with two additional Philippine isolates, Ca65 and JMB840610, and one Indian isolate B157, three more loci, tentatively designated

Pi12(t), Pi10(t) and *Pi157*, conferring complete resistance were identified (Inukai et al. 1996; Naqvi and Chattoo 1996). *Pi12(t)* confers complete resistance to isolate Ca65 and was localized on chromosome 2 (Inukai et al. unpublished). *Pi10(t)* and *Pi157* confer complete resistance to the Philippine isolates JMB840610 and the Indian isolate B157, respectively. RAPD markers *RRF6* and *RRH18* were found to be tightly linked with the locus *Pi10(t) (RRF6,* 2.8 ± 0.9 cM; *RRH18,* 3.9 ± 1.2 cM), and mapped to chromosome 5 (Fig. 1) (Naqvi, et al. 1995; Naqvi and Chattoo,1996). RFLP markers were used for mapping the *Pi157* locus (Naqvi and Chattoo 1996). Cosegregation linkage analysis indicated that *RG341* was tightly linked (3.0 ± 0.6 cM) to the *Pi157* locus on chromosome 12 (Fig. 1).

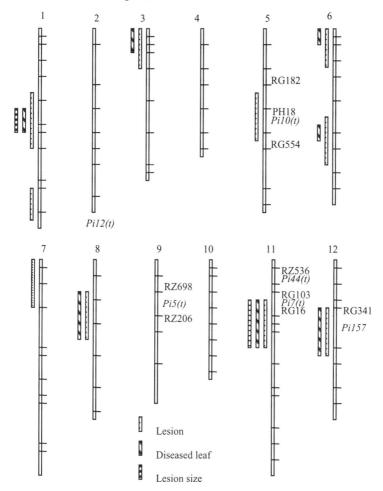

Fig. 1. Chromosomal locations of blast resistance genes identified in Moroberekan. The linkage information of *Pi12(t)* on chromosome 2 is not available; the locus is listed under the chromosome 2. Bars at the left side of the chromosomes are the regions where QTLs conferring partial resistance are located. The map is adapted from Wang et al. 1994; Naqvi and Chattoo 1996 and Chen et al. 1997.

In characterizing the QTL effect on diverse blast isolates, two lines, RIL 23 and RIL276 were selected. Based on the previous study (Wang et al. 1994; Inukai et al. 1997), RIL23 and RIL276 each carry two QTLs for partial resistance, one on chromosome 6, near *RG64*, and another on chromosome 1, near CDO920. We found that RIL23 and RIL276 both confer partial resistance to 26 isolates representing six lineages of *P. grisea* (Chen 1993). In addition, both lines were qualitatively resistant to a set of isolates of lineage 44 (Chen et al. 1996). This raised the question whether resistance to lineage 44 in RIL276 and RIL23 was conferred by an additional major gene or by the QTL, which functions as a qualitative effect in lineage 44.

To identify and map the locus in these lines conferring resistance to isolates of lineage 44, an F_2 population consisting of 100 individuals was developed from a cross between RIL276 and CO39. The F_2 was advanced into F_3 families for phenotypic analysis. Genetic analysis of the F_2 and F_3 populations using isolate C9240-1 of lineage 44, showed that resistance in RIL276 to lineage 44 isolates was governed by a single major locus. This locus was tentatively designated *Pi44(t)* for its qualitative resistance to isolates of lineage 44 of *P. grisea* (Chen et al. 1999).

2.2.   Further Analysis of Resistance loci *Pi5(t)* and *Pi7(t)*

Since lines carrying *Pi5(t)* and *Pi7(t)* showed high levels of resistance in two field trials in Indonesia, in the Philippines, and in greenhouse tests, these loci were hypothesized to play a major role in durable resistance in Moroberekan and were therefore targeted for positional cloning.

Based on the previous study (Wang et al. 1994), *Pi5(t)* was located on chromosome 4, flanked by RFLP markers *RG788* and *RG864*, and *Pi7(t)* was mapped to chromosome 11, bracketed by *RG103A* and *RG16*. This study indicated that three recombinant inbred lines (RIL206, RIL249 and RIL260) carried *Pi5(t)* and two lines (RIL125 and RIL29) carried *Pi7(t)*. Therefore, these lines were selected for further mapping of the *Pi5(t)*, and *Pi7(t)* loci (Chen et al. 1997).

Candidate lines for *Pi5(t)* and *Pi7(t)* (RIL206, RIL249, RIL260, RIL29 and RIL125) were crossed with CO39. Five F_2 populations, each consisting of about 50 progeny, were obtained from the crosses. The F_2 progeny and their F_3 families were inoculated with the isolate PO6-6 at 21 days after sowing (17 to 20 seedlings per family). Disease was evaluated 7 days after inoculation. DNA from 17 to 20 seedlings of each F_3 family was extracted to represent the F_2 individuals. Bulked DNA segregant analysis by AFLP was applied to identify DNA markers associated with the resistance to isolate PO6-6. The AFLP analysis was essentially following the protocol developed by Zabeau and Vos (1993).

All five F_2 populations segregated for resistance to blast isolate PO6-6 in a 3:1 ratio, indicating a single locus conferred resistance to isolate PO6-6. The F_2 inoculation results were confirmed in two replicated inoculations of F_3 families with the same isolate.

*Eco*RI and *Mse*I AFLP primers were surveyed for DNA markers associated with the resistance loci in two resistant and two susceptible bulks for each of the five F_2 populations. DNA markers associated with resistance in the bulks were confirmed in the F_2 individuals. Out of 350 primer pairs surveyed, 10 and 11 AFLP markers were identified to be associated with the resistance in the RIL206-derived and RIL249-derived F_2 populations, respectively. Out of 1024 primer pairs surveyed, 12 markers were identified to be associated with the resistance in the RIL260-derived F_2 population. One linked marker showed polymorphism in both the RIL206-derived and RIL249-derived populations, and three markers were associated with resistance in both the RIL249-and RIL260-derived populations. These results suggest that the resistance loci in the three lines were likely to be the same or tightly linked. Though six markers were co-segregating with the resistance in one of the three small populations, the exact genetic distance needs to be determined in a larger mapping population.

963 and 579 AFLP primer combinations were screened for markers associated with blast resistance in the two *Pi7(t)* candidate lines, RIL29 and RIL125, respectively. Two primer pairs gave polymorphic markers linked with resistance in RIL29 to isolate PO6-6. One primer pair gave a polymorphic marker (S04/G03) associated with resistance in RIL125. Unexpectedly, this marker (S04/G03) also co-segregated with resistance in RIL260 which carries *Pi5(t)* (Wang et al. 1994). Phenotypic analysis using California blast isolate CAL-1 showed that RIL125 was moderately resistant to CAL-1, to the same degree as RIL260, RIL249 and RIL206. In contrast, RIL29 was completely resistant to CAL-1. This result suggests that RIL125 may carry the same locus as that present in RIL260.

Primer pairs generating the linked markers were used to analyze the parents of a mapping population Black Gora/Labell (provided by D.J. Mackill; Redona and Mackill 1996). Eight of these primer pairs (S04/G03, S10/F07; S08/G24, S10/M04A, S10/M04B, B11/G19, R01/G23A, R01/G23B) gave rise to polymorphism between Black Gora and Labell. Segregation of the 80 F_2 progeny of Black Gora/Labell at these loci fits a 3:1 ratio. Linkage analysis with Mapmaker and Joinmap showed that these eight markers from the RIL249-derived and RIL260-derived populations were placed on chromosome 9 (Fig. 1). These results suggest that the loci conferring resistance to PO6-6 in RIL249 and RIL260 were most likely the same or tightly linked. Because all the markers linked to the locus conferring resistance to isolate PO6-6 mapped to chromosome 9 rather than chromosome 4, we tentatively considered this locus as *Pi5(t)*. Of the two *Pi7(t)* linked markers in RIL29, one was monomorphic in the two mapping populations and another was mapped to chromosome 11, near *RG103A* (Chen and Ronald, unpublished), which confirmed the previous report of the *Pi7(t)* map position (Wang et al. 1994).

The location of *Pi5(t)* in the lines RIL125, RIL206, RIL249 and RIL260 on chromosome 9 differed from the previous report in which *Pi5(t)* was mapped to chromosome 4 (Wang et al. 1994). However, the chromosomal location of *Pi5(t)* in this study agreed with the finding of Inukai and Aya (1997), who found that resistance to isolate PO6-6 in the F_2 population of RIL249 / CO39 did not map to chromosome 4. The discrepancy between the location of *Pi5(t)* in this report and the previous study could be attributable to the skewed RI population used in the previous study, which could lead to pseudo-linkage, or that Moroberekan could carry an additional locus other than *Pi5(t)* and *Pi7(t)* resistant to PO6-6. Reanalysis of the *Pi5(t)* candidate lines used in the previous study (Wang et al. 1994) is needed to clarify the discrepancy.

For fine mapping of the *Pi5(t)* locus, the RIL260 derived segregating population was retained for further analysis. A F_4 segregating population was developed consisting of over 2,000 individuals. AFLP markers R01/G23A and S04/G03, which bracketed the *Pi5(t)* locus to a 2.1 *cM* region in the small F_2 segregating population, were used to identify the rare recombinants in the F_4 population. Fine mapping of the *Pi5(t)* locus is now under way (Chen and Ronald, unpublished).

3.   **Allelism relationship between resistance genes in moroberekan and known resistance genes**

More than 20 major loci conferring resistance to rice blast have been identified and mapped through classical genetics or molecular marker technology (McCouch et al. 1994). Some of the genes were present in near-isogenic lines (Mackill and Bonman 1992; Inukai et al. 1994), which makes allelism analysis among resistance genes possible. Inukai et al. (1996) initiated a study on allelism analysis among the newly identified resistance genes in Moroberekan and known genes.

    To test the allelic relationship between *Pi5(t)* and *Pi7(t)*, they made crosses between RIL29 and RIL249, RIL29 and C101LAC (carrying *Pi1* on chromosome 11), and RIL249 and C104PKT (carrying *Pi3*). The F_2 of the RIL29/RIL249 and RIL29/C101LAC crosses were inoculated with isolate PO6-6, and with isolate P03-82-51 for cross of RIL249 and C104PKT. Results showed that *Pi5(t)* in RIL249 and *Pi7(t)* in RIL29 were non-allelic; *Pi7(t)* was allelic or tightly linked to *Pi1*, and *Pi5(t)* in RIL249 was allelic or closely linked to *Pi3*. The reaction pattern of *Pi12(t)* to Japanese isolates was similar to that of the *Pib*. Allelism tests showed that *Pi12(t)* appeared to be allelic or closely linked to the *Pib* locus (Inukai et al. unpublished). The reaction pattern *Pi44(t)* in RIL276 was distinct from that of the existing CO39 NILs and the *Pi5(t)* and *Pi7(t)* (Chen and Nelson, unpublished). Since *Pi1*, *Pi7(t)* and the multi-allele locus, *Pik* were all located on the distal segment of chromosome 11, the allelism relationship among these genes needs to be determined.

    Because resistance genes *Pib* and *Pi12(t)* were clustered in the chromosome 2 and *Pi1*, *Pi7(t)*, *Pi44(t)*, *and Pik* were in the same region of the chromosome 11, allelism analysis may not be able to distinguish the two genes in the cluster. Molecular cloning of the genes will elucidate fine genetic structure of these loci.

4.   **Phenotypic characteriazation of resistance in Moroberekan**

The durability of blast resistance in Moroberekan has been well recognized. It has been postulated that durable resistance in Moroberekan is attributable to a complex of genes with both partial and complete resistance (Wang et al. 1994). At least six major genes and ten QTLs have been so far identified and mapped to different chromosomes. Phenotypic characterization of individual major genes and QTLs in different genetic backgrounds against diverse blast isolates would provide insight into the individual contribution of the genes in the durability of resistance to blast in Moroberekan.

Such information would be useful for designing desirable resistance genotypes in breeding programs.

In the study of Wang et al. (1994), RI lines carried *Pi5(t), Pi7(t)* and QTL(s) identified using isolate PO6-6 under greenhouse conditions were tested against field blast populations at two blast screening sites in different countries (Cavinti in the Philippines and Sitiung in Indonesia). Diseased leaf area was evaluated for each line. RI lines with *Pi5(t)* and *Pi7(t)* generally had low levels of disease in the two testing sites. Lines carrying more QTLs were generally more resistant to those carrying none or only one QTL.

Field performance of the lines carrying *Pi5(t)* and *Pi7(t)* suggest that *Pi5(t)* and *Pi7(t)* confer broad resistance spectra against blast isolates, because genetically diverse blast isolates were present in the Cavinti screening site where the environmental conditions favor the blast epidemic (Chen et al. 1995). However, greenhouse testing is necessary to confirm the resistance spectrum of the individual loci. Lines carrying *Pi-5(t)* were resistant to at least 6 races belonging to 4 lineages in the Philippines, while it was susceptible to California isolates. Lines carrying *Pi7(t)* were resistant to at least 6 races belonging to 4 lineages in the Philippines and to twelve California isolates tested in both CO39 and M202 genetic backgrounds (Chen and Ronald, unpublished).

Pi44(t) is only resistant to a set of isolates of lineage 44 from the Philippines (Chen 1993) and some Japanese isolates (Inukai, personal communication). Though *Pi44(t)* showed a narrower resistance spectrum, it complements the resistance of *Pi2* (Chen et al. 1996). The combination of *Pi44(t), Piz* and other genes may provide more broad and durable resistance. Little is known about the resistance spectrum of *Pi10(t), Pi12(t)* and *Pi157*.

For a more detailed characterization of the QTL effect on diverse blast isolates, 26 isolates representing six lineages of *P. grisea* were inoculated on RIL23 and RIL276, which carry two QTLs on chromosome 6, near *RG64* and on chromosome 1, near CDO920 (Wang et al. 1994; Inukai et al. 1996). Lesion numbers on these two lines were consistently reduced in all interactions, compared with those on the susceptible parent CO39 (Chen 1993). These two lines also showed low disease levels in field tests (Wang et al. 1994). These results demonstrated the QTL identified using a single isolate (PO6-6) in the greenhouse could reliably predict the line performance in the field and in greenhouse assays with diverse isolates.

To understand the effect of genetic background on gene performance, we crossed lines carrying *Pi5(t)* (RIL125, RIL206, RIL249 and RIL260) and RIL29 (*Pi7(t)*) with a blast-susceptible California cultivar M202. F_1 plants were inoculated with California isolate CAL-1. The F_1 of the RIL125/M202, RIL206/M202, RIL249/M202 and RIL260/M202 were susceptible, whereas RIL29/M202 F_1 showed high levels of resistance. These results suggest that the moderate resistance to California isolates observed in the *Pi5(t)* lines was conferred by the genes from CO39, rather than *Pi5(t)*, and that *Pi7(t)* confers high levels of resistance in the M202 background.

5.   Conclusions and considerations

The use of molecular markers has facilitated progress towards understanding the genetic basis of durable resistance to rice blast in Moroberekan. The identification of at least six major genes and 10 QTLs in Moroberekan suggests that durability of the resistance to blast is a function of the combination of major genes for qualitative resistance and QTLs for partial or quantitative resistance. In the past several decades, breeding efforts have been mainly devoted to incorporation of single major genes into rice cultivars. However, this approach is often counteracted by rapid evolution of the pathogen populations with a subsequent breakdown of the resistance. The integration of major genes and QTLs into breeding programs may aid in the development of durable blast resistance cultivars. Though pyramiding genes would be difficult through the conventional breeding methodology, it is now possible to combine genes by molecular marker-assisted breeding (Huang et al. 1997).

6.   References

Ahn, S.W. (1994) International collaboration on breeding for resistance to rice blast, in Zeigler, R.S., Leong, S.A. and Teng, P.S. (eds.), Rice Blast Disease, CBA and IRRI, Wallingford, Oxon, UK, pp. 137-153.

Bidaux, J.M. (1978) Screening for horizontal resistance to rice blast (*Pyricularia oryzae*), in Buddenhagen, I. W. And Persley, G.J. (eds.), Rice in Africa. Academic Press, London, pp. 159-174.

Chen, D.-H. (1993) Population structure of *Pyricularia grisea* at two screening sites and quantitative characterization of major and minor resistance genes. PhD Thesis, University of the Philippines at Los Baños, Laguna, Philippines.

Chen, D., Zeigler, R.S., Leung, H., Nelson, R.J. (1995) Population structure of *Pyricularia grisea* at two screening sites in the Philippines. Phytopathology 85: 1011-1-20.

Chen, D.H., Zeigler, R.S., Ahn, S.W., Nelson, R.J. (1996) Phenotypic characterization of the rice blast resistance gene *Pi2(t)*. Plant Dis. 80: 52-56.

Chen, D.-H., Wang, G.L., and Ronald. P.C. (1997) Location of the rice blast resistance locus *Pi-5(t)* in Moroberekan by AFLP bulked segregant analysis. Rice Genetics Newsletter 14: 95-98.

Chen D.-H., M. dela Viña, T. Inukai, D.J. Mackill, P.C. Ronald[1] and R.J. Nelson (1999) Molecular mapping of the blast resistance gene, *Pi44(t)*, in a line derived from a durably resistant rice cultivar. Theor. Appl. Genet. (in press)

Fomba, S.N. and Taylor, D.R. (1994) Rice blast in West Africa: Its nature and control, in Zeigler, R.S., Leong, S.A. and Teng, P.S. (eds.), *Rice Blast Disease,* CBA and IRRI, Wallingford, Oxon, UK, pp. 343-355.

Higash, T. and Kushibuchi, K. (1978) Genetic analysis of field resistance to leaf blast (*Pyricularia oryzae*) in rice. Jap. J. of Breed. 28: 277-286.

Higash, T. and Saito, S. (1985) Linkage groups of field resistance genes of upland rice variety Sensho to leaf blast caused by *Pyricularia oryzae*. Jap. J. of Breed. 35: 438-448.

Huang, N., Angeles, E.R., Domingo, J., Magpantay, G., Singh, S., Zhang, G.,Kumaravadivel, N.,

Bennett, J., and Khush, G.S. (1997). Pyramiding of bacterial blight resistance genes in rice: marker-assisted selection using RFLP and PCR. Theor. Appl. Genet. 95: 313-320.

Inukai, T., Nelson, R.J., Zeigler, R.S., Sarkarung, S., Mackill, D.J., Bonman, J.M., Takamure, I., Kinoshita, T. (1994) Allelism of blast resistance genes in near-isogenic lines of rice. Phytopathology 84: 1278-1283.

Inukai, T. and Aya, S. (1997) Determination of chromosomal location of blast resistance gene in West African upland cultivar Moroberekan. Breed. Sci. 47 (Suppl.1), 268

Inukai, T., Nelson, R.J., Zeigler, R.S., Sarkarung, S., Mackill, D.J., Bonman, J.M., Takamure, I., Kinoshita, T. (1996) Development of pre-isogenic lines for rice blast resistance by marker aided selection from a recombinant inbred population. Theor. Appl. Genet. 93: 560-567.

Johnson, R. (1981) Durable resistance: definition of, genetic control, and attainment in plant breeding. Phytopathology 71: 567-568.

Mackill, D.J. and Bonman, J.M. (1992) Inheritance of blast resistance in near-isogenic lines of rice. Phytopathology 82: 746-749.

McCouch S.R., Nelson R.J., Tohme J., Zeigler RS (1994) Mapping of blast resistance genes in rice, in Zeigler, R.S., Leong, S.A., Teng, P.S. (eds.) Rice Blast Disease, CBA and IRRI, Wallingford, Oxon, UK, pp. 137-153.

Naqvi, N.I. and Chattoo, B.B. (1996) Molecular genetic analysis and sequence characterized amplified region-assisted selection of blast resistance in rice, in [IRRI] International Rice Research Institute, Rice Genetics III. Proceedings of the Third International Rice Genetics Symposium, 16-20 Oct 1995. Manila (Philippines): IRRI. pp 507-572.

Naqvi, N.I., Bonman, J.M., Mackill, D.J., Nelson, R.J., Chattoo, B.B. (1995) Identification of RAPD markers linked to a major blast resistance gene in rice. Molecul. Breed. 1: 341-348.

Parlevliet, J.E. (1988) Identification and evaluation of quantitative resistance, in Leonard, K.J., and Fry, W.G. (eds.), Plant Disease Epidemiology, Genetics, Resistance, and Management, Vol. 2. McGraw-Hill, New York, pp. 377

Redoña, E.D. and Mackill, D.J. (1996) Mapping quantitative trait loci for seedling vigor in rice using RFLPs. Theor. Appl. Genet. 92: 395-402.

Wang, G.L., Mackill, D.J., Bonman, J.M., McCouch, S.R., Champoux, M.C., Nelson, R.J. (1994) RFLP mapping of genes conferring complete and partial resistance to blast in a durably resistant rice cultivar. Genetics 136: 1421-1434.

Zabeau, M. and Vos, P. (1993) Selective restriction fragment amplification : a general method for DNA fingerprinting. European Patent Application No. 0534858 A1. European patent Office, Paris.

Zeigler, R.S., Cuoc, L.X., Scott, R.P., Bernardo, M.A., Chen, D.H., Valent, B., Nelson, R.J. (1995) The relationship between lineage and virulence in *Pyricularia grisea* in the Philippines. Phytopathology 85: 443-451.

5. Mapping quantitative trait loci[*]

STEVEN J. KNAPP

Department of Crop and Soil Science, Oregon State University,
Corvallis, OR 97331, U.S.A. <Steven.J.Knapp@orst.edu>

Contents

1.   Introduction

Different alleles at *quantitative trait loci* (QTL) cause genetic differences between individuals and families for quantitative traits (Bulmer 1980; Falconer 1981). QTL genotypes cannot be determined by inspecting the distributions of trait phenotypes alone. This is one of the fundamental problems of quantitative genetics. Historically important quantitative genetic parameters, e.g., additive genetic variance and heritability, summarize differences between alleles at QTL but do not shed light on the genetics of QTL. Methods for mapping QTL are needed to achieve this. QTL are mapped by using genetic markers linked to QTL to draw inferences about differences between alleles at QTL.

Several parameter estimation methods, mostly variants of two classes of *interval mapping* methods, have been described for mapping QTL using matings between inbred lines (Carbonell et al. 1992; Haley and Knott 1992; Jansen 1992; Jensen 1989; Knapp et al. 1990; Knapp et al. 1991; Knott and Haley 1992; Lander and Botstein 1989; Lou and Kearsey 1989; Martinez and Curnow 1993; Simpson 1989; Van Ooijen 1992; Weller 1986). While interval mapping should not be narrowly defined, it has most often meant using marker-brackets or flanking markers as independent variables to build a genetic model for testing they hypothesis of no QTL against the hypothesis of one QTL within a marker-bracket (differentiating between two genetic models). A great many other genetic models can be hypothesized and tested, e.g., see Knapp (1991) and Martinez and Curnow (1993), and these use the same principles.

Least squares (Haley and Knott 1992; Knapp et al. 1990) and maximum likelihood (Carbonell et al. 1992; Jensen 1989; Lander and Botstein 1989; Lou and Kearsey 1989;

[*]Reprinted without change from the first edition.

R.L. Phillips and I.K. Vasil (eds.), DNA-Based Markers in Plants, 59 - 99.
© 2001 *Kluwer Academic Publishers. Printed in the Netherlands.*

Simpson 1989) methods are the two chief classes of interval mapping methods. The fundamentals of maximum likelihood methods for interval mapping have been extensively reviewed (Lander and Botstein 1989; Knott and Haley 1992; Van Ooijen 1992; Weller 1993). As Haley and Knott (1992) show, the power of least squares and ML methods is equal for matings between inbred lines.

ML interval mapping methods have been criticized for failing to find QTL when multiple QTL are segregating (Martinez and Curnow 1993). Martinez and Curnow (1993), for example, show that interval mapping (testing the hypothesis of one QTL versus no QTL) can fail when two linked QTL are segregating. This is not surprising since what is needed is a test of the hypothesis of two linked QTL versus one QTL or no QTL. The problem has nothing to do with the parameter estimation method *per se*. Rather it has to do with the inadequacy of the underlying genetic model, or the failure to test hypotheses about different genetic models.

The distinction between the genetic model and the parameter estimation method has to be kept clear. Every parameter estimation method is prone to failure when the stated genetic model is wrong. The challenge is to model the genetics adequately and to develop statistics for differentiating between genetic models. Unless software is developed to do this, most investigators are not likely to test hypotheses about different multilocus genetic models.

Least squares methods are more suited to handling a wide range of parameter estimation problems than ML methods, such as estimating the parameters of multiple QTL, efficiently searching genomes for multiple QTL, and estimating QTL parameters from multiple environment experiments done using replicated experiment designs, e.g., randomized complete blocks. At least the computations are more straightforward. The problem of handing different experiment and environment designs has not been thoroughly addressed for any QTL mapping method (Jansen 1992; Knapp and Bridges 1990; Soller and Beckmann 1990). The problem of how to most effectively execute genome searches is not completely settled either, although the fundamentals are clear (Knapp et al. 1993).

Genetic models for mapping QTL are functions of frequencies of observed marker phenotypes and hypothesized QTL phenotypes. They are defined by the pedigree of the individuals, lines, or families, the number of quantitative trait loci, whether or not any of the QTL are linked, and the genetic map of the population. Several genetic models are routinely tested when mapping QTL, although the ultimate objective of any experiment is to find the most satisfactory multilocus genetic model for a given trait. The joint frequencies of marker and hypothesized QTL genotypes must be defined to estimate the parameters of multilocus models. This becomes very cumbersome as the number of loci increases. This problem is closely tied to the problem of who to search a genome and how to get unbiased test statistics (Knapp et al. 1993).

An important group of mating designs for plant breeders uses inbred lines as parents. Some of the progeny types which can be developed from matings between inbred lines are F_2, F_3, $F_{2:4}$, $F_{3:4}$, $F_{2:5}$, $F_{3:5}$, F4:5, backcross (BC), BC_1S_1, doubled haploid (DH), recombinant inbred (RI), and an assortment of testcross progeny. Genetic models and methods for mapping QTL have been defined for some of these progeny types, but many gaps remain. Defining these genetic models is nonetheless straightforward for matings between inbred lines.

2. Experiment design

Most of the QTL mapping methods and software developed thus far can be used for experiments where unreplicated progeny are tested or for experiments where replicated progeny are tested and completely randomized (CR) experiment designs are used. MAPMAKER-QTL (Lander and Botstein 1989), for example, cannot be used to estimate QTL parameters from a randomized complete blocks experiment design or from multiple environment experiments without ignoring blocks or environments. Nor can linear regression interval mapping methods (Haley and Knott 1992; Curnow and Martinez 1992) be used for these experiments, although they can be extended without any difficulty. There are two separate problems to address. One is developing methods to handle a range of experiment, environment, and mating designs. This is what we address by using linear least squares interval mapping (LIM). The other is developing software to implement these methods.

The acronym LIM is used for the entire group of linear least squares interval mapping methods whether parameters are estimated using linear regression *per se* (Haley and Knott 1992; Curnow and Martinez 1992), regression on dummy variables with QTL genotype coefficients used to define linear contrasts among marker genotype means, as is done throughout this paper.

Several QTL mapping advances have been described where parameter estimation was done using standard statistics software, e.g., GENSTAT or SAS; however, none of it has been automated (Knapp 1989; Knapp et al. 1993; Jensen 1991). The greatest needs for plant breeders are methods and software for estimating QTL and other genetic parameters for balanced and unbalanced linear models for different experiment, mating, and environment designs. LIM is undoubtedly the most effective way to handle these estimation problems.

LIM methods are developed below for virtually any experiment or environment design by bringing together linear least squares (Haley and Knott 1992) and linear model theory for unbalanced linear models (Knapp et al. 1993). The example used is the randomized complete blocks experiment design. Suppose an experiment is done where lines of some sort, e.g., doubled haploid or F_3 lines, are tested in randomized complete blocks in one environment (Table 1). The linear model

$$y_{ij} = \mu + b_i + g_j + e_{ij}$$

can be used to estimate the usual quantitative genetic parameters, e.g., line means and between line variances, where y_{ij} is the ijth observation of the quantitative trait, is the population mean, b_i is the effect of the ith block, g_j is the effect of the jth line and e_{ij} is the random error for the jth line in the ith block. The effects of lines and blocks are random (Table 1). The objective of such an experiment is usually to select the most outstanding lines. Other objectives might be to test the null hypothesis of no between line variance ($H_o: \sigma^2_G = 0$) and to estimate heritabilities and expected selection gains. Methods for estimated the parameters of (1) are straightforward.

Suppose, for example, doubled haploid lines are tested. The genetic variance between doubled haploid lines is

$$\sigma^2{}_G = 2\sigma^2{}_A = [E(M_G - E(M_E)] \, / \, r \, ,$$

while the line-mean heritability for selection among doubled haploid lines is

$$H = \frac{\sigma^2{}_G}{\sigma^2{}_G + \sigma^2{}_E / r} = \frac{2\sigma^2{}_A}{2\sigma^2{}_A + \sigma^2{}_E / r} = 1 - \frac{E(M_E)}{E(M_G)}$$

where is the additive genetic variance and epistasis is ignored (Table 1). Without a genetic map and marker phenotypes for several loci dispersed throughout the genome, this is where the analysis of this experiment might end. The addition of a genetic map and molecular marker phenotypes for several loci, however, creates the basis for estimating the parameters of genes or QTL underlying differences between lines, for which additional genetic models and estimation methods are needed.

Table 1.   Degrees of freedom, Type III sum of squares, and expected mean squares for lines tested in randomized complete blocks in one environment where the effects of QTL genotypes (QTL) are fixed and the effects of lines and bloks are random and factors other than QTL genotypes are balanced.

Factor	Degrees of freedom[a]	Sum of squares	Expected mean square[b]
Block	$df_R = b - 1$	$R\,[b|\mu, g]$	$E(M_B) = \sigma^2{}_E + N\sigma^2{}_B$
Line (G)	$df_G = N - 1$	$R\,[g|\mu, b]$	$E(M_G) = \sigma^2{}_E + r\sigma^2{}_G$
QTL (Q)	$df_Q = q - 1$	$R\,[q|\mu, b]$	$E(M_Q) = \sigma^2{}_E + r\sigma^2{}_{G:Q} + \phi^2{}_Q$
G:Q	$df_{G:Q} = N - 1$	$R\,[g(q)|\mu, g, b]$	$E(M_{G:Q}) = \sigma^2{}_E + r\sigma^2{}_{G:Q}$
Residual	$df_E = (N-1)(b-1)$	$R\,[e|\mu, b, g]$	$E(M_E) = \sigma^2{}_E$

[a] q is the number of QTL genotypes, $N = \Sigma^q_{i=1} \, n_i$ is the number of lines where n_i is the number of lines of the ith QTL genotype, and b is the number of bloks.

[b] $\sigma^2{}_E$ is the error variance, $\sigma^2{}_{G:Q}$ is the line nested in QTL genotype variance, $\sigma^2{}_G$ is the between line variance, $\phi^2{}_Q$ is the variance of fixed effects of QTL genotypes, and

$$\bar{n} = \frac{N - \dfrac{\Sigma n_i^2}{N}}{q - 1}$$

Classical quantitative genetic parameters can be defined as functions of marker or QTL parameters by using linear models. The effects of lines, for example, can be defined as a function of the effects of their QTL (Knapp and Bridges 1990). Rewriting (1) to show this gives

$$y_{ikj} = \mu + b_i + q_k + g(q)_{kj} + e_{ikj}$$

where y_{ikj} is the *ikj*th observation of the quantitative trait, is the population mean, b_i is the effect of the *i*th block, q_k is the effect of the *k*th QTL genotype, $g(q)_{kj}$ is the effect of the *j*th line nested in the *k*th QTL genotype and e_{ikj} is the random error for the *j*th line of the *k*th QTL genotype in the *i*th block. The effects of QTL genotypes are fixed, while the effects of other factors are random.

The effects of lines are the sum of the effects of QTL genotypes and lines nested in QTL genotypes ($g_j = q_k + g(q)_{kj}$). The latter are the effects between lines which are left over after estimating the effects between QTL genotypes. They are the effects of all of the genes which are not part of the model. This is easily seen by examining the expected mean squares. The sum of squares between lines (S_G) from (1) is the sum of squares between QTL genotypes (SQ) and between lines nested in QTL genotypes ($S_{G:Q}$) from (2) (Knapp and Bridges 1990); thus,

$$E(M_G) = \sigma^2_e + r\sigma^2_G = \frac{df_Q\, E(M_Q) + df_{G:Q}\, E(M_{G:Q})}{df_Q + df_{G:Q}} = \sigma^2_E + r\sigma^2_{G:Q} + \frac{r\bar{n}\,(q-1)}{N-1}\sigma^2_Q$$

$$(3)$$

where σ^2_Q is the genetic variance between QTL genotypes.

The QTL genotypes of (2) cannot be observed so the frequencies of hypothesized QTL genotypes must be inferred from the marker phenotypes to estimate QTL genotype effects (q_i). This is the essential feature of QTL mapping. A specific genetic model must be used to illustrate how this is done using least squares, but the principles work for virtually any genetic model. To review the theory and illustrate LIM, a genetic model for doubled haploid lines is used where a QTL is hypothesized toile between two linked marker loci A and B and there is no interference, i.e., the coefficient of coincidence (γ) is equal to 1.0 (Table 2). Genetic models for the doubled haploid mating design are well known (Knapp et al. 1990; Knapp 1991).

Gentic models for QTL mapping problems are nonlinear; however, they can be made linear by fixing the recombination frequencies between the QTL and the markers (θ_{AQ} and θ_{BQ}) (Haley and Knott 1992; Knapp et al. 1990; Lander and Botstein 1989). Under null interference ($\gamma = 1$),

$$\theta_{AB} = \theta_{AQ} + \theta_{BQ} - 2\theta_{AQ}\theta_{BQ},$$

$$\theta_{AQ} = \frac{\theta_{BQ} - \theta_{AB}}{2\theta_{BQ} - 1}$$

and

$$\theta_{BQ} = \frac{\theta_{AQ} - \theta_{AB}}{2\theta_{AQ} - 1}$$

Table 2.   Expected marker genotypes means (μ_m) for doubled haploid lines under no interference where A and B are condominant marker loci with alleles A and a and B and b, respectively, Q is a hypothesized quantitative trait locus with alleles Q and q, the locus order is AQB, θ_{AB} is the recombination frequency between A and Q, μ_{QQ} is the mean of QQ genotypes at the QTL, μ_{qq} is the mean of qq genotypes at the QTL, and *m* indexes marker genotypes.

Marker genotype		Expected marker genotype mean
A	B	
AA	BB	$\mu_1 = \dfrac{(1 - \theta_{AQ} - \theta_{BQ} + \theta_{AQ}\theta_{BQ})\mu_{QQ} + \theta_{AQ}\theta_{BQ}\mu_{qq}}{1 - \theta_{AQ} - \theta_{BQ} + 2\theta_{AQ}\theta_{BQ}}$
AA	bb	$\mu_2 = \dfrac{(\theta_{BQ} - \theta_{AQ}\theta_{BQ})\mu_{QQ} + (\theta_{AQ} - \theta_{AQ}\theta_{BQ})\mu_{qq}}{\theta_{AQ} + \theta_{BQ} - 2\theta_{AQ}\theta_{BQ}}$
aa	BB	$\mu_3 = \dfrac{(\theta_{AQ} - \theta_{AQ}\theta_{BQ})\mu_{QQ} + (\theta_{BQ} - \theta_{AQ}\theta_{BQ})\mu_{qq}}{\theta_{AQ} + \theta_{BQ} - 2\theta_{AQ}\theta_{BQ}}$
aa	bb	$\mu_4 = \dfrac{\theta_{AQ}\theta_{BQ}\mu_{QQ} + (1 - \theta_{AQ} - \theta_{BQ} + \theta_{AQ}\theta_{BQ})\mu_{qq}}{1 - \theta_{AQ} - \theta_{BQ} + 2\theta_{AQ}\theta_{BQ}}$

The recombination frequency between the marker loci (θ_{AB}) is estimated by usual methods (Ott 1990) and thereafter fixed, and θ_{AQ} and θ_{BQ} are fixed for some distance from the markers, then the other parameters (the means of the QTL genotypes or linear functions of those means) are estimated. Recombination frequencies between marker and quantitative trait loci are factored out of the model by doing this. Interval mapping is done by estimating test statistics (likelihood ratios of F-statistics) for different θ_{AQ}, thereby fixing θ_{BQ} as well, and finding the θ_{AQ} and θ_{BQ} where the test statistic is maximized. This is grid searching the parameter space of θ_{AQ} and θ_{BQ}. For example, statistics can be estimated for one cM increments of $\hat{\theta}_{AB}$ starting at marker locus A and ending at marker locus B ($0 \leq \theta_{AQ} \leq \hat{\theta}_{AB}$) where

$$\theta_{BQ} = \frac{\theta_{AQ} - \hat{\theta}_{AB}}{2\theta_{AQ} - 1}$$

and $\hat{\theta}_{AB}$ is the maximum likelihood estimate of the recombination frequency between A and B.

The parameters of the genetic model (Table 2) can be estimated without factoring out the recombination frequencies (Knapp et al. 1990; Weller 1993). Instead of estimating QTL genotype means and test statistics for different recombination frequencies by grid searching (fixing and for several fixed points between A and B), the recombination frequencies and QTL genotype means can be estimated by minimizing or maximizing some statistic, e.g., by maximizing the likelihood or minimizing the error sum of squares. The genetic model (expected means of marker phenotypes) becomes linear when the recombination frequencies are fixed, and linear least squares methods can then be used to estimate the QTL genotype means. Differences between the various parameter estimation methods are mostly a matter of mechanics. The gain achieved by using linear models, least squares, and grid searching is that virtually any experiment and environment design can be handled using standard software.

The expected means of marker phenotypes (Table 2) can be redefined to implement LIM. Let

$$\theta_{BQ} = \frac{\theta_{AQ} - \hat{\theta}_{AB}}{2\theta_{AQ} - 1}$$, then the expected means of the marker phenotypes are

$$E(\mu_{AB/AB}) = \frac{(1 - \theta_{AB} - 2\theta_{AQ} + \theta_{AQ}\theta_{AB} + \theta^2_{AQ})\mu_{QQ} + (\theta_{AQ}\theta_{AB} - \theta^2_{AQ})\mu_{qq}}{(\theta_{AB} - 1)(2\theta_{AQ} - 1)}$$

$$= p_1\mu_{QQ} + p_2\mu_{qq} = \frac{(1 - \theta_{AB} - 2\theta_{AQ} + 2\theta^2_{AQ})\alpha}{(\theta_{AB} - 1)(2\theta_{AQ} - 1)} = x_1\alpha$$

$$E(\mu_{Ab/Ab}) = \frac{(\theta_{AQ} + \theta_{AQ}\theta_{AB} - \theta_{AB} - \theta^2_{AQ})\mu_{QQ} + (\theta_{AQ}\theta_{AB} - \theta_{AQ} + \theta^2_{AQ})\mu_{qq}}{2\theta_{AB}\theta_{AQ} - \theta_{AB}} \ ,$$

$$= p_3\mu_{QQ} + p_4\mu_{qq} = \frac{-(\theta_{AB} - 2\theta_{AQ} + 2\theta^2_{AQ})\alpha}{2\theta_{AB}\theta_{AQ} - \theta_{AB}} = -x_2\alpha$$

$$E(\mu_{aB/aB}) = \frac{(\theta_{AQ}\theta_{AB} - \theta_{AQ} + \theta^2_{AQ})\mu_{QQ} + (\theta_{AQ} + \theta_{AQ}\theta_{AB} - \theta_{AB} - \theta^2_{AQ})\mu_{qq}}{2\theta_{AB}\theta_{AQ} - \theta_{AB}} \ ,$$

$$= p_4 \mu_{QQ} + p_3 \mu_{qq} = \frac{(\theta_{AB} - 2\theta_{AQ} + 2\theta^2{}_{AQ}) \, \alpha}{2\theta_{AB} \, \theta_{AQ} - \theta_{AB}} = x_2 \alpha$$

and

$$E(\mu_{ab/ab}) = \frac{(\theta_{AQ} \, \theta_{AB} - \theta^2{}_{AQ}) \, \mu_{QQ} + (1 - \theta_{AB} - 2\theta_{AQ} + \theta_{AQ} \, \theta_{AB} + \theta^2{}_{AQ}) \, \mu_{qq}}{(\theta_{AB} - 1) \, (2\theta_{AQ} - 1)},$$

$$= p_2 \mu_{QQ} + p_1 \mu_{qq} = \frac{- (1 - \theta_{AB} - 2\theta_{AQ} + 2\theta^2{}_{AQ}) \, \alpha}{(\theta_{AB} - 1) \, (2\theta_{AQ} - 1)} = x_1 \alpha \qquad (4)$$

where $\mu_{QQ} = \alpha$, $\mu_{qq} = -\alpha$, $\mu_{QQ} - \mu_{qq} = 2\alpha$ and α is the additive effect of the QTL. Substituting $\hat\theta_{AB}$ for θ_{AB} and θ_{AQ} for $\hat\theta_{AQ}$, the expected frequencies of the QTL can be estimated where $\hat\theta_{AQ}$ is the estimated (fixed) distance of Q from A ($0 \le \hat\theta_{AQ} \le \hat\theta_{AB}$).

The expected means of marker genotypes for $\hat\theta_{AB} = 0.10$, $\hat\theta_{AQ} = 0.0527864$,

and $\hat\theta_{BQ} = \dfrac{0.0527864 - 0.10}{2(0.0527864) - 1} = 0.0527864$, for example, are

$E(\mu_{AB/AB}) = p_1 \mu_{QQ} + p_2 \mu_{qq} = 0.9969\mu_{QQ} + 0.0031\mu_{qq} = 0.9938\alpha,$

$E(\mu_{Ab/Ab}) = p_3 \mu_{QQ} + p_4 \mu_{qq} = 0.5\mu_{QQ} + 0.5\mu_{qq} = 0.0\alpha,$

$E(\mu_{aB/aB}) = p_4 \mu_{QQ} + p_3 \mu_{qq} = 0.5\mu_{QQ} + 0.5\mu_{qq} = 0.0\alpha,$

and

$$E(\mu_{ab/ab}) = p_2 \mu_{QQ} + p_1 \mu_{qq} = 0.0031\mu_{QQ} + 0.9969\mu_{qq} = -0.9938\alpha; \qquad (5)$$

thus, the expected means of marker genotypes, with the recombination frequencies fixed, are linear functions of the means of QTL genotypes.

A QTL is mapped by estimating the means of QTL genotypes for $0 \le \hat\theta_{AQ} \le \hat\theta_{AB}$, and testing the hypothesis of no QTL ($H: \mu_{QQ} = \mu_{qq}$). Evidence for a QTL between A and B exists when the null hypothesis is rejected for some $\hat\theta_{AQ}$.

QTL parameters can be estimated and hypotheses about differences between QTL genotype means can be tested by using linear contrasts between marker means where the contrast coefficients are defined for fixed θ_{AQ} and θ_{BQ} and marker phenotypes *per se* are used as independent variables. A linear model for the RCB experiment is

$$y_{ikj} = \mu + b_i = m_k = g(m)_{kj} = eijk \qquad (6)$$

where y_{ikj} is the *ikj*th observation of the quantitative trait, is the population mean, b_i is the effect of the *i*th block, m_k is the effect of the *k*th marker genotype, and e_{ikj} is the random error for the *j*th line of the *k*th marker genotype in the *i*th block (Table 3). The effects of marker genotypes are fixed, while the effects of other factors are random. The parameters of (6) can be estimated by using standard linear model methods since the marker phenotypes are known. The QTL effects of (2) can be estimated by using linear differences among the marker means of (6). The additive effect of the QTL for doubled haploid lines, for example, is estimated by

$$2\alpha = \hat{x}_A \hat{\mu}' = [\hat{x}_1 \ -\hat{x}_2 \ \hat{x}_2 \ -\hat{x}_1] \begin{bmatrix} \hat{\mu}_{AB/AB} \\ \hat{\mu}_{Ab/Ab} \\ \hat{\mu}_{aB/aB} \\ \hat{\mu}_{ab/ab} \end{bmatrix} = \hat{x}_1 \hat{\mu}_{AB/AB} - \hat{x}_2 \hat{\mu}_{Ab/Ab} + \hat{x}_2 \hat{\mu}_{aB/aB} - \hat{x}_1 \hat{\mu}_{ab/ab}$$

where $\mu = [\mu_{AB/AB} \ \mu_{Ab/Ab} \ \mu_{aB/aB} \ \mu_{ab/ab}]$ is the vector of marker phenotype means and $\hat{x}_A$ is the vector of coefficients for estimating the additive effect of the QTL — the coefficients are estimated by fixing the recombination frequencies.

Table 3. Degrees of freedom, Type III sum of squares, and expected mean squares for lines tested in randomized complete blocks in one environment where the effects of marker genotypes are fixed and the effects of lines and blocks are random and factors other than marker genotypes are balanced.

Factor	Degrees of freedom[a]	Sum of squares	Expected mean square[b]
Block	$df_R = b - 1$	$R[b|\mu, g]$	$E(M_B) = \sigma^2_E + N\sigma^2_B$
Line (G)	$df_G = N - 1$	$R[g|\mu, b]$	$E(M_G) = \sigma^2_E + r\sigma^2_G$
Marker (M)	$df_M = m - 1$	$R[q|\mu, b]$	$E(M_M) = \sigma^2_E + r\sigma^2_{G:M} + \phi^2_M$
G:M	$df_{G:M} = N - m$	$R[g(q)|\mu, m, b]$	$E(M_{G:M}) = \sigma^2_E + r\sigma^2_{G:M}$
Residual	$df_E = (N - 1)(b - 1)$	$R[e|\mu, b, g]$	$E(M_E) = \sigma^2_E$

[a] m is the number of QTL genotypes, $N = \sum_{i=1}^{q} n_i$ is the number of lines where n_i is the number of lines of the *i*th marker genotype, and b is the number of bloks.

[b] σ^2_E is the error variance, $\sigma^2_{G:Q}$ is the line nested in QTL genotype variance, σ^2_G is the between line variance, and θ^2_M is the variance of fixed effects of marker genotypes.

The hypothesis of no difference between QTL genotype means can be tested by using the F-statistic $F = M_Q / M_{G:M}$ since

$$\frac{E(M_Q)}{E(M_{G:M})} = \frac{\sigma^2_E + r\sigma^2_{G:M} + \phi^2_Q}{\sigma^2_E + r\sigma^2_{G:M}}$$

where M_Q is the mean square for the hypothesis being tested, which is estimated by a difference among marker phenotype means (Table 3), and the null hypothesis is rejected with a Type I error probability of α when $F > F\alpha:df_Q, df_{G:M}$. For doubled haploid lines and one QTL, the null hypothesis is $H_0{:}\mu_{QQ} = \mu_{qq}$. Because the other parameters $(\theta_{AB}, \theta_{AQ}, \text{and } \theta_{BQ})$ are fixed, the sum of squares for the single degree of freedom additive effect contrast for doubled haploid lines is equal to the sum of squares for marker genotypes, which has three degrees of freedom, as long as the genetic model for the hypothesized QTL is adequate. This works out this way for any mating design where two genotypes are observed at a given QTL, e.g., backcross, recombinant inbred, and various testcross progeny. When one QTL is hypothesized to lie between two segregating marker loci A and B, there are three genotypes at each QTL, and two degrees of freedom for differences between QTL genotype means for F_2, F_3, F_4, or other segregating generations among inbred lines. These can be estimated as linear differences between marker phenotype means, for which there is eight degrees of freedom.

A LIM example is developed for one marker-bracket (*Plc* and *iABI151*) (Kleinhofs et al. 1992) and an experiment where 150 barley doubled haploid lines were tested in randomized complete blocks experiment designs at Corvallis, Oregon and Pullman, Washington in 1991 (Hayes et al. 1993). The quantitative trait used for the example is seed yield (ka/ha). The number of replications of lines was not balanced. Fifty lines were replicated twice, while the other 100 lines were unreplicated, so the coefficients for the expected mean squares are more complicated than those described above for (6) (Table 3).

The hypothesis of no differences between marker genotype means $(H_0{:}\mu_{AB/AB} = \mu_{Ab/Ab} = \mu_{aB/aB} = \mu_{ab/ab})$ from experiments within each environment were tested using

$$F = M_M / M_{G:M} \text{ where}$$

$$\frac{E(M_M)}{E(M_{G:M})} = \frac{\sigma^2_E + 1.2\sigma^2_{G:M} + \phi^2_M}{\sigma^2_E + 1.3\sigma^2_{G:M}}$$

(Tables 4 and 5). The other coefficients for $\sigma^2_{G:M}$ of M_M and $M_{G:M}$ are not equal because the number of replications of lines are unequal. Nor are the Type I and Type III sum of squares equal, and Type III sum of squares must be used to get unbiased test statistics and parameter estimates for this problem (Searle 1971). The mean square for marker genotypes, for example, was estimated by

$$\frac{R\,[m|\mu,b]}{m - 1}$$

where $R\,[m|\mu,b]$ is the reduction in sum of squares due to fitting the effects of marker genotypes after the mean and the effects blocks (Tables 4 and 5).

The hypothesis of no differences between marker genotypes *per se* is not important, but since the marker phenotypes are known within errors of ascertainment (misscored marker phenotypes), the coefficients for the mean squares and Type III sum of squares can be directly estimated and used for Type III tests of hypotheses of differences between means of hypothesized QTL genotypes. The hypothesis of no difference between QTL genotype means $(H_0{:}\mu_{QQ} = \mu_{qq})$ from (6) was tested using $F = M_Q / M_{G:M}$ where

$$\frac{E(M_Q)}{E(M_{G:M})} = \frac{\sigma^2_E + 1.2\sigma^2_{G:M} + \phi^2_Q}{\sigma^2_E + 1.3\sigma^2_{G:M}},$$

ϕ^2_Q is the variance of fixed effects of QTL genotypes, $M_Q = \dfrac{R\,[g|\mu,\,b]}{q-1}$

the mean square for the hypothesis, and $R\,[g|\mu,\,b]$ is the reduction in sum of squares for QTL genotypes (estimated as differences between marker phenotype means) estimated after the population mean and the effects of blocks (Tables 4 and 5).

Table 4.   Degrees of freedom (DF), Type III mean squares (MS), F-statistics (F), and probabilities for F-statistics (Pr > F) for different factors affecting the seed yields (kg/ha) of barley doubled haploid lines tested at Corvallis, Oregon.

Factor	DF	MS	F	Pr > F	Expected mean square[a]
Block	1	6,179,351	5.2	0.03	$E(M_B) = \sigma^2_E + 50.0\sigma^2_B$
Line (G)	149	2,143,979	1.8	0.01	$E(M_G) = \sigma^2_E + 1.3\sigma^2_G$
Marker	3	5,764,777	2.7	0.048	$E(M_M) = \sigma^2_E + 1.2\sigma^2_{G:M} + \phi^2_M$
QTL	1	16,559,393	7.8	0.0059	$E(M_Q) = \sigma^2_E + 1.2\sigma^2_{G:M} + \phi^2_Q$
G:M	146	2,113,975	1.8	0.011	$E(M_{G:M}) = \sigma^2_E + 1.3\sigma^2_{G:M}$
Error	49	1,191,234			$E(M_E) = \sigma^2_E$

[a] σ^2_E is the error variance, $\sigma^2_{G:M}$ is the line nested in marker genotype variance, σ^2_G is the between line variance, ϕ^2_M is the variance of fixed effects of marker genotypes, and ϕ^2_Q is the variance of fixed effects of QTL genotypes estimated using differences between marker genotype means.

Table 5.   Degrees of freedom (DF), Type III mean squares (MS), F-statistics (F), and probabilities for F-statistics (Pr > F) for different factors affecting the seed yields (kg/ha) of barley doubled haploid lines tested at Pullman, Washington.

Factor	DF	MS	F	Pr > F	Expected mean square[a]
Block	1	1,599,643	2.3	0.14	$E(M_B) = \sigma^2_E + 50.0\sigma^2_B$
Line (G)	149	1,662,027	2.4	0.0004	$E(M_G) = \sigma^2_E + 1.3\sigma^2_G$
Marker	3	614,284	0.4	0.75	$E(M_M) = \sigma^2_E + 1.2\sigma^2_{G:M} + \phi^2_M$
QTL	1	774,525	0.5	0.48	$E(M_Q) = \sigma^2_E + 1.2\sigma^2_{G:M} + \phi^2_Q$
G:M	146	1,689,607	2.4	0.0003	$E(M_{G:M}) = \sigma^2_E + 1.3\sigma^2_{G:M}$
Error	49	703,162			$E(M_E) = \sigma^2_E$

[a] σ^2_E is the error variance, $\sigma^2_{G:M}$ is the line nested in marker genotype variance, σ^2_G is the between line variance, ϕ^2_M is the variance of fixed effects of marker genotypes, and ϕ^2_Q is the variance of fixed effects of QTL genotypes estimated using differences between marker genotype means.

Parameters and test statistics were estimated for every cM between *Plc* and *iABI151* (Fig. 1). The recombination frequency between *Plc* and *iABI151* was $\hat{\theta}_{AB} = 0.18$. Evidence for a QTL between *Plc* and *iABI151* for seed yield was found for lines tested at Corvallis, but not at Pullman (Fig. 1 and Tables 4 and 5). F-statistics for $H_0:\mu_{QQ} \neq \mu_{qq}$ for some $\hat{\theta}_{AQ}$ were significantly greater than for lines $F0.01:df_Q,df_{G:M}$ tested at Corvallis, but not at Pullman. The maximum difference between QTL genotype means was found for $\hat{\theta}_{AQ} = 0.09$ for Corvallis and for $\hat{\theta}_{AQ} = 0.07$ for Pullman, although the latter was not significant (Fig. 1 and Tables 4 and 5). The contrast coefficients and least square means (kg/ha) for $\hat{\theta}_{AQ} = 0.09$ for Corvallis are

$$\hat{x}_A = [0.976 - 0.110 \ 0.110 - 0.976]$$

and

$$\hat{\mu} = [\hat{\mu}_{AB/AB} \ \hat{\mu}_{Ab/Ab} \ \hat{\mu}_{aB/aB} \ \hat{\mu}_{ab/ab}] = [6{,}223.3 \ 6{,}407.9 \ 5{,}913.0 \ 5{,}589.0];$$

so the additive effect of the hypothesized QTL affecting the seed yield of lines tested at Corvallis is $(\hat{x}_A \ \mu')/2 = 168.4$ where $\mu_{QQ} - \mu_{qq} = 2\alpha$.

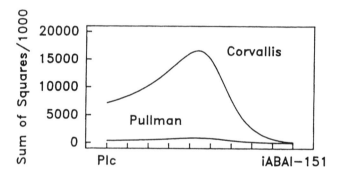

Fig. 1. Seed yield (kg/ha) sum of squares for the additive effect of a QTL between the *Plc* and *iABI151* locus for barley doubled haploid lines tested in Corvallis, Oregon and Pullman, Washington.

The parameters and test statistics for this example were estimated using PROC GLM of SAS (1992). The code for the example is

```
proc glm;
class block line;
model y = block line;
proc glm;
class block line marker;
model y = block marker line (marker);
contrast 'QTL A' marker 0.976 0.110 -0.110 -0.976;
lsmeans marker;
```

The parameters of (1) were estimated by the first `proc glm` sequence, while the parameters of (6) were estimated by the second `proc glm` sequence. The first `proc glm` sequence is not essential because the between line sum of squares can be estimated by summing the sum of squares for markers and lines nested in markers. The `contrast` statement estimates the sum of squares for the additive effect of the QTL, while the `ls-means` statement estimates the least square means for marker phenotypes.

The versatility of LIM should be clear from this example. Only the model statement needs to be rewritten for different experiment designs. The other statements (`marker`, `contrast`, `lsmeans`) are determined by the mating design, and are not affected by different experiment designs.

Table 6.   Degrees of freedom, Type III sum of squares, and expected mean squares for the analysis of the least square means of lines tested in one environment where the effects of marker genotypes are fixed and the effects of lines are random and factors other than marker genotypes are balanced.

Factor	Degrees of freedom[a]	Sum of Squares	Expected mean square[b]
Marker (M)	$df_M = m - 1$	$R\,[m|\mu, b]$	$E(M_M) = \sigma^2_{G:M} + \phi^2_M$
Line:M (G:M)	$df_{G:M} = N - m$	$R\,[g(m)|\mu, m, b]$	$E(M_{G:M}) = \sigma^2_{G:M}$

[1] m is the number of marker genotypes and $N = \sum_{i=1}^{q} n_i$ is the number of lines where n_i is the number of lines of the ith marker genotype.

[b] $\sigma^2_{G:M}$ is the line nested in QTL genotype variance and ϕ^2_M is the variance of fixed effects of marker genotypes.

The problem of estimating QTL parameters from different experiment designs or any experiment design can be simplified by analyzing the least square means of lines instead of the original observations - if the experiment design is balanced, then the least square and arithmetic means of lines are equal (Searle 1971). The experiment is first analyzed ignoring marker genotypes, but not other factors, and the least square means of lines are output and used for subsequent QTL analyses. The standard analysis of lines is done for the experiment design used. By using the least square means of lines for the QTL analysis, observations from different experiment designs can be merged, and one QTL analysis can be used for any experiment design. Pure error is eliminated from the model and the error for testing hypotheses about QTL genotypes is the line nested in QTL genotype variance (Table 7). A line-means analysis can of course be used with methods other than LIM, e.g., with maximum likelihood methods.

Software which handles only one-factor linear models, e.g., MAPMAKER-QTL (Lander and Botstein 1990), can be used to estimate QTL parameters from different experiment designs by inputting the least square means of lines, rather than by inputting the original observations, and ignoring the experiment design. These two choices are different. The former eliminates pure error variance, while the latter confounds the pure error and line nested in marker or QTL genotypes variances, and can be less powerful. Whether or not it is less powerful is a function of the number of QTL affecting the trait and how their effects are estimated (Knapp et al. 1993).

Most experiment, mating, and environment designs can be handled using PROC GLM or other software for unbalanced linear models since the only unique parts of estimation are the coefficients for testing hypotheses about differences between means of 'hypothesized QTL genotypes', and these coefficients are not affected by the experiment or environment design. Another way to implement LIM is to use these coefficients as covariates. SAS code for implementing LIM this way is

```
proc glm;
   class block marker line;
   model y = block x marker line (marker);
```

where x is the vector of coefficients for a stated θ_{AB} and θ_{AQ}. Because x is listed in the model statement, but not in the class statement, SAS uses it as a covariate. Hypotheses about QTL genotype differences are tested using Type I mean squares — the covariate absorbs 100% of the between marker genotype sums of squares when θ_{AQ} maximizes the sums of squares for the QTL effects and the data are balanced.

The estimation of covariate coefficients can be built in without difficulty. SAS code for the doubled haploid example, for example, is

```
data a;
input locusa locusb;
cards;
1 1
1 2
2 1
2 2
;
data b;
set a;
rab = 0.18;
raq = 0.09;
rbq + (raq - rab) / (2*raq - 1);
p1 = 0;
p2 = 0;
marker = 0;
if locusa = 1 and locusb = 1 then do;
  p1 = (1 - raq - rbq + raq*rbq) / (1 - raq - rbq + 2*raq*rbq);
  p2 = (raq*rbq) / (1 - raq - rbq + 2*raq*rbq);
  marker = 1;
end;
else if locusa = 1 and locusb = 2 then do;
  p1 = (rbq - raq*rbq) / (raq + rbq - 2*raq(rbq);
  p2 = (raq - raq*rbq) / (raq + rbq - 2*raq*rbq);
  marker = 2;
end;
```

```
else if locusa = 2 and locusb = 1 then do;
  p1 = (raq − raq*rbq) / (raq + rbq − 2*raq*rbq)
  p2 = (rbq − raq*rbq) / (raq + rbq − 2*raq*rbq);
  marker = 3;
end;
else if locusa = 2 and locusb = 2 then do;
  p1 = (raq*rbq) / (1 − raq − rbq + 2*raq*rbq);
  p2 = (1 − raq − rbq + raq*rbq) / (1 − raq − rbq + 2*raq*rbq);
  marker = 4;
end;
x = p1 − p2;
output;
data b;
set b;
keep locusa locusb marker x;
proc sort data = b;
by locusa locusb;
data c;
infile 'barley.dat';
input line locusa locusb block y;
proc sort data = c;
by locusa locusb;
data d;
merge b c;
by locusa locusb;
proc print data = d;
proc glm data = d;
proc glm data = d;
  class block marker line;
  model y = block x marker line (marker);
run;
```

The estimation of the covariate (x) uses the expected frequencies defined by (4).

    LIM creates a straightforward means for handling replicated progeny and the diversity of linear models arising in plant breeding experiments. This is further exemplified by multiple environment experiments, which are reviewed next.

3.   Multiple environment experiments

LIM for multiple environment experiments is handled no differently than for an experiment in one environment. Once coefficients are defined for testing hypotheses about QTL genotypes for a mating design, hypotheses about those genotypes can be tested from a virtually endless number of experiment and environment designs. An example is developed for an experiment done across locations, but it could just as easily be done for an experiment done across years, or for any other environment design. Staying with the RCB experiment design, a linear model for lines tested across locations is

$$y_{ijm} = \mu + b_i + g_j + l_m + gl_{jm} + e_{ijm} \tag{7}$$

where y_{ijm} is the ijmth observation of the quantitative trait, is the population mean, b_i is the effect of the ith block, g_j is the effect of the jth line, l_m is the effect of the mth location, gl_{jm} is the effect of the interaction between the jth line and the mth location, and e_{ijm} is the random error for the jth line in the mth location in the ith block, while a linear model for QTL genotypes tested across locations is

$$y_{ikmj} = \mu + b_i + q_k + g(q)_{kj} + l_m + ql_{km} + g(ql)_{kmj} + e_{ikmj} \tag{8}$$

where y_{ikmj} is the $ikmj$th observation of the quantitative trait, is the population mean, b_i is the effect on the ith block, q_k is the effect of the kth QTL genotype, $g(q)_{kj}$ is the effect of the jth line nested in the kth QTL genotype, l_m is the effect of the mth location, ql_{km} is the effect of the kth QTL genotype in the mth location, $g(ql)_{kmj}$ is the effect of the jth line nested in the kth QTL genotype in the mth location, and e_{ikmj} is the random error for the jth line nested in the kth QTL genotype in the mth location and the ith block (Table 7).

The sum of squares for lines is the sum of the sum of squares for QTL genotypes and lines nested in QTL genotypes, as before, and the sum of the sum of squares for lines by locations is the sum of the sum of squares for QTL genotypes by locations and for lines nested in QTL genotypes by locations; thus,

$$E(M_G) = \sigma^2_E + r\sigma^2_{GL} + rl\sigma^2_G = \frac{df_Q\, E(M_Q) + df_{G:Q}\, E(M_{G:Q})}{df_Q + df_{G:Q}}$$

$$= \sigma^2_E + r\sigma^2_{G:QL} + \frac{r\bar{n}\,(q-1)}{N-1}\,\sigma^2_{QL} + rl\sigma^2_{G:Q} + \frac{rl\bar{n}}{N-1}\, \frac{\Sigma\, q^2_k}{q-1} \tag{9}$$

and

$$E(M_G) = \sigma^2_E + r\sigma^2_{GL} = \frac{df_{QL}\, E(M_{QL}) + df_{G:QL}\, E(M_{G:QL})}{df_{QL} + df_{G:QL}}$$

$$= \sigma^2_E + r\sigma^2_{GQL} + \frac{r\bar{n}\,(q-1)}{N-1}\,\sigma^2_{QL} \tag{10}$$

where $\sigma^2_G = \sigma^2_G + \dfrac{\bar{n}}{N-1}\, \dfrac{\Sigma\, q^2_k}{q-1} = \sigma^2_{G:Q} + \phi^2_Q$ and $= \sigma^2_{GL} = \sigma^2_{G:QL} + \dfrac{\bar{n}\,(q-1)}{N-1}\,\sigma^2_{QL}$.

Table 7.   Degrees of freedom, Type III sum of squares, and expected mean squares for lines
tested in randomized complete blocks in different locations where the effects of QTL
genotypes are fixed, and the effects of lines, blocks, and locations are random and
factors other than QTL genotypes are balanced.

Factor	Degrees of freedom[a]	Sum of Squares	Expected mean square[b]
Block	$df_B = b - 1$		$E(M_B) = \sigma^2_E + lN\sigma^2_B$
Locations (L)	$df_L = l - 1$		$E(M_L) = \sigma^2_E + r\sigma^2_{GL} + rN\sigma^2_L$
Line (G)	$df_G = N - 1$	$R\,[g\|\mu, l, b, gl]$	$E(M_G) = \sigma^2_E + r\sigma^2_{GL} + rl\sigma^2_G$
QTL	$df_G = q - 1$	$R\,[g\|\mu, l, b, gl]$	$E(M_G) = \sigma^2_E + r\sigma^2_{G:QL} + rl\sigma^2_{G:Q}$ $+ rn\sigma^2_{QL} + \phi^2_Q$
G:QTL	$df_{G:Q} = N - q$	$R\,[g(q)\|\mu, l, b, gl, q]$	$E(M_{G:Q}) = \sigma^2_E + r\sigma^2_{G:QL} +$ $rl\sigma^2_{G:Q}$
G x L	$df_{GL} = (N - 1)\,(l - 1)$	$R\,[gl\|\mu, l, b, g]$	$E(M_{GL}) = \sigma^2_E + r\sigma^2_{GL}$
QTL x L	$df_{QL} = (q - 1)\,(l - 1)$	$R\,[ql\|\mu, l, b, g]$	$E(M_{QL}) = \sigma^2_E + r\sigma^2_{G:QL} + rn\sigma^2_{QL}$
G:QTL x L	$df_{G:QL} = (N - q)\,(l - 1)$	$R\,[g(ql)\|\mu, l, b, g, ql]$	$E(M_{G:QL}) = \sigma^2_E + r\sigma^2_{G:QL}$
Residual	$df_E = (Nl - 1)\,(b - 1)$	$R\,[e\|\mu, l, b, g, gl]$	$E(M_E) = \sigma^2_E$

[a] q is the number of QTL genotypes and $N = \sum_{i=1}^{q} n_i$ is the number of lines where n_i is the number of
lines of the ith QTL genotype, b is the number of blocks, and l is the number of locations.
[b] σ^2_E is the error variance, θ^2_G is between line variance, ϕ^2_Q is the variance of fixed effects of
QTL genotypes, $\sigma^2_{G:Q}$ is the line nested in QTL genotype variance, σ^2_{GL} is the line by location
variance, σ^2_{QL} is the QTL genotype by location variance, $\sigma^2_{G:QL}$ is the line nested in QTL genotype by location variance, and

$$\bar{n} = \frac{N - \dfrac{\sum n_i^2}{N}}{q - 1}$$

Since QTL genotypes are not 'observed', we proceed as before by defining a linear
model where differences between marker genotype means are used to estimate QTL
parameters. This is done by substituting the effects of QTL genotypes in (8) with the
effects of marker genotypes, viz.

$$y_{ikmj} = \mu + b_i + m_k + g(m)_{kj} + l_m + ml_{km} + g(ml)_{kmj} + e_{ikm}$$

where y_{ikmj} is the $ikmj$th observation of the quantitative trait, is the population mean, b_i
is the effect of the jth line nested in the kth marker phenotype, l_m is the effect of the mth
location, ml_{km} is the effect of the kth marker genotype in the .·h location, $g(ml)_{kmj}$ is the
effect of the jth line nested in the kth marker genotype in the n·h location, and e_{ikmj} is the
random error for the jth line nested in the kth marker ger.:·type in the mth location and
the ith block (Table 8).

The most important null hypotheses for [11] are $H_0:\sigma^2_G = 0$ and $H_0:\sigma^2_{GL} = 0$ and subsets of $H_0:\sigma^2_{ML} = 0$ and $H_0:\phi^2_M = 0$, which test hypotheses about QTL genotypes (Table 8). The first two hypotheses are used to determine whether or not there are significant differences between lines and between lines by locations. The second two hypotheses are used to determine whether or not there are significant differences between QTL genotypes and between QTL genotypes by locations. The QTL genotype by locations are random. For fixed environment factors, e.g., fertilization or irrigation rate, the effects are fixed, with consequent effects on the expected mean squares.

Table 8.   Degrees of freedom, Type III sum of squares, and expected mean squares for lines tested in randomized complete blocks across locations where the effects of QTL genotypes are fixed, and the effects of lines, blocks, and locations are random and factors other than QTL genotypes are balanced.

Factor	Degrees of freedom[a]	Sum of Squares	Expected mean square[b]
Block	$df_B = b - 1$		$E(M_B) = \sigma^2_E + lN\sigma^2_B$
Locations (L)	$df_L = l - 1$		$E(M_L) = \sigma^2_E + r\sigma^2_{GL} + rN\sigma^2_L$
Line (G)	$df_G = N - 1$	$R\,[g|\mu, l, b, gl]$	$E(M_G) = \sigma^2_E + r\sigma^2_{GL} + rl\sigma^2_G$
Marker (M)	$df_M = m - 1$	$R\,[m|\mu, l, b, gl]$	$E(M_M) = \sigma^2_E + r\sigma^2_{G:ML} + rl\sigma^2_{G:M} + rn\sigma^2_{ML} + \phi^2_M$
G:M	$df_{G:M} = N - m$	$R\,[g(m)|\mu, l, b, gl, m]$	$E(M_{G:M}) = \sigma^2_E + r\sigma^2_{G:ML} + rl\sigma^2_{G:M}$
G x L	$df_{GL} = (N - 1)\,(l - 1)$	$R\,[gl|\mu, l, b, g]$	$E(M_{GL}) = \sigma^2_E + r\sigma^2_{GL}$
M x L	$df_{ML} = (m - 1)\,(l - 1)$	$R\,[ml|\mu, l, b, g]$	$E(M_{ML}) = \sigma^2_E + r\sigma^2_{G:ML} + rn\sigma^2_{ML}$
G:M x L	$df_{G:ML} = (N - m)\,(l - 1)$	$R\,[g(ml)|\mu, l, b, g, ml]$	$E(M_{G:ML}) = \sigma^2_E + r\sigma^2_{G:ML}$
Residual	$df_E = (Nl - 1)\,(b - 1)$	$R\,[e|\mu, l, b, g, gl]$	$E(M_E) = \sigma^2_E$

[a] m is the number of QTL genotypes and $N = \sum\limits_{i=1}^{q} n_i$ is the number of lines where n_i is the number of lines of the ith QTL genotype, b is the number of blocks, and l is the number of locations.

[b] σ^2_E is the error variance, θ^2_G is between line variance, ϕ^2_M is the variance of fixed effects of marker genotypes $\sigma^2_{G:Q}$ is the line nested in QTL genotype variance, σ^2_{GL} is the line by location variance, σ^2_{ML} is the QTL genotype by location variance, $\sigma^2_{G:ML}$ is the line nested in marker genotype by location variance, and

$$\bar{n} = \frac{N - \dfrac{\Sigma n_i^2}{N}}{q - 1}$$

The hypothesis of no differences between marker genotype means $(H_0:\mu_{AB/AB} = \mu_{aB/ab} = \mu_{aB/ab} = \mu_{ab/ab})$ across locations for (11) are tested using

$$F = M_M / (M_{G:M} + M_{ML} - M_{G:ML}) = (M_M + M_{G:ML}) / (M_{G:M} + M_{ML})$$

since

$$\frac{E(M_M)}{E(M_{G:M}) + E(M_{ML}) - E(M_{G:ML})} = \frac{\sigma^2_E + r\sigma^2_{G:ML} + rl\sigma^2_{G:M} + r\bar{n}\sigma^2_{ML} + \phi^2_M}{\sigma^2_E + r\sigma^2_{G:ML} + rl\sigma^2_{G:M} + r\bar{n}\sigma^2_{ML}}$$

and specific hypotheses about differences between hypothesized QTL genotypes are tested using

$$F = M_Q / (M_{G:M} + M_{ML} - M_{G:ML}) = (M_Q + M_{G:ML}) / (M_{G:M} + M_{ML}) \qquad (12)$$

since

$$\frac{E(M_Q)}{E(M_{G:M}) + E(M_{ML}) - E(M_{G:ML})} = \frac{\sigma^2_E + r\sigma^2_{G:ML} + rl\sigma^2_{G:M} + r\bar{n}\sigma^2_{ML} + \phi^2_Q}{\sigma^2_E + r\sigma^2_{G:ML} + rl\sigma^2_{G:M} + r\bar{n}\sigma^2_{ML}}$$

where M_Q is the mean square for differences between marker genotype means and $E(M_Q)$ is the expected value of M_Q (Table 8). M_Q is estimated using the appropriate contrast or contrasts for the mating design used. These are the sort of complex F-tests which arise in most multiple environment breeding experiments (Gardner 1963). Since (12) is a linear function of sums of mean squares, the error degrees of freedom is not known and must be estimated (Searle 1970). Numerator and denominator degrees of freedom for (12) are estimated by

$$\frac{[M_Q + M_{G:ML}]^2}{M^2_Q / (df_Q) + M^2_{G:ML} / (df_{G:ML})}$$

and

$$\frac{[M_{G:M} + M_{ML}]^2}{M^2_{G:M} / (df_{G:M}) + M^2_{ML} / (df_{ML})}$$

respectively (Searle 1970).

The hypothesis of no differences between marker genotype means by locations or of no differences between QTL genotypes by locations is more straightforward. The latter is tested using

$$F = M_{QL} / M_{G:ML}$$

since

$$\frac{E\,(M_{QL})}{E\,(M_{G:ML})} = \frac{\sigma^2_E + r\sigma^2_{G:ML} + rn\sigma^2_{QL}}{\sigma^2_E + r\sigma^2_{G:ML}}$$

(Table 8).

Type III sum of squares must be used to get unbiased hypothesis tests for marker genotypes or QTL genotypes and for marker genotypes by locations or QTL genotypes by locations since marker genotypes are unbalanced and other factors might not be balanced either (Searle 1970). The mean square for marker genotypes is estimated by

$$\frac{R\,[m|\mu,\,l,\,b,\,g,\,gl]}{m - 1}$$

where $R\,[m|\mu,\,l,\,b,\,g,\,gl]$ is the reduction in sum of squares for marker genotypes estimated after the mean and the effects blocks, locations, and marker genotypes by locations (Table 8).

The LIM steps outlined for one environment are repeated to test hypotheses about QTL genotype effects across environments and QTL genotype by environment effects. It is only necessary to define contrast coefficients for the QTL genotype by environment tests since the coefficients for QTL genotypes across environments are equal to those for individual environments. This is illustrated for the doubled haploid example. The null hypothesis of no additive effect of a QTL across two locations estimated using doubled haploid lines is

$$H_0{:}\mu_{QQ1} + \mu_{QQ2} = \mu_{qq1} + \mu_{qq2}\,,$$

This effect is estimated by $\hat{\mu}_{QQ1} - \hat{\mu}_{qq1} + \hat{\mu}_{QQ2} - \hat{\mu}_{qq2}$. The null hypothesis of no additive by location effect of a QTL estimated using doubled haploid lines is

$$H_0{:}\mu_{QQ1} + \mu_{qq2} = \mu_{qq1} + \mu_{QQ2}\,.$$

This effect is estimated by $\hat{\mu}_{QQ1} - \hat{\mu}_{qq1} - \hat{\mu}_{QQ2} + \hat{\mu}_{qq2}$. To test these hypotheses, we proceed by estimating the parameters of (11), e.g., the sum of squares for each factor and the least square means for marker genotypes and marker genotypes by environments, and by defining linear differences between these means.

The parameters of (11) can be estimated by using standard methods since the marker phenotypes are known. The additive effect of the QTL for doubled haploid lines tested across locations is estimated by

$$2\alpha = \hat{x}_A\hat{\mu}' = [\hat{x}_1 \ \text{-} \ \hat{x}_2 \ \hat{x}_2 \ \text{-} \ \hat{x}_1] \begin{bmatrix} \hat{\mu}_{AB/AB} \\ \hat{\mu}_{Ab/Ab} \\ \hat{\mu}_{aB/aB} \\ \hat{\mu}_{ab/ab} \end{bmatrix} = \hat{x}_1\hat{\mu}_{AB/AB} - \hat{x}_2\hat{\mu}_{Ab/Ab} + \hat{x}_2\hat{\mu}_{aB/aB} - \hat{x}_1\hat{\mu}_{ab/ab}$$

where $\mu = [\mu_{AB/AB} \; \mu_{Ab/Ab} \; \mu_{aB/aB} \; \mu_{ab/ab}]$ are the means of marker phenotypes across locations and $\hat{x}_A$ are the estimated coefficients for estimating the additive effect of QTL — the coefficients are estimated by fixing the recombination frequencies.

Coefficients for testing the hypothesis of no QTL genotype by environment interaction are found by multiplying coefficients for testing the hypothesis of no difference between environment means by coefficients for testing the hypothesis of no difference between QTL genotype means. Coefficients for testing the hypothesis of no difference between two locations, for example, are $x_L = [1 \; -1]$. Scalar-multiplying this vector by $\hat{x}_A$ we get

$$[\hat{x}_1 \; -\hat{x}_2 \; \hat{x}_2 \; -\hat{x}_1 \; -\hat{x}_1 \; \hat{x}_2 \; -\hat{x}_2 \; \hat{x}_1]$$

where

$$[\mu_{AB/AB1} \; \mu_{Ab/Ab1} \; \mu_{aB/aB1} \; \mu_{ab/ab1} \; \mu_{AB/AB2} \; \mu_{Ab/Ab2} \; \mu_{aB/aB2} \; \mu_{ab/ab2}]$$

is the vector of marker phenotype means for the two locations. Coefficients can be defined for any number of environments in this way as shown below.

A marker-bracket is interval mapped as for one environment, but with test-statistics estimated for the effect of QTL genotypes in each environment, the mean effect of QTL genotypes across environments, and the effects of QTL genotype by environment interactions. The θ_{AQ} at which the test-statistics for each of these tests is maximum can be and often is different. The most probable distance of the QTL from either of the marker loci could be estimated by finding the maximum for the pooled hypothesis test.

Since marker genotypes and blocks were unbalanced in the barley example, the variance coefficients for the expected mean squares are functions of the numbers of observations within subsets of cells (Table 9). The analysis is messy but doable. Parameters and test statistics for (11) were estimated for every cM between *Plc* and *iABI151* (Fig. 1). The contrast coefficients and least square means for seed yields of doubled haploid lines across environments (kg/ha) for $\hat{\theta}_{AB} = 0.18$ and $\hat{\theta}_{AQ} = 0.09$ for the *Plc-iAbI151* marker-bracket are

$$\hat{x}_A = [0.976. \; -0.110 \; 0.110 \; -0.976]$$

and

$$\hat{\mu} = [\hat{\mu}_{AB/AB} \; \hat{\mu}_{Ab/Ab} \; \hat{\mu}_{aB/aB} \; \hat{\mu}_{ab/ab}] = [5,825.2 \; 6,009.6 \; 5,820.7 \; 5,580.3];$$

thus, the additive effect of the hypothesized QTL between *Plc* and *iABI151* is $(\hat{\mu}_A \hat{\mu}') / 2$ = 129.9. The hypothesis of no differences between marker genotype means across locations and no additive effect of the QTL across locations were tested by

$$F = (M_M + M_{G:ML}) / (M_{G:M} + M_{ML}) = 0.6$$

and

$$F = (M_Q + M_{G:ML}) / (M_{G:M} + M_{ML}) = 1.1 \; ,$$

respectively, where the estimated numerator degrees of freedom for the former and latter are

$$\frac{[M_M + M_{G:ML}]^2}{M^2_M / (df_M) + M^2_{G:ML} / (df_{G:ML})} = 9.4 \cong 9.0$$

and

$$\frac{[M_Q + M_{G:ML}]^2}{M^2_Q / (df_Q) + M^2_{G:ML} / (df_{G:ML})} = 1.8 \cong 2.0$$

respectively, and the estimated denominator degrees of freedom for either test are

$$\frac{[M_{G:M} + M_{ML}]^2}{M^2_{G:M} / (df_{G:M}) + M^2_{ML} / (df_{ML})} = 2.2 \cong 2.0$$

(Searle 1970). Differences between marker genotype means were not significant ($0.6 < F_{0.09:9,2}$) across locations (Table 9). The additive effect of the QTL across locations was not significant ($1.1 < F_{0.01:149, 149}$) for any θ_{AQ}(Fig. 1 and Table 9). Nor was there any evidence ($F = M_G / M_{GL} = 1.2$) for genetic variance between lines tested across locations ($1.2 < F_{0.01:149,149}$) (Table 9). There was evidence ($F = M_{GL} / M_E = 1.8$) for a line by location interaction ($1.8 > F_{0.001:149,99}$) and for a QTL genotype by location interaction ($7.3 > F_{0.001:1, 146}$) where $F = M_{QL}/M_{G:ML} = 7.3$ (Fig. 1 and Table 9).

This example illustrates how QTL parameters can be estimated for a typical multiple environment breeding experiment. The parameters and test statistics for this example were estimated using SAS (1992). The code for the example is

```
proc glm data = b;
class location block line marker;
model y = location block marker line(marker)
location*marker location*line(marker);
contrast 'QTL A' marker 0.976 0.110 -0.110 -0.976;
contrast 'QTL A x L' location*marker
0.976 0.110 -0.110 -0.976 -0.976 -0.110 0.110 0.976;
```

The only difference between this code and the code shown for the analysis of the individual environment experiments, other than the `model` statement, is the addition of the `contrast` statement for testing the hypothesis of no QTL genotype by location interaction. We showed how to get the coefficients earlier.

Additional contrast coefficients are needed if lines are tested in more than two environments. One way to handle any number of environments is to use orthogonal polynomials to define contrast coefficients for environments and to subsequently scalar-multiply the environment coefficient matrix by X_A where X_A is made by stacking e x_A vectors. The scalar product is an $(e - 1)$ x $4e$ matrix of coefficients. For three environments, for example, orthogonal polynomials for estimating contrasts between the three environment means, e.g. $[\mu_1 \ \mu_2 \ \mu_3]$, are

$$x_E = \begin{bmatrix} 1 & 0 & -1 \\ 1 & -2 & 1 \end{bmatrix},$$

so the coefficients found by scalar multiplying this matrix by

$$\dot{X}_A = \begin{bmatrix} \hat{x}_1 & -\hat{x}_2 & \hat{x}_2 & -\hat{x}_1 \\ \hat{x}_1 & -\hat{x}_2 & \hat{x}_2 & -\hat{x}_1 \end{bmatrix}$$

are

$$\begin{bmatrix} \hat{x}_1 & -\hat{x}_2 & \hat{x}_2 & -\hat{x}_1 & 0 & 0 & 0 & 0 & -\hat{x}_1 & \hat{x}_2 & -\hat{x}_2 & \hat{x}_1 \\ \hat{x}_1 & -\hat{x}_2 & \hat{x}_2 & -\hat{x}_1 & -2\hat{x}_1 & 2\hat{x}_2 & -2\hat{x}_2 & 2\hat{x}_1 & \hat{x}_1 & -\hat{x}_2 & \hat{x}_2 & -\hat{x}_1 \end{bmatrix}$$

Table 9.   Degrees of freedom (DF), Type III mean squares (MS), F-statistics (F), probabilities for F-statistics (Pr > F), and expected mean squares (EMS) for different factors affecting the seed yields (kg/ha) of barley doubled haploid lines tested at Corvallis, Oregon and Pullman, Washington.

Factor	DF	MS	F	Pr >F	EMS [a]
Location (L)	1	8,894,799	5.1	0.03	$E(M_L) = \sigma^2_E + 1.2\sigma^2_{GL} + 180.0\sigma^2_L$
Block	1	7,033,500	7.4	0.008	$E(M_B) = \sigma^2_E + 100.0\sigma^2_B$
Line (G)	149	2,054,952	1.2	0.1	$E(M_G) = \sigma^2_E + 1.3\sigma^2_{GL} + 2.6\sigma^2_G$
Marker (M)	3	2,200,611	0.6	0.8	$E(M_M) = \sigma^2_E + 1.2\sigma^2_{G:ML} + 2.3\sigma^2_{G:M} + 38.5\sigma^2_{ML} + \phi^2_M$
QTL (Q)	1	4,922,170	1.1	0.5	$E(M_Q) = \sigma^2_E + 1.2\sigma^2_{G:ML} + 2.3\sigma^2_{G:M} + 38.5\sigma^2_{ML} + \phi^2_M$
G:M	146	2,064,989	1.2	0.1	$E(M_{G:M}) = \sigma^2_E + 1.3\sigma^2_{G:ML} + 2.6\sigma^2_{G:M}$
G x L	149	1,741,562	1.8	0.0006	$E(M_{GL}) = \sigma^2_E + 1.3\sigma^2_{GL}$
M x L	3	2,064,989	1.2	0.3	$E(M_{ML}) = \sigma^2_E + 1.2\sigma^2_{G:ML} + 38.5\sigma^2_{ML}$
Q x L	1	12,514,198	7.3	0.008	$E(M_{QL}) = \sigma^2_E + 1.2\sigma^2_{G:ML} + 38.5\sigma^2_{QL}$
G:M x L	146	1,710,704	1.8	0.0009	$E(M_{G:ML}) = \sigma^2_E + 1.3\sigma^2_{G:ML}$
Error	99	945,160			$E(M_{G:ML}) = \sigma^2_E$

[a] σ^2_E is the error variance, σ^2_G is between line variance, ϕ^2_M is the variance of fixed effects of QTL genotypes, $\sigma^2_{G:M}$ is the line nested in marker genotype variance, σ^2_{GL} is the line by location variance, σ^2_{ML} is the marker genotype by location variance, and $\sigma^2_{G:ML}$ is the line nested in marker genotype by location variance.

Contrast coefficients can be defined for any environment design. For multiple location and year experiments, for example, coefficients can be defined for testing hypotheses about QTL genotype by location by year interactions by using contrasts among marker genotype by location by year means, in addition to those for testing hypotheses about QTL genotype by hear and QTL genotype by location interactions.

Suppose, for the final example, lines are tested in randomized complete blocks across locations and years. To estimate the parameters of lines only, the linear model

$$y_{ijmn} = \mu + b_i + g_j + l_m + y_n + gl_{jm} + gy_{jn} + ly_{mn} + gly_{jmn} + e_{ijmn} \tag{13}$$

can be used where y_{ijmn} is the $ijmn$th observation of the quantitative trait, is the population mean, b_i is the effect of the ith block, g_j is the effect of the jth line, l_m is the effect of the mth location, y_n is the effect of the nth year, gl_{jm} is the effect of the interaction between the jth line and the mth location, gy_{jn} is the effect of the interaction between the jth line and the nth year, ly_{mn} is the effect of the interaction between the mth location and the nth year, gly_{jmn} is the effect of the interaction between the jth line, the mth location, and the nth year, e_{ijmn} is the random error for the jth line in the mth location, nth year, and ith block, and the effects of lines, locations, years, and blocks are random (Table 10).

Rewriting (13) as a function of QTL parameters leads to

$$y_{ijkmn} = + b_i + q_k + g(q)_{kj} + l_m + ql_{km} + g(ql)_{kmj} + y_n + gy_{kn} + g(qy)knj + ly_{mn} + gly_{kmn} +$$

$$g(qly)_{kmnj} + e_{ikjmn} \tag{14}$$

where y_{ikjmn} is the $ijkmn$th observation of the quantitative trait, is the population mean, b_i is the effect of the ith block, q_k is the effect of the kth QTL genotype, $g(q)_{kj}$ is the effect of the ith line nested in the kth QTL genotype, l_m is the effect of the mth location ql_{km} is the effect of the interaction between the kth QTL genotype and the mth location, $g(ql)_{kmj}$ is the effect of the jth line nested in the kth QTL genotype and the mth location, y_n is the effect of the nth year, gy_{kn} is the effect of the interaction between the kth QTL genotype and the nth year, $g(qy)_{knj}$ is the effect of the jth line nested in the kth QTL genotype and the nth year, ly_{mn} is the effect of the interaction between the mth location and the nth year, qly_{kmn} is the effect of the interaction between the kth QTL genotype, mth location, and nth year, $g(qly)_{kmnj}$ is the effect of the jth line nested in the kth QTL genotype, mth location, and nth year, and e_{ijkmn} is the random error for the jth line in the kth QTL genotype, mth location, nth year, and ith block (Table 10). There are three QTL genotype by environment interaction variances to estimate in addition to the effects of QTL genotypes across environments. The expected mean squares for lines and between lines by environments can be expressed as a function of the QTL genotype and QTL genotype by environment expected mean squares as shown for the other examples. The pooled expected mean square for lines from (13) and (14) for example, is

$$E(M_G) = \sigma_E^2 + r\sigma_{GLY}^2 + rl\sigma_{GY}^2 + ry\sigma_{GL}^2 + rl\sigma_G^2 = \frac{df_Q E(M_Q) + df_{G:Q} E(M_{G:Q})}{df_Q + df_{G:Q}}$$

$$= \sigma_E^2 + r\sigma_{G:QLY}^2 + \frac{r\bar{n}(q-1)}{N-1}\sigma_{QLY}^2 + rl\sigma_{G:QY}^2 + \frac{rl\bar{n}(q-1)}{N-1}\sigma_{QY}^2 + ry\sigma_{G:QL}^2$$

$$+ \frac{ry\bar{n}\,(q-1)}{N-1}\sigma_{QL}^2 + rly\sigma_{G:Q}^2 + \frac{rly\bar{n}}{N-1}\left[\frac{\Sigma q_k^2}{q-1}\right];$$

Table 10. Degrees of freedom, Type III sum of squares, and expected mean squares for lines tested in randomized complete blocks across locations and years where the effects of QTL genotypes are fixed, the effects of lines, blocks, locations, and years are random, and factors other than QTL genotypes are balanced.

Factor	Degrees of freedom[a]	Sum of squares	Expected mean squares[b]
Block	$df_B = b-1$		
Locations			
(L)	$df_L = l-1$		
Years (Y)	$df_Y = y-1$		
Y × L	$df_{YL} = (l-1)(y-1)$		
Line (G)	$df_G = N-1$	$R[g\|\mu,l,y,b,gl,gy,gly]$	$E(M_G) = \sigma_E^2+r\sigma_{GLY}^2+rl\sigma_{GY}^2+ry\sigma_{GL}^2+$ $rly\sigma_G^2$
QTL	$df_Q = q-1$	$R[q\|\mu,l,y,b,ql,qy,qly]$	$E(M_Q) = \sigma_E^2+r\sigma_{G:QLY}^2+r\bar{n}\sigma_{QLY}^2+$ $rl\bar{n}\sigma_{QY}^2+rl\sigma_{G:QY}^2+ry\bar{n}\sigma_{QL}^2+ry\sigma_{G:QL}^2$ $+rly\sigma_{G:Q}^2+rly\bar{n}\phi_Q^2$
G:QTL	$df_{G:Q} = N-q$	$R[g{:}q\|\mu,l,y,b,q,ql,qy,qly]$	$E(M_{G:Q}) = \sigma_E^2+r\sigma_{G:QLY}^2+rl\sigma_{G:QY}^2+$ $ry\sigma_{G:QL}^2+rly\sigma_{G:Q}^2$
G × L	$df_{GL} = (N-1)(l-1)$	$R[gl\|\mu,l,y,b,g,gy,gly]$	$E(M_{GL}) = \sigma_E^2+r\sigma_{GLY}^2+ry\sigma_{GL}^2$
QTL × L	$df_{QL} = (q-1)(l-1)$	$R[ql\|\mu,l,y,b,q,qy,qly]$	$E(M_{QL}) = \sigma_E^2+r\sigma_{G:QLY}^2+r\bar{n}\sigma_{QLY}^2+$ $ry\sigma_{G:QL}^2+ry\bar{n}\sigma_{QL}^2$
G:QTL × L	$df_{G:QL} = (N-q)(l-1)$	$R[g{:}ql\|\mu,l,y,b,q,ql,qy,qly]$	$E(M_{G:QL}) = \sigma_E^2+r\sigma_{G:QLY}^2+ry\sigma_{G:QL}^2$
G × Y	$df_{GY} = (N-1)(y-1)$	$R[gy\|\mu,l,y,b,g,gl,gly]$	$E(M_{GY}) = \sigma_E^2+r\sigma_{GLY}^2+rl\sigma_{GY}^2$
QTL × Y	$df_{QY} = (q-1)(y-1)$	$R[qy\|\mu,l,y,b,q,ql,qly]$	$E(M_{QY}) = \sigma_E^2+r\sigma_{G:QLY}^2+r\bar{n}\sigma_{QLY}^2+$ $rl\sigma_{G:QY}^2+rl\bar{n}\sigma_{QY}^2$
G:QTL × Y	$df_{G:QY}=(N-q)(y-1)$	$R[g{:}qy\|\mu,l,y,b,q,ql,qy,qly]$	$E(M_{G:QY}) = \sigma_E^2+r\sigma_{G:QLY}^2+rl\sigma_{G:QY}^2$
G × L × Y	$df_{GLY} =$ $(N-1)(l-1)(y-1)$	$R[gly\|\mu,l,y,b,g,gy,gl]$	$E(M_{GLY}) = \sigma_E^2+r\sigma_{GLY}^2$
QTL × L × Y	$df_{QLY} =$ $(q-1)(l-1)(y-1)$	$R[gly\|\mu,l,y,b,q,qy,ql]$	$E(M_{QLY}) = \sigma_E^2+r\sigma_{G:QLY}^2+r\bar{n}\sigma_{QLY}^2$
G:QTL × L × Y	$df_{G:QLY} =$ $(N-q)(l-1)(y-1)$	$R[g{:}qly\|$ $\mu,l,y,b,q,ql,qy,qly]$	$E(M_{G:QLY}) = \sigma_E^2+r\sigma_{G:QLY}^2$
Residual	$df_E = (Nly-1)(b-1)$		$E(M_E) = \sigma_E^2$

[a] q is the number of QTL genotypes, $N = \sum_{i=1}^{q} n_i$ is the number of lines where n_i is the number of lines of the ith QTL genotype, b is the number of blocks, l is the number of locations, and y is the number of years.

[b] σ_E^2 is the error variance, σ_G^2 is the between line variance, ϕ_Q^2 is the variance of fixed effects of QTL genotypes, $\sigma_{G:Q}^2$ is the line nested in QTL genotype variance, σ_{GL}^2 is the line by location variance, σ_{QL}^2 is the QTL genotype by location variance, $\sigma_{G:QL}^2$ is the line nested in QTL genotype by location variance, σ_{GY}^2 is the line by year variance, σ_{QY}^2 is the QTL genotype by year variance, $\sigma_{G:QY}^2$ is the line nested in QTL genotype by year variance, σ_{GLY}^2 is the line by location by year variance, σ_{QLY}^2 is the QTL genotype by location by year variance, $\sigma_{G:QLY}^2$ is the line nested in QTL genotype by location by year variance, and

$$\bar{n} = \frac{N - \frac{\sum n_i^2}{N}}{q - 1}$$

hence,

$$\sigma^2_{GLY} = \sigma^2_{G:QLY} + \frac{\bar{n}(q-1)}{N-1}\sigma^2_{QLY}, \quad \sigma^2_{GY} = \sigma^2_{G:QY} + \frac{\bar{n}(q-1)}{N-1}\sigma^2_{QY},$$

$$\sigma^2_{GL} = \sigma^2_{G:QL} + \frac{\bar{n}(q-1)}{N-1}\sigma^2_{QL}, \text{ and } \sigma^2_G\sigma^2_{G:Q} + \frac{\bar{n}}{N-1}\left[\frac{\Sigma q_k^2}{q-1}\right] = \sigma^2_{G:Q} + \phi^2_Q \text{ (Table 10).}$$

The F-statistic for testing the hypothesis of no differences between QTL genotype means is

$$F = \frac{M_Q + M_{G:QL} + M_{G:QY} + M_{G:QLY}}{M_{G:Q} + M_{QL} + M_{QY} + M'_{QLY}}$$

where

$$\frac{E(M_Q) + E(M_{G:QL}) + E(M_{G:QY}) + E(M_{G:QLY})}{E(M_{G:Q}) + E(M_{QL}) + E(M_{QY}) + E(M_{QLY})}$$

$$= \frac{\sigma^2_E + r\sigma^2_{G:QLY} + r\bar{n}\sigma^2_{QLY} + rl\bar{n}\sigma^2_{QY} + rl\sigma^2_{G:QY} + ry\bar{n}\sigma^2_{QL} + ry\sigma^2_{G:QL} + rly\sigma^2_{G:Q} + \phi^2_Q}{\sigma^2_E + r\sigma^2_{G:QLY} + r\bar{n}\sigma^2_{QLY} + rl\bar{n}\sigma^2_{QY} + rl\sigma^2_{G:QY} + ry\bar{n}\sigma^2_{QL} + ry\sigma^2_{G:QL} + rly\sigma^2_{G:Q}}$$

(Table 10) with approximate numerator and denominator degrees of freedom

$$\frac{[M_Q + M_{G:QL} + M_{G:QY} + M_{G:QLY}]^2}{M^2_Q/(df_Q) + M^2_{G:QL}/(df_{G:QL}) + M^2_{G:QY}/(df_{G:QY}) + M^2_{G:QLY}/(df_{G:QLY})}$$

and

$$\frac{[M_Q + M_{QL} + M_{QY} + M_{QLY}]^2}{M^2_{G:Q}/(df_{G:Q}) + M^2_{QL}/(df_{QL}) + M^2_{QY}/(df_{QY}) + M^2_{QLY}/(df_{QLY})},$$

respectively (Satterthwaite 1943; Searle 1970). Complex F-statistics are needed for testing hypotheses about the QTL genotype by location and QTL genotype by year variances as well. This analysis is messy as is, and worsens when additional factors are unbalanced. The most expedient tack might be to analyze lines ignoring marker and QTL parameters, and then to analyze marker phenotypes (QTL) by using line by environment least square means as the dependent variable. Complex F-statistics arise either way, but the model and analysis are simplified and standardized by using line by environment least square means.

4.   Replications of lines and replications of QTL genotypes

The same size of an experiment (the total number of experimental units) is equal to Nr. Increasing r is usually less expensive than increasing N to achieve a give sample size because the marker phenotypes of each line must be assayed, and this is usually more expensive than increasing the number of replications of lines. The number of replications of QTL genotypes ($\bar{n}$) is determined by the total number of lines (N) and by the genetic model. The number of replications of lines (r) is fixed for a given experiment, while the number of replications of QTL genotypes ($\bar{n}$) and how these replications are laid out within blocks or incomplete blocks is determined by the genetic model. The number of replications of QTL genotypes fluctuates as the number of QTL parameters increases or decreases -$\bar{n}$ decreases as the number of QTL parameters increases. Any of several genetic models might be tested, and the number of genetic models which might be tested for a given experiment often exceeds the number of observations for the experiment. The spatial layout of replications of QTL genotypes within blocks differs for every genetic model and from marker-bracket to marker-bracket.

Power for mapping QTL is affected differently by replications of lines and replications of QTL genotypes (Knapp and Bridges 1991). To estimate QTL genotype means and test hypotheses about QTL genotypes, individuals or lines need not be replicated as long as marker or QTL genotypes are replicated (Knapp 1989; Knapp and Bridges 1990). Unreplicated progeny are nevertheless disadvantageous. By using replicated lines, many useful classical quantitative genetic and QTL parameters can be estimated, and a minimum number of lines need to be assayed for their marker phenotypes (for a given overall sample size). In addition, replicated lines are essential for estimating errors for testing hypotheses about QTL genotype effects and QTL genotype by environment interaction effects as shown earlier (Tables 1, 3, 4, 5, 7, 8, 9 and 10).

Increasing the number of replications of lines only increases power for mapping QTL when most of the between line variance has been explained by differences between QTL genotypes (Knapp and Bridges 1990).

$$\hat{\sigma}^2_{G:Q} \cong 0 \text{ and } \hat{\sigma}^2_G \cong \frac{\bar{n}}{N-1}\left[\frac{\Sigma q^2_k}{q-1}\right] \qquad \text{when this is achieved.}$$

When genome is searched marker-bracket by marker-bracket, the line nested in QTL genotype variance ($\hat{\sigma}^2_{G:Q}$) is rarely if every close to zero. As model building progresses, more and more QTL are added to the model and more and more QTL parameters are estimated until, ultimately, $\hat{\sigma}^2_{G:Q} \cong 0$; whereupon power is greatly increased for testing hypotheses about QTL genotypes by increasing the number of replications of lines (Knapp and Bridges 1990).

This problem is closely tied to model building and greatly impacts the gain which can be achieved from marker-assisted selection (MAS). Gains from MAS are maximized by maximizing the variance explained by marker genotypes (Lande and Thompson 1990) or by QTL genotypes, which is the goal of a QTL mapping experiment (Lander and Botstein 1989; Knapp and Bridges 1990; Knapp et al. 1992; Van Ooijen 1992). When this is achieved $\hat{\sigma}^2_{G:Q} \cong 0$ and $\hat{\sigma}^2_G \cong \hat{\phi}^2_G$ since $\hat{\sigma}^2_G = \hat{\sigma}^2_{G:Q} + \hat{\phi}^2_Q \cdot \hat{\sigma}^2_{G:Q}$ tends to zero when the parameters of the most important QTL have been estimated, and the coefficient

$$\frac{\bar{n}\,(q-1)}{N-1}$$

tends to 1.0 as q increases.

With a balanced number of replications of each QTL genotype, which can only happen for certain progeny types and genetic models, $\bar{n} = 1$ and $\sigma^2_G = \phi^2_Q$ when $N = q$ (Table 1). $\sigma^2_G = \sigma^2_{G:Q}$ when no QTL are mapped and $\sigma^2_{G:Q} = 0$ when all of the important QTL have been mapped.

Every line within a sample could have a unique QTL genotype. Whether or not they do is a function of the number of lines tested (N) and the number of QTL segregating within the population, which can be different from the number of QTL modeled (q). If a great number of QTL underlie a trait, say 10 independent loci, then $3^{10} = 59,049$ genotypes can arise within an F_2 or F_3 population, while $2^{10} = 1,024$ genotypes can arise within a DH or recombinant inbred line population. For the example of 10 independent QTL, each DH, RI, or F_3 line within most samples almost surely has a unique ten locus QTL genotype.

If the QTL genotype of each line is unique, and the parameters of $q = N$ QTL genotypes are estimated, then

$$N - 1 = q - 1, \bar{n} = \frac{N-1}{q-1} = 1, \hat{\sigma}^2_{GL} = \frac{\bar{n}q - \bar{n}}{N-1}\,\hat{\sigma}^2_{QL} = \hat{\sigma}^2_{QL},$$

and $\hat{\sigma}^2_G = \hat{\phi}^2_Q$. A complete QTL model explains a maximum of $100H$ percent of the line-mean phenotypic variance ($\hat{\sigma}^2_p$) or 100 percent of the between line variance where H is the family-mean heritability (Table 2).

$\sigma^2_{G:QL}$ is the line by location variance which is not explained by the mapped QTL. $\sigma^2_{G:Q}$ and $\sigma^2_{G:QL}$ are 'lack-of-fit' variances. Some of the QTL underlying significant QTL genotype of environment interaction effects might be different from those underlying significant QTL effects across environments. This has ramifications for searching genomes for important QTL. If the mean effects of QTL across environments are used to determine which QTL are retained in a model, then QTL underlying a line by environment interaction might be overlooked. This can be avoided by using the effects of QTL by environments in addition to the effects of QTL across environments to determine which QTL should be retained in a model.

5.   Multiple loci

One challenge for implementing QTL mapping methods for a wide range of mating designs is defining frequencies of marker and quantitative trait loci, linked and unlinked, for several loci, and then grid searching (interval mapping) the multi-dimensional distances between the markers. Joint frequencies are needed to estimate Type III sum of squares and to get unbiased hypothesis tests and parameter estimates (Knapp et al. 1993); however, this poses a special problem, and ultimately leads to a paradox. The paradox is that once several marker-brackets are added to the model, missing cells can arise (Knapp et al. 1993). Unbiased QTL effects can then only be estimated when there is no interaction between the loci. Since these interactions cannot be estimated when there are missing cells, the only recourse is to estimate the parameters and Type III sum of squares with the knowledge that the estimates could be biased. Although it is hard to imagine much bias caused by interaction (epistasis), the question of the extent of epistasis still needs to be answered, and it very well could be important for a given population or trait. MAS is probably not going to be affected much by the epistasis caused bias, but it might be enhanced by jointly estimating the intralocus (main) effects of a maximum number of QTL.

Knapp (1991), Haley and Knott (1992), and Carbonell et al. (1992) gave two-locus examples of joint frequencies for different mating types. The estimation of these frequencies for more than two loci is straightforward but can be very cumbersome. Hypotheses about additive by additive, additive by dominant, and dominant by dominant QTL genotype interactions can be tested by using contrasts between marker genotype means. The coefficients for these contrasts are defined by the joint expected frequencies of the QTL for fixed recombination frequencies.

The computing becomes burdensome as the number of marker-brackets or QTL increases. One problem posed by multilocus estimation is the number of tests needed to interval map more than one QTL simultaneously. The number for k QTL is $n_i + 1 \cdot n_{i+1} + 1 \cdot ... \cdot n_k + 1$ where n_i is the distance between the ith and ith + 1 marker loci, and statistics are estimated for every 1 cM between two markers.

The problem of estimating the parameters of multiple QTL warrants much further study. Rodolphe and Lefort (1993) and Jansen (1993) examined the problem and proposed methods for estimating the parameters of multiple QTL. The sort of advances they propose are essential for gaining a less biased and more thorough understanding of the genetics of complex traits.

6.   Selecting sources of favorable alleles and marker-assisted selection

Two of the goals of QTL mapping experiments are to find sources of favorable alleles for developing superior cultivars and hybrids and to gain the knowledge necessary for maximizing selection gains through marker-assisted selection (MAS) (Tanksley et al. 1989; Lande and Thompson 1990; Page 1991; Edwards and Page 1993; Stuber and Sisco 1992). Although the cost effectiveness of MAS is widely debated, the usefulness of QTL mapping for finding new favorable alleles is hard to dispute. Tanksley et al. (1989) stressed this, and illustrated how marker-assisted backcross breeding could be used to minimize linkage drag and greatly speed up the development of near-isogenic lines. Creating a picture of the distribution of favorable alleles between parents and progeny is obviously extremely useful (Young and Tanksley 1989a,b). Favorable allele frequencies and coupling linkages of favorable alleles have been built up through years of breeding and selection and new favorable alleles for most traits are hidden in exotic germplasm. The cost of assaying marker phenotypes seems to be the limitation to using this technology more widely, and these costs are bound to decrease as marker technology advances.

One of the most important problems in plant breeding is finding new sources of favorable alleles among donor inbred lines for improving elite inbred line cultivars or the inbred parents of elite single cross hybrids (Dudley 1987; Zanoni and Dudley 1987; Gerloff and Smith 1988a,b; Stuber et al. 1993). Several statistics have been developed and have proven useful for finding new favorable alleles among donor inbred lines for developing new single-cross hybrids superior to an elite single-cross hybrid. Dudley (1984, 1987) proposed methods for estimating the number of favorable alleles in a donor inbred line which are not in either parent of an elite single-cross hybrid (n_G). This parameter and many others can be estimated from the means of the parent and donor inbred lines and their hybrids (Dudley 1987; Zanoni and Dudley 1987; Gerloff and Smith 1988a). The necessary experiments are inexpensive and useful for selecting parents for developing new inbred lines (Dudley 1987; Zanoni and Dudley 1987; Gerloff and Smith 1988a,b). So what is lacking, or rather, what is gained by mapping QTL? Parameters such as n_G summarize populations not genes. Restrictive assumptions about the genetics and needed to estimate n_G, e.g., the effects of individual genes must be equal, the effects of the genes must be completely dominant, no epistasis or linkage, and so on. Despite this, these methods seem to work extremely well because none of the biases seem to greatly affect the conclusions (Zanoni and Dudley 1987; Misevic 1989). What QTL mapping adds, however, is knowledge about the distribution of favorable alleles between parent and donor inbred lines and their progeny, estimates of gene effects without restrictive assumptions about their genetics, and marker loci linked to the genes to be selected. Once new alleles are found, they must be introgressed from donor to elite inbred lines. To develop superior single-cross hybrids, new lines must be developed which are fixed for favorable alleles in one of the parents of the elite hybrid and for new favorable alleles in the donor inbred line. MAS can be used to introgress the new favorable alleles through backcross breeding (Tanksley et al. 1989) or through pedigree or other variants of inbred line breeding methods (Lande and Thompson 1990; Lande 1992).

The usefulness of a line as a donor of new alleles cannot necessarily be determined by its phenotype alone regardless of the cultivar objective. Historically, plant breeders have had to rely only on phenotype parameters, e.g., means and variances, and experience, but these alone do not guarantee the selection of sources of favorable alleles different from those already fixed in an elite line (Stuber and Sisco 1992). This is partly why exotic germplasm has not been widely used and the gap between 'elite and exotic' germplasm constantly widens (Goodman 1986; Troyer et al. 1989; Troyer 1990).

A straightforward theoretical example illustrates the basic problem. Suppose alleles at six quantitative trait loci (A, B, C, D, E, and F) are dispersed among several inbred lines for some trait and an elite inbred line (P_1) is homozygous for favorable alleles at five (A, B, C, D, and E) of the six loci (Table 11). A source of favorable alleles for locus F is needed to develop a line superior to P_1 (Table 11). Two of the donor lines (P_2 and P_5) are sources of favorable alleles at the F locus; however, using genotype means alone, you could not determine the usefulness of P_5 or distinguish between P_2 and P_3, which have equal genotype means. Although the means of P_2 and P_3 are equal, a line superior to P_1 cannot be developed from the P_1 x P_3 mating, whereas a line superior to P_1 can be developed from the P_1 and P_2 and the P_1 and P_5 matings. The P_1 and P_5 mating obviously poses more of a problem for developing a line superior to P_1 than the P_1 x P_2 mating. P_2 is superior to P_5 as a donor of F locus alleles because it is fixed for favorable alleles at three additional loci (C, D, and E) and P_5 is fixed for unfavorable alleles at every other locus. But how could P_5 ever be found to be a source of useful alleles without QTL mapping? Is much of the diversity sought by plant breeders hidden within exotic germplasm as exemplified by P_5?

Table 11. Hypothetical favorable (+) and unfavorable (-) allele distributions among inbred lines for one trait and set of target environments where the genotype mean is increased to 5.0 units for each favorable allele and the mean of the line homozygous for unfavorable alleles at each locus is 40.0 units.

Locus	P_1	P_2	P_3	P_4	P_5	P_6
A	++	—	++	—	—	—
B	++	—	++	—	—	—
C	++	++	++	++	—	—
D	++	++	—	—	—	—
E	++	++	++	—	—	—
F	—	++	—	—	++	—
Genotype mean	90.0	80.0	80.0	50.0	50.0	40.0

Transgressive segregates for new favorable alleles are often obscured when the heritability of the trait is low, the effect of the new favorable alleles is small, and the source of new favorable alleles is fixed for unfavorable alleles at most other loci. This seems plausible because favorable alleles with large effects are the easiest to find and fix. This is partly why it becomes harder to accumulate new favorable alleles in elite germplasm.

An example of favorable alleles being obscured by genes with large effects comes from QTL mapping experiments in barley where doubled haploid lines were developed from a cross between a feed grain and malting barley (Steptoe x Morex) (Hayes et al. 1993). Steptoe is an extremely important feed grain barley cultivar with very poor malting quality characteristics; however, Steptoe alleles were found to be favorable at two loci affecting malting quality. The effects of these QTL were small compared to effects of most of the other QTL, all of which were fixed for favorable alleles in Morex. Morex, of course, was fixed for favorable alleles at more loci affecting malting quality, but Steptoe was nevertheless found to be a source of favorable alleles at two QTL, and it could conceivably be used to develop a malting barley slightly superior to Morex (Hayes et al. 1993). The problem, of course, is to introgress those two alleles, while retaining the favorable alleles fixed in Morex. Favorable alleles for many important quantitative traits could be dispersed throughout the exotic germplasm of many of our crop species, as has been shown in tomato (Paterson et al. 1992) and many other species. QTL mapping is certainly no panacea, but it gives plant breeders an objective and powerful vehicle for finding sources of new alleles and selecting the most outstanding sources of these alleles.

The multilocus questions briefly reviewed above the addressed somewhat more fully elsewhere (Knapp et al. 1993; Dudley 1993) are of critical importance for finding genes with small effects and maximizing the efficiency of marker-assisted index selection (MAS). The response to MAS is a function of the heritability of the trait and the percentage of the additive genetic variance explained by marker (σ^2_M) or QTL (σ^2_A) genotypes (Smith 1967; Lande and Thompson 1990). Maximizing these is the goal of a QTL mapping experiment.

Lande and Thompson (1990) and Lande (1992) developed indexes for selection using phenotype and marker scores. Index scores are estimated by $I = b_y Y + b_M m$ where y is the phenotype score, m is the marker score (sum of the additive effects of the QTL or marker loci), b_y is the weight for phenotype scores, and b_M is the weight for marker scores. The weights are estimated by

$$b = P^{-1}Gd = \begin{bmatrix} \sigma^2_P & \sigma^2_M \\ \sigma^2_M & \sigma^2_M \end{bmatrix}^{-1} \begin{bmatrix} \sigma^2_A & \sigma^2_M \\ \sigma^2_M & \sigma^2_M \end{bmatrix} \begin{bmatrix} 1 \\ 0 \end{bmatrix} = \begin{bmatrix} b_Y \\ b_M \end{bmatrix} = \begin{bmatrix} \dfrac{\sigma^2_A - \sigma^2_M}{\sigma^2_P - \sigma^2_M} \\[2ex] \dfrac{\sigma^2_P - \sigma^2_G}{\sigma^2_P - \sigma^2_M} \end{bmatrix}$$

where P is the phenotypic variance-covariance matrix, G is the genotypic variance-covariance matrix, d is the vector of economic weights, σ^2_P is the phenotypic variance, σ^2_A is the additive genetic variance, and σ^2_M is the additive genetic variance associated with marker loci (Lande and Thompson 1990). The efficiency of MAS relative to conventional selection is

$$\frac{R_M}{R_P} = \frac{\sigma_P \sqrt{\sigma^4_A - 2\sigma^2_A\sigma^2_M + \sigma^2_P\sigma^2_M}}{\sigma^2_A\sqrt{\sigma^2_P - \sigma^2_M}} = \sqrt{\frac{\sigma^2_M/\sigma^2_A}{\sigma^2_A/\sigma^2_P} + \frac{(1-\sigma^2_M/\sigma^2_A)^2}{1 - (\sigma^2_A/\sigma^2_P)\,(\sigma^2_M/\sigma^2_A)}}$$

$$= \sqrt{\frac{\sigma^2_M/\sigma^2_A}{H} + \frac{(1-\sigma^2_M/\sigma^2_A)^2}{1-H(\sigma^2_M/\sigma^2_A)}}$$

where i is the selection intensity, H is the heritability, the gain from selection using trait phenotypes alone is

$$R_P = i\,\frac{\sigma^2_A}{\sigma^2_P}\,\sigma_P = i\,\frac{\sigma^2_A}{\sigma_P} = iH\sigma_P,$$

and the gain from using a MAS index is

$$R_M = i\sqrt{d^T G b} = i\sqrt{[1\ 0]\begin{bmatrix}\sigma_A^2 & \sigma_M^2 \\ \sigma_M^2 & \sigma_M^2\end{bmatrix}\begin{bmatrix}\dfrac{\sigma_A^2 - \sigma_M^2}{\sigma_P^2 - \sigma_M^2} \\[2mm] \dfrac{\sigma_P^2 - \sigma_A^2}{\sigma_P^2 - \sigma_M^2}\end{bmatrix}} = i\sqrt{\dfrac{\sigma_A^2(\sigma_A^2-\sigma_M^2)}{\sigma_P^2 - \sigma_M^2} + \dfrac{\sigma_M^2(\sigma_P^2-\sigma_A^2)}{\sigma_P^2 - \sigma_M^2}}$$

(Lande and Thompson 1990).

The efficiency of MAS relative to conventional selection increases as σ_M^2 / σ_A^2 increases and is maximum when $\sigma_M^2 / \sigma_A^2 = 1.0$. As stated earlier, it should be feasible to find marker-brackets which explain 100% of the genetic variance between lines when the genome is completely covered — not necessarily saturated — by markers. Complete coverage is achieved when every segregating QTL is bracketed by markers. Then the genetic variance between marker genotypes should be equal to the genetic variance between lines.

Suppose QTL are found to lie between ten independent marker-brackets among 150 doubled haploid lines. There are 4 genotypes for each marker bracket and $4^{10} = 1,048,576$ genotypes, so the genotype of every line is almost always going to be unique unless the pairs of markers are closely linked, then some of the genotypes might be duplicated. If the marker genotype of each line is unique, the genetic variance among the marker genotypes has to be equal to the genetic variance between lines. The response to MAS is maximized by maximizing σ_M^2 / σ_A^2. This can be achieved in any experiment by randomly selecting markers to absorb the genetic variance between lines. If the selected markers are not linked to genes underlying the trait, then MAS obviously has no effect, even though $\sigma_M^2 / \sigma_A^2 = 1.0$ is achieved. The markers selected obviously must be linked to bona fide QTL. The significance of an effect, e.g., a significant likelihood odds or F-statistic, does not always ensure this because of Type I and II errors (false positives and negatives).

How a genome is searched and how multiple QTL are mapped affects the implementation of MAS and the probabilities of Type I and II errors (Knapp et al. 1993). The efficiency of MAS increases as heritability decreases, while the probability of finding a given QTL decreases as heritability decreases for a given sample size. So maximizing σ_M^2 / σ_A^2 is theoretically harder as heritability decreases, which is when it is most critical to do so.

The probability of encountering missing cells increases as the number of marker-brackets modeled increases. Unless there are no interactions between QTL, inferences about their effects cannot be made for many multilocus models because of missing cells. This means the marker scores and σ_M^2 for MAS cannot be estimated from the multilocus model either. The compromise is to estimate the effects of minor QTL along with the effects of major QTL, because the sampling bias, if there is any, is much greater for the latter (Knapp et al. 1993).

7.   TL genotype by environment interactions

Many methods have been developed for characterizing line or cultivar by environment interactions (Baker 1988; Becker and Leon 1988; Freeman 1973; Lin et al. 1986; Rosielle and Hamblin 1981; Wescott 1986; Zobel et al. 1988). The goal of some of these methods is to select cultivars or lines which maximize selection gains for a set of target environments. The breeding of any trait whether using MAS or not, entails optimizing selection for a set of *target environments* (Comstock and Moll 1963; Gardner 1963). A set of target environments is usually defined by locations and years but might encompass more definitive fixed factors, e.g., irrigation or fertilization rates.

Genotype means across *test environments* are used to select lines, populations, hybrids, and cultivars for target environments and marker and QTL alleles for MAS across target environments. Test environments — samples of years, locations, and other factors — are selected to maximize the speed of a selection cycle while minimizing the cost of testing and maximizing selection gains for target environments. Genotype by environment interactions can cause test environments to fail to maximize selection gains for target environments, with equivalent consequences for selection with the without markers. At the extreme, this happens when differences between genotypes are observed across test environments, and there are no differences across target environments. Or when differences between genotypes are not observed across test environments, and there are differences across target environments. The consequences are either to fix unfavorable alleles, or, for MAS only, to fix alleles at QTL which have no mean effect across target environments, but which had an effect across the sample of test environments used. The root of the problem with genotype by environment interactions are differences between test and target environments. Nothing can be done about the outcome of selection if test and target environments are fixed.

The nature of line by environment interactions can be explained as function of individual QTL genotype by environment interactions. If the ranks of means of genotypes change across environments, then those genotypes manifest *crossover* genotype by environment interactions. Conversely, if differences between means of genotypes change across environments, but their ranks are constant, then those genotypes manifest *non-crossover* genotype by environment interactions. Baker (1988) proposed statistical methods (Azzalini and Cox 1984; Gail and Simon 1985) for differentiating between crossover and non-crossover interactions when testing lines or cultivars. Differentiating between crossover and non-crossover genotype by environment interactions is useful because only crossover interactions affect how cultivars are bred and deployed (Baker, 1988); however, the outcome of selection can only be changed by redefining or reselecting test or target environments when faced with cross-over or non-crossover interactions. Non-crossover interactions do not lead to fixing unfavorable alleles — at worst QTL might be selected which do not affect the trait across target environments or QTL might be missed which affect the trait across target environments. When non-crossover interactions arise, the difference between QTL genotype means across environments is usually less than within some environments.

Lines, populations, hybrids, and cultivars often manifest genotype by environment interactions, especially when tested across diverse environments. If a significant line by environment interaction is observed for a trait, then one or more of the genes underlying the trait might manifest QTL genotype by environment interactions (Tables 1, 3, 7, and 8). The consequences of genotype by environment interactions for QTL are no different than for lines. Sorting out questions about genotype by environment interactions is obviously one of the objectives of multiple environment QTL mapping experiments. The methods described earlier can be used to sort out many of the relevant questions. Is there evidence for QTL genotype by environment interactions? What is the nature of a significant QTL genotype by environment interaction? Is there a subset of environments for which a QTL exhibits no QTL genotype by environment interaction?

Additional methods are needed to determine the nature of QTL genotype by environment interactions. These methods are hardly necessary for practicing MAS. Only the means of QTL genotypes across test environments need to be estimated to select QTL for MAS. This only fails if every QTL manifests crossover interaction and the test environments did not uncover these interactions, both of which are very unlikely for carefully selected test and target environments. Suppose, for example, a QTL manifests a crossover interaction which is not observed among the QTL genotype by test environment means, but which leads to no overall effect of the QTL within target environments, then putting selection pressure against this QTL is equivalent to selecting a neutral locus. This diminishes selection response by decreasing selection intensity, as do many errors (Edwards and Page 1993; Page 1991).

Differentiating between non-crossover and crossover QTL genotype by environment interactions might be important for optimizing MAS. Crossover QTL genotype by environment interactions could affect the outcome of MAS, whereas non-crossover interactions should be of no consequence to the outcome of MAS. Understanding and characterizing the nature of QTL genotype by environment interactions is useful for optimizing MAS or conventional selection.

Test statistics have been developed to gain insights into the nature of line by environment interactions (Baker 1988), and these can be used to gain insights into the nature of QTL genotype by environment interactions. A crossover QTL genotype by environment interaction exists when, for two QTL genotypes in two environments, the inequality

$$\mu_{qq1} - \mu_{QQ1} > 0, \; \mu_{qq2} - \mu_{QQ2} < 0$$

or

$$\mu_{qq1} - \mu_{QQ1} < 0, \; \mu_{qq2} - \mu_{QQ2} > 0$$

are satisfied (Azzalini and Cox, 1984) where μ_{QQi} and μ_{qqi} are means of the QQ and qq genotypes in the ith environment. Azzalini and Cox (1984) developed a statistic for testing the null hypothesis of no crossover interaction. The null hypothesis of no crossover interaction for two QTL genotypes tested in two environments is rejected when the inequalities

$$\hat{\mu}_{qq1} - \hat{\mu}_{QQ1} \geq t_a \hat{\sigma}_{\overline{M}}, \; \hat{\mu}_{qq2} - \hat{\mu}_{QQ2} \leq -t_a \hat{\sigma}_{\overline{M}}$$

or

$$\hat{\mu}_{qq1} - \hat{\mu}_{QQ1} \leq t_a \hat{\sigma}_{\overline{M}}, \; \hat{\mu}_{qq2} - \hat{\mu}_{QQ2} \geq -t_a \hat{\sigma}_{\overline{M}}$$

are satisfied where

$$t_\alpha = \Phi\{[\frac{(-2)\log(1-\alpha)}{q(q-1)e(e-1)}]^{1/2}\} = \Phi(x),$$

α is the probability of a Type I error, $\Phi(x)$ is the standard normal cumulative distribution function, $\hat{\sigma}^2_M$ is the standard error of a difference between marker genotype by environment means, q is the number of QTL genotypes, and e is the number of environments. These inequalities were defined by Azzalini and Cox (1984) as a *quadruple*. Every quadruple is then tested for a set of genotypes and environments. The null hypothesis of no crossover interaction is rejected when the inequalities are satisfied. An expedient way to implement this test is to test the quadruple for the two QTL genotypes with the largest mean difference with different signs. If either inequality is satisfied, then the null hypothesis is rejected.

There was a significant QTL genotype by environment interaction for the *Plc-iABI151* marker-bracker (Table 9 and Fig. 2). The additive effects of the QTL genotypes for Corvallis and Pullman were $\hat{\mu}_{QQ} - \hat{\mu}_{qq} = 336.8$ and $\hat{\mu}_{QQ} - \hat{\mu}_{QQ} - \hat{\mu}_{qq} = -88.9$. The ranks of the QQ and qq genotypes changed across environments, and these changes were statistically significant (Table 9 and Fig. 2). The standard error of a marker by location mean was $\hat{\sigma}_M = 113.8$,

$$t_\alpha = t_{0.01} = \Phi\{ [\frac{(-2)\log(1-0.01)}{2(2-1)2(2-1)}]^{1/2}\} = \Phi(0.0005) = 2.57,$$

and $t_{0.01} \hat{\sigma}^2_M = 2.57 (113.8) = 292.5$. The quadruple for testing the hypothesis of no crossover QTL genotype by environment interaction is $336.8 \geq 292.5$ and $-88.9 > -292.5$, so we fail to reject the hypothesis of no crossover interaction because the magnitude of the difference between QQ and qq genotypes for Pullman was less than the test statistic. While the interaction between QTL genotypes and locations was significant and the estimated ranks of QTL genotype means were different at Corvallis and Pullman, the evidence does not exist to support a crossover interaction. This is consistent with what we found for individual location estimates. The QTL had no effect at Pullman, whereas it had a significant effect at Corvallis (Tables 3 and 4 and Fig. 1).

The number of quadruples which must be tested increases as the number of environments increases. The picture can become quite muddy. Some of the interactions between QTL genotypes and environments could be crossover and non-crossover for any locus or set of environments. This test (Azzalini and Cox 1984) at least shows how the nature of QTL genotype by environment interactions can be investigated.

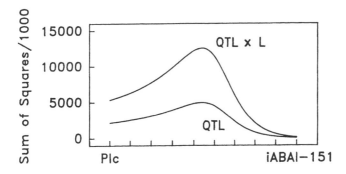

Fig. 2. Seed yield (kg/ha) sum of squares for the additive (QTL) and additive by location (QTL x L) effect of a QTL between the *Plc* and *iABI151* locus for barley doubled haploid lines tested in Corvallis, Oregon and Pullman, Washington.

8. Acknowledgements

This work was funded by USDA grants (#90-34213-5191, #91-37300-6529, and #92-37300-7652) and an Alexander von Humbolt Fellowship. I am grateful to Prof. Dr. Jeff Schell, Dr. Reinhard Toepfer, and Dr. Norbert Martini at the Max Planck Institute, Cologne, Germany where some of this work was done, and to Dr. William Beavis (Pioneer Hi-Bred International), Dr. Robert Stucker (University of Minnesota), and Dr. Chris Haley (University of Edinburgh) for critiquing the original manuscript.

9. References

Azzalini, A., and Cox, D.R. (1984) Two new tests associated with analysis of variance. J.R. Statist. Soc. B. 46: 335-343

Baker, R.J. (1988a) Tests for crossover genotype-environment interactions. Can. J. Plant Sci. 68: 405-410.

Baker, R.J. (1988b) Analysis of genotype-environment interactions in crops. ISI Atlas Sci.: Animal and Plant Sci. 1: 1-4.

Becker, H.C. and Leon, J. (1988) Stability analysis in plant breeding. Plant Breed. 101: 1-23.

Bulmer, M.G. (1980) The Mathematical Theory of Quantitative Genetics. Oxford University Press, Oxford.

Carbonell, E.A., Gerig, T.M., Balansard, E. and Asins, M.J. (1992) Interval mapping in the analysis of nonadditive quantitative trait loci. Biometrics 48: 305-315.

Carbonell, E.A., Asins, M.J., Baselga, M., Balansard, E., and Gerig, T.M. (1993) Power studies in the estimation of genetic parameters and the localization of quantitative trait loci for backcross and doubled haploid populations. Theor. Appl. Genet. 86: 411-416.

Comstock, R.E. and Moll. R.H. (1963) Genotype-environment interactions. In: W.D. Hanson and H.F. Robinson (eds.), Statistical Genetics and Plant Breeding. NAS-NRC Publ. 982.

Doebley, J. and Stec, A. (1991) Genetic analysis of the morphological differences between maize and teosinte. Genetics 129: 285-295.

Doebley, J., Stec, A., Wendel, J. and Edwards, M. (1990) Genetic and morphological analysis of a maize-teosinte F_2 population: implications for the origin of maize. Proc. Natl. Acad. Sci. U.S.A. 87: 9888-9892.

Dudley, J.W. (1984) A method of identifying lines for use in improving parents of a single-cross. Crop Sci. 24: 355-357.

Dudley, J.W. (1987) Modifications of methods for identifying inbred lines useful for improving parents of elite single crosses. Crop Sci. 27: 944-947.

Dudley, J.W. (1993) Molecular markers in plant improvement: manipulation of genes affecting quantitative traits. Crop Sci. 33: 660-668.

Edwards, M.D. and Page, N.J. (1993) Evaluation of marker-assisted selection through computer simulation. Theor. Appl. Genet. (in press).

Edwards, M.D., Stuber, C.W. and Wendel, J.F. (1987) Molecular marker facilitated investigations of quantitative trait loci in maize. I. Number, genomic distribution, and type of gene action. Genetics 116: 113-125.

Falconer, D.S. (1981) Introduction to Quantitative Genetics. Longman, New York.

Freeman, G.H. (1973) Statistical methods for the analysis of genotype-environment interactions. heredity 31: 339-354.

Freund, R.J., Little, R.C. and Spector, P.C. (1986) SAS˝ System for Linear Models, SAS, Cary, North Carolina.

Gail, M. and Simon, R. (1985) Testing for qualitative interaction between treatment effects and patient subsets. Biometrics 41: 361-372.

Gardner, C.O. (1963) Estimates of genetic parameters in cross-fertilizing plants and their implications in plant breeding. In: W.D. Hanson and H.F. Robinson (eds.), Statistical Genetics and Plant Breeding. NAS-NRC Publ. 982.

Gerloff, J. and Smith, O.S. (1988) Choice of method for identifying germplasm with superior alleles. I. Theoretical results. Theor. Appl. Genet. 76: 209-216.

Gerloff, J. and Smith, O.S. (1988) Choice of method for identifying germplasm with superior alleles. I. Computer simulation results. Theor. Appl. Genet. 76: 217-227.

Haley, C.S. and Knott, S.A. (1992) A simple regression method for mapping quantitative trait loci in line crosses using flanking markers. Heredity 69: 315-324.

Hanson, W.D. (1963) Heritability. In: W.D. Hanson and H.F. Robinson (eds.), Statistical Genetics and Plant Breeding. NAS-NRC Publ. 982.

Hayes, P.M., Liu, B.-H., Knapp, S.J., Chen, F., Jones, B., Blake, T., Frankowiak, J., Rasmusson, D., Sorrells, M., Ullrich, S.E., Wesenberg, D., and Kleinhofs, A. (1993) Quantitative trait locus effects and environmental interaction in a sample of North American barley germplasm. Theor. Appl. Genet. 87: 392-401.

Helentjaris, T., Slocum, M., Wright, S., Schaefer, A. and Nienhuis, J. (1986) Construction of genetic linkage maps in maize and tomato using restriction fragment length polymorphisms. Theor. Appl. Genet. 72: 761-769.

Jansen, R. (1992) A general mixture model for mapping quantitative trait loci by using molecular markers. Theor. Appl. Genet. 85: 252-260.

Jansen, R. (1993a) Interval mapping multiple quantitative trait loci. Genetics 135: 205-211.

Jansen, R. (1993b) Maximum likelihood in a generalized linear finite mixture model by using the EG algorithm. Biometrics (in press).

Jensen, J. (1989) Estimation of recombination parameters between a quantitative trait locus (QTL) and two marker gene loci. Theor. Appl. Genet. 78: 613-618.

Kleinhofs, A., Kilian, A., Saghai Maroof, M.A., Biyashev, R.M., Hayes, P.M., Chen, F.Q., Lapitan, N., Fenwick, A., Blake, T.K., Kanazin,V., Ananiev, E., Dahleen, L., Kudrna, D., Bollinger, J., Knapp, S.J., Liu, B.-H., Sorrells, M., Heun, M., Frankowiak, J.D., Hoffman, D., Skadsen, R. and Steffenson, B.J. (1993) A molecular, isozyme, and morphological map of the barley (*Hordeum vulgare*) genome. Theor. Appl. Genet. 86: 705-712.

Knapp, S.J. (1989) Quasi-Medelian genetic analysis of quantitative trait loci using molecular marker linkage maps. In: G. Roebbelen, (ed.), Proc. XIIth Eucarpia Congress (European Plant Breeding Congress), Gπttingen, Germany.

Knapp, S.J. (1991) Using molecular markers to map multiple quantitative trait loci: Models for backcross, recombinant inbred, and doubled haploid progeny. Theor. Appl. Genet. 81: 333-338.

Knapp, S.J. and Bridges, W.C. (1990) Using molecular markers to estimate quantitative trait locus parameters: Power and genetic variances for unreplicated and replicated progeny. Genetics 126: 769-777.

Knapp, S.J., Bridges, W.C. and Birkes, D. (1990) Mapping quantitative trait loci using molecular marker linkage maps. Theor. Appl. Genet. 79: 583-592.

Knapp, S.J., Bridges, W.C. and Liu, B.-H. (1992) Mapping quantitative trait loci using nonsimultaneous and simultaneous estimators and hypothesis tests. In: J.S. Beckmann and T.S. Osborn (eds.), Plant Genomes: Methods for Genetic and Physical Mapping, pp. 209-237. Kluwer Academic Publishers, Dordrecht, The Netherlands.

Knott, S.A. and Haley, C.S. (1992) Aspects of maximum likelihood methods for the mapping of quantitative trait loci in line crosses. Genet. Res. 60: 139-151.

Lande, R. (1992) Marker-assisted selection in relation to traditional methods of plant breeding. In: H.T. Stalker and J.P. Murphy (eds.), Plant Breeding in the 1990s. proc. Symp. Plant Breeding in the 1990s, C.A.B. International, Wallingford, U.K.

Lande, R. and Thompson, R. (1990) Efficiency of marker-assisted selection in the improvement of quantitative traits. Genetics 124: 743-756.

Lander, E.S. and Botstein, D. (1989) Mapping Mendelian factors underlying quantitative traits using RFLP linkage maps. Genetics 121: 185-199.

Landry, B.S., Kesseli, R.V., Fararra, B., and Michelmore, R.W. (1987) A genetic map of lettuce (*Lactuca sativa* L.) with restriction fragment length polymorphisms, isozymes, disease resistance, and morphological markers. Genetics 116: 331-337.

Lin, C.S., Binns, M.R. and Lefkovitch, L.P. (1986) Stability analysis: Where do we stand? Crop Sci. 26: 894-900.

Lou, Z.W. and Kearsey, M.J. (1989) Maximum likelihood estimation of linkage between a marker gene and a quantitative trait locus. Heredity 66: 117-124.

Martinez, O. and Curnow, R.N. (1992) Estimating the locations and the sizes of the effects of quantitative trait loci using flanking markers. Theor. Appl. Genet. 85: 480-488.

McLachlan, G.J. and Basford, K.E. (1988) Mixture Models: Inference and Applications to Clustering. Marcel Dekker, New York.

Misevic, D. (1988) Evaluation of three test statistics used to identify maize inbred lines with new favorable alleles not present in elite single crosses. Theor. Appl. Genet. 77: 402-408.

Nienhuis, J., Helentjaris, T., Slocum, M., Ruggero, B. and Schaefer, A. (1987) Restriction fragment length polymorphism analysis of loci associated with insect resistance in tomato. Crop Sci. 27: 797-803.

Ooijen, J.W. (1992) Accuracy of mapping quantitative trait loci in autogamous species. Theor. Appl. Genet. 84: 803-811.

Osborn, T.C., Alexander, D.C. and Fobes, F. (1987) Identification of restriction fragment length polymorphisms linked to genes controlling soluble solids content in tomato fruit. Theor. Appl. Genet. 73: 350-356.

Page, N.P. (1991) A computer simulation evaluation of the utility of marker-assisted selection. Ph.D. thesis, University of Minnesota, St. Paul, MN.

Paterson, A.H., Lander, E.S. Hewitt, J.D., Paterson, S., Lincoln, S.E. and Tanksley, S.D. (1988) Resolution of quantitative traits into Mendelian factors by using a complete linkage map of restriction fragment length polymorphisms. Nature 335: 721-726.

Paterson, A.H., Damon, S., Hewitt, J.D., Zamir, D., Rabinowitch, H.D., Lincoln,S.E., Lander, E.S. and Tanksley, S.D. (1991) Mendelian factors underlying quantitative traits in tomato: comparison across species, generations, and environments. Genetics 127: 181-197.

Rhodolphe, F. and Lefort, M. (1993) A multi-marker model for detecting chromosomal segments displaying QTL activity. Genetics 134: 1276-1277.

Rosielle, A.A. and Hamblin, J. (1981) Theoretical aspects of selection for yield in stress and non-stress environments. Crop Sci. 21: 943-946.

Satterthwaite (1946) An approximate distribution of estimates of variance components. Biometrics Bull. 2: 110-114.

Searle, S. (1971) Linear Models. John Wiley and Sons, NewYork.

Shukla, G.K. (1972) Some statistical aspects of partitioning genotype-environmental components of variability. Heredity 29: 237-245.

Simpson, S.P. (1989) Detection of linkage between quantitative trait loci and restriction fragment length polymorphisms using inbred lines. Theor. Appl. Genet. 77:815-819.

Smith, C. (1967) Improvement of metric traits through specific genetic loci. Anim. Prod. 9:349-358.

Stuber, W.C. and Sisco, P.H. (1991) Marker-facilitated transfer of QTL alleles between elite inbred lines and responses in hybrids. Ann. Corn Sorghum Res. Conf. 46: 104-133.

Stuber, C.W., Edwards, M.D. and Wendel, J.F. (1987) Molecular-marker facilitated investigations of quantitative trait loci in maize. II. Factors influencing yield and its component traits. Crop Sci. 27: 639-648.

Stuber, C.W., Lincoln, S.E., Wolff, D.W., Helentjaris, T. and Lander, E.S. (1992). Identification of genetic factors contributing to heterosis in a hybrid from two elite maize inbred lines using molecular markers. Genetic 132: 823-839.

Soller, M. and Beckmann, J.S. (1990) Marker-based mapping of quantitative trait loci using replicated progenies. Theor. Appl. Genet. 80: 205-208.

Titterington, D.M., Smith, A.F.M. and Makov, U.E. (1985) Statistical Analysis of Finite Mixture Distributions. Wiley and Sons, New York.

Troyer, A.F. (1990) A retrospective view of corn genetic resources. J. Hered. 81: 17-24

Troyer, A.F., Openshaw, S.J. and Knittle, K.H. (1988) Measurement of genetic diversity among popular commercial corn hybrids. Crop Sci. 28: 481-485.

van Ooijen, J.W. (1992) Accuracy of mapping quantitative trait loci in autogamous species. Theor. Appl. Genet. 84: 803-811.

Weller, J.I. (1986) Maximum likelihood techniques for the mapping and analysis of quantitative trait loci with the aid of genetic markers. Biometrics 42: 627-640.

Westcott, B. (1986) Some methods of analyzing genotype-environment interaction. Heredity 56: 243-253.

Young, N.D. and Tanksley, S.D. (1981a) Restriction fragment length polymorphism maps and the concept of graphical genotypes. Theor. Appl. Genet. 77: 95-101.

Young, N.D. and Tanksley, S.D. (1939b) RFLP analysis of the size of chromosomal segments retained around the Tm-2 locus of tomato during backcrossing. Theor. Appl. Genet. 77: 353-359.

Zanoni, U. and Dudley, J.W. (1989) Comparison of different methods of identifying inbreds useful for improving elite maize hybrids. Crop Sci. 29: 577-582.

Zhang, W. and Smith, C. (1992) Computer simulation of marker-assisted selection utilizing linkage disequilibrium. Theor. Appl. Genet. 83: 813-820.

Zobel, R.W., Wright, M.J. and Gauch, H.G.Jr. (1988) Statistical analysis of a yield trial. Crop Sci. 80: 388-393.

6.   Comparative mapping of plant chromosomes

ANDREW H. PATERSON[1], *and* JEFFREY L. BENNETZEN[2]
[1]Department of Crop and Soil Science, University of Georgia, Athens GA 30602
<paterson@uga.edu>
[2]Department of Biological Sciences, Purdue University, West Lafayette IN 47907-1392

Contents

1.   Introduction

The two major subclasses of flowering plants, monocots and dicots, are thought to have diverged from a common ancestor about 135-200 million years ago, yet a host of morphological and functional similarities remain apparent even among diverse taxa. Vavilov's "law of homologous series in variation" (Vavilov 1922) was an early recognition of fundamental similarity between different plant species at the phenotypic level.

Over the past two decades, molecular genetics has provided the tools required to investigate similarities and differences in the hereditary information of taxa that cannot interbreed. Many investigations in diverse taxa generally point to two broad, yet distinct, messages that will be the subject of the remainder of this chapter:

1. The small but essential portion of most plant genomes encoding genes evolves relatively slowly, with corresponding genes retaining recognizable DNA sequences and parallel order along the chromosomes of taxa that have been reproductively-isolated for millions of years.

2. The large portion of most plant genomes encoding repetitive DNA element families evolves relatively quickly, including many species-specific repetitive DNA families absent from closely-related taxa, and exhibits rapidly-changing arrangement due partly if not largely to sequence mobility.

R.L. Phillips and I.K. Vasil (eds.), DNA-Based Markers in Plants, 101 - 114.
© *2001 Kluwer Academic Publishers. Printed in the Netherlands.*

2.   Conservation of gene content and order, determined by RFLP analysis

Comparative mapping is based upon analysis of "orthologous" genetic loci in different plant species. To be "orthologous," genetic loci must be directly derived from a common ancestral locus. Orthology is similar to allelism — except that it refers to corresponding loci in taxa which cannot hybridize sexually, thus a direct test of allelism by classical means is impossible.

Comparative mapping relies heavily upon cloned DNA sequences that occur at only one, or very few, places in the respective genomes of the investigated taxa. If there is only one copy of a DNA sequence per genome, then orthology is relatively simple to establish. DNA sequences that encode a protein product (i.e., genes) are especially well-suited to comparative analysis, because functional constraints tend to preserve the sequences of genes and therefore common probes can be used to detect corresponding genes in different taxa. The classical RFLP technique has been the method of choice for comparative genetic mapping in plants, because it is amenable to the use of heterologous DNA probes to detect DNA polymorphisms at corresponding loci in different taxa. Progress in assembling physical maps for plant chromosomes, comprised of contiguous sets of recombinant bacterial artificial chromosomes (BACs), yeast artificial chromosomes (YACs) or other large DNA clones, will significantly improve the level of detail of comparative data for plant genomes.

Since RFLP probes are often chosen because they are expressed (i.e., cDNAs), single copy and/or unmethylated within the genome (Burr et al. 1988), they often correspond to genes. One of the earliest and most important observations of comparative RFLP maps was that an RFLP probe from one plant species almost always detected a corresponding DNA from any other plant species within the same family. This indicated that plant genomes had, by this criterion, the same genes. Copy numbers for some gene families often vary between species, but it does not appear that wholly different gene contents separated plant species. Hence, the very different biologies of different plants is apparently not due to novel genes, but to novel allelic specificities (Bennetzen and Freeling 1993; 1997).

3.   Colinearity in five families of angiosperms

Five different taxonomic families of plants, each including major crops, have been selected to illustrate particular aspects of comparative mapping. These include monocots and dicots; diploids, disomic polyploids, and polysomic polyploids; autogamous and allogamous crops; tropical and temperate crops; and in a few instances, uncultivated experimental models. Within each of these five plant families (and presumably others), a recurring picture of conservation of gene order, punctuated by occasional chromosome rearrangement, bodes well for cross-utilization of genetic information in closely-related taxa. The accuracy of molecular maps is supported by their congruity with predictions from classical genetics regarding gene and genome duplication (maize, soybean, cotton, *Brassica*), subgenomic origin and composition (cotton), and similarity in gene order of closely-related taxa (mungbean and cowpea, tomato and potato). The selected plant taxa are discussed in alphabetical order.

3.1. A. Brassicaceae

The Brassicaceae comprises 360 genera, organized into 13 tribes (Shultz 1936; Al-Shehbaz 1973). The most economically important species are in the genus *Brassica* (tribe Brassiceae), which includes six species that are grown worldwide for a variety of uses. Three of the species are considered diploid (*B. rapa*, A genome, $2n = 2x = 20$; *B. oleracea*, C genome, $2n = 2x = 18$; *B. nigra*, C genome, $2n = 2x = 16$), and three are amphiploid derivatives of the diploids (*B. juncea*, AB, $2n = 4x = 36$; *B. napus*, AC, $2n = 4x = 38$; *B. carinata*, BC, $2n = 4x = 34$). Extensive genetic maps have been reported in four different *Brassica* species (*B. oleracea, B. rapa, B. nigra, B. napus*; Osborn and Paterson 1994).

While some large regions of *Brassica* chromosomes revealed common gene order in different species, both inversions and translocations were found to differentiate among the diploid *Brassica* genomes (cf. Slocum 1989; Slocum et al. 1990; McGrath and Quiros 1991; Teutonico and Osborn 1994). Tetraploid *Brassica* genomes show considerable correspondence with the diploid genomes, but also significant differences. Lydiate et al. (1993) have developed an RFLP map for cultivated tetraploid *B. napus*. Cultivated strains have essentially the same gene order as wild *B. napus*, but chromosomal rearrangements subsequent to polyploid formation have differentiated the tetraploid sub-genomes from their corresponding diploid genomes.

A factor that complicates genetic analysis of *Brassica* is that most DNA probes detect two, three, or more loci. In comparison of different *Brassica* taxa, this makes it difficult to distinguish orthologous (defined above) from paralogous loci (the results of duplication with subsequent descent). The finding by many investigators that most extant forms of *Brassica* are highly duplicated (even "diploid" forms), poses a complication to understanding genome organization of *Brassica*.

One possible solution to this problem derives from comparative mapping. *Arabidopsis thaliana*, n = 5 (tribe Sisymbrieae), a preferred model system in plant develment, biochemistry, physiology, and genetics, is a close relative of plants within the genus *Brassica*. Although extensive chromosomal rearrangements have occurred since the divergence of *B. oleracea* and *A. thaliana*, islands of conserved organization are discernible. A detailed comparative map of *Brassica oleracea* and *Arabidopsis thaliana* has been established based largely on RFLP mapping of *Arabidopsis* cDNAs (i.e, expressed sequence tags or ESTs) (Kowalski et al. 1994; Lan et al. 2000). Based on conservative criteria for inferring synteny, "one to one correspondence" between *Brassica* and *Arabidopsis* chromosomes accounted for 57% of comparative loci. At least 19 chromosomal rearrangements differentiate *B. oleracea* and *A. thaliana* orthologs. Chromosomal duplication in the *B. oleracea* genome was strongly suggested by parallel gene arrangements on different chromosomes that accounted for 42% of mapped duplicate loci. At least 22 chromosomal rearrangements differentiate putative *B. oleracea* homologs from one another. Triplication of some *Brassica* chromatin, and duplication of some *Arabidopsis* chromatin, was suggested by data that could not be accounted for by the one-on-one and duplication models, respectively. The relative orders of DNA markers along homoeologous chromosomal regions in *Brassica* permit one to infer whether specific chromosomal rearrangements predate, or postdate, duplication of *Brassica* chromosomes. The extensive rearrangement of putatively homoeologous *Brassica* chromosomes makes it difficult to classify many individual rearrangements as inversions or translocation. However, structural polymorphisms for corresponding chromosomes within individual *Brassica* species (especially *B. oleracea*) tend to be inversions (McGrath and Quiros 1991; Lan et al. submitted), suggesting that inversions occur (or at least persist) more frequently than translocations.

3.2. B. Poaceae

Most of the world's food and feed crops are members of the family Poaceae, including rice, wheat, maize, sorghum, sugarcane, barley, oat, rye, millet, and others. A recurring message of conservation across large chromosomal tracts emerges from many investigations of diverse Poaceae spanning 65 million years of evolution (cf. Bennetzen and Freeling 1993). Pioneering efforts in comparative mapping of maize and sorghum (Hulbert et al. 1990) have been supported by more detailed studies (Whitkus et al. 1992; Berhan et al. 1993; Binelli et al. 1993; Chittenden et al. 1994; Periera et al. 1994), and supplemented by detailed investigations of the comparative organization of maize and rice (Ahn and Tanksley 1993), wheat and rice (Kurata et al. 1994); and maize, wheat, and rice (Ahn et al. 1993). A host of laboratories have published many additional studies to encompass many other cultivated Poaceae, with particular emphasis on detailed analyses of interrelationships among the homoeologous chromosome sets of the Triticeae and their relatives (cf. Naranjo et al. 1987; Chao et al. 1989; Liu et al. 1991, 1992; Devos et al. 1992, 1993, 1995; Xie et al. 1993; Namuth et al. 1994; Hohmann et al. 1995; Marino et al. 1995; Mickelson-Young et al. 1995; Nelson et al. 1995 a-c; Van Deynze et al. 1995). Curiously, even in the relatively 'conservative' Poaceae, particular lineages appear to be rapidly-evolving. The genomes of *Secale* (rye) and *Triticum* (wheat) appear to differ by about 13 chromosomal rearrangements after only about 6 million years of divergence (Devos et al. 1992), a rate of reshuffling that is more than twice the average of 9 taxa compared to date (Paterson et al. 1996), and exceeded only by the Brassica-Arabidopsis lineage.

Comparative mapping has provided new information on the somewhat controversial issue of ancient chromosomal duplication in the Poaceae (particularly sorghum). While it has long been acknowledged that wheat is a polyploid, and recently it has become clear that maize is an ancient tetraploid (Helentjaris et al. 1988; Ahn and Tanksley 1993; Gaut and Doebley 1997), evidence for other taxa such as sorghum was equivocal (Whitkus et al. 1992). Development of "complete" (i.e., number of linkage groups = number of chromosomes) genetic maps of sorghum in two different populations (Chittenden et al. 1994; Periera et al. 1994) has demonstrated that at least some regions of the sorghum genome are indeed duplicated. A prominent "grass genome circle" model indicates that sorghum maps primarily as a diploid (Moore et al. 1995; Gale and Devos 1998) — however, these comparisons rely on relatively low density maps in populations with low levels of polymorphism that often preclude mapping of duplicated loci.

Their close relationship, high degree of colinearity, and cross-hybridization of genic DNA probes, all suggest the use of the small genome of *Sorghum* to guide molecular mapping and positional cloning in the closely-related genus *Saccharum* (sugarcane), one of the world's most valuable crops (Ming et al. 1998). At least five *Saccharum* homologous groups (HGs) appear to correspond largely if not completely to single *Sorghum* chromosomes (linkage groups A, C, G, H, and I). In no case does any HG correspond clearly to more than two linkage groups (LGs), suggesting that no more than four or five major inter-chromosomal translocations have occurred since the *Saccharum* - *Sorghum* divergence (involving homologs of sorghum LGs B, E, and J).

The consequences of polyploid formation in plants are also exemplified by comparison of *Saccharum* (sugarcane) and *Sorghum*. In as little as five million years since *Saccharum* and *Sorghum* diverged from a common ancestor (Al-Janabi et al. 1994b), *Saccharum* species have reached gametic chromosome numbers ranging from 18-85 or more. Data available to date reveal only one inter-chromosomal (apparent translocation) and two intra-chromosomal (apparent inversion) rearrangements that distinguished both *Saccharum spontaneum* and *S. officinarum* from *Sorghum*. Chromosome structural polymorphism within the *Saccharum* genus was suggested by the discovery of eleven intra-chromosomal rearrangements that distinguished *S. spontaneum* and *S. officinarum*. (Ming et al. 1998), like *Brassica* pointing to a relatively high tolerance of plant genomes for intra-chromosomal rearrangement.

3.3. C. Fabaceae

The Fabaceae is the third-largest family of flowering plants, and like the Poaceae, makes a disproportionately large contribution to the sustenance of humankind. The Fabaceae includes major legumes and oilseeds such as soybean (*Glycine max*, $2n = 4x = 40$) peanut (*Arachis hypogaea*, $2n = 4x = 40$), mung bean (*Vigna radiata*, $2n = 2x = 22$), chickpea (*Cicer arietinum*, $2n = 2x = 16$), and lentil (*Lens culinaris*, $2n = 2x = 14$), as well as vegetable crops such as common bean (*Phaseolus vulgaris*, $2n = 2x = 22$), and pea (*Pisum sativum*, $2n = 2x = 14$), and finally forages such as alfalfa (*Medicago sativa*, $2n = 4x = 32$). The Fabaceae are distinguished from other major crops in that about 90% of legume species fix atmospheric nitrogen in a relationship with *Rhizobium* bacteria — thus supplying themselves with one of the most costly fertilizer elements.

Detailed genetic maps have been assembled in at least eight genera of the Fabaceae (*Arachis, Glycine, Lens, Medicago, Phaseolus, Pisum, Vicia,* and *Vigna-*). A pioneering report of comparison between lentil and pea documented common gene order across at least 40% of the genomes (Weeden et al. 1992), and suggested conservation with many chromosomal regions in *Vicia*, based on isozyme loci (Torres et al. 1993). Not surprisingly, mungbean and cowpea, different species within the genus *Vicia*, also exhibit a high degree of linkage conservation (Menacio-Hautea et al. 1993).

Soybean is an allopolyploid and exhibits a high level of duplication within the genome (Funke et al. 1993). Progress in establishing homoeology has been hampered by a relatively low level of DNA polymorphisms among tetraploid *Glycine* species, but is proceeding apace, and duplicated paralogous chromosomal regions are quickly being found. Early results from analysis of duplicated DNA marker loci suggest that homoeologous genomic regions may be extensively rearranged, as DNA probes linked in one chromosomal region tend to detect duplicate loci which are widely-distributed (Keim et al. 1990; Shoemaker et al. 1992).

A concerted effort to align the genomes of many legumes is presently underway (N.D. Young, pers. comm). The completion of this task is likely to quickly yield a better understanding of the organization and evolution of these important crops.

3.4. D. Malvaceae

Cultivated tetraploids ($2n = 4x = 52$) of the cotton genus, *Gossypium*, provide the world's leading natural fiber and are also major oilseed crops. All five tetraploid species exhibit disomic chromosome pairing (Kimber 1961). Chromosome pairing in interspecific crosses between diploid and tetraploid cottons suggests that tetraploids contain two distinct genomes, which resemble the extant A genome of *G. herbaceum* ($2n = 2x = 26$) and D genome of *G. raimondii* Ulbrich ($2n = 2x = 26$), respectively.

A detailed RFLP map of cotton (Reinisch et al. 1994) reveals the genomic origins of, and homoeologies among, most of the linkage groups in tetraploid cotton, as well as the nature and frequency of rearrangements which distinguish homoeologs. As was true for *Brassica*, cotton shows an abundance of duplicated DNA marker loci. About 20% of DNA probes segregated for RFLPs at two or more loci, providing a framework for identifying homoeologous chromosomes in tetraploid cottons. Homoeologous regions for most tetraploid cotton chromosomes have now been identified (Reinisch et al. 1994; A.H.P., pers. comm.). Most homoeologous pairs are distinguished by one or more inversions (Reinisch et al. 1994; A.H.P., pers. comm.). In only two cases did putative homoeologs show parallel DNA marker order over the entire region of the chromosome for which homoeology could be inferred, although large blocks of sequence conservation were evident in most cases. Four pairs clearly differed by at least one inversion, and two additional pairs differed by at least two inversions. At least one pair showed a translocation, while several additional pairs may have represented either translocations or simply small linkage groups which had not yet linked up. Detection of rearrangements among the homoeologs of cotton is not surprising, in light of the 6 to 11 million years of divergence of the diploid 'subgenomes' from a common ancestor (Wendel 1989).

3.5. E. Solanaceae

The Solanaceae includes several economically important plant species, specifically tomato (*Lycopersicon esculentum*, $2n = 2x = 24$), pepper (*Capsicum annuum*, $2n = 2x = 24$), potato (*Solanum tuberosum*, $2n = 2x = 24$), eggplant (*Solanum melongena*, $2n = 2x = 24$), and tobacco (*Nicotiana tabacum*, $2n = 4x = 48$). The Solanaceae were among the first plant taxa to which comparative mapping was applied. Within the genus *Lycopersicon*, there is no evidence that any species have differences in gene order. Numerous inves–tigators have studied crosses between and among species, finding consistent genetic maps that differed primarily by the extent of recombination in localized regions.

A detailed comparative map of potato and tomato has been published (Bonierbale et al. 1988; Tanksley et al. 1992). The genomes of potato and tomato differ by only five chromosomal rearrangements. The rearrangements involve paracentric inversions of the short arms of chromosomes 5, 9, 11, and 12, and the long arm of chromosome 10, each indicating a single break at or near the centromere. This further reinforces the relatively high propensity (or tolerance) of plants for intra-chromosomal rearrangement.

In contrast, the genomes of tomato and pepper are more extensively rearranged, with a minimum of 15 chromosome breaks found since divergence from a common ancestor (Prince et al. 1993). Although the genetic map of pepper remains only fragmentary, it is clear that pepper has taken a remarkably different evolutionary path than other well-studied genera of the Solanaceae.

4. Comparisons across taxonomic families, toward a unified map of higher plant chromosomes

As exemplified above, closely-related (confamilial) genera often retain large chromosomal tracts in which gene order is co-linear, punctuated by structural mutations such as inversions and translocations. As evidenced from Brassica, sugarcane, and tomato, plants have a higher propensity (or tolerance) for intra-chromosomal rearrangements such as inversions than inter-chromosomal rearrangements (translocations). Several examples point to the possibility that conservation of gene order might extrapolate to more distantly-related taxa. Such a possibility is supported in a general sense by the observation that nine pairs of taxa, for which there exist both comparative genetic maps and plausible estimates of divergence time, showed an average of 0.14 (±0.06) structural mutations per chromosome per million years of divergence (Paterson et al. 1996) — this suggests that for taxa such as monocots and dicots which diverged from a common ancestor about 130-200 million years ago, 43-58% of chromosomal tracts 3 cM may remain co-linear. To test the hypothesis that regions of co-linearity may be discernible in distantly-related taxa, common DNA probes were mapped in *Arabidopsis, Sorghum, Gossypium* (cotton), and B*rassica* (broccoli). The close relationship among the crucifers *Arabidopsis* and *Brassica* (see above) enabled us to treat these genera as having a unified map. Genes that were closely-linked in the crucifers were usually also closely-linked in sorghum and/or cotton. Eight pairs of genes, linked at 3 cM in the crucifers, could be mapped in sorghum — of these, seven pairs (87.5%) were linked in sorghum, at distances of 1.4, 3.3, 9.0, 12.5, 13.5, 26.0, and 46.7 cM, respectively. Among the fewer genes mapped in cotton, three pairs were linked at 3 cM or less in the crucifers — the first two pairs of loci were linked at 5.8 and 41.1 cM in cotton. The last pair was unlinked. The fewer conserved gene blocks found in cotton than sorghum is presumed to be an artifact of the lower density of comparative markers (avg 38 cM spacing, versus 19 cM in sorghum). As few as 150-200 large rearrangements may distinguish the genomes of *Arabidopsis* and *Sorghum* : 500 corresponding loci could yield a detailed comparative map. Results from sorghum extrapolate to rice, maize or other taxa, since the grass genomes are already well-integrated. By phylogenetic analysis, ancestral versus derived gene orders might be discerned, revealing the course of chromosome evolution (cf. Kowalski et al. 1994).

5.   Local genomic colinearity

Comparisons of genetic maps by RFLP technology is limited in sensitivity, partly by the number of orthologous probes available but mainly by population sizes relative to the frequency of recombination. Most RFLP maps contain only a few hundred comparable markers (at most), hence yielding the ability to detect colinearity at the 5 cM to 10 cM level. This low sensitivity could miss many small rearrangements, including deletions, duplications and inversions. In contrast, comparative restriction map or sequence analysis of orthologous genomic segments of 100 kb or so cloned into BAC or YAC vectors can provide highly detailed comparisons, albeit for only a small portion of any genome.

Initial comparative studies of the restriction maps of YAC or BAC clones and genetic maps based on RFLPs suggested that a high level of genic colinearity was usually present within the grasses, although exceptions were also observed (Dunford et al. 1995; Kilian et al. 1995). More recently, genomic sequence comparisons have shown extensive colinearity between the rice and sorghum genomes (Chen et al. 1998) and between the maize and sorghum genomes (Tikhonov et al. 1999). However, once again, several exceptions were observed. In comparisons between maize, sorghum and rice, only the genes exhibit conserved content and order. The sequences between genes are highly variable. In particular, the larger maize genome contains large amounts of intragenic DNA, most of which is composed of retrotransposons inserted within each other (Chen et al. 1997; 1998; SanMiguel et al. 1996; Tikhonov et al. 1999). These intragenic retrotransposons (IRPs) show a preference for insertion between genes (SanMiguel et al. 1996) and have amplified within the last 2-6 million years to comprise over 50% of the total maize nuclear genome (SanMiguel et al. 1998; SanMiguel and Bennetzen 1998). Hence, the IRPs mostly account for the greater distance between genes in large genome grasses compared to the intergene distances in small genome grasses (Chen et al. 1997; Tikhonov et al. 1999). Clearly, these mobile DNAs are able to insert into genomic regions without usually affecting local gene content or order.

Aside from the unconserved intragenic regions, some exceptions to colinearity have also been observed with genes in otherwise orthologous regions. For instance, orthologous regions of the maize, sorghum and rice genomes all exhibit tight genetic and physical linkage of *sh2* and *a1* homologues, but these two loci are not linked in *Arabidopsis* (Bennetzen et al. 1998). The *adh1*-homologous regions of maize and sorghum show extensive colinearity, but exhibit no colinearity with the *adh1*-orthologous region of rice (Tikhonov et al. 1999). Moreover, this *adh1*-region also exhibits little or no colinearity with a similar sequenced region of the *Arabidopsis* genome (Tikhonov et al. 1999). Even in a comparison between maize and sorghum, two closely related species that last shared a common ancestor about 15-20 million years ago (Gaut and Doebley 1997), some exceptions to colinearity were observed in *adh1*-homologous regions. A twelve gene segment (about 70 kb) of sorghum DNA differed from a nine gene segment of maize DNA (about 220 kb) by two apparent deletions (Tikhonov et al. 1999). However, copies of the three genes missing from the *adh1* region were detected in other parts of the maize genome, suggesting that these deletions were tolerated because of the tetraploid nature of maize (Gaut and Doebley 1997; Tikhonov et al. 1999)

6. Synthesis

As recently as ten years ago, extensive commonality in plant gene content and order seemed an unlikely proposition, given the tremendous variability in plant development, biochemistry and morphology. However, comparative genetic maps based on RFLPs now have shown that this tremendous level of variation is more due to allelic variation than it is to huge differences in gene content (cf. Bennetzen and Freeling 1993; 1997). Unfortunately, belief in genomic colinearity within particularly plant families has rapidly moved from an unjustified rejection to an equally unjustified wholesale acceptance. Most comparative genetic maps continue to contain fairly few markers, and investigators tend to feature the colinearities rather than the deviations from colinearity. Further, virtually all comparative maps lack the level of polymorphism that allows mapping of all of the bands for any given probe, thus giving rise to confusion regarding orthology and parology. Finally, even genomes that show extensive genomic colinearity by most criteria still may lack colinearity for certain classes of genes, as has been suggested for nucleolar organizers (Dubcovsky and Dvorak 1995) and disease resistance loci (Leister et al. 1998). Hence, despite the many cases and exceptional potential value of genomic colinearity in plants, any given region cannot be assumed to be colinear at the outset of an investigation. Colinear RFLP maps and a high degree of relatedness are good predictors of local colinearity, but only analysis of the targeted local region can determine whether microcolinearity is actually present.

The two broad messages from comparative genome analysis are remarkably incongruous — how is parallel gene order preserved along the chromosomes of taxa that have been reproductively-isolated for millions of years, simultaneously with rapidly-changing arrangement of intergenic DNA? Clearly, the vast majority of chromosomal breaks are resolved in the rejoining of "sticky ends" to a common inserted DNA molecule (such as a transposable element). In fact, recent data have suggested that transposable elements may play an active role in "repair" of chromosomal breaks (Teng et al. 1996). A fundamental question that remains to be resolved is the extent to which gene order is an evolutionary force — that is, whether gene order is preserved by selection, or simply evolves at a stochastic rate that is determined by the fidelity of chromosomal replication. Predicted lengths of "smallest conserved evolutionary unit sequences (SCEUS-O'Brien et al. 1993) provide a null hypothesis for identifying unusual features of particular genomes. Conserved "gene blocks" that are larger than expected to persist by chance might reflect fitness advantages of particular gene arrangements or structural features. Taxa that show particularly high rates of chromosome structural rearrangement such as *Brassica* and *Secale* (cf. Paterson et al. 1996) might be fertile systems for investigating the structural mutation process. The degree to which structural mutations are either a cause or a consequence of speciation might be clarified.

It is not yet clear what should be concluded from the discovery of frequent deviations from microcolinearity (conserved local gene order), in some genomes that exhibit extensive 'macrocolinearity' at the level of RFLP maps. Are small rearrangements, such as inversions of a few centiMorgans or less, particularly common? This type of rearrangement could preserve macrocolinearity while interrupting microcolinearity.

Is there any relationship between the frequencies of large and small rearrangements within a genome? What are the molecular mechanisms that underlie small and/or large rearrangements? Why do some taxa have more frequent rearrangements than others? Could selection for conserved gene order be more important in some lineages than others, or might there be selection in some families for rearrangements that would increase the level of infertility in wide crosses? Alternatively, some species may have evolved in highly-fragmented environments that may be conducive to fixation of frequency-dependent mutations such as chromosome structural rearrangements (cf. Kowalski et al. 1994), or environments that routinely expose them to higher levels of chromosome-breaking and transposable element-inducing mutagens (e.g., intense solar radiation). The availability of genomic DNA sequences for large chromosomal segments of selected monocot and dicot taxa will provide a powerful new data set for further investigating these questions.

The applied significance of conserved gene order is also noteworthy. Widespread use of heterologous DNA markers has accelerated genome analysis in many crops, and extended it from leading crops to many minor crops. Physical maps and genomic sequence for facile models such as *Arabidopsis,* aligned to genetic maps of major crops, may aid in the cloning of agriculturally-important genes or QTLs. Thousands of genetically-mapped mutants of *Arabidopsis*, maize, rice, and other taxa might be united into a central tool for comparative study of plant development. Mutants unique to one taxon may facilitate molecular dissection of processes that are invariant in other taxa.

8.   References

Ahn, S. and Tanksley, S.D. (1993) Comparative linkage maps of rice and maize genomes. Proc. Natl. Acad. Sci. USA 90: 7980-7984.

Ahn, S., Anderson, J.A., Sorrells, M.E. and Tanksley, S.D. (1993) Homoeologous relationships of rice, wheat and maize chromosomes. Mol. Gen. Genet. 241: 483-490.

Al-Janabi, S.M., Honeycutt, R.J., Peterson C. and Sobral, B.W.S. (1994) Phylogenetic analysis of organellar DNA sequences in the Andropogoneae: *Saccharum*. Theor. Appl. Genet. 88: 933-944.

Al-Shehbaz, I.A. (1973) The biosystematics of the genus *Thelypodium* (Cruciferae). Contrib. Gray Herb. Harv. Univ. 204: 3-148.

Bennetzen, J.L. and Freeling, M. (1993) Grasses as a single genetic system: genome composition, collinearity and compatibility. Trends Genet. 9: 259-261.

Bennetzen, J.L. and Freeling, M. (1997) The unified grass genome: synergy in synteny. Genome Res. 7: 301-306.

Bennetzen, J.L., SanMiguel, P., Chen, M., Tikhonov, A., Francki M. and Avramova Z. (1998) Grass genomes. Proc. Natl. Acad. Sci. USA 95: 1975-1978.

Berhan, A.M., Hulbert, S.H., Butler, L.G. and Bennetzen, J.L. (1993) Structure and evolution of the genomes of *Sorghum bicolor* and *Zea mays*. Theor. Appl. Genet. 86: 598-604.

Binelli, G.L. Gianfrancesci, M.E.Pè, Taramino, G., Busso, C., Stenhouse, J. and Ottaviano, E. (1993) Similarity of maize and sorghum genomes as revealed by maize RFLP probes.

Theor. Appl. Genet. 84: 10-16.

Bonierbale, M.D., Plaisted, R.L. and Tanksley, S.D. (1988) RFLP maps based on a common set of clones reveal modes of chromosomal evolution in potato and tomato. Genetics 120: 1095-1103.

Burr, B., Burr., F.A., Thompson, K.H., Albertsen, M.C., Stuber, C.W. (1988) Gene mapping with recombinant inbreds in maize. Genetics 118: 519-526.

Chao, S., Sharp, P.J., Worland, A.J. et al. (1989) RFLP-based genetic maps of wheat homoeologous group 7 chromosomes. Theor. Appl. Genet. 78: 495-504.

Chen, M., SanMiguel, P. and Bennetzen, J.L. (1998) Sequence organization and conservation in *sh2/a1*-homologous regions of sorghum and rice. Genetics 148: 435-443.

Chen, M., SanMiguel, P., de Oliveira, A.C., Woo, S.-S., Zhang, H., Wing, R.A. and Bennetzen, J.L. (1997) Microcolinearity in the *sh2*-homologous regions of the maize, rice and sorghum genomes. Proc. Natl. Acad. Sci. USA 94: 3431-3435.

Chittenden, L.M., Schertz, K.F., Lin, Y-R., Wing, R.A., Paterson, A.H. (1994) A detailed RFLP map of *Sorghum bicolor* x *S. propinquum* suitable for high-density mapping suggests ancestral duplication of chromosomes or chromosomal segments. Theor. Appl. Genet. 87: 925-933.

Devos, K.M., Atkinson, M.D., Chinoy, C.N. et al. (1993) RFLP-based genetic map of the homoeologous group 2 chromosomes of wheat, rye and barley. Theor. Appl. Genet. 85: 784-792.

Devos, K.M., Atkinson, M.D., Chinoy, C.N. et al. (1992) RFLP-based genetic map of the homoeologous group 3 chromosomes of wheat and rye. Theor. Appl. Genet. 83: 931-939.

Devos, K.M., Dubcovsky, J., Dubcovsky, J. et al. (1995) Structural evolution of wheat chromosomes 4A, 5A, and 7B and its impact on recombination. Theor. Appl. Genet. 91: 282-288.

Devos, K., Atkinson, M.D., Chinoy, C.N., Liu, C.J. and Gale, M.D. (1992) Chromosomal rearrangements in rye genome relative to that of wheat. Theor. appl. Genet. 83: 931-939.

Dunford, R.P., Kurata, N., Laurie, D.A., Money, T.A., Minobe, Y. and Moore, G. (1995) Conservation of fine scale DNA marker order in the genomes of rice and the Triticeae. Nucl. Acids. Res. 23: 2724-2728.

Dubcovsky, J. and Dvorak, J. (1995) Ribosomal RNA multigene loci: nomads of the Triticeae genomes. Genetics 140: 1367-1377.

Funke, R.P.A. Kolchinsky, P.M. Gresshoff. (1993) Physical mapping of a region in the soybean (*Glycine max*) genome containing duplicated sequences. Plant Molec. Biol. 22: 437-446.

Gale, M.D. and K.M. Devos (1998) Plant comparative genetics after 10 years. Science 282: 656-659.

Gaut, B.S. and Doebley, J.F. (1997) DNA sequence evidence for the segmental allotetraploid origin of maize. Proc. Natl. Acad. Sci. USA 94: 6809-6814.

Helentjaris, T.D. Weber, S. Wright. (1988) Identification of the genomic locations of duplicate nucleotide sequences in maize by analysis of restriction fragment length polymorphisms. Genetics 118: 353-363.

Hohmann, U., Graner, A. and Endo, T.R.(1995) Comparison of wheat physical maps with barley linkage maps for group 7 chromosomes. Theor. Appl. Genet. 91: 618-626.

Hulbert, S.H., Richter, T.E., Axtell, J.D. and Bennetzen, J.L. (1990) Genetic mapping and characterization of sorghum and related crops by means of maize DNA probes. Proc. Natl. Acad. Sci. (USA) 87: 4251-4255.

Keim, P.B., Diers, W., Olson, T.C., Shoemaker, R.C. (1990) RFLP mapping in soybean: association between marker loci and variation in quantitative traits. Genetics 126: 735-742.

Kilian, A., Kudrna, D.A., Kleinhofs, A., Yano, M., Kurata, N., Steffenson, B. and Sasaki, T. (1995) Rice-barley synteny and its application to saturation mapping of the barley *Rpg1* region. Nucl. Acids Res. 23: 2729-2733.

Kimber, G. (1961) Basis of the diploid-like meiotic behavior of polyploid cotton. Nature 191: 98-99.

Kowalski, S.P., Lan, T-H., Feldmann, K.A. and Paterson, A.H. (1994) Comparative mapping of *Arabidopsis thaliana* and *Brassica oleracea* chromosomes reveals islands of conserved organization. Genetics 138: 499-510.

Kurata, N., Moore, G., Nagamura, Y., Foote, T., Yano, M., Minobe, Y. and Gale, M. (1994) Conversation of genome structure between rice and wheat. Bio/Technology 12: 276-278.

Leister, D., Kurth, J., Laurie, D.A., Yano, M., Sasaki, T., Devos, K., Graner, A. and Sculze-Leifert P. (1998) Rapid reorganization of resistance gene homologues in cereal chromosomes. Proc. Natl. Acad. Sci. USA 95: 370-375.

Liu, C.J., Atkinson, M.D., Chinoy, C.N. et al. (1992) Nonhomoeologous translocations between group 4, 5, and 7 chromosomes within wheat and rye. Theor. Appl. Genet. 83: 305-312.

Liu, Y.G., Tsunewaki, K. (1991) Restriction fragment length polymorphism (RFLP) analysis in wheat. II. Linkage maps of the RFLP sites in common wheat. Jpn. J. Genet. (1991) 66: 617-633.

Lydiate, D.A. Sharpe, U. Lagercrantz and I. Parkin. 1993. Mapping the *Brassica* genome. Outlook Agr. 2: 85-92.

Marino C.L., Nelson J.C., Lu Y.H. et al. (1995) RFLP-based linkage maps of the homoeologous group 6 chromosomes of hexaploid wheat (*Triticum aestivum* L. em. Thell). Genome. submitted.

McGrath, J.M. and C.E. Quiros. (1991) Inheritance of isozyme and RFLP markers in *Brassica campestris* and the comparison with *B. oleracea*. Theor. Appl. Genet. 82: 668.

Menacio-Hautea, D.C.A. Fatokum, L. Kumar, D. Danesh, N.D. Young. 1993. Comparative genome analysis of mungbean (*Vigna radiata* L. Wilczek) and cowpea (*V. unguiculata*) using RFLP analysis. Theor. Appl. Genet. 86:797-810.

Mickelson-Young, L., Endo, T.R. and Gill, B.S. (1995) A cytogenetic ladder-map of the wheat homoeologous group-4 chromosomes. Theor. Appl. Genet. 90: 1007-1011.

Ming, R., Liu, S-C., Lin, Y-R., da Silva, J., Wilson, W., Braga, D., Van Deynze, A., Wenslaff, T.E., Wu, K.K., Moore, P.H., Burnquist, W., Irvine, J.E., Sorrells, M.E., Paterson, A.H. (1998) Alignment of the *Sorghum* and *Saccharum*: chromosomes: Comparative genome organization and evolution of a polysomic polyploid genus and its diploid cousin. Genet. 150: 1663-1682.

Moore, G., Devos, K.M., Wang, Z. and Gale, M.D. (1995) Grasses, line up and form a circle. Curr. Biol. 5:737-739.

Namuth, D.M., Lapitan, N.L.V., Gill, K.S. et al. (1994) Comparative RFLP mapping of *Hordeum vulgare* and *Triticum tauschii*. Theor. Appl. Genet. 89: 865-872.

Naranjo, T., Roca, P., Goicoechea, P.G. et al. (1987) Arm homoeology of wheat and rye chromosomes. Genome 29: 873-882.

Nelson, J.C., Van Deynze, A.E., Autrique, E. et al. (1995) Molecular mapping of wheat. Homoeologous group 2. Genome 38: 116-124.

Nelson, J.C., Van Deynze, A.E., Autrique, E. et al. Molecular mapping of wheat. Homoeologous group 3. Genome 1995; 38: 125-133.

Nelson, J.C., Sorrells, M.E., Van Deynze, A.E. et al. (1995) Molecular mapping of wheat. Major genes and rearrangements in homoeologous groups 4, 5, and 7. Genetics 141: 721-731.

O'Brien, S.J., Womack, J.E., Lyons, L.A., Moore, K.J., Jenkins, N.A. and Copeland, N.G. (1993) Anchored reference loci for comparative genome mapping in mammals. Nature Genetics 3, 103-112.

Osborn, T. and Paterson A.H. (1994) The Crucifereae. Contribution to "The USDA Plant Genome Research Program", J. Miksche et al. Chapter 1 in Adv. Agronomy, Vol. 56.

Paterson, A.H., Lan, T.H., Reischmann, K.P., Chang, C., Lin, Y.R., Liu, S.C., Burow, M.D., Kowalski, S.P., Katsar, C.S., DelMonte, T.A., Feldmann, K.A., Schertz, K.F., Wendel, J.F. (1996) Toward a unified map of higher plant chromosomes, transcending the monocot-dicot divergence. Nature Genetics 14: 380-382.

Periera, M.G., Lee, M., Bramel-Cox, P., Woodman, W., Doebley J. and Whitkus, R. (1994) Construction of an RFLP map in sorghum and comparative mapping in maize. Genome 37: 236-243.

Pereira, M.G. and Lee, M. (1995) Identification of genomic regions affecting plant height in sormghum and maize. Theor. Appl. Genet. 90: 380-388.

Prince, J.P., Pochard, E., Tanksley, S.D. (1993) Construction of a molecular linkage map of pepper, and a comparison of synteny with tomato. Genome 36: 404-417.

Reinisch, A.R., Dong, J.-M., Brubaker, C., Stelly, D., Wendel, J. and Paterson, A.H. (1994) An RFLP map of cotton (*Gossypium hirsutum* x *G. barbadense*): chromosome organization and evolution in a disomic polyploid genome. Genetics 138: 829-847.

SanMiguel, P. and Bennetzen, J.L. (1998) Evidence that a recent increase in maize genome size was caused by the massive amplification of intergene retrotransposons. Annals Bot. 82: 37-44.

SanMiguel, P., Gaut, B. and Bennetzen, J.L. (1998) The paleontology of intergene retrotransposons in maize. Nature Genetics 20: 43-45.

SanMiguel, P., Tikhonov, A., Jin, Y.-K., Motchoulskaia, N., Zakharov, D., Melake-Berhan, A., Springer, P.S., Edwards, K.J., Lee, M., Avramova, Z. and Bennetzen, J.L. (1996) Nested retrotransposons in the intergenic regions of the maize genome. Science 274: 765-768.

Shultz, O.E. (1936) pp. 227-658 in Die Natürlichen Pflanzenfamilien, edited by A. Engler and H. Harms, no. 2, vol. 17b. Shoemaker, R.R. Guffy, L. Lorenzen, J. Specht. 1992. Molecular genetic mapping of soybean: map utilization. Crop Sci. 32: 1091-1098.

Slocum, M.K. (1989) Analyzing the genomic structure of *Brassica* species and subspecies using RFLP analysis, p. 73-80: In T. Helentjaris and B. Burr (eds.), Development and Application of Molecular Markers to Problems in Plant Genetics. Cold Spring Harbor Laboratory Press, Cold Spring Harbor, NY.

Slocum, M.K, Figdore, S.S., Kennard, W.C., Suzuki, J.Y., and Osborn, T.C. (1990) Linkage arrangement of restriction fragment length polymorphism loci in *Brassica oleracea*. Theor. Appl. Genet. 80: 57-64.

Tanksley, S., Ganal, D.M.W., Prince, J.P., DeVicente, M.C., Bonierbale, M.W., Broun, P., Fulton, T.M., Giovannoni, J.J., Grandillo, S., Martin, G.B., Messeguer, R., Miller, J.C., Miller, L., Paterson, A.H., Pineda, O., Roder, M.S., Wing, R.A., Wu W. and Young N.D. (1992) High density molecular linkage maps of the tomato and potato genomes. Genetics 132: 1141-1160.

Teng, S-C., Kim, B. and Gabriel, A. 1996. Retrotransposon reverse-transcriptase-mediated repair of chromosomal breaks. Nature 383: 641-644.

Teutonico, R.A. and Osborn, T.C. (1994) Mapping of RFLP and qualitative trait loci in *Brassica rapa,* and comparison to linkage maps of *B. napus, B. oleracea,* and *Arabidopsis thaliana.* Theor. Appl. Genet. 89: 885-894.

Tikhonov, A.P., SanMiguel, P.J., Nakajima, Y., Gorenstein, N.D., Bennetzen, J.L. and Avramova Z. (1999) Colinearity and its exceptions in orthologous *adh* regions of maize and sorghum. Proc. Natl. Acad. Sci. USA 96: 7409-7414.

Torres, A.M., Weeden, N.F., Martin, A. (1993) Linkage among isozyme, RFLP, and RAPD markers in *Vicia faba.* Theor. Appl. Genet. 85: 937-945.

Van Deynze A.E., Dubcovsky, J., Gill, K.S. et al. (1995) Molecular-genetic maps for chromosome 1 in *Triticeae* species and their relation to chromosomes in rice and oats. Genome 38: 47-59.

Vavilov, N.I. (1922) The law of homologous series in variation. J. Genet. 12 (1): 47-89

Weeden, N.L., Muehlbauer, F.J. and Ladizinsky, G. (1992) Extensive conservation of linkage relationships between pea and lentil genetic maps. J. Hered. 83: 123-129.

Wendel, J.F. (1989) New World cottons contain Old World cytoplasm. Proc. Natl. Acad. Sci. USA 86: 4132-4136.

Whitkus, R., Doebley, J. and Lee, M. (1992) Comparative genome mapping of sorghum and maize. Genetics 132: 119-130.

Xie, D.X., Devos, K.M., Moore, G. et al. (1993) RFLP-based genetic maps of the homoeologous group 5 chromosomes of bread wheat (*Triticum aestivum* L.). Theor. Appl. Genet. 87: 70-74.

7.   Breeding multigenic traits

CHARLES W. STUBER

U.S. Department of Agriculture, Agricultural Research Service Department of Genetics,
North Carolina State University Raleigh, North Carolina 27695-7614 - <cstuber@ncsu.edu>

Contents

1.   Introduction

Increases in the productivity of food, feed, and fiber in domesticated crop plants can be attributed to the collective plant breeding efforts of man over many millennia. These increases (many have been dramatic) have resulted from artificial selection, either conscious or unconscious, on the phenotypes of the targeted species. Until the 20th century, plant breeding was largely an art with little or no knowledge of genetic principles. Although plant improvement since the rediscovery of Mendel's principles has involved both art and science, the contributions of science will undoubtedly assume a much greater role as new technology becomes more widely used and as additional gains in agricultural productivity are required at an ever-increasing pace.

If we consider the increase in maize grain yield production in the United States, for example, we note that the gain for the time span from 1930 to 1980 averaged nearly 100 kg ha^{-1} yr^{-1} with about three-fourths of that gain attributed to genetic improvement (Duvick 1984; Russell 1984). The remainder is attributed to changes in cultural practices such as increased rates of mineral fertilizers and the use of herbicides for weed control and pesticides for control of insects and diseases. In a more recent comparison of 36 widely grown maize hybrids adapted to central Iowa and released at intervals from 1934 to 1991, Duvick (1997) reported that yielding ability had increased at a linear rate of about 74 kg ha^{-1} yr^{-1}. He concluded that increased grain yielding ability of these widely successful hybrids is due primarily to improved tolerance of abiotic and biotic stresses, coupled with the maintenance of the ability to maximize yield per plant under non-stress growing conditions. By scaling down the yields obtained in his trials to the level of on-farm yields and by making hybrid comparisons on side-

R.L. Phillips and I.K. Vasil (eds.), DNA-Based Markers in Plants, 115 - 137.

by-side trials, Duvick now estimated that all of the yield gain could be attributed to genetic improvement.

Because there are limited opportunities for gains resulting from changes in cultural practices (particularly in the United States and other developed countries), future gains in the productivity of most crops may depend almost entirely on genetic improvements. In fact, environmental concerns may cause a reduction in the use of agricultural chemicals and fertilizers. Therefore, plant breeders will need to develop and apply new technology (such as DNA-based markers) at a faster pace to more effectively improve crop plants for the ever increasing global human population as well as for the changes in consumer preferences.

Most economically important plant traits, such as grain or forage yield, are classified as multigenic or quantitative. Even traits considered to be more simply inherited, such as disease resistance, may be 'semi-quantitative' for which trait expression is governed by several genes (e.g., a major gene plus several modifiers). The challenge to strategically use new technology (such as DNA-based markers) to increase the contribution of 'science' to the 'art + science' equation for plant improvement therefore applies to most, if not all, traits of importance in plant breeding programs.

Historically, one of the first questions in quantitative genetics was whether the inheritance of these continuously distributed traits was Mendelian (Comstock 1978). Obviously, the answer to this question has major implications in the consideration of the use of markers for plant breeding programs. Evidence reported by East and Hayes (1911), Emerson and East (1913), and others contributed to the rejection of the 'blending' inheritance hypothesis and to the conclusion that Mendelian principles apply to quantitative as well as to qualitative traits. Over the past 80 years, both plant and animal geneticists have obtained convincing evidence for the shaping of the general model that embraces the multiple-factor hypothesis for quantitative traits (with genes located in chromosomes and hence sometimes linked, and incomplete heritability because of the contribution of environmental factors to total phenotypic variation).

2.   Challenges facing plant breeders

Plant breeders are faced with numerous challenges in their improvement programs, many of which might be met with the development and application of new technology. Some of these challenges that have stimulated research in the application of marker technology relate to:
1. understanding the basis of heterosis and prediction of hybrid performance,
2. identification of useful genetic factors in divergent populations or lines (such as exotic accessions),
3. introgression of desired genetic factors into elite breeding lines,
4. improvement of recurrent selection programs based on phenotypic responses.
5. understanding and adjusting for genotype by environment interaction.

In order to meet these challenges, the primary emphasis has been on the development of new tools, such as DNA-based markers, with the major focus on the improvement of breeding precision and efficiency.

In spite of the fact that numerous investigations have been conducted on the inheritance of multigenic traits (using primarily classical biometrical methods), plant breeders typically have little information on: (1) the number of genetic factors (loci) influencing the expression of the traits, (2) the chromosomal location of these loci, (3) the relative size of the contribution of individual loci to trait expression. (4) pleiotropic effects, (5) epistatic interactions among genetic factors, and (6) variation of expression of individual factors in different environments. In order to achieve the maximum benefit from marker-based procedures for the manipulation and improvement of multigenic traits, an increased understanding of the genetic bases underlying quantitative trait variation will be necessary.

Marker-based technology already is providing scientists with a powerful approach for identifying and mapping quantitative trait loci (QTLs) and should ultimately lead to the development of a better understanding of genetic phenomena such as epistasis, pleiotropy, and heterosis. A number of recent investigations (particularly in maize, tomato, and rice) are providing some clues to understanding such phenomena (e.g., Edwards et al. 1987; Stuber et al. 1987; Paterson et al. 1990, 1991; Abler et al. 1991; Koester 1992; Edwards et al. 1992; Stuber et al. 1992; Eshed and Zamir 1995; Li et al. 1997). It must be acknowledged, however, that studies such as these have identified and mapped only rather large chromosomal segments (in most cases probably 20 to 30 cM long). Although results from such studies may be adequate for many plant breeding endeavors, novel approaches will be necessary to identify individual genes and quantify individual gene action and interactions among genes.

3. Hybrid predictions

Heterosis (or hybrid vigor) is a major reason for the success of the commercial maize industry as well as for the success of breeding efforts in many other crop and horticultural plants. Development of inbred lines suitable for use in production of superior hybrids is very costly and requires many years in traditional breeding programs. Much of the developmental effort is devoted to field testing of newly created lines in various single-cross combinations to identify those lines with superior combining ability.

The search for a reliable method for predicting hybrid performance without generating and testing hundreds or thousands of single-cross combinations has been the goal of numerous marker studies, particularly in maize. For example, in several earlier marker studies in maize, correlations between isozyme allelic diversity and grain yield were estimated in single-cross hybrids derived from commercially used lines (Hunter and Kannenberg 1971; Heidrich-Sobrina and Cordeiro 1975; Gonella and Peterson 1978; and Hadjinov et al. 1982). With 11 or fewer isozyme marker loci and 15 or fewer inbred parental lines, the estimated correlations between isozyme allelic diversity and specific combining ability were low and nonsignificant in these studies. Even in a much larger study in which 100 maize hybrids derived from 37 elite lines were used to evaluate associations of hybrid yield performance with allelic diversity at 31 isozyme loci, an R-square value of only 0.36 was reported by Smith and Smith (1989).

Also, in another recent study, no association was found between hybrid grain yield and isozyme diversity in a study of six enzyme marker loci in 75 F_1 rice (*Oryza sativa* L.) hybrids (Peng et al. 1988).

Frei et al. (1986a) reported results from a study of 114 single-cross maize hybrids with a somewhat different orientation from those discussed above. In this study, the merits of using isozyme marker diversity were compared with the merits of using pedigree diversity for predicting hybrid yield performances. Genotypes of 21 isozyme loci were used to classify inbred line pairs into similar and dissimilar groups. These isozyme diversity groups were then further subdivided into similar and dissimilar pedigree classes based on commonality of pedigree background between line pairs. Comparisons of the isozyme diversity groups showed that average grain yield of hybrids with dissimilar isozymes was significantly higher (10% greater) than hybrids with similar isozyme genotypes. However, hybrid yields for the dissimilar pedigree class averaged about 37% more than for the hybrids in the similar pedigree class. Although isozyme dissimilarity was significantly associated with higher yield in the single-cross hybrids tested, the investigators concluded that the useful predictive value of these markers would be limited primarily to lines with similar pedigrees. In another isozyme diversity/hybrid performance study in maize, Lamkey et al. (1987) investigated F_1 hybrids among 24 high-yielding and 21 low-yielding lines from a group of 247 inbred lines derived using single-seed descent in the Iowa Stiff Stalk Synthetic (BSSS) population. Allelic differences at 11 isozyme marker loci were not predictive of hybrid performance in comparisons among crosses of high- and low-yielding lines.

The above studies suggest that isozyme genotypes provide limited value in the prediction of hybrid performance in crops such as maize and rice. Several factors may contribute to this somewhat disappointing conclusion. For example, the low number of isozyme loci assayed in most of the studies would effectively mark only a small fraction of the genome. Therefore, only a limited proportion of the genetic factors contributing to the hybrid response would be sampled. More importantly, it is unlikely that these marker loci affect the phenotypic expression of the targeted quantitative trait directly; rather they serve to identify adjacent (linked) chromosomal segments. Certainly allelic differences at marker loci *do not* assure allelic differences at linked QTLs. For a limited number of markers to be useful as predictors for hybrid performance, the effects of QTL 'alleles' linked to specific marker alleles must be ascertained.

Also, it should be noted that the type of gene action associated with specific QTLs will affect the predictive value of linked marker loci. In maize populations developed from crosses of two inbred lines, it has been shown that the number of heterozygous marker loci is positively correlated with grain yield of F_2 plants or backcross families (Edwards et al. 1987; Stuber et al. 1992). These results corroborated other data that implicates dominant (or even overdominant) types of gene action as the predominate contributor to the expression of grain yield in maize. In such cases, marker allele diversity that reflects linked QTL 'allele' diversity should be predictive of grain yield responses. However, for traits governed largely by additive gene action (and this type of gene action might prevail for some loci affecting grain yield) the heterozygous QTL genotype would not be the most favorable. Again, as stated in the preceding paragraph, effective prediction of hybrid performance based on markers requires knowledge of QTLs linked to the markers.

Furthermore, it should be stressed that the level of linkage integrity must not be overlooked in the consideration of markers for hybrid predictions. For example, if the proposed hybrids are derived from lines produced from a randomly mated population, or if the lines comprise some subset of publicly available inbreds, then the associations between marker alleles and QTL alleles might be expected to be essentially random, i.e., near linkage equilibrium. For marker-based procedures to be successful for predictive or selective purposes for complexly inherited traits, such as grain yield, the genome should be well saturated with uniformly spaced markers and/or a high level of linkage disequilibrium must exist.

In the study discussed previously by Smith and Smith (1989), associations of grain yield with diversity of RFLP genotypes also were measured in the more than 100 hybrids derived from 37 elite maize inbred lines. Plots of F_1 grain yield against RFLP diversity, based on 230 marker loci, showed an R-square value of 0.87. This value presents a striking contrast between the use of 230 RFLP marker loci versus 31 isozyme loci (with an R-square value of 0.36) for the prediction of hybrid performance. However, it is important to note that even with 230 RFLP markers, yields varied from 8150 to 10,660 kg ha^{-1} (130 to 170 bushels/acre) for the subset of hybrids with the maximum detected 'distance' (0.70 to 0.80 on a scale ranging from 0.00 to 0.80) between the parental lines. Most breeders would be working with similar subsets of largely unrelated lines for which, again, marker diversity alone does not appear to be very satisfactory for predictive purposes.

In another maize study, Melchinger et al. (1990) compared RFLP genotypes at 82 marker loci with field data on 67 hybrids reported earlier by Darrah and Hallauer (1972). Twenty inbred lines were involved in the parentage of the hybrids. They concluded that associations of hybrid yield, heterosis, and specific combining ability with multilocus heterozygosity of RFLP loci generally were too weak to be useful as a supplementary tool for predicting yield performance of crosses between unrelated lines. In addition, they concluded that for unrelated lines, genetic distance measures based on a large number of RFLPs uniformly distributed throughout the genome are not markedly superior to those based on a small number of isozymes for predicting hybrid yield. Thus, their results show that better marker coverage alone will not increase predictive power substantially. Melchinger et al. further state that 'it seems necessary to employ specific markers for those segments that significantly affect the expression of heterosis for grain yield'.

Dudley et al. (1992) reported a study for which the major objectives were to evaluate methods of using molecular marker data to: (1) identify parents useful for improving a single-cross hybrid and (2) compare marker genotypic means measured at the inbred level to those measured at the hybrid level. Genotypic data from 14 isozyme and 52 RFLP marker loci were compared with field performance data from a diallel mating design of 14 maize inbreds in their investigations. They found that marker genotypic differences measured in inbreds were positively correlated with differences measured in hybrid backgrounds; however, these correlations were only slightly higher than those between phenotypic midparent and hybrid values. Their findings suggested that genotypic differences may be useful for preliminary selection of loci and alleles for possible improvement of hybrids but probably will not accurately predict final performance of a hybrid. They concluded that number of unique alleles in a donor line was not a good measure for identifying lines that have value for improving a single cross and stated that 'uniqueness of alleles does not necessarily indicate the presence of a favorable QTL'. Thus, results from this study corroborated results from many earlier attempts to correlate marker allele diversity of parental lines with hybrid performance.

Bernardo (1994) evaluated the use of a best linear unbiased prediction of single cross performances based on (i) RFLP data on the parental inbreds and (ii) yield data on a related set of 54 single crosses. Sets of n predictor hybrids (where n = 10, 15, 20, 25, or 30) were chosen at random, and pooled correlations between predicted yields and observed yields of the remaining (54 - n) hybrids ranged from 0.65 to 0.80 (r^2 = 0.42 to 0.64). Although Bernardo concluded that single-cross yield can be predicted effectively based on parental RFLP data and yields of a related set of hybrids, these results are no better for hybrid predictions than those discussed above by Smith and Smith (1989).

The use of marker-aided prediction of advanced generation combining ability based on data from early generation testcrossing was evaluated by Johnson and Mumm (1996). Based on a total marker score generated from regressing F_3 line testcross yields on the corresponding F_3 line RFLP genotypes, F_5 testcross yields were predicted with more accuracy than using only F_3 testcross yields as predictors. It was concluded that marker-aided prediction of advanced generation performance from early generation testcrosses was effective. These results were in agreement with those reported by Eathington et al. (1995).

4.   Marker allele frequency changes

An earlier approach in the development of markers for plant breeding purposes was based on the association of isozyme allelic frequencies with phenotypic changes of targeted quantitative traits in long term recurrent selection experiments. Studies in this area were designed not only to search for the presence of marker associations with quantitative traits but also to ascertain the strength of these associations; mapping QTLs came later in development of breeding strategies based on marker technology.

Changes in allelic frequencies at a large number of isozyme marker loci were monitored over different cycles of long-term recurrent selection in several populations of maize (Stuber and Moll 1972; Stuber et al. 1980). Statistically significant changes were reported at eight marker loci which were greater than would be expected from drift acting alone. More importantly, these frequency changes were highly correlated with phenotypic changes in the selected trait, grain yield. Associations between isozyme marker genotypes and several agronomic and morphological traits in maize recurrent selection experiments also have been reported by Pollak et al. (1984), Kahler (1985), and Guse et al. (1988).

Results from studies such as these led to the hypothesis that manipulation of allelic frequencies at selected isozyme marker loci should produce significant responses in the correlated trait. An investigation to test this hypothesis was based on the results from the study reported by Stuber et al. (1980). This investigation, which was conducted in an open-pollinated maize population, showed that selections based solely on manipulations of allelic frequencies at seven isozyme loci significantly increased grain yield (Stuber et al. 1982). Ear number, a trait highly correlated with grain production, also increased based on these manipulations. Frei et al. (1986b) conducted a somewhat similar study in a population generated from a composite of elite inbred lines. They reported that manipulations of isozyme allelic frequencies produced responses nearly equal to responses based on phenotypic selection in the same population.

Although these marker-allele manipulation studies produced statistically significant results, the findings were not dramatic. In both of the studies listed above, the level of linkage disequilibrium between marker loci and QTLs was probably low because the target populations had been subjected to several generations of random mating prior to the investigation. This undoubtedly reduced the effectiveness of the marker loci for manipulating the associated quantitative traits. Although these studies could be considered to be only mildly successful, the results were sufficiently positive to stimulate further investigations using larger numbers of markers in more structured types of populations. The impetus for further study and development of marker-technology related to the breeding of quantitative traits is probably the major contribution that can be ascribed to these earlier attempts.

5.   QTL identification and manipulation in populations

Mapping methods for identifying and locating QTLs are discussed very completely in Chapter 6 of this Volume. In addition, Kearsay and Farquhar (1998) have briefly reviewed methods currently available for QTL analysis in segregating populations and summarized some of the conclusions arising from such analyses in plant populations. They also have provided a summary of QTL properties from 176 trial-trait combinations in plants and details of the publications summarized can be found by accessing: http://www.biology.bham.ac.uk/qtl-rev-papers/.

Although Kearsey and Farquhar (1998) summarized a large number of studies focused on the identification and mapping of QTLs, there is very little documented evidence on the manipulation of quantitative traits using marker-facilitated procedures. In this section, I will focus the discussion on the identification and manipulation of QTLs in maize and largely on our research program in Raleigh, North Carolina.

In the marker-facilitated research conducted in the maize genetics program at Raleigh, North Carolina, QTLs have been identified and mapped in more than 20 populations (F_2, F_3, and recombinant inbred) derived from eight elite inbred lines and five inbred lines with a partial exotic (Latin American, expected to be 50%) component (Edwards et al. 1987; Stuber et al. 1987; Abler et al. 1991; Edwards et al. 1992; Stuber et al. 1992; Koester et al. 1993; Ragot et al. 1995; Graham et al. 1997; and unpublished data). Both isozyme and RFLP marker loci were used in these studies, although the earlier studies used only isozymes. Measurements recorded on individual plants in the field experiments included dimensions, weights, and counts of numerous vegetative and reproductive plant parts as well as silking and pollen shedding dates.

Findings from these studies showed that QTLs affecting most of the quantitative traits evaluated were generally distributed throughout the genome, however, certain chromosomal regions appeared to contribute greater effects than others to trait expression. In the earlier studies not all of the chromosome regions were well marked, and presumably major factors also may have been segregating in regions of the genome devoid of marker loci in these studies.

In two of the early F_2 studies, nearly 1900 plants were genotyped and evaluated for more than 80 quantitative traits in each population. One of these F_2 populations was derived from the cross of CO159 with Tx303 (denoted as COTX); the other from the cross of T232 with CM37 (denoted as CMT). Data from these F_2 populations provided a much stronger case for marker-based selection for manipulating quantitative traits than that from the earlier isozyme allelic frequency studies. To evaluate the efficacy of marker-facilitated manipulation, data from the COTX and the CMT F_2 population studies were used (Stuber and Edwards 1986). Selections were based solely on the genotypes of the F_2 plants evaluated in the mapping studies; evaluations of selection response were then made on progenies of these open-pollinated F_2 plants. Phenotypic (mass) selection, based solely on the phenotypic expression of each F_2 plant also was conducted for comparison with the marker-facilitated selection responses. The evaluations of the selection responses were made at three locations in North Carolina in the year after the individual F_2 plants were studied.

In these marker-based selection studies, several traits were manipulated and a breeding value was determined for each of 15 markers for every F_2 plant (about 1900 in each population) for each trait. This breeding value was determined by calculating one-half the difference between the quantitative trait means of the two homozygous classes for the marker locus. Plants homozygous for the favorable marker allele were assigned a plus breeding value for that marker locus; those plants homozygous for the unfavorable allele were given a minus value. Marker heterozygotes received a value of zero. A composite breeding score was then calculated for each plant/trait combination by totaling the individual breeding values for the 15 isozyme marker loci. Based on these composite scores, individual plants were chosen to provide divergent selection groups (positive and negative) for each trait, which included grain yield, ear number, and ear height.

In Table 1 it can be noted that for the CMT F_2 population, the mean of the increased-yield (genotypic-positive) entry was about 40% greater than the mean for the decreased-yield (genotypic-negative) entry. Also, the mean yield of the positive entry was about 20% greater than the mean of the unselected check (a sample of the open-pollinated population from the same F_2). It can also be noted that the response to phenotypic (mass) selection was nearly identical to the genotypic response. In the COTX population, responses to genotypic selection were similar to those for CMT, but were not so striking. This was expected because considerably more variation was accounted for by the marker loci for grain yield in CMT than in COTX. Ear height and ear number are highly correlated with grain yield and the responses of these two traits to selection for yield reflected this close association (Table 1).

Table 1.   Means for three traits following divergent genotypic (marker-locus) and phenotypic (mass) selection for grain yield in CMT (T232 x CM37) F_2 population.

Selection criterion	Trait means		
	Grain yield (g/plant)	Ear height	Ear number (cm)
Genotypic-positive	151.2	73.5	1.48
Genotypic-negative	107.7	47.1	1.20
Phenotypic-positive	151.7	68.5	1.43
Phenotypic-negative	122.4	57.8	1.28
Check-F_2 randomly mated	127.2	59.2	1.35
S.E.$_d$[a]	6.4	2.2	0.04

[a]   Standard error of mean difference.

To further analyze the marker-assisted selection response, marker allelic frequencies were calculated in each of the selected populations. In the CMT population, frequencies of the favorable alleles at the 15 marker loci averaged 0.38 greater in the population selected for increased (genotypic-positive) yield than in the one selected for decreased (genotypic-negative) yield (Edwards and Stuber, unpublished). The divergent selection showed differences in allelic frequencies ranging from 0.02 to 0.73 for individual loci. As expected, loci showing the greatest differences were those that accounted for the largest proportion of the phenotypic variation for grain yield.

Frequency differences between the divergent populations derived from phenotypic (mass) selection averaged 0.13, about one-third the 0.38 value for the genotypic selection in the CMT populations even though the overall selection responses were similar. These findings imply that loci in the unmarked regions of the genome, which would not be expected to respond in the marker-based scheme, contributed to the phenotypic selection response. Results in the COTX populations were similar to those for CMT, but magnitudes of frequency differences were less in COTX, which corresponds to the smaller response to selection as noted above.

From this investigation, it was concluded that marker-facilitated selection (based on 15 isozyme marker loci which probably represent no more than 30 to 40% of the genome) was as effective as phenotypic selection which would be expected to involve the entire genome. Furthermore, the results imply that a significant increase in the relative effectiveness of marker-based selection could be reasonably expected if the entire genome were marked with uniformly distributed loci (e.g., every 10 to 20 cM).

Edwards and Johnson, (1994) reported an investigation designed to develop improved inbred lines using QTL mapping information from two elite sweet corn breeding populations. A selection index involving 34 traits was created and index performance was then correlated with marker loci to determine which loci were associated with index performance. Significant gains in hybrid performance was evident in crosses of lines that had been generated by selection that was based only on marker genotypic information. They concluded that marker-facilitated selection allowed simultaneous gains for a number of traits, many of which require processing in a processing plant and are difficult and expensive to characterize.

Although results from these population selection studies conclusively demonstrated that quantitative traits, such as grain yield, can be manipulated using only genotypic (marker) data, Lande and Thompson (1990) have provided the theory and shown analytically that the maximum rate of improvement may be obtained by integrating both phenotypic and marker data. In their investigation several selection indices were derived that maximize the rate of improvement in quantitative characters under several schemes of marker-assisted phenotypic selection (including the use of phenotypic data on relatives). They also analyzed statistical limitations on the efficiency of marker-assisted selection, which included the precision of the estimated associations between marker loci and QTLs as well as sampling errors in estimating weighting coefficients in the selection indices.

Findings from the Lande and Thompson (1990) investigation showed that (on a single trait) the potential selection efficiency by using a combination of molecular and phenotypic information (relative to standard methods of phenotypic selection) depends on the heritability of the trait, the proportion of additive genetic variance associated with the marker loci, and the selection scheme. The relative efficiency of marker-assisted selection (MAS) is greatest for characters with low heritability if a large fraction of the additive genetic variance is associated with the marker loci. Limitations that may affect the potential utility of marker-assisted selection in applied breeding programs include: (1) the level of linkage disequilibria in the populations which affects the number of marker loci needed, (2) sample sizes needed to detect QTLs for traits with low heritability, and (3) sampling errors in the estimation of relative weights in the selection indices.

Following the analytical approach of Lande and Thompson (1990), which focused on first generation selection, the efficiency of MAS over several successive generations has been investigated using computer simulations (Zhang and Smith 1992, 1993; Gimelfarb and Lande 1994a, b, 1995; Wittaker et al. 1995). Results from these studies showed that MAS could be more efficient than purely phenotypic selection in quite large populations and for traits with relatively low heritabilities. The simulations also showed that additional genetic gain provided by MAS, when compared with purely phenotypic selection, rapidly decreased when several successive cycles of selection had occurred, and that MAS may become less efficient than phenotypic selection in the long term. This situation becomes more acute when the effects associated with markers are not re-evaluated at each generation.

More recently, Hospital et al. (1997) conducted computer simulations to study the efficiency of MAS based on an index combining the phenotypic value and the molecular score of each individual in the targeted population. In this case, the molecular score is computed from the effects attributed to markers by multiple regression of phenotype on marker genotype. Their results were consistent with earlier studies in that they also found that MAS may become less efficient than phenotypic selection in the long term. This is because the rate of fixation of unfavorable alleles at QTLs with small effects is higher under MAS than under phenotypic selection, and could be a consequence of the strong selection applied to QTLs with large effects under MAS in early generations. Hospital et al. (1997) point out, however, that this problem may be of little consequence in a practical breeding program because it takes place after a number of generations that is greater than the length of most breeding programs. They also indicate MAS 'is of interest' when it is compared with purely phenotypic selection over several successive generations in a breeding program involving an alternation of generations with and without phenotypic selection, if heritability is high. In this situation, the effects attributed to markers are better estimated in the phenotypic evaluation step, so that selection on markers-only without phenotypic evaluation is then efficient in the next generation, even for small population sizes. In addition, the cost of MAS in this context is greatly reduced.

Although recent advances in molecular genetics have promised to revolutionize agricultural practices, Lande and Thompson state 'There are, however, several reasons why molecular genetics can never replace traditional methods of agricultural improvement, but instead should be integrated to obtain the maximum improvement in the economic value of domesticated populations'. Their analytical results, as well as the more recent computer simulations and the limited empirical results, however, are encouraging and support the use of DNA-based markers to achieve substantial increases in the efficiency of artificial selection.

6. Heterosis and marker-facilitated enhancement of heterosis

As stated previously heterosis (or hybrid vigor) has been a major contributor to the success of the commercial maize industry and is a major component of the breeding strategies in many crops and horticultural plants. The term, heterosis, was coined by G.H. Schull and first proposed in 1914 (see Hayes 1952) and normally is defined in terms of F_1 hybrid superiority over some measure the performance of one or both parents. Possible genetic explanations for this phenomenon include true overdominance (i.e., single loci for which two alleles have the property that the heterozygote is truly superior to either homozygote), pseudo-overdominance (i.e., closely linked loci at which alleles having dominant or partially dominant advantageous effects are in repulsion linkage phase), or possibly certain types of epistasis.

An extensive field and laboratory investigation (which will be referred to as the 'heterosis' study) was reported by Stuber et al. (1992) in which a major goal was to obtain data that might lay a foundation for understanding the basis of this important phenomenon in maize. In this study, 76 marker loci (nine isozyme and 67 RFLP) that represented from 90 to 95% of the genome were used to identify and map QTLs contributing to heterosis in the cross between the elite inbred lines, B73 and Mo17. Experimental materials were derived by backcrossing 264 F_3 lines (developed by single-seed descent) to each of the two parental inbreds. Phenotypic data were recorded on the backcross progenies in six diverse environments (four in North Carolina, one in Illinois, and one in Iowa). Evaluations of phenotypic effects associated with these QTLs showed that for the trait, grain yield, the heterozygote had a higher phenotype than the respective homozygote (with only one exception), suggesting not only overdominant gene action (or pseudo-overdominance), but also that these detected QTLs contributed significantly to the expression of heterosis. There was little evidence for genotype (QTL) by environment interaction even though the mean yield levels for the environments sampled varied from 3950 to 7240 kg ha^{-1} (63 to 116 bushels per acre).

The goal of a companion study conducted with the 'heterosis' investigations was to use molecular markers to identify and locate genetic factors in two other elite inbred lines, Oh43 and Tx303, that would be expected to enhance the heterotic response for grain yield in the B73 x Mo17 single-cross hybrid (Stuber and Sisco 1991; Stuber 1994). Experimental materials were developed by hybridizing the two lines, Oh43 and Tx303, and selfing for two generations to create 216 single-seed descent F_3 lines. One plant from each of these lines was testcrossed to both B73 and Mo17, and then selfed to provide progeny for marker genotyping. Phenotypic data were recorded on the testcross progenies in the same six environments used for the 'heterosis' study. By making appropriate comparisons among phenotypic means for marker classes (from testcross progenies in this study and backcross progenies in the 'heterosis' study), six chromosomal segments were identified in Tx303 and another six in Oh43 that (if transferred into B73 and Mo17, respectively) would be expected to enhance the B73 x Mo17 hybrid response for grain yield (Figs. 1 and 2).

Three backcross generations (two marker-facilitated) were used for the transfer (introgression) of identified chromosomal segments into the target lines, B73 and Mo17. Marker analyses in the BC_2 and BC_3 families were used to select individuals for the succeeding backcross and selfing generations, respectively. Individuals were selected if they had the desired marker genotype in the vicinity of the donor segment to be transferred and the recipient line's marker genotype in the remainder of the genome. At least one marker was monitored on each of the 20 chromosome arms and usually two markers were genotyped in the vicinity of the donor segment being transferred. After the third backcross, selected plants were selfed for two generations to fix the desired donor segment. Marker genotyping followed similar procedures in the two backcross generations and the two selfing generations. However, if a marker locus became fixed in a line, that marker was not evaluated in that line in the succeeding generation and only segregating loci were analyzed. This reduced the laboratory analyses considerably.

After the second selfing generation, 141 BC_3S_2 modified B73 lines were identified for testcrossing to the original Mo17. Likewise, 116 BC_3S_2 modified Mo17 lines were targeted for testcrossing to the original B73. These 257 testcross hybrids were evaluated in replicated field plots at three locations in North Carolina (Stuber and Sisco 1991; Stuber 1994). Of the modified B73 x original Mo17 testcrosses, 45 (32%) yielded more than the check hybrid (normal B73 x normal Mo17) by at least one standard deviation. Only 15 (11%) yielded less than the check. Evaluations of modified Mo17 x original B73 testcrosses showed that 51 (44%) yielded more grain than the normal check hybrid and only 10 (<9%) yielded less than the check hybrid. The highest yielding testcross hybrids exceeded the check by 8 to 11% (565 to 753 kg ha^{-1}).

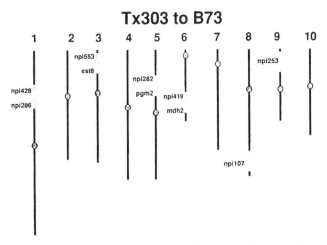

Figure 1. Chromosomal locations of segments identified in the inbred line T x 303 for transfer into inbred line B73.

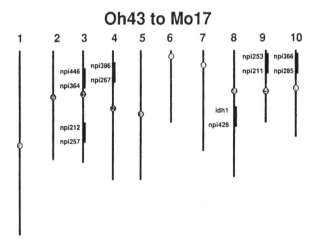

Figure 2. Chromosomal locations of segments identified in the inbred line Oh43 for transfer into inbred line Mo17.

Based on these initial evaluations, the better performing modified lines were selected for intercrossing and were designated as 'enhanced' lines. Fifteen 'enhanced' B73 lines were chosen for crossing with 18 'enhanced' Mo17 lines producing 93 hybrids that were evaluated in replicated field trials in North Carolina in 1993 (Stuber 1994, 1998). At a planting density of about 71560 plants ha^{-1}, only 15 of the test hybrids produced less grain than either of two check hybrids (normal B73 x normal Mo17 and a high yielding commercial hybrid, Pioneer 3165). However, six exceeded the checks by two standard deviations or more. The two highest yielding 'enhanced' B73 x 'enhanced' Mo17 hybrids exceed the checks by more than 15% (1381 to 1506 kg ha^{-1}).

These 'enhanced' B73 x 'enhanced' Mo17 hybrids were evaluated again in 1994, with some additional reciprocal crosses, increasing the number of test hybrids to 149 (Stuber 1994, 1998). Of these hybrids, 20 exceeded the checks (same as used in 1993) by two standard deviations or more, with the three best exceeding the checks by 12 to 15% (1130 to 1506 kg ha^{-1}). Although the rankings changed slightly over the two years of testing, the average yields over both years for the best yielding hybrids exceeded the check hybrids by 8 to 10% (628 to 1004 kg ha^{-1}). More importantly, the parental lines that showed superior general combining ability in 1993 also showed similar performance in 1994. Evaluations of the highest yielding 'enhanced' B73 x 'enhanced' Mo17 hybrids were again conducted in 1995 and 1997 in North Carolina and results corroborated those of the previous years. In addition, nine of these hybrids were evaluated at four locations in central Iowa by Pioneer Hi-Bred International in 1995. Of the nine, three exceeded the grain yields of the best Pioneer commercial check hybrids by 377 to 628 kg ha^{-1}. These results were surprising because all of the selections were based on evaluations in North Carolina.

Although the results from this marker-facilitated introgression experiment are striking, some might have anticipated a higher success rate. It should be noted, however, that a restricted amount of resources were available for the study. Only about 600 backcross families for each of the two lines could be evaluated. Therefore, not all of the targeted segments could be transferred and most of the 'enhanced' lines have only one to three introduced segments. Also, by following only one marker per chromosome arm to monitor the recovery of the recipient genome in the non-targeted regions, it is likely that some of the 'enhanced' lines have incorporated unidentified regions from the donor lines that might include deleterious genes. In addition, the 'tightness' of the linkage of the targeted segments with the associated marker loci could not be assessed. Therefore, deviations in the negative direction from the check could be due to a loss of linkage integrity in some cases.

Nevertheless, as with any applied plant breeding program, availability of resources places limitations on the size of the program and certain compromises must be made. This situation affects the magnitude and success of breeding programs based on marker techniques just as it does in programs using traditional approaches.

Although comparisons of the above responses with those that might be obtained with traditional backcrossing and selfing would be informative, resources were not available to conduct a concurrent traditional study. If a traditional backcrossing program had been conducted without any field testing until the final testcrosses were made, it seems reasonable to expect that the number of testcrosses inferior to the check would be equal to (or perhaps, even greater than) the number of superior yielding testcrosses. Personal communications with corn breeders have indicated that attempts to improve B73 for traits such as yield using traditional methods have frequently met with failure. Also, with a traditional breeding approach, the time required would probably be at least double to achieve the level of homozygosity that was achieved with three backcrossing generations and two selfing generations using the DNA-based marker approach.

Results from the transfer of the targeted segments from Tx303 into B73 and from Oh43 into Mo17 have demonstrated that marker-facilitated backcrossing can be successfully employed to manipulate and improve complex traits such as grain yield in maize. Not all of the six targeted segments have been successfully transferred into a single modified B73 or modified Mo17 line. There appears to be some indication that there may be no advantage in transferring more than two to four segments. In fact, there is some indication that there could be a disadvantage. Increasing the number of transferred segments may be replacing the recipient genome with an excessive amount of linked donor chromosomal segments that could cause a deleterious effect. Also, epistatic interactions between a larger number of introgressed segments may result in a negative effect. In addition, favorable epistatic complexes in coupling phase (e.g., between recurrent parent alleles) could be disrupted. Further evaluations are necessary to determine the effects of larger numbers of transferred segments.

7.   Marker facilitated introgression (backcrossing)

The preceding section outlined an investigation that demonstrated the use of marker-facilitated backcrossing, however, the procedures used were probably more complex than will be encountered in most plant breeding strategies. Several of the important traits that must be manipulated by plant breeders are more simply inherited than grain yield but still may involve the expression of several genes. For example, disease resistance frequently is controlled by only a few genetic factors. However, for many diseases resistance is considered to be a multigenic ('semi-quantitative') trait. For example, Bubeck et al. (1992) have shown that resistance to gray leaf spot in maize is based on four or five genes.

Normally, the first steps in a marker-based introgression program are the identification and mapping of the genes (more realistically, chromosomal segments) targeted for transfer to the desired line or strain. Experimental procedures for these steps have already been outlined in previous chapters. As suggested by Kearsey and Farquhar (1998), breeders may not need to know the locations of their targeted QTLs with very great accuracy. They will mainly be interested in incorporating (into elite lines) those QTLs which have a large effect, and which may have been missed by conventional selection procedures. Once the appropriate analyses have been performed to identify the genes of interest in the resource (perhaps, exotic) strain, as well as linkages to resource-specific marker alleles, repeated backcrossing to the recipient line or cultivar — choosing in each cycle only those backcross progeny with desired linked marker alleles — will provide effective introgression of the desired genes of interest into the recipient line. As was demonstrated in the previous section, marker-assisted selection against unwanted chromosomal regions from the donor (reducing linkage drag) will expedite the introgression process.

Table 2 (taken from Beckmann and Soller 1986) shows the frequency of a favorable allele following one to six backcross generations, with and without selection for a linked marker allele, including selection for a pair of bracketing marker alleles. It can be noted that the frequency of the introgressed favorable allele after three generations of backcrossing is 0.66 for the single marker and 0.85 for bracketing markers (with recombination of 0.40 between markers). These frequencies are in striking contrast to the 0.06 with no markers after three backcrosses and only 0.01 after six backcrosses. Also, with marker-assisted introgression, frequencies for the introgressed alleles are sufficiently high that two, three or even more alleles could be readily introduced and brought to fixation in a given breeding cycle. As stated by Beckmann and Soller (1986), 'Without marker assistance, a great many backcross products will have to be screened for the introduced trait, even in the case of one introduced allele, due to the extreme rarity of backcross products carrying desired exotic alleles'.

Table 2.   Frequency of a favorable allele after a given number of backcross generations, with and without selection for a linked marker allele or for a pair of markers bracketing the favorable allele, and with and without marker-assisted selection against the remaining donor genome (taken from Beckmann and Soller 1986).

Number of backcross generations	Marker-assisted selection				
	For favorable allele			Against remainder of donor genome	
	None	Single marker[a]	Marker bracket[b]	None	Full marker coverage[c]
	(Frequency of favorable allele)			(Proportion of recipient genome recovered)	
1	0.25	0.81	0.92	0.75	0.85
2	0.12	0.73	0.88	0.88	0.99
3	0.06	0.66	0.85	0.94	1.00
4	0.03	0.59	0.82	0.97	1.00
5	0.02	0.53	0.78	0.98	1.00
6	0.01	0.48	0.75	0.99	1.00

[a] Proportion of recombination between marker allele and linked favorable allele = 0.10.
[b] Proportion of recombination between the two markers of the bracket = 0.40.
[c] Assuming two markers per chromosome.

8.   Using near-isogenic lines (NILs) as a breeding tool

Although the results discussed in Section 6. of this chapter showed that the enhancement of lines B73 and Mo17 was successful, the procedure for the development of the 'enhanced' lines (NILs) was very inefficient and would not be recommended for a practical breeding program. That procedure depended on the identification of the targeted segments (containing the putative QTLs) prior to transfer to the recipient lines. In our maize research program at Raleigh, NC, we have outlined, and have tested, a marker-based breeding scheme for systematically generating superior lines without any prior identification of QTLs in the donor source(s). The identification and mapping of QTLs in the donor is a bonus obtained when the derived NILs are evaluated.

Choice of the donor usually will be based on prior knowledge of its likely potential for providing superior genetic factors, and, in maize, may involve appropriate heterotic relationships.

The procedure involves the generation of a series of NILs by sequentially replacing segments of an elite line (the recipient genome that is targeted for improvement) with corresponding segments from the donor genome. The objective is to generate a set of NILs containing, collectively, the complete genome of the donor source, with each NIL containing a different chromosomal segment from the donor. Marker-facilitated backcrossing, followed by marker-facilitated selfing to fix introgressed segments, is used to monitor the transfer of the targeted segments from the donor and to recover the recipient genotype in the remainder of the genome. The number of backcrosses required will depend on the number of evaluations that can be made in the marker laboratory. As few as two backcross and one selfing generation will suffice if the laboratory resources are adequate to handle the required number of plant samples.

In maize, the NILs are then crossed to an appropriate tester(s) to create hybrid test-cross progeny that are evaluated in replicated field trials (with appropriate checks) for the desired traits. (The NILs would be tested per se for crops such as soybean and wheat.) The superior performing testcrosses will be presumed to have received donor segments that contain favorable QTLs. Thus, QTLs are mapped by function, which should be an excellent criterion for QTL detection. The breeding scheme not only creates 'enhanced' elite lines that are essentially identical to the original line, but it also provides for the identification and mapping of QTLs as a fringe benefit with no additional cost. Obviously, the scheme is based on having a reasonably good marker map with alternate alleles in the donor and recipient lines.

This breeding strategy should be an excellent procedure for tapping into the potential of exotic germplasm. Furbeck (1993) used this procedure to develop a set of 149 NILs using the elite line, Mo17, as the recipient. An exotic population, derived from the Brazilian racial collection Cristal (MGIII) and the Peruvian collection Arizona (AYA41), was used as the donor. Fig. 3 shows some of the significant introgressed segments and traits involved. Using this procedure, positive segments from the exotic (such as the segment associated with *Phi1* on chromosome 1) are immediately available in an adapted background (Mo17) for further breeding use. Also, and equally important, segments with negative effects — such as those associated with *Dia1* and *Dia2*, both -596 kg ha^{-1} (-9.5 bushels/acre) — are eliminated. Moreover, because each segment was incorporated independently, the detection of positive effects is not biased by adjacent negative segments — e.g., *Phi1*, +659 kg ha^{-1} (+10.5 bushels/acre), and *Dia2*, -596 kg ha^{-1} (-9.5 bushels/acre), on chromosome 1. These adjacent segments would likely segregate together in a traditional breeding program and would effectively cancel each other.

A somewhat similar marker-facilitated backcrossing scheme was used by Brown et al. (1989) to transfer isozyme-marked segments from wild barley (*Hordeum spontaneum*) into an elite barley (*H. vulgare*) cultivar. Each of the 84 NILs was then made homozygous for a single isozyme-marked segment with two selfing generations. After evaluating these lines, per se, in the field, they also concluded that this was a useful approach for identifying QTLs for improving yield in divergent germplasm.

Eshed and Zamir (1995) have also very effectively used the NIL breeding procedure to extract favorable genetic factors from a wild species of tomato.

A major advantage of this NIL approach is that once a favorable QTL has been identified, it is already fixed in the elite recipient line and the breeding work is essentially completed. In addition, lines with favorable QTL alleles can be easily maintained and then used for pyramiding several favorable QTL alleles into a single line. A possible disadvantage of this approach is that favorable epistatic complexes between QTLs may not be identified. However, there is little experimental evidence documenting the occurrence of such epistatic interactions.

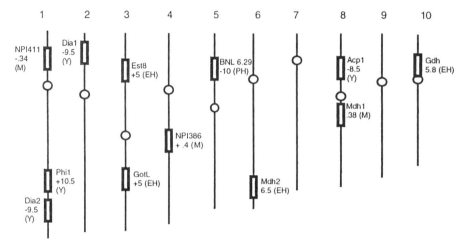

Figure 3.   Maize chromosome map showing locations of positive and negative effects on several traits derived from segments transferred from an exotic population (Cristal x Arizona) into the elite inbred line Mo17. Effects measured in testcrosses to inbred line B73. Y = yield (bushels/acre); M = moisture (%); EH = ear height (cm); PH = plant height (cm).

9.   Conclusions

The effectiveness of molecular-marker technology for identifying and mapping QTLs has been widely demonstrated in numerous crop plants. Also, the positive results from a limited number of marker-facilitated selection and introgression studies are encouraging for transferring desired genes between breeding lines and, thus, increasing the precision and efficiency of plant breeding. It is hoped that markers also should expedite the acquisition of important genes from exotic populations or from wild species.

In maize, for example, most mapping studies have been conducted in populations derived from domestic lines. Efforts in exotic maize populations have been less effective and frequently have met with considerable frustration (Koester 1992). This has been particularly true in the use of RFLPs because of the large number of marker variants and multiple banding patterns that have been difficult to interpret. Differentiating between multiple alleles at a single marker locus versus alleles associated with duplicate loci frequently is nearly impossible. Hopefully, with new marker technology (e.g., PCR-based techniques), this limitation will be overcome and the use of exotic germplasm can be tapped as a vast source of new genes in maize breeding, as well as in other crops.

In most QTL identification studies, rather stringent probability levels have been set so that there is a low risk in making Type I errors (i.e., false positives). Thus, only QTLs with major effects are identified as being significant. It should be pointed out that these would be the QTLs with high heritabilities, are easily manipulated using traditional breeding practices, and may already be fixed in many breeding lines. It may prove to be more productive, therefore, to use marker technology as a means for placing greater emphasis on those QTLs (or chromosomal regions) that show only relatively minor effects.

There have been some attempts to reduce the size of the regions identified as containing major QTLs through 'fine-mapping' such as the study reported by Paterson et al. (1990). Many have envisioned this approach as an initial step in identifying single genes that might ultimately be manipulated using transformation (recombinant DNA) technology. One might speculate, however, that this approach could be counter-productive if these major QTLs contain a number of genes that have evolved as highly integrated complexes over many cycles of selection. For example, the region on the short arm of chromosome 5, bracketed by isozyme markers *Amp3* and *Pgm2*, has been found to have a very significant effect on grain yield in several maize populations (Stuber 1992; Stuber et al. 1992). This region has been targeted for fine-mapping and it was found that there were at least two smaller QTLs (Graham et al. 1997). Although earlier analyses (Stuber et al. 1992) suggested that the major QTL identified in this region acted in an overdominant fashion, effects at these two smaller QTLs appear to act in a dominant manner, and are in repulsion phase linkage. Thus, the effects at these two QTLs support the dominance theory of heterosis in this chromosomal region. It will be of interest to determine whether chromosomal segments such as this can be further improved with marker technology (e.g., by placing the favorable dominant factors in coupling phase linkage in one, or both, of the parental lines). In practical breeding programs, manipulation of large segments (such as the one identified in maize on chromosome 5) may be simpler, and more effective in the short term, than extracting and manipulating individual genes.

It should be stressed that little is known regarding the stability of QTL alleles when transferred to different genetic backgrounds and when evaluated in varying environments. Tanksley and Hewitt (1988) illustrated the potential dangers of establishing breeding programs based on associations of markers with quantitative traits prior to evaluations of the identified factors in appropriate genetic backgrounds. Although there is some evidence for the interaction of QTLs with environments, results tend to be contradictory. Stuber et al. (1992) found little evidence for such interaction in maize. In a major follow-up study in a population generated from the cross of maize inbred lines B73 and Mo17 (Stuber et al. 1996; LeDeaux and Stuber 1997; Stuber 1998; and unpublished data), there was little evidence for QTL by environment interaction even when very severe stress conditions were imposed on the field evaluations. These data would suggest that breeders and geneticists can rely on mapping data from favorable environments for breeding materials adapted to stress environments. This suggestion should be viewed with caution, however, because the results may reflect the fact that the parental lines (used in the study) have been selected for stability over a wide range of environments, and it may not be prudent to extrapolate to more widely variable and more divergent materials.

Each new investigation using DNA-based marker technology as a tool for plant geneticists and plant breeders adds evidence to the projected role of markers not only for identifying useful genes (or chromosomal segments) in various germplasm sources but also for transferring these genes into desired cultivars or lines. As laboratory analyses become more automated, the cost (now one of the major deterrents in the use of marker technology) will decrease and the use of DNA-based markers for improving multigenic traits will be greatly expanded in the future.

10. References

Abler, B.S.B., Edwards, M.D. and Stuber, C.W. (1991) Isoenzymatic identification of quantitative trait loci in crosses of elite maize hybrids. Crop Sci. 31: 267-274.

Beckmann, J.S. and Soller, M. (1986) Restriction fragment length polymorphisms in plant genetic improvement. Oxford Surv. Plant Mol. Cell Biol. 3: 197-250.

Bernardo, R. (1994) Prediction of maize single-cross performance using RFLPs and information from related hybrids. Crop Sci. 34: 20-25.

Brown, A.H.D., Munday, J. and Oram, R.N. (1989) Use of isozyme-marked segments from wild barley (*Hordeum spontaneum*) in barley breeding. Plant Breed. 100: 280-288.

Bubeck, D.M., Goodman, M.M., Beavis, W.D. and Grant, D. (1993) Quantitative trait loci controlling resistance to gray leaf spot in maize. Crop Sci. 32: 838-847.

Comstock, R.E. (1978) Quantitative genetics in maize breeding. In: D. Walden (ed.), Maize Breeding and Genetics, pp. 191-206. Wiley, New York.

Darrah, L.L. and Hallauer, A.R. (1972) Genetic effects estimated from generation means in four diallel sets of maize inbreds. Crop Sci. 12: 615-621.

Dudley, J.W., Saghai Maroof, M.A. and Rufener, G.K. (1992) Molecular marker information and selection of parents in corn breeding programs. Crop Sci. 32: 301-304.

Duvick, D.N. (1984) Genetic contributions to yield gains of U.S. hybrid maize, 1930 and 1980. In: W.R. Fehr (ed.), Genetic Contributions to Yield Gains of Five Major Crop Plants, pp. 15-47. Crop Science Society of America, Madison, WI.

Duvick, D.N. (1997) What is yield? In: G.O. Edmeades, B. Banziger, H.R. Mickelson and C.B. Pena-Valdivia (eds.), Proc. of Symposium on Developing Drought- and Low N-Tolerant Maize, CIMMYT, El Batan, Mexico, pp. 332-335.

East, E.M. and Hayes, H.K. (1911) Inheritance in maize. Conn. Agr. Expt. Sta. Bull. 167.

Eathington, S.R., Dudley, J.W. and Rufener, G.K. (1995) Evaluation of early-generation selection utilizing marker-QTL associations and phenotypic data. Proc. 31st Annual Illinois Corn Breeders' School. University of Illinois at Urbana-Champaign.

Edwards, M. and Johnson, L. (1994) RFLPs for rapid recurrent selection. Proc.of Symposium on Analysis of Molecular Marker Data. Am. Soc. Hort. Sci. and Crop Sci. Soc. Am., Corvallis, OR, pp. 33-40.

Edwards, M.D., Stuber, C.W. and Wendel, J.F. (1987) Molecular-marker-facilitated investigations of quantitative-trait loci in maize. I. Numbers, genomic distribution, and types of gene action. Genetics 116: 113-125.

Edwards, M.D., Helentjaris, T., Wright, S. and Stuber, C.W. (1992) Molecular-marker-facilitated investigations of quantitative trait loci in maize. IV. Analysis based on genome saturation with isozyme and restriction fragment length polymorphism markers. Theor. Appl. Genet. 83: 765-774.

Emerson, R.A. and East, E.M. (1913) The inheritance of quantitative characters in maize. Nebr. Agr. Expt. Sta. Res. Bull. 2.

Eshed, Y. and Zamir, D. (1995) An introgression line population of *Lycopersicon pennellii* in the cultivated tomato enables the identification and fine mapping of yield-associated QTL. Genetics 141: 1147-1162.

Frei, O.M., Stuber, C.W. and Goodman, M.M. (1986a) Use of allozymes and genetic markers for predicting performance in maize single cross hybrids. Crop Sci. 26: 37-42.

Frei, O.M., Stuber, C.W. and Goodman, M.M. (1986b) Yield manipulation from selection on allozyme genotypes in a composite of elite corn lines. Crop Sci. 26: 917-921.

Furbeck, S.M. (1993) The development and evaluation of molecular-marker derived near isogenic lines to study quantitative traits in maize. Ph.D. thesis, North Carolina State University, Raleigh (Diss. Abstr. DA9330287).

Gimelfarb A. and Lande R. (1994a) Simulation of marker-assisted selection in hybrid populations. Genet. Res. 63: 39-47.

Gimelfarb A. and Lande R. (1994b) Simulation of marker-assisted selection for non-additive traits. Genet. Res. 64: 127-136.

Gimelfarb A. and Lande R. (1995) Marker-assisted selection and marker-QTL associations in hybrid populations. Theor. Appl. Genet. 91: 522-528.

Gonella, J.A. and Peterson, P.A. (1978) Isozyme relatedness of inbred lines of maize and performance of their hybrids. Maydica 23: 55-61.

Graham, G.I., Wolff, D.W. and Stuber, C.W. (1997) Characterization of a yield quantitative trait locus on chromosome five of maize by fine mapping. Crop Sci. 37: 1601-1610.

Guse, R.A., Coors, J.G., Drolsom, P.N. and Tracy, W.F. (1988) Isozyme marker loci associated with cold tolerance and maturity in maize. Theor. Appl. Genet. 76: 398-404.

Hadjinov, M.I., Scherbak, V.S., Benko, N.I., Gusev, V.P., Sukhorzheuskaya, T.B. and Voronova, L.P. (1982) Interrelationships between isozyme diversity and combining ability in maize lines. Maydica 27: 135-149.

Hayes, H.K. (1952) Development of the heterosis concept. In: J. Gowen (ed.), Heterosis, pp. 49-65. Iowa State College Press, Ames, IA.

Heidrich-Sobrinho, E. and Cordeiro, A.R. (1975) Codominant isoenzymatic alleles as markers of genetic diversity correlated with heterosis in maize (*Zea mays* L.). Theor. Appl. Genet. 46: 197-199.

Hospital, F., Moreau, L., Lacoudre, F., Charcosset, A. and Gallais, A. (1997) More on the efficiency of marker-assisted selection. Theor. Appl. Genet. 95: 1181-1189.

Hunter, R.B. and Kannenberg, L.W. (1971) Isozyme characterization of corn (*Zea mays* L.) inbreds and its relation to single cross hybrid performance. Can. J. Genet. Cytol. 13: 649-655.

Johnson, G.R. and Mumm, R.H. (1996) Marker assisted maize breeding. Proc. 51st Annual Corn and Sorghum Industry Research Conf., American Seed Trade Assoc. 51: 75-84.

Kahler, A.L. (1985) Associations between enzyme marker loci and agronomic traits in maize. Proc. 40th Annual Corn and Sorghum Industry Research Conf., American Seed Trade Assoc. 40: 66-89.

Kearsey, M.J. and Farquhar, A.G.L. (1998) QTL analysis in plants; where are we now? Heredity 80: 137-142.

Koester, R.P. (1992) Identification of quantitative trait loci controlling maturity and plant height using molecular markers in maize (*Zea mays* L.). Ph.D. thesis, North Carolina State University, Raleigh, NC.

Koester, R.P., Sisco, P.H. and Stuber C.W. (1993) Identification of quantitative trait loci controlling days to flowering and plant height in two near isogenic lines of maize. Crop Sci. 33: 1209-1216.

Lande, R. and Thompson, R. (1990) Efficiency of marker-assisted selection in the improvement of quantitative traits. Genetics 124: 743-756.

Lamkey, K.R., Hallauer, A.R. and Kahler, A.L. (1987) Allelic Differences at enzyme loci and hybrid performance in maize. J. Hered. 78: 231-234.

LeDeaux, J.R. and Stuber, C.W. (1997) Mapping heterosis QTLs in maize grown under various stress conditions. Symposium on The Genetics and Exploitation of Heterosis in Crops. Abstracts p. 40-41.

Li, Z., Pinson, S.R.M., Park, W.D., Paterson, A.H. and Stansel, J.W. (1997) Epistasis for three grain yield components in rice (*Oryza sativa* L.). Genetics 145: 453-465.

Melchinger, A.E., Lee, M., Lamkey, K.R. and Woodman, W.L. (1990) Genetic diversity for restriction fragment length polymorphisms: Relation to estimated effects in maize inbreds. Crop Sci. 30: 1033-1040.

Paterson, A.H., Deverna, J.W., Lanini, B. and Tanksley, S.D. (1990) Fine mapping of quantitative trait loci using selected overlapping recombinant chromosomes, in an interspecies cross of tomato. Genetics 124: 735-742.

Paterson, A.H., Damon, S., Hewitt, J.D., Zamir, D., Rabinowitch, H.D., Lincoln, S.E., Lander, E.S. and Tanksley, S.D. (1991) Mendelian factors underlying quantitative traits in tomato: Comparison across species, generations, and environments. Genetics 127: 181-197.

Peng, J.Y., Glaszmann, J.C. and Virmani, S.S. (1988) Heterosis and isozyme divergence in indica rice. Crop Sci. 28: 561-563.

Pollak, L.M., Gardner, C.O. and Parkhurst, A.M. (1984) Relationships between enzyme marker loci and morphological traits in two mass selectd maize populatiosn. Crop Sci. 24: 1174-1179.

Ragot, M., Sisco, P.H., Hoisington, D.A. and Stuber, C.W. (1995) Molecular-marker-mediated characterization of favorable exotic alleles at quantitative trait loci in maize. Crop Sci. 35: 1306-1315

Russell, W.A. (1984) Agronomic performance of maize cultivars representing different eras of maize breeding. Maydica 29: 375-390.

Smith, J.S.C. and Smith, O.S. (1989) The use of morphological, biochemical, and genetic characteristics to measure distance and to test for minimum distance between inbred lines of maize (*Zea mays* L.). (Mimeo of paper presented at UPOV Workshop, Versailles, France,

October 1989). Pioneer Hi-Bred International, Inc., Johnston, IA.

Stuber, C.W. and Edwards M.D. (1986) Genotypic selection for improvement of quantitative traits in corn using molecular marker loci. Proc. 41st Annual Corn and Sorghum Industry Research. Conf., American Seed Trade Assoc. 41: 70-83.

Stuber, C.W. and Moll, R.H. (1972) Frequency changes of isozyme alleles in a selection experiment for grain yield in maize (*Zea mays* L.). Crop Sci. 12: 337-340.

Stuber, C.W. and Sisco, P.H. (1991) Marker-facilitated transfer of QTL alleles between elite inbred lines and responses in hybrids. Proc. 46th Annual Corn and Sorghum Industry Research Conf., American Seed Trade Assoc. 46: 104-113.

Stuber, C.W., Moll, R.H., Goodman, M.M., Schaffer, H.E. and Weir, B.S. (1980) Allozyme frequency changes associated with selection for increased grain yield in maize (*Zea mays* L.). Genetics 95: 225-236.

Stuber, C.W., Goodman, M.M. and Moll, R.H. (1982) Improvement of yield and ear number resulting from selection at allozyme loci in a maize population. Crop Sci. 22: 737-740.

Stuber, C.W., Edwards, M.D. and Wendel, J.F. (1987) Molecular marker-facilitated investigations of quantitative trait loci in maize. II. Factors influencing yield and its component traits. Crop Sci. 27: 639-648.

Stuber, C.W. (1992) Biochemical and molecular markers in plant breeding. In: J. Janick (ed.) Plant Breeding Reviews 9: 37-61.

Stuber, C.W., Lincoln, S.E., Wolff, D.W., Helentjaris, T. and Lander, E.S. (1992) Identification of genetic factors contributing to heterosis in a hybrid from two elite maize inbred lines using molecular markers. Genetics 132: 823-839.

Stuber, C.W. (1994) Success in the use of molecular markers for yield enhancement in corn. Proc. 49th Annual Corn and Sorghum Industry Res. Conf., American Seed Trade Assoc. 49: 232-238.

Stuber, C.W. (1998) Case history in crop improvement: Yield heterosis in maize. In: Molecular Analysis of Complex Traits, A.H. Paterson (Ed.) CRC Press, Inc. pp 197-206.

Stuber, C.W., Graham, G. and Senior, M.L. (1996) Effects of environmental stresses on mapping heterosis QTLs in maize. Plant Genome IV Abstr. p. 23.

Tanksley, S.D. and Hewitt, J. (1988) Use of molecular markers in breeding for soluble solids content in tomato — a re-examination . Theor. Appl. Genet. 75: 811-823.

Wittaker, J.C., Curnow, R.N., Haley, C.S. and Thompson, R. (1995) Using marker-maps in marker-assisted selection. Genet. Res. 66: 255-265.

Zhang, W. and Smith, C. (1992) Computer simulation of marker-assisted selection utilizing linkage disequilibrium. Theor. Appl. Genet. 83: 813-820.

Zhang, W. and Smith, C. (1993) Simulation of marker-assisted selection utilizing linkage disequilibrium: the effects of several additional factors. Theor. Appl. Genet. 86: 492-496.

8. Information systems approaches to support discovery in agricultural genomics

BRUNO W. S. SOBRAL, MARK E. WAUGH, *and* WILLIAM D. BEAVIS
Virginia Bioinformatics Institute (0477) 1750 Kraft Drive
Corporate Research Center Bldg. 10, Suite 1400 Blacksburg, VA 24061
<sobral@vt.edu>

Contents

1. Introduction

Reductionist approaches to biological questions have provided important insights to genetic mechanisms and genome structures. These approaches have also led to the engineering of technologies that generate data at rates that far exceed our ability to comprehend their meaning. If biologists are going to address the fundamental questions concerning the complexity that underlies growth, development and phenotypic variability, then there is an unprecedented need to acquire, understand, manipulate, and exploit high-value genomic information. Further, for genomics to deliver the promise of using biological processes to make agricultural systems more efficient, sustainable, and environmentally friendly, various types of biological data will need to be integrated to reveal the functional relationships between DNA, RNA, proteins, environment and phenotypes. Such integrative approaches will rely heavily on development of information systems (IS) and data analysis methods that will require the same rigor and

R.L. Phillips and I.K. Vasil (eds.), DNA-Based Markers in Plants, 139 - 166.
© 2001 *Kluwer Academic Publishers. Printed in the Netherlands.*

resources that have been applied to development of laboratory experimental protocols. In addition, software and process engineering will need to be applied to information systems development much in the same way as engineering principles have been applied to data production in biological laboratories, resulting in a shift from cottage industry to industrial scale biological data factories.

Systems approaches to understanding complex biological systems are required to support our rapid rate of progress at grinding up organisms and acquiring data on various parts of those systems (processes as well as components). It is not at all clear that simply storing all the data from studying the parts will enable system-level questions to be addressed. In fact, the current situation with most biological databases suggests that if a systems approach is not taken, the result will be the drowning of the biological community in mountains of data that are not easily accessed and utilized in the pursuit of fundamental and applied biological questions. The information overload or 'data poisoning' as we call it here is already evident in private sector agribusiness where such large-scale data sets have been developed.

Databases provide the scientific community with *memory*. This is similar to how scientific literature currently provides this function to the research community. However, with the increase in the size and complexity of data sets, it seems more and more likely that this memory will need to have more electronic components if we are to avoid data poisoning.

To support US genomic data production in crops and livestock species, a number of species or organismal databases (SDBs) were developed in the early 1990s. Now that the first challenges of storage of mostly genetic-map-based data have been met with SDBs, it is important to recognize that digital information can quickly lose its meaning. This occurs because, during the process of encoding and massaging, the biological or scientific rules can be hidden in application programs understood only by their developers (the 'black box' syndrome).

In much the same way as in complex biological systems, the context of digital information is the environment in which the information resides, how it is used, and what other information relates to it. For any given user, the more context available, the more easily that user will be able to decipher and apply digital information. Similarly, the more context that is made available for any new information, the more easily that information will be understood and applied.

Context is not limited to the category (i.e., is it expression data or phenotypic data) of the information. In many cases it is necessary to define the associated rules that give information meaning. In genomics these rules define the behavior associated with types of data stored in the system. For example, there are rules that define what information is needed to create a genetic map, and what are the valid states of genetic maps.

Because of genomics, parallel gene expression analysis technologies, proteomics, metabolomics, and as yet undiscovered (and unnamed) high-throughput technologies to produce biological data, we have succeeded in breaking organisms into their component parts and then generating large amounts of data concerning those components (or processes in which they are involved). However, to increase our understanding of biological systems, a systems approach must be taken to reassemble and interpret those data meaningfully. Thus, Integrated Biological Information Systems (IBIS) will be needed.

Among many other requirements, IBIS should be designed to provide a *nervous system* for multidisciplinary teams to work together effectively. Knowledge workers in agricultural genomics are currently suffering from a deluge of data and this situation will only become worse with the full application of new technologies to key agricultural species, their relatives, and key model systems such as *Arabidopsis thaliana*. Not only are there more data, but data are being stored in (and thus need to be acquired from) increasingly varied and heterogeneous sources.

Epistemologically, there are two main challenges in problem solving: failure to use known information and introduction of unnecessary constraints (Rubinstein 1975). "The moment there were two different operating systems or two different languages or two different network interfaces, unnecessary constraints were introduced. These constraints led to failure to use known information" (Ryan 1996). Fortunately, IS are evolving to provide an integrative role in the business world, thus providing the agricultural and biological research community with useful information technologies, products and experience.

It is time for public IBIS to be developed such that researchers from diverse scientific disciplines can effectively manage and integrate information from heterogeneous, distributed DBs. Integration of biological data from SDBs will be required to enable the power of discovery and the research and resource synergy associated with comparative biology. But IBIS will have to be more than simply information archives. Ideally, IBIS should enable scientific hypothesis generation and discovery *in silico*, thereby pushing the limits of discovery and innovation in genomics. In effect, IBIS are needed to facilitate our understanding of the relationship between genotype and phenotype. An agriculturally oriented IBIS should capture data from agriculturally relevant organisms and model systems. In addition, *IBIS should have a cohesive interface permitting scientists to store, access, view and analyze diverse types of biological data and enable queries across diverse data resources in a manner that is transparent to the user*. In this sense, IBIS interfaces might look and feel like an Internet browser, through which a user connects and accesses diverse types of data distributed worldwide.

The recent funding of the National Science Foundation's (NSF) Plant Genome Initiative by the US Congress will result in a major ramp-up in the scope and volume of production of agriculturally significant biological data. In addition, the NSF Request for Proposals (RFP) was novel to the US plant research community in its structure, as it is aimed at financing primarily collaborative, multidisciplinary teams of plant scientists. Thus NSF's new program is stimulating a positive and much-needed change in the way plant scientists interact with each other, by rewarding them for extensive collaboration across institutions and disciplines. Unfortunately, with the exeption of the NSF BDI there has not been a similar interest by most federal funding agencies that fund biological research with respect to data management. This is despite significant lessons that have been extracted from the Human Genome Project and from agribusinesses that have invested heavily in massive data production only to find that they use about 1% of the resulting data. As a result of increased funding levels, many leading plant research groups are pursuing massive data production without adequate support for data management, analysis, sharing, and publication. Many of the proposals funded by

NSF in the 1997 RFP contemplate creation of web pages or use of single graduate or post-doctoral students to handle the outputs from large-scale data generation, using a SDB approach. If this trend continues, it is clear that there will be a massive phenomenon of 'data poisoning'. As a result of data poisoning, much of the data will not be effectively used in downstream applications that yield useful new technologies and products for society. Thus, there is a clear need to think about data management with a long-term perspective in mind.

To respond to some of these challenges, in fulfillment of its mission (http://www.ncgr.org), in April of 1998 the National Center for Genome Resources (NCGR) convened a Scientific Meeting on IBIS from April 17-19 1998 in Santa Fe New Mexico. Thirteen agricultural genomics scientists from the public and private sectors were invited and attended. The meeting had four major objectives:

1. Establish types of data to be supported/included;
2. Establish key 'use cases' of the system;
3. Establish key interfaces; and
4. Constitute a rotating Scientific Advisory Committee for interaction and evaluation of progress and requirements.

Based on that meeting and subsequent interactions with diverse members of the scientific community, as well as the development of proof-of-principle projects, a number of results have been obtained. Herein some of those results are reported and some of the lessons learned are discussed.

2.   Information technology and terminology

As with other types of data, biological data have a broad scope in meaning. A database is a structured collection of data, which are intentionally brought together and made persistent and suited for querying (deCaluwe et al. 1997).

Knowledge may be viewed as the combination of facts and their logical inter-relationships. Knowledge can be stored in a database; in fact IBIS should be designed to be able to evolve toward knowledge management.

Generically, database needs can be categorized as data management, object management, and knowledge management (Cattell 1992, Ullman 1989, deCaluwe 1997). Ideally, IBIS should address all three of these needs. Typically, data management is the only aspect addressed by most file (generally hierarchical) and relational database management systems (DBMS). A legacy system is any system currently in production (Ryan 1996).

In object-oriented programming, objects are central to the programming paradigm. Objects are used to represent real world or abstract entities. Objects consist of encapsulated data structures (deCaluwe et al. 1997); thus an object-oriented software component presents a well-defined interface to whomever or whatever wants to use it. Objects organize data, have behavioral capacities and can help infuse data with meaning (Ryan 1996). Objects can cooperate with each other without having been specifically designed to do so. This provides potential for the system to evolve in response to changing needs and data producing methods of biological researchers.

Object management provides the ability to manage complex data structures, often from multiple database sources. Knowledge management provides the ability to integrate objects and rules in order to draw inferences about relationships among objects. Relationships discovered among biological objects may be used as testable hypotheses for laboratory/field experiments, with results feeding back into the system. By combining iterative object-oriented software development principles with iterative data acquisition and hypothesis building within the system by biological researchers, IBIS may be able to evolve from data to knowledge management.

3.   Some requirements

3.1.   Supporting distributed collaborators in real time

Competitive US public funding of Plant Genome research is being administered through NSF's Plant Genome Initiative and the USDA National Research Initiative (NRI). Most plant genomic projects are collaborative, typically involving multi-disciplinary teams and geographically distributed members residing at multiple institutions. Thus, for the foreseeable future, distributed data generation and data curation should be expected. Indeed, most of the results of public funding have been stored and curated in distributed legacy databases (Table 1).

Although it is generally recognized that integrating the information generated by the various projects is desirable (Mary Clutter, NSF, personal communication), there has been no desire to centralize data curation for the varied projects. In addition, existing SDB resources are simply too valuable to throw away and rewrite from scratch. Most SDBs that support agricultural research in the US now publish data through WWW servers. However, integration is not yet possible because WWW browsers do not process queries across multiple DBs. Furthermore, the existing systems are composed of hierarchical, relational and object-oriented database architectures. Because of the geographic and disciplinary distances and the heterogeneity of DB architectures among collaborators and DB resources, there is a very large unmet need for "real time" support for these new genome projects.

From the biologist's (in a broad sense, to include breeders as well as molecular and organismal biologists, and biochemists, et al.) perspective there are a number of exploratory functions and activities that can be articulated and will need to be supported early in the development process:

- Biologists should be able to submit research results to multiple DBs in a single transaction;
- Biologists should be able to enter data into DBs in a manner that is intuitive for their respective discipline and that is cohesive with their normal work flow (i.e., the additional overhead of getting data into IBIS should be non-existent or at least minimized to a large degree);
- Manual browsing, where users will tend to look at objects and explore their general properties interactively;

- Sweeping searches, extending to major portions of the IBIS (some comparative queries would likely fall into this category, for example);
- Searches extending across the existing systems;
- Creation of personal, community, biological process, or organismally specific data subsets, requiring the capacity to make persistent objects that were selected; and
- Creation of new objects through compounding of existing ones.

These activities suggest that IBIS will need to provide:
- Integration of different data types within an organism,
- Support for data modeling components via visualization tools, and
- The ability to compare genomic information across organisms and data types.

IBIS's data integration model requires a software architecture providing persistence for objects and object-oriented applications on top of distributed existing data resources. Considering the very short cycles of innovation observed in object-oriented technologies (Keller et al. 1996) and in the engineering of new high-throughput laboratory technologies, modularization will be important to assure creation of an evolutionary system. Interfaces close to standards enable the exchange of custom-made parts with commercially available software once better solutions appear. Clearly, a benefit from a component-based design approach and development and application of standards is that collaborators working on the same project (or related projects) may view the same information from multiple sources, at the same time and in the same way.

As an example, consider the development of a prototypic system to support distributed genome projects in real time (Waugh et al. 2000). This prototype was developed by NCGR for the *Phytophthora* Genome Initiative (PGI — www.ncgr.org/research/pgi). PGI is a distributed, international genome project focused on the biology of interactions between *Phytophthora* and its hosts. To collaborating scientists, the actual location of data should be invisible, i.e., encapsulated, so that it seems as if it is locally curated. None-the-less decisions need to be made concerning the physical location of data and how it should be accessed.

Distributed genomics efforts, in which IBIS will provide a 'virtual laboratory' (NSF RFP for Plant Genome Initiative), have a key characteristic: there are many interactions among individuals, research teams, databases, analysis tools, and visualization tools. Consequently, important features of IBIS must include:
- Responsiveness to change and capability for evolutionary growth;
- Provision of a unified infrastructure for software developers' needs;
- End-user-control (i.e., systems should mold to user's needs rather than forcing users to adapt to fixed templates);
- Provision of mechanisms to capture scientific knowledge in ways that are essentially technology-independent;
- Ability to easily transfer data from one architecture to another, because of legacy systems;
- Multi-platform availability and interoperatibility of IBIS's components (as seen by the end user); and
- Simplified migration to future platforms (for DBs, etc.).

Table 1. Existing database resources.

DBName	Organism common name	Scientific name	DBMS	Data/Comments
Arabidopsis		*Arabidopsis thaliana*		Highly curated interactive genetic and physical maps; nucleotide and protein sequence information; sequence analysis software genotype and phenotypes of a large collection of mutants and wild types; links to stock centers; citations.
Alfagenes	Alfalfa	*Medicago sativa*	AceDB v: 4.1	Data #records: Cultivar 364, Environment 128, Germplasm 6352, Locus 499, Map 33, Metabolite 28 (medicarpin pathway), Pathway 10 (Medicarpin) Restriction Endonuc. 344, Sequence 195 (all known), Species 62, Trait 148, citations
BeanGenes	Beans	*Phaseolus* and *Vigna* species	AceDB v: 4.1	Molecular maps of *P. vulgaris* (RFLP; RFLP/RAPD); RFLP maps of mungbean and cowpea; loci, probe and gene information as well as cultivar information and Pathology.
CoolGenes	peas, lentil, chickpea, sweetpea, fava bean	*Pisum, Lens, Cicer, Lathyrus,* and *Vicia faba*	AceDB v: 3.0	Gene maps, information on pathology and diseases, gene symbols and typeline information.
CottonDB	Cotton	*Gossypium hirsutum*	AceDB v: 4.1	Data #Records: Papers: 1915, Pathology: 286, Trait Scores: 28168 (Germplasm yield, fiber, seed data), Collection: 8, Authors: 0655, Loci: 933, Sequence: 48, Journal: 897, Traité Study:1029, Maps: 79 (observed and molecular markers), Polymorphism: 87, Germplasm: 2823, Species: 42
GrainGenes	Wheat, oats, rye, barley, sugercane, cross-species		AceDB 4.5	Genetic and cytogenetic maps, genomic probes, nucleotide sequences, genes, alleles and gene products, associated phenotypes, quantitative traits and QTLs, genotypes and pedigrees of cultivars, genetic stocks, and other germplasms, pathologies and the corresponding pathogens, insects, and abiotic stresses, a taxonomy of the *Triticeae* and *Avena*

Table 1.   Continued.

DBName	Organism common name	Scientific name	DBMS	Data/Comments
MaizeDB	Corn	*Zea mays*	Genera software supporting gateways to Sybase RDMS	418 genetic maps, 7703 mapped loci, including 1108 genes, 2924 probed sites; recombination (1592), map score data (5681); 93(recombinant inbreds) and umc 95 (immortal F2) maps; 6011 probes (anchor/core markers for map bins; 2479 genetic/cytogenetic stocks, 6931 stock pedigrees; 21,401 locus variations; 12,252 bibliographic references; 3625 addresses of maize researchers; 45 QTL experiments
MilletGenes	Pearl millet and related millets.	*Panicum, Setaria, Echinochloa, Pennisetum* and *Paspalum; Eleusine; Eragrostis*	AceDB v: 4.3	Probe information, End-sequences of RFLP probes, RFLP and STS polymorphism data, Autoradiograph and gel images, Segregation data, Genetic maps for 8 crosses and their consensus map, QTL data, Morphological data.
RiceGenes	Rice	*Oryza sativa*	AceDB v: 4.5 version being ported to a RDBMS	7 rice genetic maps; 3 rice comparative maps (maize [1], oat [1] and the Triticeae[1]); 2.500+ probes (including the anchor set used for comparative mapping of grasses); 3,000+ sequences; 200+ rice QTLs ; 120+ maize QTLs; 1,500+ references 100 rice variety releases
RGP Japanese Rice DB	Rice	*Oryza sativa*	HTML/flatfile	High-density rice genetic map,1998; rice genetic map in Nature Genetics, 1994; Genome Map Information (in progress) of rice (RiceDB); Genotype data for RI lines (Asominori X IR24) with 375 RFLP markers; RFLP analysis of japonica rice varieties; Rice Physical Mapping Data from RGP; Genotype data (BC1F5) for a Backcross Inbred Lines (BILs) derived from Nipponbare/Kasalath // Nipponbare with 245 RFLP markers
KRGRP Genome	Rice	*Oryza sativa*	AceDB 4.5.1	Academia Sinica Map; Cornell Map; Japan (Korean NIAR-STAFF Tsukuba Map; Korea Map - Rice Rice NIAST, RDA; Morphological Map; 5,100 ESTs Research as of

Table 1. Continued.

DBName	Organism common name	Scientific name	DBMS	Data/Comments
				August 1, 1997; nucleotide sequences, Program) deduced amino acid sequence and restriction map of complete chloroplast genome from 12 species including Rice, Tobacco, Maize
RoseDB	Apple, peach, cherry, plum, almond, strawberry, and rose	*Malus, Prunus*, and other *Rosaceae*	AceDB v: 4.6	Marker segregation and linkage maps for apple cultivars: Rome Beauty, White Angel, Wijcik McIntosh, NY75441-58, and NY75441-67; ~1000 markers are listed, most are PCR-based, but morphological, isozyme, and RFLP loci are also included. Gel images are available on line for most markers. Intermap comparisons of marker positions and marker order are provided
SolGenes		*Solanaceae* sp.	AceDB v: 4.5	Solanaceae Genome Database, contains information about potatoes, tomatoes, and peppers
SorghumDB	Sorghum	*Sorghum bicolor*	AceDB v: 4.3	118 maps; also contains pathology, gene, race, genetic germplasm and stock information as well as 179 sequences
SoyBase	Soybean		AceDB v: 4.5	Map Collection- 14 (Classical genetic map and 6 molecular marker maps); Loci- 2607; Probes- 444; Genes- 397; QTL- 328; Germplasm- 30,000: Reaction/Pathway- 445 (Clickable diagrams of metabolic pathways covering 524 enzymes and 640 metabolites); Enzyme- 810 (EC number, purification, clones, physical properties and species and cultivars studied); Traits- 526; Pathology; 19 Information on soybean diseases including causative organism, Storage Protein- 17; Nodulin- 91
TreeBASE	many	many	ACI 4th Dimension database software	Relational database of phylogenetic information. Contains published phylogenetic trees and data matrices; bibliographic information on phylogenetic studies, and some details on taxa, characters, algorithms used, and analyses performed

Table 1.   Continued.

DBName	Organism common name	Scientific name	DBMS	Data/Comments
TreeGenes	Multiple: forest trees		AceDB v: 4.7	Genetic maps, DNA sequence, germplasm, and other related information for a large number of forest tree species
ChlamyDB		*Chlamydomnas reinhardtii*	AceDB v: 3.0	Genetic and molecular maps of Chlamydomonas reinhardtii; information on genetic loci, mutant alleles, sequenced genes; descriptions of strains in the Chlamydomonas Genetics Center collection; plasmids in the Chlamydomonas Genetics Center collection; links to GenBank for sequenced genes Chlamydomonas, Volvox, Dunaliella, and other Chlamy cousins; colleagues' names and addresses; Chlamydomonas accessions in other algal collections; more than 2800 bibliographic citations
Saccharomyces Genome Database	Yeast	*Saccharomyces cerevisiae*	AceDB v: 4.5 being ported to Oracle RDBMS	Contains a variety of genomic and biological information and is maintained and updated by SGD curators. The SGD also maintains the S. cerevisiae Gene Name Registry, a complete list of all gene names used in S. cerevisiae
SubtiList		*Bacillus subtilis*	UNIX Sybase	Provides a clean dataset of non-redundant DNA sequences of B. subtilis (strain 168), associated to relevant annotations and protein sequences
UMN Biocatalysis/ Biodegradation Database	NA	NA Store	Object store	Microbial biocatalytic reactions and biodegradation pathways primarily for xenobiotic, chemical compounds
ReBase	NA	NA	?	Restriction enzymes, methylases, the micro-organisms from which they have been isolated

Table 1. Continued.

DBName	Organism common name	Scientific name	DBMS	Data/Comments
DOGS	many	many	flatfile	Database Of Genome Sizes. The ultimate goal is to compile a list of all the known organisms and their respective genome sizes. Both the completed and estimated genomes are listed.
Mendel	Plants	NA	AceDB v: 4.1	Gene Name; Gene synonym; Accession number; Gene product name and synonym;Information on expression (in a few cases at present); Coding sequence co-ordinates; SwissProt and PID numbers
CIMMYT Maize Germplasm Database	Corn	*Zea mays*	CD-ROM, also on diskette	Maize germplasm database
MGCSC Catalog	Corn	*Zea mays*		Catalog of stocks available from Maize Genetics Cooperation Stock Center
CIP Germplasm Database	Potato		Diskette; when complete, will be available on the Internet	
Germplasm Resource Information Network (GRIN)	multiple	multiple	?	Plant and animal germplasm information
Grass Genera of the World	multiple	multiple	Generated from a DELTA database running under MS-Windows	Morphological, anatomical and physiological descriptions of over 800 grass genera

Table 1.   Continued.

DBName	Organism common name	Scientific name	DBMS	Data/Comments
PLANTS National Database	many	many	Oracle	Standardized plant names, symbols and other plant attribute information
Indices Nominum Supragenericorum Plantarum Vacsularium Project Database	vascular plants		Star/WEB database	Index of vascular plant names
PhytochemDB - plant chemicals	many	many	AceDB v: 4.1	Plant chemical data, including quantity, taxonomic occurence, and chemical activity
Phytochemical and Ethnobotanical	many	many	Relational: SQL-queryable database	Chemicals and activities in a particular plant; high concentration chemicals; chemicals with one activity; ethnobotanical uses; GRIN Accessions. Databases
PathoGenes Plant Viruses Online	many multiple	many multiple	AceDB v: 3 Delta/HTML	Fungal pathogens of small-grain cereals Host range; transmission and control; geographical distribution; physical, chemical and genomic properties; taxonomy and relationships; and selected literature references; database accession numbers (up to Gb[90] and Em[44]) of the genomic sequences of viruses and of satellite RNAs
The Potyvirus Sequence Database (PSD)	Potyvirus		ftp	Genbank format flatfiles of potyvirus sequences

3.2. Data inclusion and curation

In addition to software development, a second major effort required to make and sustain an IBIS is data curation by biologists. To transform data into knowledge requires experience in the areas associated with the IBIS (very broadly speaking, biology and mathematics).

An early challenge is to create a system whereby users can be fairly confident that the most important data needed can be accessed. Thus, it will be important to identify SDBs that store the data of relevance. Of course, it will take considerable wisdom to decide which data are in fact relevant. Questions to be answered include:

1. Which databases to access? Which columns, objects, files, etc., in those DBs? How to map from one schema to the next?
2. What types of curatorial activities need to be associated with identification of data that belong in IBISs?
3. How to serve the needs of users interested in all the data vs. those interested in subsets of the data? On which subset themes should focus be directed and how will curation and long-term publication occur into the future?

3.3. Interfaces

As with other types of users of information systems, to the biologist, the interface is the software. Graphical User Interfaces (GUIs) make IS easy to access and use because they contain metaphors that make the computer appear to match what the user has in mind when accessing or analyzing information. Unfortunately, many GUIs provide a shallow sense of having what is required. However, when the user wants to access information not located in the DB or local machine, they must search through hot-linked DBs, figure out the structure and type of data, run programs to access the data and find relevant programs to acquire, reformat, analyze and visualize the data. Ritter (1994) provided an early prototypic example of an integrated system. In that case, many of the challenges described were revealed. In an IBIS's all relevant data will need to be integrated with applications in a transparent fashion.

Biological knowledge workers often do not desire detailed information about specific applications. It is important to allow users to simply point at the objects they want to access and the needed applications will load automatically and transparently. This design feature will allow users to focus on their questions rather than on the particulars of the programs and applications they are running. Also, different users will have different requirements and personalities. Ideally, IBIS users should be able to organize their graphical desktops in a manner that is best suited to their needs. It is this customization of the interface that makes the computer interface appear friendlier and easier to use.

3.4. Collaborative construction of an IBIS

To build IBIS data producing centers, data consumers and information technology centers will need to interact in collaborative programs to define and evolve data models, model the work flows and discovery processes that scientists apply, and build the necessary software systems and tools. In addition, biological scientists will be required for ongoing data curation and continuous feedback on systems evolution. Building systems for those who produce and use the data, with their input and continued feedback, increases the likelihood that systems will be successfully used. Such a building strategy involves the research community in the developmental path of an IBIS, providing an active voice in the evolution of the system. To help assure collaborative development of public IBIS will require:

- Deployment in the public domain to support public research;
- Use and integration of existing public (national and international) biological data wherever possible;
- Support for distributed curation, data submission, and distributed annotation, as required;
- Integration through implementation of object-oriented components (for data acquisition, storage, and analysis/visualization); and
- Provision of adequate security for subsets of collaborators.

4.   Implementation

To some extent, IS add to problems of data interpretation. For example, databases filled with DNA sequences are hard to understand without the biological rules that give them meaning. Unfortunately, these rules tend to be built into application programs used to access data. Biological rules hidden in application programs are difficult to integrate in a client-server environment. These rules must be broken out of the program and attached more effectively to the data, thus providing meaning to the data. A database without rules is like a book or encyclopedia that only contains nouns. Books and encyclopedias are much more readable and useful when verbs are included.

The result of tying biological rules with biological data is a set of interfaces that are exported by the server. IBIS must make use of valuable SDBs (Table 1) and provide enhancements to give an object-oriented look and feel to them. A three-tier approach seems to be gaining in popularity elsewhere (Chaudhri and Loomis 1998). The object-oriented Client connects to an object-oriented Server, which then ties to an existing data source. Communication between the legacy data source may be through simple database calls, or may use Remote Procedure Calls (RPC) of application code on the legacy machine(s). Using RPC may require reworking some of the code of the legacy system but may be less work than rewriting the entire systems to run native on an object-oriented server. The IBIS server is thus elevated to the genomic object repository.

There are several architectures that might provide a fully functional IBIS. We are currently considering a three-tiered architecture to support these requirements (Fig. 1).

The application server is both a client and a server in this architecture. In general, the terms "client" and "server" become meaningful more as "roles" than as labels for particular hardware or software components in complex distributed architectures. The communications between client and server in the three-tier architecture, and indeed for any distributed architecture, can be implemented using any networking technology.

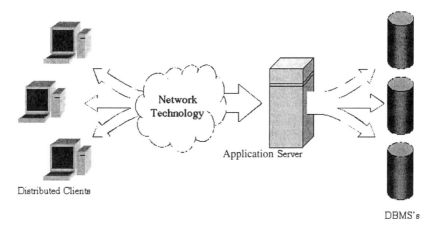

Figure 1.   Basic three-tier architecture.

CORBA, COM, and Java's RMI are examples of higher-level networking technologies, i.e. ones that hide the lowest-level code from programmers, but they are only the latest in a long line.

What has worked elsewhere is an architecture composed of: the user interface, the query support component, and the data warehouse or a federation of databases (which includes the encapsulated SDBs). The prototype successfully tested by Ritter (1994) used a virtual database. Generally speaking, these components have the following tasks associated with them:

- User interface: helps to create queries and control the flow of extracted data. Complex queries might be built from component queries using a drag-and-drop interface, with components represented as icons on the desktop, which can be connected. This is an evolution from text-based queries and is sometimes called a 'cartoon-based' query interface. Whole desktops should be amenable to saving for later use, and users might build customized macros for shared use.
- Query support layer: responsible for parsing, tokenizing, and transforming the query string, and returning feedback after accessing the data (warehouse, federation, or virtual).
- Data warehouse, federated DB, or virtual DB: contains the actual data and corresponding metadata for IBIS.

Another challenge to integrating SDBs is that they were designed to meet species-centric and discipline-centric needs. As a result there are semantic and nomenclature inconsistencies that compound the technical issues associated with heterogeneous architectures. It has been argued that no amount of connectivity can compensate for semantic and nomenclature inconsistencies. While it is true that there is a need for greater communication and discussion among the various communities supporting the SDBs, resolution of the various semantic and nomenclature inconsistencies by developing a varibioum ontology need not be a pre-requisite to the integration of genomic information. It should be possible to reveal associations between genomic objects based on sequence information, syntenic map locations, expression patterns, etc., and then leave it to the researcher to decide if synonymous use of different names or vice-versa has occurred (for a practical example see Paterson et al. 1995).

Clearly, to implement any feasible architecture will require recognition of both technical and social challenges. Software engineers can cite numerous cases where, despite expenditures of millions of dollars, projects failed because such issues were not addressed early in the development process. We need first to recognize that software development requires the same rigor and planning as development of laboratory protocols. We also need to recognize that merely posting research results on the WWW in a virtual 'scrap-book' will not be sufficient to effectively address the fundamental questions of complexity in biological systems. Intellectually, we have acknowledged these issues, but this intellectual acknowledgement has not yet translated into research proposals with adequate resources devoted to informatics. One of the most challenging issues for software development to support biological research is that groups wishing to offer their expertise must do so through collaborative agreements with data-producing groups. On one hand, this is highly desirable because of the multidisciplinary nature of the problem. On the other hand, this means that software engineering is tied to specific projects rather than being able to tackle the big picture. Not surprisingly, many excellent biological research groups do not fully appreciate the scale of the informatics problem and thus may not be proposing effective solutions. We would not expect computer scientists or software engineers to provide solutions to biological data interpretation and research design. Until funding mechanisms explicitly address the challenges, there will continue to be difficult and unnecessary hurdles and expenses for the biological research community to get what they need.

4.1. Technical strategies

The key technological platform on which an IBIS can be constructed is Distributed Object Technology (DOT). DOT "is the synergistic application of object-oriented principles to distributed, network-oriented application development and implementation" (Ryan 1996). It is an outgrowth of the idea that objects and networks were made for each other. DOT ensures that data and functions can be encapsulated into objects. "Encapsulation provides information hiding in which software units protect themselves by hiding the internal details from the outside world" (Khana 1994). Encapsulation makes it possible to change distributed object systems component by component without needing to rewrite the entire system (Ryan 1996). This 'evolutionary' nature of software development using DOT is much like selective breeding in biological systems, selection is thought to act on components of biological systems that we wish to optimize for agricultural needs. When client software developers want to make new functionality available to the end user, they simply add the method call to their client and alter the user interface to enable the user to access the new functionality. The primary development for DOT is Object-Oriented Technology (OOT).

The current standards and technologies that will permit DOT implementation of IBIS's include:

1. Common Object Request Broker Architecture (CORBA): Defined by the Object Management Group (OMG) attempts to foster cooperation between systems suppliers, software vendors and users. Activities related to genomics and bioinformatics that are in some way tied into OMG standards include:

   - European Bioinformatics Institute (EBI), who have developed several CORBA-based servers (see CORBA at EBI home page: http://corba.ebi.ac.uk/)
   - BBSRC: Roslin Institute, making heavy use of CORBA in their server-side development relating to "ArkDB" (pig, chicken, sheep, cattle, horse, cat, and tilapia data), see http://www.ri.bbsrc.ac.uk/bioinformatics/research.html.
   - NetGenics SYNERGY system is a well-developed system based on CORBA. Their 'white paper' on the use of CORBA in bioinformatics is at http://www.netgenics.com/white_paper.html.
   - OMG's Life Sciences Research (http://www.omg.org/homepages/lsr) is a task force of the OMG specifically focused on applying CORBA in the life sciences. They have working groups covering many specific areas, including sequence analysis, maps, and gene expression. There are around 100 members from industry, government, and academia (only 50 of them are members of the OMG and have voting privileges).

2. Object Linking and Embedding/Component Object Model (OLE/COM): Invented by Microsoft to serve as a CORBA substitute.

3. World Wide Web (WWW): accessed through Hypertext Transfer Protocol (HTTP), Hypertext Markup Language (HTML), Common Gateway Interface (CGI), and Java (a powerful programming language developed by Sun Microsystems that allows developers to write the program once and deploy it on any platform).

The above have been described as "centers of gravity for the distributed object universe" (Ryan 1996).

The second major development for DOT has been the evolution of client/server technology (Fig. 2). The diagram depicts what might be called a "classical" or "2-tier" client-server architecture. The main drawback of such an arrangement is that the client needs to be more memory- and compute-intensive (in fact, this kind of architecture is sometimes called a "fat-client" architecture) than in a 3-tier architecture because it cannot rely on an application server to share some of the tasks that are common among all clients. Another problem is that clients and servers are "tightly coupled", i.e. heavily interdependent, and thus each needs to be altered when the other changes. This can be a problem since clients are typically hard to update because they are often installed and maintained at numerous sites all over the internet.

The combination of intelligent client workstations and high-speed intelligent networks has brought new opportunities for an IBIS development. In client/server technology, objects are applied to distributed applications via GUI-based clients, networks and application servers. User interfaces have gained much from object technology. GUI systems perform their functions effectively because of object-based technology. The evolution of intelligent client and GUI enables developers to deliver greater customization under end-user control. Improved ease-of-use because of windows and icons with customization is a result of object technology application to desktop clients. These developments point toward customizable user-interfaces for different end-users (contrast, say, breeders with molecular biologists). Of course, many people would argue that it would be possible to achieve many of the new functionalities without OOT, though most would agree that OOT has helped to improve GUIs.

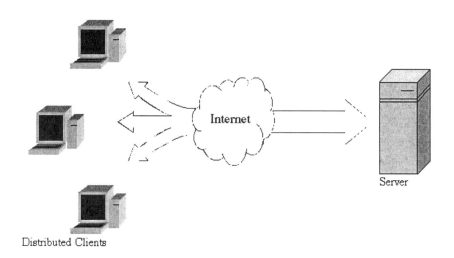

Figure 2. Basic client/server arrangement.

Standards in DOT (CORBA, OLE, and web browsers and servers) simplify integration and usability of multiple GUI applications. Standards are leading the way for simple GUIs to become object-oriented GUIs. Object-oriented GUIs will hide the implementation of data formats and access methods. An object-oriented client is an object-oriented GUI with location transparency, i.e., all data appear local and integrate more naturally.

A strategy to consider for implementing an IBIS is to develop it in parts (components), *with an understanding of the whole system in mind.* It is logical to separate IBIS's development into component parts that, together, make up the IBIS, and make it a general requirement for components that these parts effectively function as a whole as the system evolves. This parallels the situation that we face with biological data generated on specific components of complex biological systems.

We have pointed out that a component-based approach can be achieved through the application of DOT. Because DOT provides encapsulation of the data with object interfaces the internal data organization becomes invisible to application programs that access it. Encapsulation also occurs with application programs. By hiding implementation, later developments and improvements can be achieved without requiring changes to the components' interface or the human procedures or software used to access it.

This will ensure that an IBIS is capable of evolving with the rapidly changing environment of biological research. A new OOT system can be constructed by reconfiguring existing object-oriented components. This is cheaper than developing a system with entirely new components.

Also, OOT will provide developers of IBIS with the ability to reuse components. Mary Loomis, Director of Software Technology at Hewlett-Packard (HP) Laboratories, to quote, "object technology helps in building smaller components that can also leverage other components' capabilities." (Ryan 1996). Improved reusability means that IBIS engineers can spend more time developing new components and less time maintaining old ones.

Fortunately, the nature of OOT is compatible with the goal of a systems approach to making use of biological information. Recall that objects represent real or abstract entities. This is similar to the biologist's world in which organisms and processes occur. Object-oriented programming makes use of inheritance in its ability to design objects by 'inheriting' attributes from existing objects. This is a very biological approach to software engineering. Additionally, biologists tend to favor thinking about biological information within a hierarchical framework and have long used hierarchical classifications for living organisms or developmental patterns within organisms. To date, genomics has not developed a hierarchical taxonomy, but as patterns of organization become apparent, it is likely that IBIS will require the development of classes of genome objects that take advantage of the biological ordering from the general to the specific.

CORBA, OLE and the WWW provide new standards for implementation of DOT to IBIS. Various CORBA-based software companies are offering links from OLE/COM to their tools, which should allow bridging for integrated OLE/COM - CORBA-based networks without needing to integrate the two models themselves. Web technologies are less mature: HTTP and HTML are not really object-oriented; CGI has performance issues and does not implement RPC. However, the web provides quite useful encapsulation (hiding) of network data and processes. And the Web is enhanced by its diversity, including products from individuals to major corporations. Java, Microsoft's ActiveX, Netscape Frames and Plug-Ins are examples of web technologies that are improving client browsing and server data access.

To summarize, IBIS can be built using DOT because correctly implemented object technology can perform more efficiently and effectively than simple file systems and relational databases (Ryan 1996). The Web, CORBA, and OLE provide standards for implementation. When users and developers on object-oriented clients are empowered to specify what they want, without consideration for how to access it or where it is located, they can spend more time on scientific questions rather than working on the implementation of their information system.

4.2. From needs to plan

The kind of specialization introduced by client/server technology is amplified in a DOT environment. Client development teams focus on rapid prototyping GUI development, user interface standards, and optimization, whereas server developers focus more on effective representation of objects, effective database access, exporting reusable interfaces, etc. In a DOT environment, there should also be a focus on effective reuse of distributed objects. More and more applications will use combinations of already existing components. Client developers thus eventually focus on combining components from a library of components. In this sense it will be important to develop and make available class libraries for genomic components.

Integration of any new technology poses new difficulties and challenges. Cultural changes may need to occur to enable the most effective implementation and use of DOT. It will be important to explore new ideas and technologies as IBIS are being developed. It is not realistic to expect that there will be a clean separation of data resources for different generations of software development technologies (Keller et al. 1998) and technological renovation will have to be weighed against solid advantages for an existing IBIS. It also will be important to assure application and vendor independence. In practice, application independence addresses issues that expedite system production by decoupling applications from the database. Early in the development of IBIS it will be necessary to identify areas that would be best to standardize immediately and allow developers to use the library base as the foundation for subsequent code (and avoid the 'multiple solution syndrome'). Application code should not be affected by changes in the database design or implementation. Therefore, application code should not call the database directly. Vendor independence provides a safeguard mechanism for future evolution of the vendor's products as well as of IBIS. Ideally, all database products would conform to standards allowing future swaps of DBMS without significant effort. In practice, it is likely that an IBIS's database component(s) will need to be encapsulated to allow easy database replacement.

5.   Ongoing evaluation of new technologies, user groups, and prototypes

Software engineering groups will need resources for evaluation and testing of potentially useful products for an IBIS implementation, in a 'proof of principle' mode. A deeper understanding with a concomitant higher level of confidence concerning potential solutions may only come through hands-on experience and training. Prototyping as a major design tool provides potentially the best real-world test of any proposed approaches to be used in IBIS, much in the same way as preliminary laboratory experiments are frequently required to validate a laboratory approach. Prototyping is fundamentally important to IBIS's ultimate design success because it is likely to lead to key requirements being identified and tested at a much earlier stage. It is very important that the evaluation process be given sufficient time and funding so that there is adequate assessment via prototypes of the main issues for IBIS.

Implementation of IBIS will rely heavily on the interaction between software engineering staff and users of the systems. A key aspect of object modeling is the application of *use cases*. This was one goal of the April 17-19 (1998) first scientific advisory meeting at NCGR. However, it was determined that smaller, more focused expert groups should be employed to target specific use cases for components of IBIS. A component approach should be more effective, though there will always be a need to tie the resulting information back into the data model of IBIS. The goal of use cases is to capture how the user interacts with the information system. The user is the agent, generating input events. Input events are then acted upon through a system operation, generating output events.

The best way to generate use cases is through interviews as the system is being built. Further refinement of requirements from use cases will be achieved through prototypes that show the user interface on a GUI. Although such a prototype has no back end, the user may perform interactions on it and see what would be the results of certain inputs. Once the use cases have been adequately explored and developed, system operations to support those user interfaces can be defined. In parallel, users work with server modelers to create the genome objects that will support the client applications. Client and server components of the application should be created simultaneously.

The goal of the client developers is usability and making the information available to users to meet their requirements. In this model, the goal of the server developers is to create genome objects that will last and be useful for future applications.

The strategy of implementing IBIS in parts suggests the use of prototypes. Prototypes are significant aids in understanding the value of technology. A software engineering philosophy critical for the long-term evolution of IBIS is to develop scenarios with the end-user community and play those scenarios through prototypes. These scenarios can then evolve iteratively.

5.1. Some key technical challenges

There are critical technical challenges for IBIS implementation that have not been thoroughly addressed herein because it was not our goal to do so. However, we do wish to emphasize a few. In addition, there are likely to be other challenges we cannot envision at this stage.

As we have seen, there are many data types and many heterogeneous data sources, and there is a need to execute a parallel development of pieces and whole of IBIS. Additionally, IBIS must be flexible and able to evolve in response to change. These realities suggest an approach to system engineering that includes:

- Conceptual definition of components and subcomponents of IBIS;
- Definition and evolution of data models for systems integration;
- Definition and iterative refinement of key use-cases for IBIS;
- Significant investment in heterogeneous database research; and
- Definition of architectural models for IBIS.

Particular issues for the integrated system's engineering include the fact that we observe two kinds of heterogeneity that specifically relate to IBIS. The first heterogeneity occurs within data types, i.e., multiple data resources for each data type, with distinct solutions for data modeling. The second type of heterogeneity is observed across data types, i.e., each 'type' of data, whether genomic sequences or phenotypic information, are gathered and modeled differently and without prior thought on integrative use of the data. Additionally, software engineers are faced with non-standard DBMS. Such DBMS typically suffer from limited query languages, unreliability and poor performance (all of which affect scalability of those systems in response to increased data production).

There also are two levels of integration problems to be faced. A lower level problem relates to protocols for communication among generic components. For this, as we have seen, there are standards emerging. At a higher level, though, there is a need for specification of the nature or particular, domain-specific objects (such as maps, chromosomes, biochemical pathways, etc.). The latter is being addressed by OMG's LSR, for example.

6.   Example prototype systems

6.1.   Real-time support for distributed genome projects

6.1.1.   *The Phytophthora Genome Initiative (PGI)*

Federal funding for agricultural genomics is increasing in the US and worldwide. International collaborations for genomic sequencing of major agricultural species and model organisms seem to be the paradigm that is gaining acceptance (humans, *Arabidopsis*, yeast, and rice, for example). To support increased federal funding and a distributed paradigm for genomic projects, there is a need to deploy web-accessible solutions for information handling. In response to this growing need, the international *Phytophthora* Genome Initiative (PGI) (http://www.ncgr.org/research/pgi) is being used as a development/testing/implementation ground for such a system. This system is essentially re-usable for other distributed genome collaborations as needed. The goal is to offer to biological research collaborators a significant reduction in the amount of repetitive tasks that their staff needs to execute, thus freeing time for thinking about biological questions. The system is available entirely over the web, so that an Internet browser is sufficient for complete system access. The system also is designed to be as unintrusive as possible with respect to people needing to upload data into it for analysis.

Initially, the PGI system has focused on supporting DNA sequence data acquisition from PE/ABD automated sequencers. The raw ABD files are sent via ftp directly to NCGR by the sequencing center. To store and analyze sequence data, we used a relational database and developed an automated analysis "pipeline" (Fig. 3). The relational database was adapted from the Genome Sequence Database (GSDB), version 1.0 (Harger et al. 1998). A small subset of the GSDB schema was used to store sequence and feature annotation. In addition, several tables were used to monitor activities ecapture state information (the "Tracker" database).

In the analysis pipeline, raw data are automatically processed after transfer via ftp to NCGR. Inputs to the pipeline are raw sequences and chromatogram traces from an ABI377 AutoSequencer (in the case of PGI) or 3700 in other cases. Outputs of the pipeline are annotated sequences and reports that summarize analysis results. Access to the database is available via the world wide web. In general, output from one step of the analysis pipeline serves as input to the next step.

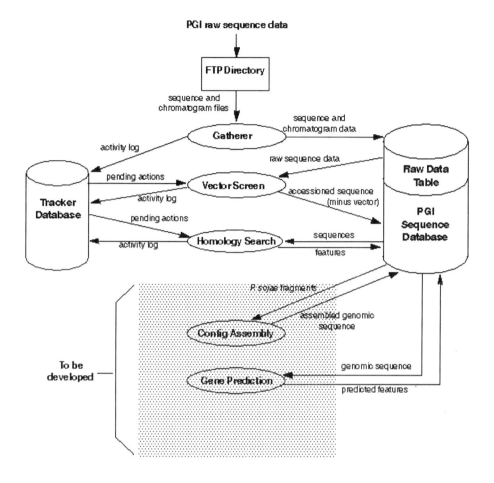

Figure 3. The *Phytophthora* Genome Initiative 'data pipeline'.

Steps in the analysis pipeline presently include vector screening, similarity searching and motif searching. Derivitives of PGI including MGI & NPC also support automated contig assembly and in the case of EST data, redundancy checking. Future plans include support for gene prediction integrated comparisons of genetic and physical maps.

A "Gatherer" program examines incoming files for obvious errors before adding the data to the database. Raw sequences are subsequently screened for cloning vector sequences and bacterial sequence contamination. Sequence regions with high similarity to known vector sequences or bacterial sequence are removed before further analysis. The insert sequence is saved to the database, provided that the sequence is not entirely composed of vector, and that there are no chimerical vector regions within the insert sequence.

After vector sequences have been removed, insert sequences are examined for similarity to other sequences whose function is known. To search for similarity to other sequences, we developed an interface to BLAST (Altschul et al. 1997) which adds database connectivity and parses BLAST output into data objects. Each sequence is queried against current versions of NCBI's 'nr' non-redundant amino acid reference library [ftp://ftp.ncbi.nlm.nih.gov/blast/db] and a non-redundant set of nucleotide sequences compiled from GSDB. Similarity regions that result from BLAST analysis are added to the database as features linked to the query sequence. The BLAST output is also stored intact in the database. PGI uses the default BLAST parameters, and does not store any similarity regions with expected values above 0.01.

To the *Phytophthora* biologist, what PGI has delivered is the ability to have real-time access to results from sequencing without needing to allocate the local laboratory resources into DNA sequence data manipulation and analysis, thus allowing increased devotion to biological questions using the power of a genomic approach. PGI also has contributed to connecting and strengthening the *Phytophthora* research community worldwide.

6.1.2. *The plant R genes DB*

In addition to significant efforts in software engineering IBIS will require significant data curation. Curation of biological data requires biological scientists just as development of IBIS requires software engineers. Discovery of biological relationships will require hypothesis testing. Thus, it is important to create avenues through which willing biological scientists can easily imbue IBIS with their knowledge and experience. In that context, NCGR has developed and deployed sabbatical and post-doctoral programs to enable biological scientists to develop and deploy curated data sets and analysis algorithms of interest to their specific communities.

An example sabbatical project was developed at NCGR with Nevin Young (University of Minnesota), on the theme of plant disease resistance (R) genes. More than 570 publicly available R-gene and R-gene-like sequences were acquired, stored, and analyzed. Nucleotide Binding Site (NBS) motifs revealed two categories in NBS-LRR type R genes. One category was absent or very rare in rice. After data acquisition and analysis, we predict 550-600 R-gene-like sequences will be discovered in *Arabidopsis*.

The PRGDB prototype was made public at the 1999 Plant and Animal Genome meeting and then available through NCGR's web site www.ncgr.org/ersearch/rgenes. Future plans include a user group workshop to discuss future implementation plans and requirements and integration of biochemical and expression data and curation/update/import tools.

6.1.3. *The Plant Specific Metabolic Archive (PlaSMA)*

The Plant Secondary Metabolic Archive (PlaSMA) was a collaborative project with the National Biotechnology Information Facility. Thematically, PlaSMA is focused on plant secondary metabolic data. From an informatics perspective, PlaSMA is a relational database that supports acquisition, storage, and viewing of plant secondary metabolic data. The PLaSMA model for data input and curation is one of using distributed biochemists focused on their pathways of expertise, in response to the requirements of that community.

PlaSMA is a relational database in which a pathway is represented as a grouping of component reactions. The reactions are shared among different pathways. This facilitates component-wise curation and a high degree of consistency. PlaSMA provides great depth of reaction information. Much of the depth of the literature is stored. Multiple sets of experimental parameters can be stored for each reaction with links to the original references. PlaSMA does not limit its scope to the representation of "generic" pathways where reaction and pathway data are based on consensus observations across taxonomic boundaries. PlaSMA provides links from enzymes and their genes to external databases such as GSDB, PDB, SWISS-PROT, and PID.

PlaSMA employs a reusable object model for database components. Components central to metabolic pathways are represented and stored as reusable objects. Examples of reusable objects include reactions, enzymes, and compounds.

An important output of the collaboration with biochemists has been clearly shown through the development of the manual data input device, the Plant Secondary Metabolism Input Device (PlaSMID, Fig. 4).

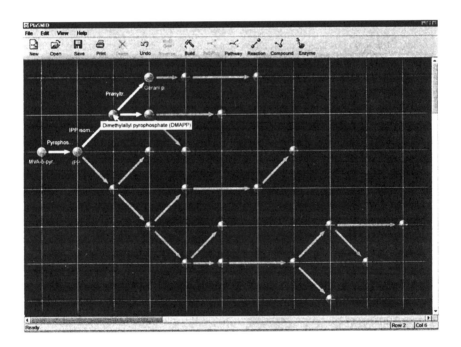

Figure 4. A pathway diagram being populated with specific reactions.

PlaSMID allows biochemists to input data into the system by drawing biochemica pathways on the screen and then using them to input the data by calling up form and is currently being used in a prototypical manner with NBIF collaborators. New requirements have surfaced since the prototypes were developed and the next iteration, PlaSMID v. beta 3 improvements will include added ability to edit reusable objects such as reactions, enzymes and compounds. Furthermore, built in safeguards to prevent 'breaking' existing reactions or pathways will be implemented. These become paramount as the number of people working on data entry increases. Finally, implementation of a more logical approach to kinetic and reaction parameter handling that better reflects the underlying biology is also under development.

Like GSDB, PlaSMA provides an underlying architecture for manipulation of a specific type of biological data. As more sequence data accumulate, it has become clear that other types of data need to be integrated to predict function in many cases. Biochemistry provides one important route to functional prediction. Thus, the underlying technologies developed for PlaSMA can be integrated into IBIS.

To enable a move toward function, future PlaSMA development plans include the development of querying and viewing tools. Though beyond the scope of the current collaboration, an important role for biochemical data is to offer it to enable integration with other biochemical components, such as metabolic engineering and modeling modules. Of course, integration with other data types is also of paramount importance.

7.   Conclusions

In the 21st century biologists will have the fruits of many genome projects. These will primarily be in the form of large data sets that require biological knowledge and interpretation to be useful. This cannot be achieved without IBIS. Rapid information/biological reagent access will require repositories serving many users and:
- Hardware: high-speed connectivity to Internet-accessible resources;
- Information systems: public, fully integrated exploratory toolkit(s) online; and
- skills to access, manipulate, and analyze information.

The shift from cottage industry to large-scale biology is causing 'data-poisoning'. Fortunately, Information Technologies are sufficiently mature to be able to offer solutions to relieve data poisoning. However, to implement these solutions, systems development must engage collaborations between biologists, computer scientists, and engineers. Furthermore, funding must be directed at enabling software providers to deploy suitable solutions.

In addition to software development to support IBIS, curation of systems must be done by those with biological knowledge. Thus, sufficient funding needs to be allocated toward curation by biologists and curators need to be provided with software tools that are useful and high-throughput, to aid in their task.

Finally, we should not build systems for single users anymore, and IBIS should be developed openly, to support public and private research efforts and provide frameworks to simplify integration across those boundaries as well.

8.   Acknowledgements

Many thanks are due to various scientists and engineers who have contributed their time towards development of the ideas reflected herein, as well as the opportunity we all have to 'stand on the shoulders of giants'. In particular, we thank the participants of the April 98 scientific meeting in Santa Fe: Robert Farber (LANL), Robert Pecherer (NCGR), RW Doerge (Purdue University), Edie Paul (GeneFlow), Ernest Retzel (University of Minnesota), Scott Tingey (DuPont), Tom Blake (Montana State University), Benjamin Bowen (Pioneer Hi-Bred International), Ronald Sederoff (North Carolina State University), Daniel Grimmaeli (CIMMYT), Christine Paszko (Accelerated Technology Laboratories), Lisa Lorenzen (Pioneer Hi-Bred International), Richard Michelmore (University of California Davis), Lynn Clark (Iowa State University), David Frisch (Clemson University), Allan Dickerman (NCGR), and Rob Pecherer (NCGR). We also thank Peter Hraber (NCGR) and Adam Siepel (NCGR) for assistance throughout this endeavor and critical readings.

9.   References

Altschul, S.F., Madden, T.L., Schaffer, A.A., Zhang, J., Zhang, Z., Miller, W., Lipman, D.J. (1997) Gapped BLAST and PSI-BLAST: A new generation of protein database search programs. Nucleic Acids Res. 25(17): 3389-3402.

Cattell, R.G.G. (1992) Object Data Management, Object-Oriented and Extended Relational Database Systems, Addison-Wesley Publishing Co.

Chaudhri, A.B. and Loomis, M. (eds.) (1998), Object Databases in Practice. Hewlett-Packard Professional Books

deCaluwe, R., van Gyseghem, N., Cross, V., (1997) Basic notions and rationale of the integration of uncertainty management and object-oriented databases, in deCaluwe R (ed.), Fuzzy and Uncertain Object-Oriented Databases: Concepts and Models, World Scientific Publishing Co.

Harger, C., Sukupski, M., Bingham, J., Farmer, A., Hoisie, S., Hraber, P., Kiphart, D., Krakowski, L., McLeod, M., Schwertfeger, J., Seluja, G., Siepel, A., Singh, G., Stamper, D., Steadman, P., Thayer, N., Thompson, R., Wargo, P., Waugh, M., Zhuang, J.J., Schad, P.A. (1998) The Genome Sequence DataBase (GSDB): improving data quality and data access. Nucleic Acids Res. 26 (1): 21-26.

Keller, W., Mitterbauer, C. and Wagner, K. (1998) Object-oriented data integration: Running several generations of database technology in parallel, in Chaudhri AB, Loomis M (eds.), Object Databases in Practice. Hewlett-Packard Professional Books.

Khana, R. (ed.) (1994) Distributed Computing, Prentice Hall.

Paterson, A.H., Lin, Y.-R., Li, Z., Schertz, K., Doebley, J.F., Pinson, S.R.M., Liu, S.-C., Stansel, J.W. and Irvine, J.E. (1995) Convergent domestication of cereal crops by independent mutations at corresponding genetic loci. Science 269: 1714-1718.

Ritter, O. (1994) The Integrated Genomic Database (IGD), in Suhai, S. (ed.), Computational Methods in Genome Research. Plenum Press.

Rubinstein, M.F. (1975) Patterns of Problem Solving. Englewood Cliffs, N.J.

Ryan, T.W. (1996) Distributed Object Technology, Hewlett-Packard Professional Books.

Ullman, J. (1989) Principles of Database and Knowledge-Based Systems, Volumes 1 and 2, Computer Science Press, Rockville, Maryland.

Waugh, M., Hraber, P., Weller, J., Wu, Y., Chen, G., Inman, J., Kiphart, D. and Sobral, B. (2000). The Phytophthora Genome Initiative Database: informatics and analysis for distributed pathogenomic research. Nucleic Acids Res. 25(1): 87-90.

9. Introduction: molecular marker maps of major crop species

RONALD L. PHILLIPS[1], *and* INDRA K. VASIL[2]
[1] *Department of Agronomy and Plant Genetics, University of Minnesota,*
St. Paul, MN 55108-6026, U.S.A. <phill005@umn.edu>
[2] *Laboratory of Plant Cell and Molecular Biology, University of Florida,*
Gainesville, FL 32611-0690, U.S.A.

Contents

---

1. Molecular marker maps

Initially genetic maps of higher plants were based almost entirely on morphological and biochemical traits. These maps have been replaced and/or supplemented with DNA-based marker maps formulated on the use of powerful new molecular techniques.

The new high precision maps can be developed with comparative ease and rapidity. They have a much higher density of markers which allows revelation of more and more restricted segments of the genome. One of the many revolutionary aspects of this technology is that linkage between molecular markers and traits of interest often can be detected in a single cross. The ability to hybridise probe after probe to the DNA of the same individuals of a segregating population allows one to pursue the analysis until linkage becomes evident. With morphological and biochemical markers used previously, a separate cross was required to test linkage with each new marker. It was seldom that more than three markers could be tested for linkage with the trait of interest in a single cross because of viability problems. With the techniques described in this volume, a new gene can be placed on the linkage map within a few days instead of the much longer time required with the previous techniques.

Recombinant inbred lines and other means to immortalize segregating lines derived from a cross coupled with appropriate software programs facilitate the building of linkage maps. Other approaches also can streamline the detection of linkage, such as the use of backcross derived lines, bulked segregant analysis, random mating, or aneuploids. Linkage maps now include the DNA-based markers as well as many qualitative and quantitative trait loci. Such maps will be extended over time to provide an impressive and useful genic array displayed across the entire genome. Maps including cDNAs and other cloned genes provide additional dimensions for determining expressed gene relationships.

The genetic dissection of quantitative traits is made feasible by DNA-based markers which have no phenotypic manifestations. Genomic regions can be recognized which

167

R.L. Phillips and I.K. Vasil (eds.), DNA-Based Markers in Plants, 167 - 168.
© *2001 Kluwer Academic Publishers. Printed in the Netherlands.*

control large portions of the phenotypic variation of important traits. The pyramiding of genes influencing a single trait will lead to many new insights and useful strains.

As new technologies develop that allow the automation of marker analysis, the applications in diagnostics, breeding, proprietary protection, and cloning of genes will no doubt increase. The ability to analyze large numbers of individuals at modest cost will enhance the attractiveness of molecular markers in these fields and lead to their routine use. Map-based cloning requires a high density marker map around the targeted region. The increasing interest in plant genome analysis is leading to even more saturated maps useful in the isolation of genes. This will be helpful not only in marker assisted breeding, but also in introducing desirable genes into crops of interest by genetic transformation.

In the following chapters, a group of leading researchers provide the latest version of DNA-based marker maps for a variety of important crops. The progress made during the past five years has been truly phenomenal.

By this date, several maps exist for most species either due to the use of various populations or marker types. In some cases these maps have been joined to give a more robust map. The mapping of telomeres on some maps clearly designates the termini of chromosome arms. In several instances, maps are available reflecting recombination in the male flowers versus female flowers. Many of the markers for the various species show segregation distortion, being a much more broadly-observed phenomenon than might have been expected. These regions sometimes reflect the presence of gametophyte factors or self-incompatibility loci. There is even some speculation that the regions showing segregation distortion are important in regard to inbreeding depression and/or heterosis.

Evolutionary information is inferred from the extensive mapping performed in most of the species. Ancient polyploidy is proposed to explain the genomic structure in species previously regarded as diploid. The extent of duplications in most species is impressive and knowledge of duplicated regions can be used to map duplicated loci in an impressively efficient manner.

10. Molecular marker analyses in alfalfa and related species

E. C. BRUMMER[1], M. K. SLEDGE [2],
J. H. BOUTON[2], and G. KOCHERT[2,3]
[1] Iowa State University, Agronomy Department, Ames, IA 50011
[2] University of Georgia, Department of Crop and Soil Science, Athens, GA 30602
[3] University of Georgia, Department of Botany, Athens, GA 30602
<kochert@dogwood.botany.uga.edu>

Contents

1.   Introduction

Alfalfa, *Medicago sativa* L., a highly productive forage species, grows throughout the temperate regions of the world and produces premium quality hay, silage, and pasturage. Alfalfa originated near the Caspian Sea in northern Iran and northeastern Turkey; its cultivation spread throughout the Mediterranean region and into Germany, France, and China by the time of the Roman Empire (Bolton 1962). Today, alfalfa is raised on all continents and is currently cultivated on more than 32 million hectares worldwide (Michaud et al. 1988). The genus *Medicago* is highly diverse, including not only perennial, cultivated alfalfa but also several annual forage species and many other species that undoubtedly represent potentially useful germplasm sources for further improvement of the cultivated species.

Alfalfa produces high dry matter yields with high crude protein and total digestible nutrients. Yields as high as 24 Mg/ha have been attained under nonirrigated conditions (Barnes et al. 1988). All the *Medicago* species fix atmospheric nitrogen, which helps improve the soil for future crops while concurrently eliminating the need for inorganic N fertilizer. In addition to these direct agronomic benefits, alfalfa is a useful model plant for the study of forage systems and autotetraploid genetic theory (Barnes et al. 1988). The annual species *M. truncatula* Gaertn. is now the model system of choice for studying plant-*Rhizobium* interactions (Barker et al. 1990).

Several excellent discussions of the *Medicago* taxonomy have been published (Lesins and Gillies 1972; Lesins and Lesins 1979; Quiros and Bauchan 1988; Small and Jomphe 1989). Between fifty- six (McCoy and Bingham 1988) and 83 (Small and Jomphe 1989) species are currently recognized within the genus. The basic chromosome number of most *Medicago* species, including cultivated alfalfa, is x = 8 although several

169

R.L. Phillips and I.K. Vasil (eds.), DNA-Based Markers in Plants, 169 - 180.
© 2001 Kluwer Academic Publishers. Printed in the Netherlands.

species have x = 7, apparently as a result of chromosomal rearrangements (Lesins and Lesins 1979). Diploid, tetraploid, and hexaploid species are known. Both annual, autogamous and perennial, allogamous species are present within the genus. Cultivated alfalfa is perennial and autotetraploid (2n = 4x = 32), but a breeding population, Cultivated Alfalfa at the Diploid Level (CADL), has been developed for research purposes (Bingham and McCoy 1979).

Medicago sativa is actually a complex of related diploid and tetraploid subspecies that are interfertile and have the same karyotype (Quiros and Bauchan 1988). Included within the complex are four subspecies: *Medicago sativa* subsp. *falcata* (2x and 4x), subsp. *sativa* (4x), subsp. *coerulea* (2x), and subsp. *glutinosa* (4x), as well as a variety of hybrid subspecies. *M.sativa* subsp. *coerulea* is the diploid form of subsp. *sativa* (Quiros and Bauchan 1988). All subspecies have purple or variegated flowers except yellow flowered subsp. *falcata*. Cultivated alfalfa consists predominantly of subsp. *sativa* germplasm, but other subspecies, particularly subsp. *falcata*, have been introgressed. Endemic to colder regions, subsp. *falcata* has provided winterhardiness to cultivated alfalfa, though it has also contributed slower regrowth and a more prostrate growth habit than pure subsp. *sativa* (Lesins and Lesins 1979).

As in most major agronomic crops, the application of abundant genetic markers in *Medicago* has generated new perspectives on the evolution, breeding, and genetics of the genus. Molecular markers have been applied in four main areas: (1) analysis of interspecific hybridization, (2) evaluation of genetic variation within and among populations, (3) development of genetic linkage maps, and (4) correspondence of parental genetic similarity with hybrid or synthetic yield.

2.   Interspecific hybridization

Attempts have been made to hybridize *M. sativa* with most other *Medicago* species (Lesins and Lesins 1979; McCoy and Bingham 1988). Twelve perennial species have been successfully hybridized with *M. sativa* although some of the hybridizations required the use of ovule-embryo culture in order to recover the interspecific progeny (McCoy and Smith 1986). A plant derived from an alfalfa x *M. scutellata* cross is the only known hybrid between the annual and perennial species (Sangduen et al. 1982). Molecular markers have been used to study interspecific chromosome pairing (McCoy et al. 1991) and to monitor introgression of wild germplasm to alfalfa (McCoy and Echt 1993).

In cases where normal crossing procedures fail to produce interspecific progeny, somatic hybridization can be used (Arcioni et al. 1997). Molecular markers provide a quick and accurate means to confirm the hybrid nature of progeny (Crea et al. 1997). In somatic hybrids between alfalfa and the shrub *M. arborea*, Nenz et al. (1996) found widespread genome reorganization using RFLP markers, not unexpected given the large evolutionary separation between the species. A highly repeated DNA sequence that is dispersed among chromosomes in alfalfa but absent in *M. arborea* may be useful to track these rearrangements (Calderini et al. 1997). Dynamic changes in chromosome numbers and ribosomal gene spacer lengths have also been observed in somatic hybrids within the *M. sativa* complex (Cluster et al. 1996). Taken together, these reports show the usefulness of molecular markers in hybrid analysis.

3. Molecular marker variability

Abundant variability exists in alfalfa, enabling breeders to successfully select virtually any agronomic trait of interest. More extensive morphological variation is found among the species within *Medicago*, with many traits (e.g. pod and leaf shape) varying significantly (Lesins and Lesins 1979; Small and Jomphe 1989). Isozyme variability, including the existence of multiple alleles at a locus, has been demonstrated for alfalfa (Quiros 1982). Restriction fragment length polymorphism (RFLP) markers have shown that large amounts of genetic variability are present within and among plants in various diploid and tetraploid populations (Brummer et al. 1991; Crochemore et al. 1996; Ghérardi et al. 1998; Kidwell et al. 1994a; Pupilli et al. 1996; Yu and Pauls 1993b). The high level of variation is not unexpected for an outcrossing species with very low self-fertility.

Despite the high intra-population genetic diversity, these studies have been able to use markers to effectively differentiate among cultivars or populations of different origin. Opinions on the best method to use for discriminating among heterogenous populations differs. Yu and Pauls (1993b) used RAPD markers on bulked DNA samples from 10 plants per population and found that DNA based clustering accurately reflected the known relatedness of the cultivars. Kidwell et al. (1994a) analyzed genetic diversity with RFLP markers on both individual plants and population bulks. They found similarities in clustering with either method but cautioned that information could be lost with bulking because rare alleles were not detected by the hybridization, a problem also encountered with RAPD marker amplification (Yu and Pauls 1993b). Ghérardi et al. (1998) developed a methodology to assess population differences based on analysis of 30 individuals per population. Given the almost certain loss of rare allele data expected with bulking, their method may provide the best estimates of interpopulation genetic distances.

Subspecies *falcata*, and germplasm derived from it, clearly differentiates from subsp. *sativa* (Brummer et al. 1991; Crochemore et al. 1996; Kidwell et al. 1994a). *Crochemore (1996) also found that even though some wild subsp. sativa* accessions had similar patterns of growth as subsp. *falcata*, they belonged to different germplasm groups based on molecular marker data. Differences among other populations are not as striking, but non-dormant germplasm (African or Middle Eastern origin) generally groups apart from semi-dormant, European-derived populations, though some overlap exists (Crochemore et al. 1996; Kidwell et al. 1994a). The non-dormant Peruvian germplasm, which had been introduced into South America by Spanish settlers (Barnes et al. 1977), was distinct from most other germplasm (Kidwell et al. 1994a). Further analyses of population differentiation may enable the construction and maintenance of heterotic groups that could be used for alfalfa yield improvement (Brummer 1999).

Molecular markers were used to differentiate accessions and species of six annual *Medicago* species (Brummer et al. 1995). Variation within and among accessions of a single species was identified, but most accessions of a given species clustered together. Species relatedness reflected traditional taxonomic status. A thorough examination of the population structure of *M. truncatula* was conducted using RAPD markers (Bonnin et al. 1996a; Bonnin et al. 1996b). They reported a restricted gene flow among populations and subpopulations, but intra-population genetic variation was higher than expected. Little population genetic analysis has been conducted on natural populations of *M. sativa*, and further studies are needed on both the perennial and annual species.

4.   Map development and use

Five genetic linkage maps for alfalfa have been published, four in diploid populations (Brummer et al. 1993; Echt et al. 1994; Kiss et al. 1993; Kiss et al. 1998; Tavoletti et al. 1996) and one in a tetraploid population (Brouwer and Osborn 1999). RFLP markers form the structural framework for all the maps, but other markers have been mapped in some populations (Table 1). Nine simple sequence repeat (SSR) markers, the first developed and mapped in alfalfa, have been added to the University of Georgia map (Diwan et al. 1997; Diwan et al. 1999) (Fig. 1). Though difficult to develop, SSR markers appear to be ideal for alfalfa because multiple alleles can be readily observed, they are PCR-based but locus-specific, and they are relatively simple to assay (Diwan et al. 1999).

Table 1.   Ploidy, population structure, map length, percent loci exhibiting segregation distortion, and number and type of loci represented on the five published alfalfa maps.

Reference	Ploidy (2n)	Pop.	Seg. Dist. %	Map Length (cM)	Number Mapped			
					RFLP	RAPD	SSR	Isozyme
Brummer et al., 1993; this chapter	2x	F$_2$	48%	467	143		9	
Kiss et al., 1993, 1998	2x	F2	~50%	523	213a	608		12
Echt et al., 1994	2x	BC	34%	553/603	76	68		
Tavoletti et al., 1996	2x	F1	9%	234/261	105			
Brouwer and Osborn, 1999	4x	BC	4%	443	91			

[a] Includes 80 known genes.

Some correspondence among the extant maps can be seen based on mapping of common markers (Table 2). Linkage group 4 of Brummer et al. (1993) split into two groups (4a and 4b, Fig. 1) when we added more markers. However, Kiss et al. (1993) showed evidence in their population that the two groups should remain linked (Table 2).

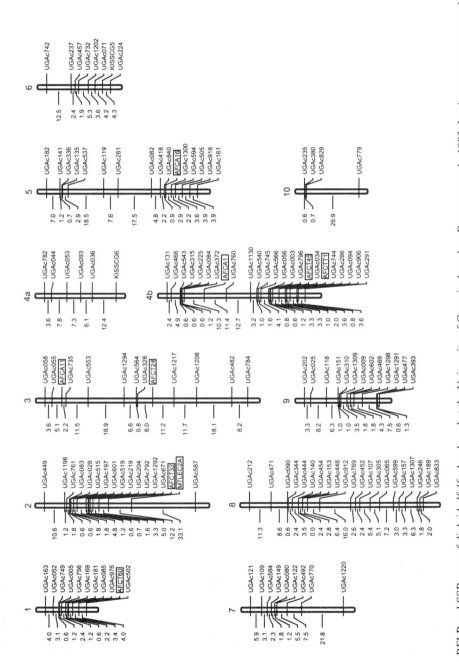

Figure 1. RFLP and SSR map of diploid alfalfa developed at the University of Georgia, based on Brummer et al., 1993. Loci names are presented to the right of linkage groups; boxes are drawn around SSR markers. Map distances (Kosambi cM) are to the left.

Many of the RFLP probes used in all the mapping projects detected more than one locus (>50% in Brouwer and Osborn 1999; Brummer et al. 1993), complicating attempts to relate linkage groups among the various maps. A project currently underway at Iowa State will provide extensive information to fully integrate the extant maps (Shah et al. 1999; Brummer, unpub. results).

Table 2.   Correspondence of linkage groups among the three major mapping projects based on at least two markers in common.

Georgia	Wisconsin	Hungary
1	3	
2		3
3	1	
4a		4
4b	8	4a
5	6,1	6a
6	3	
7	5	
8	7b	
9	5	
10	8a	

a Only one marker mapped in common.

b Correspondence based on the fact that no markers for group 8 (GA) or group 7 (WI) were polymorphic in this population.

In diploids, constructing maps in F_2 populations is straightforward because the F_1 hybrid has only two alleles per locus; heterozygosity in the parents does not affect the F_2. However, mapping a population derived from a cross of two non-inbred plants becomes more complex — recombination in each parent must be considered. Therefore, in these crosses, parent specific maps are developed that can be joined together by markers mapped in common (Ritter et al. 1990). The backcross and F_1 maps developed in alfalfa have been constructed in this manner (Echt et al. 1994; Tavoletti et al. 1996).

Mapping in tetrasomic tetraploid populations has added complexity. Maps of each individual homologue within each parent are created, and the four homologous linkage maps are merged based on common loci to produce a single map of a given chromosome (da Silva and Sorrells 1996; Grivet et al. 1996). The most commonly and easily analyzed marker in these populations are single dose restriction fragments, SDRF (Wu et al. 1992), or more generically, single dose alleles, SDA (Diwan et al. 1999). A particular marker allele present on only one of the four homologues (i.e. a single dose allele) in only one parent will segregate 1:1 for presence:absence in an F_1 population. Double and triple dose markers can be described similarly; DDA will segregate 5:1 in an F_1; TDA will not segregate in the F_1 of a tetrasomic tetraploid without double reduction. Linking homologues can be done either by mapping multiple SDA per marker or by linking DDA to two homologues directly (da Silva and Sorrells 1996). Linkage groups are assumed to be homologues if SDA alleles from multiple markers map on each one; SDA from only one marker that map to two linkage groups may represent a locus duplication that may or may not fall on a homologue (Grivet et al. 1996). Yu and Pauls (1993c) derived maximum likelihood formulae for various dominant marker linkages in tetraploid populations. Brouwer and Osborn (1999) developed the first complete map of tetraploid alfalfa. Several embryonic tetraploid maps have also been reported (Diwan et al. 1999; Yu and Pauls 1993c; Shah et al. 1999).

The length of the alfalfa genome is about 500 cM based on the five estimates of current maps (excluding Tavoletti et al. 1996, due to few markers). The genome size of alfalfa has been estimated at 0.86 pg using flow cytometry (Blondon et al. 1994). Based on this estimate, one cM on the alfalfa map corresponds to approximately 1600 kb.

Segregation distortion was observed for some loci on all the maps. The diploid F_2 population maps suffered severely, with nearly 50% of markers showing skewed segregation, often toward excess heterozygotes (Brummer et al. 1993; Kiss et al. 1993). The map constructed in a backcross population showed 34% distortion (Echt et al. 1994), but an F_1 mapping population had only 9% (Tavoletti et al. 1996). Thus, populations with higher levels of inbreeding exhibited increased segregation distortion. Alfalfa suffers severely with inbreeding due to the exposure of deleterious recessive alleles suggesting that maintaining heterozygosity, either multiple alleles at a locus or linked dominant alleles in repulsion, is necessary for survival (Bingham et al. 1994). The tetraploid backcross map showed little segregation distortion, probably due to the limited exposure of homozygotes during inbreeding a tetrasomic plant. In general, loci with segregation distortion clustered on several linkage groups, making those regions potential targets for studying the control of heterosis and inbreeding depression.

Maps are being developed primarily to identify and manipulate chromosomal regions (and eventually genes) associated with traits of agronomic interest. Five traits controlled by single genes have been mapped to date. Kiss et al. (1993) have mapped two phenotypic mutants, dwarf and sticky leaf, and two loci controlling flower color, anthocyanin and xanthophyll. Brouwer and Osborn (1997) located the unifoliate leaf/ cauliflower head mutation to linkage group 4 on the Wisconsin map. A RAPD marker has been identified with a weak linkage to somatic embryogenesis, but a complete map was not developed (Yu and Pauls 1993a).

Mapping quantitative trait loci (QTL) for agronomically important traits has only recently been attempted. Preliminary results identifying QTL for aluminum tolerance have been reported in a diploid population (Sledge et al. 1997). Mapping winterhardiness and fall dormancy (Brouwer et al. 1998) and forage yield and plant height (Shah et al. 1999) mark the first attempts at using molecular markers to target specific agronomically important traits in tetraploid germplasm.

With the ascension of *M. truncatula* to model system status, genetic maps are being constructed in a second *Medicago* species (Thoquet et al. 1999). Over 260 RAPD, AFLP, isozyme, and SSR markers have been added to this map. Preliminary results indicate that the *M. truncatula* genome is syntenous with *M. sativa* (Cook 1999). Hopefully, mapping information gained from one species will be equally useful in the other.

5.   Marker diversity and yield

Tetrasomic inheritance, severe inbreeding depression, high levels of genetic variability, and the commercialization of synthetic cultivars restrict approaches toward integrating molecular markers into a breeding program. However, markers may be useful as an aid to identify superior parents for population or cultivar development. Conceptually, the genetic similarity or distance among a set of potential parents could be determined from marker data and the most divergent genotypes chosen as parents of a synthetic cultivar. Many studies have shown that genetic diversity does not accurately predict progeny performance in many cases (e.g. Godshalk et al. 1990; Lee et al. 1989), mainly because randomly chosen markers introduce too much noise into the analysis either by not being linked to the genes of interest or by too many genes of interest not being linked to the markers (Bernardo 1992; Charcosset et al. 1991).

However, Kidwell et al. (1994b) found an association between genetic distance and forage yield in tetraploid alfalfa, raising hopes that markers could be used for parental selection in an alfalfa breeding program. A subsequent study using commercial breeding material did not find any clear pattern among populations developed from different numbers of parents selected based on either RFLP similarity or dissimilarity (Kidwell et al. 1999). The lack of clear trends may be due to the selection of random markers, as discussed above, or to the limited genetic base from which the plants were sampled. Using the non-dormant germplasm CUF101, Sledge et al. (1998) screened AFLP markers on a population of 120 individuals to select sub-populations of 12 and 24 parents based on genetic distance. They then compared the performance of synthetics developed from 12, 24, or 120 genotypes. Though their yield data are not available, they did not notice any shift in disease response in these populations as the parental pool narrowed from 120 to 12.

Using markers known to be linked to genes controlling the traits of interest (e.g. yield) to select parents from a broad genetic base may represent a clearer strategy to successfully identify superior genotypes. A plan to identify potential heterotic groups based on marker diversity has been proposed (Brummer 1999). In this scheme, selection within each group would be conducted independently. Superior plants from each group could be increased in isolation to produce foundation seed, which could then be mixed in the production field resulting in partially hybrid certified seed. This method could effect greater yield progress than has been demonstrated previously, but strict separation among heterotic groups would need to be maintained.

Molecular markers will undoubtedly continue to enlighten our understanding of the *Medicago* genome. Future molecular marker studies in alfalfa need to focus on several areas: (1) population structure, (2) selection response, (3) genotype by environment interaction, and (4) marker-assisted selection with multiple alleles. Combined with phenotypic data, marker analyses may lead to substantial improvement in selection methods and breeding strategies.

6. References

Arcioni, S., Damiani, F., Mariani, A. and Pupilli, F. (1997) Somatic hybridization and embryo rescue for the introduction of wild germplasm. In: B.D. McKersie and D.C.W. Brown (eds.), Biotechnology and the improvement of forage legumes, pp. 61-89. CAB International, New York.

Barker, D.G., Bianchi, S., Blondon, F., Dattée, Y., Duc, G., Essad, S., Flament, P., Gallusci, P., Génier, G., Guy, P., Muel, X., Tourneur, J., Dénrié, J. and Huguet, T. (1990) *Medicago truncatula*, a model plant for studying the molecular genetics of the *Rhizobium*-legume symbiosis. Plant Mol. Biol. Rep. 8: 40-49.

Barnes, D.K., Bingham, E.T., Murphy, R.P., Hunt, O.J., Beard, D.F., Skrdla, W.H. and Teuber, L.R. (1977) Alfalfa germplasm in the United States: Genetic vulnerability, use, improvement, and maintenance. USDA Tech. Bull. No. 1571. USDA-ARS, Washington.

Barnes, D.K., Golpen, B.P. and Baylor, J.E. (1988) Highlights in the USA and Canada. In: A.A. Hanson, D.K. Barnes and R.R. Hill, Jr. (eds.), Alfalfa and alfalfa improvement, pp. 1-24. ASA-CSSA-SSSA, Madison, WI.

Bernardo, R. (1992) Relationship between single-cross performance and molecular marker heterozygosity. Theor. Appl. Genet. 83: 628-634.

Bingham, E.T., Groose, R.W., Woodfield, D.R. and Kidwell, K.K. (1994) Complementary gene interactions in alfalfa are greater in autotetraploids than diploids. Crop Sci. 34: 823-829.

Bingham, E.T. and McCoy, T.J. (1979) Cultivated alfalfa at the diploid level: origin, reproductive stability, and yield of seed and forage. Crop Sci. 19: 97-100.

Blondon, F., Marie, D. and Brown, S. (1994) Genome size and base composition in *Medicago sativa* and *M. truncatula* species. Genome 37: 264-270.

Bolton, J.L. (1962) Alfalfa: Botany, cultivation, and utilization. Interscience, New York.

Bonnin, I., Huguet, T., Gherardi, M., Prosperi, J.M. and Olivieri, I. (1996a) High level of polymorphism and spatial structure in a selfing plant species, *Medicago truncatula* (Leguminosae), shown using RAPD markers. Am. J. Bot. 83: 843-855.

Bonnin, I., Prosperi, J.M. and Olivieri, I. (1996b) Genetic markers and quantitative genetic variation in *Medicago truncatula* (Leguminosae): a comparative analysis of population structure. Genetics 143: 1795-1805.

Brouwer, D.J., Duke, S. and Osborn, T.C. (1998) Identifying genetic regions affecting winter survival, fall dormancy, and freezing tolerance in tetraploid alfalfa (*Medicago sativa*L.). Intl. Plant Animal Genome VI: 139. San Diego, CA. 18-22 Jan.

Brouwer, D.J. and Osborn, T.C. (1997) Identification of RFLP markers linked to the unifoliate leaf, cauliflower head mutation in alfalfa. J. Hered. 88: 150-152.

Brouwer, D.J. and Osborn, T.C. (1999) A molecular marker linkage map of tetraploid alfalfa (*Medicago sativa* L.). Theor. Appl. Genet.

Brummer, E.C. (1999) Capturing heterosis in forage crop cultivar development. Crop Sci. 39: Jul-Aug [in press].

Brummer, E.C., Bouton, J.H. and Kochert, G. (1993) Development of an RFLP map in diploid alfalfa. Theor. Appl. Genet. 86: 329-332.

Brummer, E.C., Bouton, J.H. and Kochert, G. (1995) Analysis of annual *Medicago* species using RAPD markers. Genome 38: 362-367.

Brummer, E.C., Kochert, G. and Bouton, J.H. (1991) RFLP variation in diploid and tetraploid alfalfa. Theor. Appl. Genet. 83: 89-96.

Calderini, O., Pupilli, F., Paolocci, F. and Arcioni, S. (1997) A repetitive and species-specific sequence as a tool for detecting the genome contribution in somatic hybrids of the genus *Medicago*. Theor. Appl. Genet. 95: 734-740.

Charcosset, A., Lefort-Buson, M. and Gallais, A. (1991) Relationship between heterosis and heterozygosity at marker loci: a theoretical computation. Theor. Appl. Genet. 81: 571-575.

Cluster, P.D., Calderini, O., Pupilli, F., Crea, F., Damiani, F. and Arcioni, S. (1996) The fate of ribosomal genes in three interspecific somatic hybrids of *Medicago sativa*: three different outcomess including the rapid amplification of new spacer-length variants. Theor. Appl. Genet. 93: 801-808.

Cook, D.R. (1999) *Medicago truncatula* as a nodal species for comparative and functional legume genomics. Proc. Plant and Animal Genome VII: 57. San Diego. 17-21 Jan. Sherago, Int., NY.

Crea, F., Calderini, O., Nenz, E., Cluster, P.D., Damiani, F. and Arcioni, S. (1997) Chromosomal and molecular rearrangements in somatic hybrids between tetraploid *Medicago sativa* and diploid *Medicago falcata*. Theor. Appl. Genet. 95: 1112-1118.

Crochemore, M.-L., Huyghe, C., Kerlan, M.C., Durand, F. and Julier, B. (1996) Partitioning and distribution of RAPD variation in a set of populations of the *Medicago sativa* complex. Agronomie 16: 421-432.

da Silva, J.A.G. and Sorrells, M.E. (1996) Linkage analysis in polyploids using molecular markers. In: P.P. Jauhar (ed.) Methods of genome analysis in plants, pp. 211-228. CRC Press, Boca Raton, FL.

Diwan, N., Bhagwat, A.A., Bauchan, G.B. and Cregan, P.B. (1997) Simple sequence repeat DNA markers in alfalfa and perennial and annual *Medicago* species. Genome 40: 887-895.

Diwan, N., Bouton, J.H., Kochert, G. and Cregan, P.B. (1999) Mapping simple sequence repeats (SSR) DNA markers in diploid and tetraploid alfalfa. Theor. Appl. Genet. [in press].

Echt, C.S., Kidwell, K.K., Knapp, S.J., Osborn, T.C. and McCoy, T.J. (1994) Linkage mapping in diploid alfalfa *Medicago sativa* L. Genome 106: 123-137.

Ghérardi, M., Mangin, B., Goffinet, B., Bonnet, D. and Huguet, T. (1998) A method to measure genetic distance between allogamous populations of alfalfa (*Medicago sativa*) using RAPD molecular markers. Theor. Appl. Genet. 96: 406-412.

Godshalk, E.B., Lee, M. and Lamkey, K.R. (1990) Relationship of restriction fragment length polymorphisms to single-cross hybrid performance in maize. Theor. Appl. Genet. 80: 273-280.

Grivet, L., D'Hont, A., Roques, D., Feldmann, P., Lanaud, C. and Glaszmann, J.C. (1996) RFLP mapping in cultivated sugarcane (*Saccharum* spp.): Genome organization in a highly polyploid and aneuploid interspecific hybrid. Genetics 142: 987-1000.

Kidwell, K.K., Austin, D.F. and Osborn, T.C. (1994a) RFLP evaluation of nine *Medicago* accessions representing the original germplasm sources for North American alfalfa cultivars. Crop Sci. 34: 230-236.

Kidwell, K.K., Bingham, E.T., Woodfield, D.R. and Osborn, T.C. (1994b) Relationships among genetic distance, forage yield and heterozygosity in isogenic diploid and tetraploid alfalfa populations. Theor. Appl. Genet. 89: 323-328.

Kidwell, K.K., Hartweck, L.M., Yandell, B.S., Crump, P.M., Brummer, J.E., Moutray, J. and Osborn, T.C. (1999) Forage yields of alfalfa populations derived from parents selected on the basis of molecular marker diversity. Crop Sci. 39: 223-227.

Kiss, G.B., Csan·di, G., K·lm·n, K., Kaló, P. and ...krész, L. (1993) Construction of a basic genetic map for alfalfa using RFLP, RAPD, isozyme, and morphological markers. Mol. Gen. Genet. 238: 129-137.

Kiss, G.B., Kaló, P., Kiss, P., Felföldi, K., Kereszt, A. and Endre, G. (1998) Construction of an improved genetic map of diploid alfalfa (*Medicago sativa*) using a novel linkage analysis for chromosomal regions exhibiting distorted segregation. In: C. Elmerich, A. Kondorosi and W.E. Newton (eds.), Biological nitrogen fixation for the 21st century, pp. 313. Kluwer, Dordrecht.

Lee, M., Godshalk, E.B., Lamkey, K.R. and Woodman, W.W. (1989) Association of restriction fragment length polymorphisms among maize inbreds with agronomic performance of their crosses. Crop Sci. 29: 1067-1071.

Lesins, K. and Gillies, C.B. (1972) Taxonomy and cytogenetics of *Medicago*. In: C.H. Hanson (ed.) Alfalfa science and technology, pp. 53-86. ASA-CSSA-SSSA, Madison, WI.

Lesins, K.A. and Lesins, I. (1979) Genus *Medicago* (Leguminosae): A taxogenetic study. Kluwer, Dordrecht, The Netherlands.

McCoy, T.J. and Bingham, E.T. (1988) Cytology and cytogenetics of alfalfa. In: A.A. Hanson, D.K. Barnes and R.R. Hill, Jr. (eds.), Alfalfa and alfalfa improvement, pp. 737-776. ASA-CSSA-SSSA, Madison, WI.

McCoy, T.J. and Echt, C.S. (1993) Potential of trispecies bridge crosses and random amplified polymorphic DNA markers for introgression of *Medicago daghestanica* and *M. pironae* germplasm into alfalfa (*M. sativa*). Genome 36: 594-601.

McCoy, T.J., Echt, C.S. and Mancino, L.C. (1991) Segregation of molecular markers supports an allotetraploid structure for *Medicago sativa* x *Medicago papillosa* interspecific hybrid. Genome 34: 574-578.

McCoy, T.J. and Smith, L.Y. (1986) Interspecific hybridization of perennial *Medicago* species using ovule-embryo culture. Theor. Appl. Genet. 71: 772-783.

Michaud, R., Lehman, W.F. and Rumbaugh, M.D. (1988) World distribution and historical development. In: A.A. Hanson, D.K. Barnes and R.R. Hill, Jr. (eds.), Alfalfa and alfalfa improvement, pp. 25-91. ASA-CSSA-SSSA, Madison, WI.

Nenz, E., Pupilli, F., Damiani, F. and Arcioni, S. (1996) Somatic hybrid plants between the forage legumes *Medicago sativa* L. and *Medicago arborea* L. Theor. Appl. Genet. 93: 183-189.

Pupilli, F., Businelli, S., Paolocci, F., Scotti, C., Damiani, F. and Arcioni, S. (1996) Extent of RFLP variability in tetraploid populations of alfalfa, *Medicago sativa*. Plant Breed. 115: 106-112.

Quiros, C.F. (1982) Tetrasomic inheritance for multiple alleles in alfalfa. Genetics 101: 117-127.

Quiros, C.F. and Bauchan, G.R. (1988) The genus *Medicago* and the origin of the *Medicago sativa* complex. In: A.A. Hanson, D.K. Barnes and R.R. Hill, Jr. (eds.), Alfalfa and alfalfa improvement, pp. 93-124. ASA-CSSA-SSSA, Madison, WI.

Ritter, E., Gebhardt, C. and Salamini, F. (1990) Estimation of recombination frequencies and construction of RFLP linkage maps in plants from crosses between heterozygous parents. Genetics 125: 645-654.

Sangduen, N., Sorensen, E.L. and Liang, G.H. (1982) A perennial x annual *Medicago* cross. Can. J. Genet. Cytol. 24: 361-365.

Shah, M.M., Luth, D., Brummer, E.C., Council, C.L. and Kunz., R.C. (1999) Molecular mapping of QTLs for yield heterosis in tetraploid alfalfa. Proc. Plant and Animal Genome VII Conf.: http://www.intl-pag.org/pag/7/abstracts/ag7020.html. San Diego. 17-21 Jan. Sherago

International, NY.

Sledge, M.K., Bouton, J.H. and Kochert, G. (1998) Molecular marker diversity as a means of selecting parents for synthetic cultivars. Proc. North Amer. Alfalfa Impr. Conf.: 73-74. Bozeman, MT. 1-6 Aug.

Sledge, M.K., Bouton, J.H., Tamulonis, J., Parrott, W.A. and Kochert, G. (1997) Aluminum tolerance QTL in diploid alfalfa. Proc. XVIII Int. Grassland Congr.: 4.9-4.10. Winnepeg, Canada. 8-19 June.

Small, E. and Jomphe, M. (1989) A synopsis of the genus *Medicago* (Leguminosae). Can. J. Bot. 67: 3260-3294.

Tavoletti, S., Veronesi, F. and Osborn, T.C. (1996) RFLP linkage map of an alfalfa meiotic mutant based on an F_1 population. J. Hered. 87: 167-170.

Thoquet, P., Gherardi, M., Kereszt, A., Tirichine, L., Prosperi, J.-M. and Huguet, T. (1999) Genetic mapping of the Mediterranean model legume *Medicago truncatula*. Proc. Plant and Animal Genome VII: 135. San Diego. 17-21 Jan. Sherago Int., NY.

Wu, K.K., Burnquist, W., Sorrells, M.E., Tew, T.L., Moore, P.H. and Tanksley, S.D. (1992) The detection and estimation of linkage in polyploids using single-dose restriction fragments. Theor. Appl. Genet. 83: 294-300.

Yu, K.F. and Pauls, K.P. (1993a) Identification of a RAPD marker associated with somatic embryogenesis in alfalfa. Plant Mol. Biol. 22: 269-277.

Yu, K.F. and Pauls, K.P. (1993b) Rapid estimation of genetic relatedness among heterogeneous populations of alfalfa by random amplification of bulked genomic DNA samples. Theor. Appl. Genet. 86: 788-794.

Yu, K.F. and Pauls, K.P. (1993c) Segregation of random amplified polymorphic DNA markers and strategies for molecular mapping in tetraploid alfalfa. Genome 36: 844-851.

11. An integrated RFLP map of *Arabidopsis thaliana* *

HOWARD M. GOODMAN[1], SUSAN HANLEY[1], SAM CARTINHOUR[1],
J. MICHAEL CHERRY[1], BRIAN HAUGE[1], ELLIOT MEYEROWITZ[2],
LEONARD MEDRANO[2], SHERRY KEMPIN[2],
PIET STAMM[3], *and* MAARTEN KOORNNEEF[3]

[1] *Department of Genetics, Harvard Medical School and Molecular Biology,*
Massachusetts General Hospital, Boston, MA 02114, U.S.A.
<Goodman@frodo.mgh.harvard.edu>
[2] *Division of Biology 156-29, California Institute of Technology, Pasadena, CA 91124, U.S.A.*
[3] *Department of Genetics, Agricultural University of Wageningen,*
Dreijenlaan 2, 6703 HA Wageningen, The Netherlands

Contents

1. Introduction

Over the past several years, *Arabidopsis thaliana* has gained increasing popularity as a model system for the study of plant biology. Its short life cycle, small size and large seed output make it well suited for classical genetic analysis (reviewed in Meyerowitz 1987). Mutations have been described affecting a wide range of fundamental developmental and metabolic processes (reviewed in Estelle and Somerville 1986). A genetic linkage map consisting of some 90 loci has been assembled (Koornneef 1987) and an increasing number of cloned genes are available. In addition, *Arabidopsis* is ideally suited for physical mapping studies since it has a very small genome (approximately 100,000 kb; Hauge and Goodman, unpub. Result) and a remarkably low content of interspersed repetitive DNA (Pruitt and Meyerowitz 1986). The availability of a complete physical map of the *Arabidopsis* genome will greatly simplify the cloning of any gene based solely on its mutant phenotype and genetic map location. For the map to be of any utility it is necessary to align the physical map with the classical genetic linkage map via an RFLP map.

The current RFLP map of *Arabidopsis thaliana* is presented in Table 1, see pages 183-186. Using methods essentially equivalent to the two previously published RFLP maps (Chang et al. 1988; Nam et al. 1989) the number of markers in each data set have been increased and the independent data sets have been mathematically integrated into a single map using the computer program 'Joinmap' (Stam 1993). It should be noted, however, that both the resolution of the RFLP map and the alignment of the RFLPs with the genetic linkage map are somewhat limited. Therefore, it is currently difficult to assign with confidence a precise position on the classical genetic map for a given RFLP probe since

*Reprinted without change from the first edition.

181

R.L. Phillips and I.K. Vasil (eds.), DNA-Based Markers in Plants, 181 - 186.

only a few genetic markers per chromosome were used to align the RFLP map with the genetic linkage map. In addition, due to the necessarily limited number of progeny used for the segregation analysis, we estimate that the resolution of the RFLP map is only on the order of 2 cM. In our early mapping efforts we primarily worked with randomly selected bacteriophage (Chang et al. 1988) or cosmid clones (Nam et al. 1989). Subsequent efforts have mainly been devoted to mapping cloned genes obtained from various laboratories as indicated in Table 1. The majority of genes that have been mapped have not been genetically identified. In other words their cognate mutation is not known. In a few cases both the gene and the genetic locus is known. For example, *ag* (agamous) (Yanofsky et al. 1990), *ap3* (apetala 3) (Jack et al. 1992), and *tt-3* and *tt-5* (transparent testa 3 and 5) (Shirley et al. 1992) have been cloned. The mapping of these loci provide immediate contact points between the physical map and the classical genetic linkage map. As new loci are mapped with respect to the RFLPs, additional contact points will be established and the map will be refined.

References

Chang, C., Bowman, J.L., DeJohn, A.W., Lander, E.S. and Meyerowitz, E.M. (1988) Restriction fragment length polymorphism linkage map for *Arabidopsis thaliana*. Proc. Natl. Acad. Sci. U.S.A. 85: 6856-6860.

Cherry, J.M., Cartinhour, S. and Goodman H.M. (1992) AAtDB, an *Arabidopsis thaliana* database. Plant Mol. Biol. Rep. 10: 308-309, 409-410.

Estelle, M.A. and Somerville, C.R. (1986) The mutants of *Arabidopsis*. Trends Genet. 2: 89-93.

Hauge, B.M., Hanley, S.M., Cartinhour, S., Cherry, J.M., Goodman, H.M., Koornneef, M., Stam, P., Chang, C., Kempin, S., Medrano, L. and Meyerowitz, E.M. (1993) An integrated genetic/RFLP map of the *Arabidopsis thaliana* genome. Plant J. 3: 745-754.

Jack, T., Brockman, L.L. and Meyerowitz, E.M. (1992) The Homeotic gene APETALA3 of *Arabidopsis thaliana* encodes a MADS box and is expressed in petals and stamens. Cell 68: 683-697.

Koornneef, M. (1987) Linkage map of *Arabidopsis thaliana* (2n = 10). In: S.J. O'Brien (ed.), Genetic Maps, pp. 742-745. Cold Spring Harbor Laboratory Press, Cold Spring Harbor, NY.

Meyerowitz, E.M. (1987) *Arabidopsis thaliana*. Ann. Rev. Genet. 21: 93-111.

Nam, H.-G., Giraudat, J., den Boer, B., Moonan, F., Loos, W.D.B., Hauge, B.M. and Goodman, H.M. (1989) Restriction fragment length polymorphism linkage map of *Arabidopsis thaliana*. Plant Cell 1: 699-705.

Pruitt, R.E. and Meyerowitz, E.M. (1986) Characterization of the genome of *Arabidopsis thaliana*. J. Mol. Biol. 187: 169-183.

Shirley, B., Hanley, S. and Goodman, H.M. (1992) Effects of ionizing radiation on a plant genome: analysis of two Arabidopsis *transparent testa* mutations. Plant Cell 4: 333-347.

Stam, P. (1993) Construction of integrated genetic linkage maps by means of a new computer package: JOINMAP. Plant J. 3: 739-744.

Yanofsky, M.F., Ma, H., Bowman, J.L., Drews, G.N., Feldmann, K.A. and Meyerowitz, E.M. (1990) The protein encoded by *Arabidopsis* homeotic gene agamous resembles transcription factors. Nature 346: 35-39.

Table 1.   Classical genetic markers are indicated in italics (Koornneef, 1987) and classical genetic markers which have also been cloned and RFLP mapped are indicated in bold type. These are as follows: *chr 1*: *an*, angustifolia; **chl3**, chlorate resistant; *chl*, chlorina; *apl*, apetala; *g/2*, glabra; *c/v1*, clavata; **ga2**, gibberellin requiring; *chr2*: **hy3**, long hypocotyls; *cp2*, compacta; *er*, erecta; *as*, asymmetric leaves and lobed leaves; *py*, pyrimidine requiring; *cer8*, eceriferum; *chr3*: *hy2*, long hypocotyls; *dwf*, auxin resistant dwarf; **abi3**, abscisic acid insensitive; *gll*, glabra; **ap3**, apetala; **tt5**, transparent testa; *chr4*: **gal**, gibberellin requiring; *bp*, brevipedicellus; **ag**, agamous; *cer2*, eceriferum; *ap2*, apetala; and *chr5*: **tt4**, transparent testa; *pi*, pistillata; *ttg*, transparent testa glabra; **tt3**, transparent testa; *tz*, thiozole requiring; *biol*, biotin auxotroph. All other markers in plain text have been cloned and RFLP mapped, but not correlated with a mapped mutation. RFLP markers with names of the form m-# are random bacteriophage clones from the Meyerowitz laboratory (Chang et al. 1988) previously designated either bAt-# or LEM-#, names of the form pCITd-#, pCIT-# or pCITN7 — are random cosmid clones from the Meyerowitz laboratory, and names of the form g-# are random cosmid clones from the Goodman laboratory (Nam et al. 1989). The sources of the other markers can either be found in the publically available database AatDB (Cherry et al. 1993) or will be published elsewhere (Hauge et al. 1993). The numbers under the cM heading indicate the position on the chromosome using the Kosambi mapping function in Joinmap.

CHR1	cM	CHR2	cM	CHR3	cM	CHR4	cM	CHR5	cM
pvv4	0	pCITd112a	0	myb	0	g6844	0	pCITd94	0
g21491	6	m246	10.8	pCITd39	2.6	g3843	4	g21488	1.2
an	14.5	pCIT1291	13.2	GAP-C	2.7	U2.9snRNA	6	ubq6121	2.3
m488	23.7	g4133	14.7	m302	3.5	gal-14	7.5	pAtT80	3.9
g5972	25.1	g4532	17	CD06119	4.2	**gal**	11.3	m562	5.4
pCITd91	25.4	m497	17.7	m262	5.4	**bp**	13.5	m447	7.3
IPhAraI	26.2	g4553	18.9	m472	7.6	auxinBP	15.6	g3715	8.2
GTPbp	27.7	al4G4	31.8	g3838	8.9	g2616	16.2	CD05629	10.4
g4715	28	pCITd100	34.6	m583	9.4	m506	17.5	m217	10.9
la8	28.1	**hy3**	34.9	peaf	10.5	m456	18.8	KG-31	13.3
g19821	28.6	m216	36.4	g4119	11.7	m448	22.2	U2.5snRNA	13.4
m322	28.8	m104	39.2	g17341	12.5	PK-87	23.8	g3837	14.2
g19857	29.7	PK-20	39.4	m243	12.6	pCITf3	28.6	ASA-1	14.5

H.M. Goodman, S. Hanley, S. Cartinhour, J.M. Cherry, B. Hauge, E. Meyerowitz, L. Medrano, S. Kempin, P. Stamm, and M. Kornneef

Table 1. Continued.

CHR1	cM	CHR2	cM	CHR3	cM	CHR4	cM	CHR5	cM
g5957	29.7	m605	40.6	g4523	13.3	pCITd23	29.1	PBS811	16.3
m333	29.8	**cp2**	42.4	hsp70-9	14.7	m518	31.4	m224	17.4
m241	31.7	m465	44.4	*hy2*	14.8	m210	37.4	g6830	18
m219	33.3	m251	46.7	g4547	16.1	g6837	38.8	pCIT1243	19.9
g21497	35.1	Gpal	55.4	g2488-a	16.2	m326	40.1	g3021	20
g2358	38.6	g6842	56.2	pCIT7P	18.3	m580	41.5	**tt4**	20.1
pCITN7-31	39.2	*er*	57.1	m228	19.1	g10086	42.1	pCITd37	21.3
g3786	39.4	PBS707	60.8	m317	19.3	g4565-a	43.8	g5962	22.3
m235	44.6	m220	62.2	KG-17	19.8	m226	44.3	g4568	23.8
g2395	45.8	m283	63.3	m560	22.4	pCITd71	45.9	pCIT718	25.5
g3829	47.3	ASA-2	63.6	*dwf*	22.5	**ag**	48.7	**AR**119	25.8
g17286	50.7	g21502	65.7	GS.KB6	22.6	g4539	49.9	pCITf16	27
m215	51.2	m323	65.8	g4708	23.1	g3845	50.9	pGATC-11	27
m310	51.5	m429	68.7	pCIT1240	24.1	g19833	53	g4111	27.2
g19834	53.9	g6191	69.2	g6220	26.3	g19838	53.2	g4560	27.5
m271	54.5	g17288	69.5	y6-L31	27.7	g3883	54	g2632	28.2
m201	55.5	g6825	69.5	m255	28.7	m557	54.9	*pi*	30.3
m321	55.9	*as*	69.7	m105	29.6	06455CD	57.3	g4556	30.4
m402	57	*py*	69.7	**abi3**	30.4	*cer2*	58.7	g6843	32.3
m335	57.4	g13808	70	BWS12	31.9	pCITd104	62.7	m291	34.2
m254	57.7	m551	71	m433	32.8	KG-32	63.2	*ttg*	34.4
rpHS-1	58.2	g4514	73.8	g17287	34.8	m272	63.7	KG-10	35.4
m253	58.6	g1789	74	KG-23	37	g8300	64.1	g6856	37.2
m299	62.4	pATC4	75.3	GAP-A	37.1	TSB2	64.3	CD06455f	43.5

Table 1.   Continued.

CHR1	cM	CHR2	cM	CHR3	cM	CHR4	cM	CHR5	cM
GAP-B	67.5	m336	80.2	g2440	39.7	g17340	64.5	GS-L1	45.3
pAtT12-1	67.8	cer8	83.6	g/l	47.3	g4513	64.6	phyC	48.7
chl3	67.9	pCIT4241	86.9	pAt3-89.1	56.7	m600	64.8	GS-R1	51.8
g512av	71.4	pCITN7-26	100.2	g4117	63.4	pCITd76	56.5	m247	52.3
chl	72.1			m281b	64.1	PG11	67.4	m423	54.8
m213	72.9			g2534	64.2	g3088	69.8	m422	54.9
m281a	73.3			m249	65.2	Wyac23H12	71.7	g4028	57.6
m280	81.9			g4564-b	65.9	m214	72	pCITd90	59.5
KG-20	83.2			PB3	68	ap2	73	**tt3**	60.9
m1511b	83.9			**csr**	69.5	pCITd99	72.6	pAt5-91.5	61.5
m1511a	84.9			CAB11	69.7	pCIT1212	75.3	sAT2105	62.6
g4026	85			m409	70.6	g4551	80	m225	65.4
BWS15	87.6			m576	72	g2486	80.3	PBS813a	72.3
m305	89.7			m457	73.2	la5	83.6	m331	73.3
g6836	90.1			g4014	73.8	g3265	84.5	m268	73.8
g2488-b	90.1			apZL	74.7	g3713	85.1	pAD1.7	73.9
g17336	90.4			g17811	77.8	DHS1	88.4	KG-8	75
m315	91.3			**ap3**	78.2	DHS2	92.1	m435	80.5
m421	92.3			Syac1G12	78.2			g4130	80.7
g4121	92.6			g4125	83.2			TSB1	80.8
g4552	92.8			g2606	84.4			m233	82.8
pau10-1	93.5			m115	85.1			*tz*	84.8
pau1-1	94.3			**tt5**	85.9			PBS813b	84.8
g11447	95.1			U2.3snRNA	86.7			g3844	85.9

H.M. Goodman, S. Hanley, S. Cartinhour, J.M. Cherry, B. Hauge, E. Meyerowitz, L. Medrano, S. Kempin, P. Stamm, and M. Kornneef

Table 1. Continued.

CHR1	cM	CHR2	cM	CHR3	cM	CHR4	cM	CHR5	cM
pCITd117	95.8			m339	87.5			3878	86.9
apl	96.9			g2778	88.5			biol	87.3
m252	102			pCIT1210	88.8			g3791	87.9
K-24	103.1			XDB-G	89.3			pCIT4242	89.1
g4721	103.5			m460	89.6			m211	91.2
m453	103.6			m424	92.6			g17337	91.4
Syac1E3	104.1			g19826	97.8			g2455	91.6
TIP	104.4			g5966	102			g2368	93.5
pCIT-N7-24	106.5							pCITd110	95.2
g6836	108.2							pCITd123	96.9
gl2	108.5							g4510	98
m532	109.3							pCITd77	99.4
m3012	111.2							m555	100.2
NIA1	111.4							lox8	101.1
m237	111.6							AB5-13	102.5
g16066	111.8								
clvl	112.6								
ga2	115								
m132	115.7								

12. An integrated map of the barley genome

ANDRIS KLEINHOFS[1] *and* ANDREAS GRANER[2]
[1]Departments of Crop and Soil Sciences and Genetics and Cell Biology, Washington State University, Pullman, Washington 99164-6420, U.S.A. <andyk@wsu.edu>
[2]Institute for Plant Genetics and Crop Plant Research (IPK), Corrensstr 3, D-06466 Gatersleben, Germany

Contents

1.   Introduction

Barley, *Hordeum vulgare* L., is among the oldest cultivated crops dating back to 5000 to 7000 years B.C. (Clark 1967; Harlan 1976; Harlan 1979) and perhaps much older (Wendorf et al. 1979). Barley has also been a favorite genetic experimental organism since the rediscovery of Mendel's laws of heredity (Tschermak 1901; cited from Smith 1951). The widespread use of barley is attributable to its diploid nature (2n = 2x = 14), self fertility, large chromosomes (6-8mm), high degree of natural and easily inducible variation, ease of hybridization, wide adaptability, and relatively limited space requirements. The early literature on the genetics of barley and related information has been reviewed (Smith 1951; Nilan 1964; Nilan 1974; Briggs 1978) and more recently in Rasmusson (Rasmusson 1985) and Shewry (Shewry 1992). A comprehensive listing and description of the genes known at the time can be found in Sogaard and von Wettstein-Knowles (Sogaard and von Wettstein-Knowles 1987), updated by von Wettstein-Knowles (von Wettstein-Knowles 1992). Periodic publications such as the Proceedings of International Barley Genetic Symposia (Lamberts et al. 1964; Nilan 1971; Gaul 1976; Asher et al. 1981; Yasuda and Konishi 1987; Munck 1992; Scoles and Rossnagel 1966) and the annual Barley Genetics Newsletter (since 1971) summarize and update the barley genetics literature.

Cultivated barley (*Hordeum vulgare* subsp. *vulgare*) has seven pairs of distinct chromosomes designated by Arabic numbers 1-7 (Nilan 1964; Ramage 1985). The five chromosomes without satellites are designated 1 to 5 based on their relative lengths measured at mitotic metaphase, with chromosome 1 being the longest and chromosome

R.L. Phillips and I.K. Vasil (eds.), DNA-Based Markers in Plants, 187 - 199.

5 the shortest. The two chromosomes with satellites are designated 6 and 7, with chromosome 6 having the larger satellite and being shorter than chromosome 7 which has the smaller satellite. The initial association of linkage groups with specific chromosomes was accomplished using translocations (Burnham and Hagberg 1956) and trisomics (Tsuchiya 1960). The ranking of the lengths of the three longest chromosomes has been reported to be different from the original assignments in some studies; however, to avoid confusion, the originally used numbers have been retained by barley geneticists. Each of the seven barley chromosomes can be identified by its distinctive Giemsa C- and N-banding pattern (reviewed in Ramage 1985; Linde-Laursen and Jensen 1992). Based on the few reports of barley pachytene chromosome analyses (Sarvella et al. 1958; Singh and Tsuchiya 1975), the relative lengths and arm ratios agree reasonably well with those published for mitotic metaphase chromosomes.

There is now convincing evidence, based on comparative mapping, that barley chromosomes 1, 2, 3, 4, 5, 6 and 7 are homoeologous to wheat chromosomes 7H, 2H, 3H, 4H, 1H, 6H and 5H, respectively. This conclusion is supported by the occurrence of similar morphological, biochemical and molecular markers on the homoeologous chromosomes and the ability of the barley chromosomes to substitute for the equivalent wheat chromosomes in substitution lines, reviewed in (Shepherd and Islam 1992). Recently, it has been recommended that barley chromosomes be designated according to their homoeologous relationships with chromosomes of other Triticeae species (Linde-Laursen 1997). In this report we use the barley chromosome designations with the Triticeae designations in parentheses in order to avoid confusion with previous literature and the need to convert all databases to the new designations. When using the Triticeae nomenclature it is extremely important to include the genome H symbol in order to avoid confusion in the literature.

The barley genome is large, $1C = 4.9 - 5.3X \ 10^9$ bp (Bennett and Smith 1976; Arumuganathan and Earle 1991). Several advantages, however, make barley the organism of choice among the large genome cereals for molecular mapping. These include the ability to develop doubled haploid (DH) lines by several methods, the availability of numerous mutants and cytogenetic stocks, particularly the barley-wheat addition lines, minimal space requirements, and the recent development of large insert libraries.

Barley doubled haploid lines can be developed by techniques which sample either female or male gametes. One method is based on the observations that when *Hordeum vulgare* female flowers are pollinated with *Hordeum bulbosum*, initial fertilization and zygote formation takes place, but the *H. bulbosum* chromosomes are subsequently eliminated in the hybrid embryos (Kasha and Kao 1970). The haploid embryos can be rescued and homozygous plants developed by tissue culture and chromosome doubling techniques (Chen and Hayes 1989). The resulting DH lines are random samples of female gametes. Alternatively, in vitro regeneration from anthers or isolated microspores can be used to develop DH lines that represent random samples of male gametes (Clapham 1973; Sunderland and Xu 1982). Comparative mapping with populations developed by the two methods showed that there is increased recombination with the anther culture (male recombination) method, particularly in the telomeric regions. (Devaux et al. 1995). The anther culture method also resulted in more segregation distortion and methylation compared to the bulbosum method (Devaux et al. 1993).

The availability of barley-wheat disomic (Islam 1980) and ditelosomic (Islam 1983) addition lines further facilitates barley genetic studies. These lines contain a normal complement of Chinese Spring wheat chromosomes and a cv. Betzes barley chromosome or chromosome arm. All individual barley chromosome addition lines, except chromosome 5, are fertile and reasonably stable. These lines have been used to organize individual markers and linkage groups into chromosomes and identify centromere locations (Graner et al. 1991; Heun et al. 1991; Kleinhofs et al. 1993).

Two barley YAC libraries, 2X and 4X, respectively (Kleine et al. 1993; Simons et al. 1997) and a 6.3X barley BAC library (Yu et al. 2000) have been developed. The broad availability of the BAC library is expected to greatly facilitate barley physical mapping and map-based gene cloning.

2.   Maps

The first barley RFLP map published was for chromosome 6 (Kleinhofs et al. 1988), followed by a partial map of the whole genome incorporating RFLP, morphological, isozyme and PCR markers (Shin et al. 1990). Extensive and numerous molecular maps of the barley genome have been generated (Graner et al. 1991; Heun et al. 1991; Kleinhofs et al. 1993; Becker et al. 1995; Kasha et al. 1995; Langridge et al. 1995; Meszaros and Hayes 1997; Waugh et al. 1997; Qi et al. 1998) (and updates in the GrainGenes database http://wheat.pw.usda.gov). These maps and others (Hinze et al. 1991; Barua et al. 1993; Devos et al. 1993; Giese et al. 1994; Becker and Heun 1995; Komatsuda et al. 1995; Laurie et al. 1995; Bezant et al. 1996; Laurie et al. 1996; Schonfeld et al. 1996; Ellis et al. 1997), jointly represent over two thousand molecular markers mapped to the barley genome. Unfortunately, even to date, only limited common markers have been placed on the different maps. This means that it is difficult to accurately merge all maps into one extensive map of the barley genome at this time. The most extensive effort to merge barley molecular marker maps by generating a common platform comprising about 100 common markers has been made by our laboratories with the Steptoe x Morex and Igri x Franka maps (unpublished). These data have been used to merge some maps using "JoinMap" (Langridge et al. 1995; Qi et al. 1996). Here we take a different approach. Using the Steptoe x Morex map as a base, we have divided the barley genome in approximately 10 cM intervals to develop a "BIN" map (Fig. 1). This map allows the placement of many markers mapped on different maps in their appropriate BINS. Despite its comparatively low resolution, the BIN map will suffice most demands in terms of the rapid identification of molecular markers for a defined chromosome region. The marker descriptions and their BIN locations are incorporated in an EXCEL spreadsheet, which can be downloaded from (http://barleygenomics.wsu.edu/). The data can also be viewed at the same URL under "Barley Molecular Marker Database".

The EXCEL spreadsheet consists of 26 columns which are described in the spreadsheet. Briefly, column A identifies the BIN marker i.e. the marker closest to the short arm telomere in that BIN, columns B and C provide the locus name and previously used names, respectively. Other very significant columns are F for chromosome location,

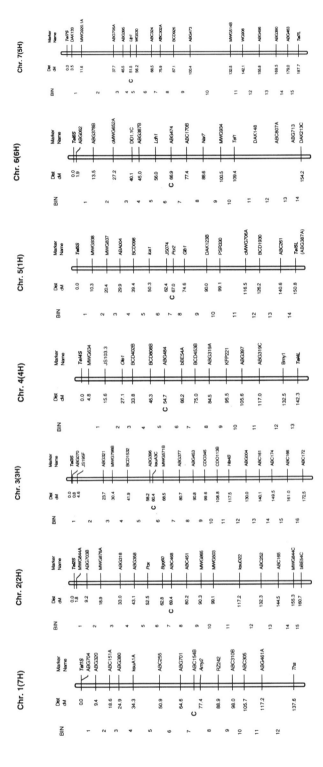

Figure 1. Barley chromosome 1-7 BIN maps. The BIN markers are shown and each BIN is identified by a number. Distances in cM are based on the Kosambi formula and are cumulative starting with 0 at the short arm telomere. Mapped telomere markers and approximate cetromere markers are included. Due to the lack of markers in the Steptoe x Morex chromosome 6L map, markers from the Harrington x Morex map are used. These are not precisely mapped and thus their approximate positions are indicated.

G for BIN location, H for the mapping population, and Y for reference. The other columns provide more detailed information about the locus or probe.

The EXCEL spreadsheet includes markers mapped by our groups and most other markers in the literature that could be reasonably accurately placed on the BIN map. The major map data integrated are from the Steptoe x Morex (SM), Igri x Franka (IF), and Proctor x Nudinka (PN) maps. To date, we have not integrated the majority of the mapped AFLP markers (Qi et al. 1998; Waugh et al. 1997) or the majority of the mapped morphological loci (Franckowiak 1997). Currently (July, '98), the EXCEL program contains information on 1,531 loci, including 1,325 RFLP (591 cDNA and 734 gDNA), 24 RAPD, 30 AFLP, 13 STS, 23 disease resistance, 51 morphological, 8 physiological, 13 isozyme, and 44 SSR loci. The information is updated routinely.

2.1. Centromeres and telomeres

Barley centromeres are readily mapped due to the availability of the ditelosomic addition lines (Islam 1983). Markers are mapped on specific chromosome arms until the switch from one arm to the other is found (Kleinhofs et al. 1993). The barley centromere locations are shown on the BIN map (Fig. 1). Physical mapping of barley chromosomes has also localized the centromeres (Kunzel and Korzun 1996; Kunzel et al. 2000).

All short arms and 4(4H), 5(1H), and 7(5H) long arms, of the 14 barley telomeres have been mapped by combined RFLP/PCR methods using telomere associated sequences (Kilian and Kleinhofs 1992; Kilian et al. 1999) (Fig. 1). Seven putative telomeric markers were also mapped using polymorphisms generated by Pulsed Field Gel Electrophoresis (PFGE) and hybridization with telomere repeat sequences from Arabidopsis and an HvRT repeat sequence from barley (Roder et al. 1993). The telomeric markers mapped by PFGE for short arm of chromosome 2(2H) and 5(1H) are in good agreement with the telomeric markers mapped by RFLP/PCR. The telomeric markers for long arm of chromosome 1(7H) and 3(3H) agrees with the most distal markers on the extensive Steptoe x Morex map. These are included in the BIN map (Fig. 1). The telomeric markers for 4L(4HL) and 6L(6HL) mapped proximal to the telomeric or most distal markers on the more extensive maps and therefore are not included in Fig. 1. The telomeric marker, *TelXba.c* for chromosome 4S (4HS) identified by Roder et al. (Roder et al. 1993) mapped 20 cM distal from the telomeric marker identified by Kilian and co-workers (1998). However, the *TelXba.c* marker data was highly distorted and only 48 DH lines were scored. Thus, we believe that the *Tel4S* locus identified by Kilian is probably more accurate.

In general, there is a good agreement between the two data sets and the discrepancies in map distances can be attributed to different recombination frequencies with different parents and population generation methods. Combining the results of Kilian et al. (1999) and Roder et al. (Roder et al. 1993), only barley chromosome 2L(2HL) is without a telomeric marker and the genetic distance of the barley genome is established at about 1,500 cM, varying slightly with different parents.

2.2. Physical maps

Understanding of the relationship between genetic and physical distances is crucial to the understanding of the genome. The earliest comparisons of genetic and physical distances were based on comparing C-bands to marker genes (Linde-Laursen 1979; Linde-Laursen 1982) and in situ hybridization of ribosomal RNA (Appels et al. 1980), B-hordein (Clark et al. 1989) and ribosomal 5S RNA (Kolchinsky et al. 1990) loci to RFLP markers (Kleinhofs et al. 1988; Kleinhofs et al. 1993). Integration of the *in situ* hybridization physical and genetic maps have been made (Fukui and Kakeda 1990; Pedersen et al. 1995).

Physical mapping of the *Hor1* and *2* loci was accomplished by Pulsed Field Gel Electrophoresis (PAGE) (Sorensen 1989; Siedler and Graner 1991).

A comprehensive comparison of the barley physical and genetic RFLP maps has been completed (Kunzel et al. 2000). This work involves PCR mapping of single copy RFLP probes on micro-isolated individual translocation and control chromosomes. The data is in agreement with earlier studies showing suppressed recombination in the chromosome proximal regions, but the quantitative data show significant variation among individual chromosomes. The physical/genetic distances varied between 0.1 to 298 Mb/cM.

Detailed comparison of physical and genetic distances have been made at the *mlo rpg4* and *Mla* loci (DeScenzo and Wise 1996; Simons et al. 1997). These data clearly illustrate the extreme variation in the barley physical/genetic distances even within short chromosome regions.

2.3. High resolution maps

High resolution maps have been prepared for several barley genome regions. The earliest such maps involved the *lig* (previously *li*), *mlo*, and *wax* (previously *wx* or *glx*) loci (Jorgensen and Jensen 1979; Konishi 1981; Rosichan et al. 1979). Molecular marker based high resolution maps have been produced for the disease resistance loci *Rpg1*, *Mla*, *rpg4* and *mlo* (DeScenzo et al. 1994; Kilian et al. 1997; Simons et al. 1997; Wei et al., 1999; Druka et al., 2000).

2.4. Comparative maps

It is now reasonably well established that the barley genome contains few, if any, rearrangements, at least among the parents used for extensive mapping. We have been used to map approximately 100 common probes on the SM and IF maps without finding any contradictions in their locations. Probes identifying a single locus have been mapped on as many as 5 different maps and always appear to detect in the same position on all maps, as judged by their position relative to other markers. There are a few reports of single copy probes not mapping to the same position in different crosses. Although real rearrangements can not be completely excluded, human error is the most likely explanation in the majority of cases. A few sequences have been identified whose map location seems to have changed in relatively recent times. These are based on probes that may produce a hybridization band in one parent that is missing in another parent. In most cases, no allelic bands are found, although other bands hybridizing to the same probe may occur representing a different locus. Some examples are BCD351, ABG078, and *Cat2*. The explanation for these apparent deletion and/or insertion events is not known.

The barley genome is highly collinear, with few minor inversions, with that of the wheat genomes A, B, and D (Devos et al. 1993; Namuth et al. 1994; Van Deynze et al. 1995; Dubcovsky et al. 1996). Thus, the comparisons of the wheat genomes with other species apply to the barley genome as well. Colinearity of the barley and rice genomes have been reported in general (Saghai Maroof et al. 1996) and in specific regions (Dunford et al. 1995; Kilian et al. 1997; Han et al. 1998). These studies indicate that there are regions of very high gene order collinearity between barley and rice. However, there are also rearrangements and probable transpositions (Han et al., 1999). Nevertheless, rice has proven to be a very good model and source of probes for saturation of specific barley chromosome regions.

3.   Markers

3.1. RFLP

Restriction Fragment Length Polymorphism (RFLP) markers remain the cornerstone of most molecular maps due to their reliability and transferability. Probes from many different sources have been used to map the barley genome and are described in detail in the EXCEL spreadsheet, to Dr. Arnis Druka for developing the Barley Molecular Marker Database. The RFLP markers are either detected by anonymous cDNA or genomic DNA probes or cloned known-function probes (mostly cDNA). Probes from other species such as wheat, *Triticum tauschii*, oat, rice, sorghum and maize hybridize well with barley DNA and have also been used. The known-function gene sequences are designated consistent with previously used or suggested gene symbols or, if none, a symbol is assigned following established guidelines (Barley Genetics Newsletter vol. 26). In cases where it was not possible to determine which previously used locus designation corresponded to the locus uncovered by the RFLP probe, a new locus number was used.

Many different gene and marker loci designations are in use, consequently, we have tried to incorporate all known previous designations of the listed loci in column C of the EXCEL spreadsheet.

3.2. PCR

Polymerase Chain Reaction (PCR) based markers have revolutionized methods for detecting polymorphisms. These include RAPD (Random Amplified Polymorphic DNA), SSR (Simple Sequence Repeat, also known as microsatellites), STS (Sequence Tagged Sites including all types of amplification based on specific pairs of primers flanking a defined sequence), AFLP (Amplified Fragment Length Polymorphism), and SAMPL

(Selective Amplification of Polymorphic Loci). In spite of some limitations, these markers provide rapid and less expensive means for generating large amounts of data.

4.   Summary

Barley molecular map development has been explosive during the past few years. As of this writing, there are over two thousand molecular markers mapped to the seven barley chromosomes and the number is increasing daily. There are doubled haploid, therefore "immortal", populations available for future mapping. These should serve the barley genetics community for many years to come. Additional work is needed to identify all telomeric markers, merge the different maps and correlate the molecular maps with classical and physical maps. The current and future developments should facilitate the application of map data to barley breeding, map-based cloning and the use of barley as a genetic organism for further basic studies.

5.   Acknowledgements

Research in A. Kleinhofs' laboratory is supported in part by CSRS Special Grant Agreement No. 90-34213-5190, USDA-NRI grant 9600794, and the Washington Barley Commission. Research in A. Graner's laboratory is supported by EU-Grant BIO4 CT97 2220. Special thanks are due to Dr. David Matthews and laboratory supervisor David Kudrna for developing the EXCEL spreadsheet and to all our laboratory members for their extensive barley genome mapping efforts.

6.   References

Appels, R., Gerlach, W.L., Dennis, E.S., Swift, H. and Peacock, W.J. (1980). Molecular and chromosomal organization of DNA sequences coding for the ribosomal RNAs in cereals. Chromosoma, 78: 293-311.

Arumuganathan, K. and Earle, E.D. (1991). Nuclear DNA content of some important plant species. Plant Mol. Biol. Rep., 9: 208-218.

Asher, M.J.C., Ellis, R.P., Hayter, A.M. and Whitehouse, R.N.H. (Eds.). (1981). Barley Genetics IV. Edinburgh: Edinburgh Univ. Press.

Barua, U.M., Chalmers, K.J., Thomas, W.T.B., Hackett, C.A., Lea, V., Jack, P., Forster, B.P., Waugh, R. and Powell, W. (1993). Molecular mapping of genes determining height, time to heading, and growth habit in barley (*Hordeum vulgare*). Genome, 36: 1080-1087.

Becker, J. and Heun, M. (1995). Barley microsatellites: allele variation and mapping. Plant Mol. Biol., 27: 835-845.

Becker, J., Vos, P., Kuiper, M., Salamini, F. and Heun, M. (1995). Combined mapping of AFLP and RFLP markers in barley. Mol. Gen. Genet., 249: 65-73.

Bennett, M.D. and Smith, J.B. (1976). Nuclear DNA amounts in angiosperms. Phil. Trans. R. Soc. London (Biol.), 274: 227-274.

Bezant, J., Laurie, D., Pratchett, N. and Chojecki, J. (1996). Marker regression mapping of QTL controlling flowering time and plant height in a spring barley (Hordeum vulgare L.) cross. Heredity, 77: 64-73.

Briggs, D.E. (1978). Barley. London: Chapman and Hall.

Burnham, C.R. and Hagberg, A. (1956). Cytogenetic notes on chromosomal interchanges in barley. Hereditas, 42: 467-482.

Chen, F. and Hayes, P.M. (1989). A comparison of Hordeum bulbosum-mediated haploid production efficiency in barley using in vitro floret and tiller culture. Theor. Appl. Genet., 77: 701-704.

Clapham, D. (1973). Haploid Hordeum plants from anthers in vitro. Z. Pflanzenzuchtg., 69: 142-155.

Clark, H.H. (1967). The origin and early history of the cultivated barleys. Agric. Hist. Rev., 15: 1-18.

Clark, M., Karp, A. and Archer, S. (1989). Physical mapping of the B-hordein loci on barley chromosome 5 by in situ hybridization. Genome, 32: 925-929.

DeScenzo, R.A. and Wise, R.P. (1996). Variation in the ratio of physical to genetic distance in intervals adjacent to the Mla locus on barley chromosome 1H. Mol. Gen. Genet., 251: 472-482.

DeScenzo, R.A., Wise, R.P. and Mahadevappa, M. (1994). High-resolution mapping of the Hor1/Mla/Hor2 region on chromosome 5S in barley. Molec. Plant-Microbe Interactions, 7: 657.

Devaux, P., Kilian, A. and Kleinhofs, A. (1993). Anther culture and Hordeum bulbosum-derived barley doubled haploids: mutations and methylation. Mol. Gen. Genet., 241: 674-679.

Devaux, P., Kilian, A. and Kleinhofs, A. (1995). Comparative mapping of the barley genome with male and female recombination-derived, doubled haploid populations. Mol. Gen. Genet., 249: 600-608.

Devos, K.M., Millan, T. and Gale, M.D. (1993). Comparative RFLP maps of homoeologous group-2 chromosomes of wheat, rye and barley. Theor. Appl. Genet., 85: 784-792.

Druka, A., Kudrna, D., Han, F., Kilian, A., Steffenson, B., Frisch, D., Tomkins, J., Wing, R. and Kleinhofs, A. (2000). Physical mapping of barley stem rust resistance gene rpg4. Mol. Gen. Genet. (in press).

Dubcovsky, J., Luo, M.-C., Zhong, G.-Y., Kilian, A., Kleinhofs, A. and Dvorak, J. (1996). Genetic map of diploid wheat, Triticum monococcum L., and its comparison with maps of Hordeum vulgare L. Genetics, 143: 983-999.

Dunford, R.P., Kurata, N., Laurie, D.A., Money, T.A., Minobe, Y. and Moore, G. (1995). Conservation of fine-scale DNA marker order in the genomes of rice and the Triticeae. Nucl. Acids Res., 23: 2724-2728.

Ellis, R.P., Forster, B.P., Waugh, R. and Bonar, N. (1997). Mapping physiological traits in barley. The New Phytologist, 137: 149.

Franckowiak, J. (1997). Revised linkage maps for morphological markers in barley, Hordeum vulgare. Barley Genetics Newsletter, 26: 9-21.

Fukui, K. and Kakeda, K. (1990). Quantitative karyotyping of barley chromosomes by image analysis methods. Genome, 33: 450-458.

Gaul, H. (Ed.). (1976). Barley genetics III. Munich: Verlag Karl Thiemig.

Giese, H., Holm-Jensen, A.G., Mathiassen, H., Kjaer, B., Rasmussen, S.K., Bay, H. and Jensen, J.

(1994). Distribution of RAPD markers on a linkage map of barley. Hereditas, 120: 267-273.

Graner, A., Jahoor, A., Schondelmaier, J., Siedler, H., Pillen, K., Fischbeck, G., Wenzel, G. and Herrmann, R.G. (1991). Construction of an RFLP map of barley. Theor. Appl. Genet., 83: 250-256.

Han, F., Kilian, A., Chen, J.P., Kudrna, D., Steffenson, B., Yamamoto, K., Matsumoto, T., Sasaki, T. and Kleinhofs, A. (1999). Sequence analysis of a rice BAC covering the syntenous barley Rpg1 region. Genome 42: 1071-1076.

Han, F., Kleinhofs, A., Ullrich, S.E., Kilian, A., Yano, M. and Sasaki, T. (1998). Synteny with rice: analysis of barley malting quality QTL and rpg4 chromosome regions. Genome, in press.

Harlan, J.R. (1976). Barley, Hordeum vulgare (Graminea-triticinae). In: N.W. Simmonds (Ed.), Evolution of Crop Plants, (pp. 93-98). London: Longman Group.

Harlan, J.R. (1979). On the origin of barley., Barley: Origin, Botany, Culture, Winter Hardiness, Genetics, Utilization, Pests., (pp. 10-36): USDA Agric. Handbook 338.

Heun, M., Kennedy, A.E., Anderson, J.A., Lapitan, N.L.V., Sorrells, M.E. and Tanksley, S.D. (1991). Construction of a restriction fragment length polymorphism map for barley (Hordeum vulgare). Genome, 34: 437-447.

Hinze, K., Thompson, R.D., Ritter, E., Salamini, F. and Schulze-Lefert, P. (1991). Restriction fragment length polymorphism-mediated targeting of the ml-o resistance locus in barley (Hordeum vulgare). Proc. Natl. Acad. Sci. USA, 88: 3691-3695.

Islam, A.K.M.R. (1980). Identification of wheat-barley addition lines with N-banding of chromosomes. Chromosoma, 76: 365-373.

Islam, A.K.M.R. (1983). Ditelosomic additions of barley chromosomes to wheat. In: S. Sakamoto (Ed.), Proc. 6th Int. Wheat Genet. Symp., Kyoto Univ. Press, Kyoto, Japan. pp. 233-238.

Jorgensen, J.H. and Jensen, H.P. (1979). Inter-allelic recombination in the ml-o locus in barley. Barley Genetics Newsletter, 9: 37-39.

Kasha, K.J. and Kao, K.N. (1970). High frequency haploid production in barley. Nature, 225: 874-876.

Kasha, K.J., Kleinhofs, A., Kilian, A., Saghai Maroof, M., Scoles, G.J., Hayes, P.M., Chen, F.Q., Xia, X., Li, X.-Z., Biyashev, R.M., Hoffman, D., Dahleen, L., Blake, T.K., Rossnagel, B.G., Steffenson, B.J., Thomas, P.L., Falk, D.E., Laroche, A., Kim, W., Molnar, S.J. and Sorrells, M.E. (1995). The North American Barley Genome Map on the cross HT and its comparison to the map on cross SM. In: Koichiro Tsunewaki, (Ed.), Plant Genome and Plastome. Their Structure and Evolution. Kodansha Scientific LT, Tokyo Press. pp 73-88.

Kilian, A., Chen, J., Han, F., Steffenson, B. and Kleinhofs, A. (1997). Towards map-based cloning of the barley stem rust resistance genes Rpg1 and rpg4 using rice as an intergenomic cloning vehicle. Plant Molecular Biology, 35: 187-195.

Kilian, A. and Kleinhofs, A. (1992). Cloning and mapping of telomere-associated sequences from Hordeum vulgare L. Mol. Gen. Genet., 235: 153-156.

Kilian, A., Kudrna, D. and Kleinhofs, A. (1998). Genetic and molecular characterization of barley chromosome telomeres. Genome (in press).

Kleine, M., Michalek, W., Graner, A., Herrmann, R.G. and Jung, C. (1993). Construction of a barley (Hordeum vulgare L.) YAC library and isolation of a Hor1-specific clone. Mol. Gen. Genet., 240: 265-272.

Kleinhofs, A., Chao, S. and Sharp, P.J. (1988). Mapping of nitrate reductase genes in barley and wheat. In: T.E. Miller and R.M.D. Koebner, eds., Proc. Seventh International Wheat

Genetics Symposium, Vol. 1 (Cambridge, England, 1988). Institute of Plant Science Research, Cambridge. pp. 541-546.

Kleinhofs, A., Kilian, A., Maroof, M.A.S., Biyashev, R.M., Hayes, P., Chen, F.Q., Lapitan, N., Fenwick, A., Blake, T.K., Kanazin, V., Ananiev, E., Dahleen, L., Kudrna, D., Bollinger, J., Knapp, S.J., Liu, B., Sorrells, M., Heun, M., Franckowiak, J.D., Hoffman, D., Skadsen, R. and Steffenson, a.B.J. (1993). A molecular, isozyme and morphological map of the barley (Hordeum vulgare) genome. Theor. Appl. Genet., 86: 705-712.

Kolchinsky, A., Kanazin, V., Yakovleva, E., Gazumyan, A., Kole, C. and Ananiev, E. (1990). 5S-RNA genes of barley are located on the second chromosome. Theor. Appl. Genet., 8: 333-336.

Komatsuda, T., Taguchi-Shiobara, F., Oka, S., Takaiwa, F., Annaka, T. and Jacobsen, H.-J. (1995). Transfer and mapping of the shoot-differentiation locus Shd1 in barley chromosome 2. Genome, 38: 1009-1014.

Konishi, T. (1981). Reverse mutation and interallelic recombination at the ligule-less locus of barley. In: R.N.H. Whitehouse, (Ed.), Barley Genetics IV. Fourth International Barley Genetics Symposium, Edinburgh, Scotland 1981. pp. 838-845.

Kunzel, G. and Korzun, L. (1996). Physical mapping of cereal chromosomes, with special emphasis on barley. In: G. Scoles and B. Rossnagel, (Eds.), V Int. Oat Conf. and VII Int. Barley Genetics Symp. Univ. Saskatchewan, Saskatoon, Sask. Canada 1996. pp. 197-206.

Kunzel, G., Korzun, L. and Meister, A. (2000). Cytologically integrated physical restriction fragment length polymorphism map for the barley genome based on translocation breakpoints. Genetics 154: 397-412.

Lamberts, H., Broekhuizen, S., Dantuma, G. and Martienssen, R.A. (Eds.). (1964). Barley Genetics I. Wageningen: Pudoc.

Langridge, P., Karakousis, A., Collins, N., Kretschmer, J. and Manning, S. (1995). A consensus linkage map of barley. Molec. Breed. 1: 389-395.

Laurie, D.A., Pratchett, N., Allen, R.L. and Hantke, S.S. (1996). RFLP mapping of the barley homeotic mutant lax-a. Theor. Appl. Genet., 93: 81-85.

Laurie, D.A., Pratchett, N., Bezant, J.H. and Snape, J.W. (1995). RFLP mapping of five major genes and eight quantitative trait loci controlling flowering time in a winter x spring barley (Hordeum vulgare L.) cross. Genome, 38: 575-585.

Linde-Laursen, I. (1979). Giemsa C-banding of barley chromosomes III. Segregation and linkage of C-bands on chromosomes 3, 6, and 7. Hereditas, 91: 73-77.

Linde-Laursen, I. (1982). Linkage map of the long arm of barley chromosome 3 using C-bands and marker genes. Heredity, 49: 27-35.

Linde-Laursen, I. (1997). Recommendations for the designation of the barley chromosomes and their arms. Barley Genetics Newsletter, 26: 1-3.

Linde-Laursen, I. and Jensen, J. (1992). Proposal for new designations of chromosome arms and positions of chromosomal markers. In: L. Munck (Ed.), Barley Genetics VI, . Copenhagen: Munksgaard Int. Pub. Ltd.

Meszaros, K. and Hayes, P.M. (1997). The Dicktoo x Morex barley mapping population. http://wheat.pw.usda.gov/ggpages/DxM/.

Munck, L. (Ed.). (1992). Barley Genetics VI. Copenhagen: Munksgaard Int. Pub. Ltd.

Namuth, D.M., Lapitan, N.L.V., Gill, K.S. and Gill, B.S. (1994). Comparative RFLP mapping of *Hordeum vulgare* and *Triticum tauschii*. Theor. Appl. Genet., 89: 865-872.

Nilan, R.A. (1964). The Cytology and Genetics of Barley, 1951-1962. Pullman: Washington State Univ. Press.

Nilan, R.A. (Ed.). (1971). Barley Genetics II. Pullman: Washington State Univ. Press.

Nilan, R.A. (1974). Barley (Hordeum vulgare). In R.C. King (Ed.), Handbook of Genetics, (pp. 93-110). New York: Plenum Press.

Pedersen, C., Giese, H. and Linde-Laursen, I. (1995). Towards an integration of the physical and the genetic chromosome maps of barley by in situ hybridization. Hereditas, 123: 77-88.

Qi, X., Stam, P. and Lindhout, P. (1996). Comparison and integration of four barley genetic maps. Genome, 39: 379-394.

Qi, X., Stam, P. and Lindhout, P. (1998). Use of locus-specific AFLP markers to construct a high-density molecular map in barley. Theor. Appl. Genet., 96: 376-384.

Ramage, R.T. (1985). Cytogenetics. In: D.C. Rasmusson (Ed.), Barley, Madison: American Society of Agronomy, pp. 127-154.

Rasmusson, D.C. (Ed.). (1985). Barley. Madison: American Society of Agronomy.

Roder, M.S., Lapitan, N.L.V., Sorrells, M.E. and Tanksley, S.D. (1993). Genetic and physical mapping of barley telomeres. Mol. Gen. Genet., 238: 294-303.

Rosichan, J., Nilan, R.A., Arenaz, P. and Kleinhofs, A. (1979). Intragenic recombination at the waxy locus in Hordeum vulgare. Barley Genetics Newsletter, 9: 79-85.

Saghai Maroof, M.A., Yang, G.P., Biyashev, R.M., Maughan, P.J. and Zhang, Q. (1996). Analysis of the barley and rice genomes by comparative RFLP linkage mapping. Theor. Appl. Genet., 92: 541-551.

Sarvella, P., Holmgren, B.J. and Nilan, R.A. (1958). Analysis of barley pachytene chromosomes. Nucleus, 1: 183-204.

Schonfeld, M., Ragni, A., Fischbeck, G. and Jahoor, A. (1996). RFLP mapping of three new loci for resistance genes to powdery mildew (Erysiphe graminis f. sp. hordei) in barley. Theor. Appl. Genet., 93: 48-56.

Scoles, G. and Rossnagel, B. (Eds.). (1966). International Oat Conference V and International Barley Genetics Symposium VII. Saskatoon: Univ. Saskatchewan Extension Press.

Shepherd, K.W. and Islam, A.K.M.R. (1992). Progress in the production of wheat-barley addition and recombination lines and their use in mapping the barley genome. In P.R. Shewry (Ed.), Barley: Genetics, Biochemistry, Molecular Biology and Biotechnology, Wallingford: CAB International, pp. 99-114.

Shewry, P.R. (Ed.). (1992). Barley: Genetics, Biochemistry, Molecular Biology and Biotechnology. Wallingford: CAB International.

Shin, J.S., Corpuz, L., Chao, S. and Blake, T.K. (1990). A partial map of the barley genome. Genome, 33: 803-808.

Siedler, H. and Graner, A. (1991). Construction of physical maps of the Hor1 locus of two barley cultivars by pulsed field gel electrophoresis. Mol. Gen. Genet., 226: 177-181.

Simons, G., van der Lee, T., Diergaarde, P., van Daelen, R., Groenendijk, J., Frijters, A., Buschges, R., Hollricher, K., Topsch, S., Schulze-Lefert, P., Salamini, F., Zabeau, M. and Vos, P. (1997). AFLP-based fine mapping of the Mlo gene to a 30-kb DNA segment of the barley genome. Genomics, 44: 61-70.

Singh, R.J. and Tsuchiya, T. (1975). Pachytene chromosomes in barley. J. Hered., 66: 165-167.

Smith, L. (1951). Cytology and genetics of barley. Bot. Rev., 17: 1-51; 133-202; 285-355.

Sogaard, B. and von Wettstein-Knowles, P. (1987). Barley: genes and chromosomes. Carlsberg

Res. Commun., 52: 123-196.

Sorensen, M.B. (1989). Mapping of the Hor2 locus in barley by pulsed field gel electrophoresis. Carlsberg Res. Commun., 54: 109-120.

Sunderland, N. and Xu, Z.H. (1982). Shed pollen culture in Hordeum vulgare. J. Exp. Bot., 33: 1086-1095.

Tsuchiya, T. (1960). Cytogenetic studies of trisomics in barley. Japanese. J. Bot., 17: 177-213.

Van Deynze, A.E., Nelson, J.C., Yglesias, E.S., Harrington, S.E., Braga, D.P., McCouch, S.R. and Sorrells, M.E. (1995). Comparative mapping in grasses. Wheat relationships. Mol. Gen. Genet., 248: 744-754.

von Wettstein-Knowles, P. (1992). Cloned and mapped genes: Current status. In P.R. Shewry (Ed.), Barley: Genetics, Biochemistry, Molecular Biology and Biotechnology., Wallingford: C.A.B. International, pp. 73-98.

Waugh, R., Bonar, N., Baird, E., Thomas, B., Graner, A., Hayes, P. and Powell, W. (1997). Homology of AFLP products in three mapping populations of barley. Mol. Gen. Genet., 255: 311-321.

Wei, F., Gobelman-Werner, K., Morroll, S.M., Kurth, J., Mao, L., Wing, R., Leister, D., Schulze-Lefert, P. and Wise, R.P. (1999). The Mla (powdery mildew) resistance cluster is associated with three NBS-LRR gene families and suppressed recombination within a 240-kb DNA interval on chromosome 5S (1HS) of barley. Genetics 153: 1929-1948.

Wendorf, F., Schild, R., El Hadidi, N., Close, A.E., Kobusiewicz, M., Wieckowska, H., Issawi, B. and Haas, H. (1979). Use of barley in the Egyptian late Paleolithic. Science, 205: 1341-1347.

Yasuda, S. and Konishi, T. (Eds.). (1987). Barley Genetics V. Okayama: Sanyo Press.

Yu, Y., Tomkins, J.P., Waugh, R., Frisch, D.A., Kudrna, D., Kleinhofs, A., Brueggeman, R.S., Muehlbauer, G.J. and Wing, R.A. (2000). Development and characterization of a bacterial artificial chromosome library for barley (Hordeum vulgare). Theor. Appl. Genet. (in press).

13.   DNA-based marker maps of *Brassica*

CARLOS F. QUIROS

Department of Vegetable Crops, University of California, Davis, CA 95616 <cfquiros@ucdavis.edu>

Contents

R.L. Phillips and I.K. Vasil (eds.), DNA-Based Markers in Plants, 201 - 237.

1. Introduction

Brassica crops have great economic importance worldwide as vegetables and oil seed crops. Vegetable brassicas have an annual value of over $ 1 billion in the US. The production of these crops is likely to increase in the future considering their rise in popularity as sources of chemoprotecting agents (Fahey et al. 1997) and fiber. As oilseed crops, 15% (Carr and McDonald, 1991) of the edible oil produced in the world comes from rapeseed. The development of canola has resulted in the improvement of *Brassica* oil seed quality, increasing production mostly in Canada and Europe. Therefore, the development of basic genetic information in brassicas and its application to breeding is justified due to the increasing importance of these crops.

The diploid species of *Brassica* range in genomic numbers from n=7 to n=12. The three diploid cultivated species are: 1) *B. nigra* (2n=2x=16, B genome) black mustard, 2) *B. oleracea* (2n=2x=18, C genome) cabbage group and 3) *B. rapa* (syn. *campestris*, 2n=2x=20, A genome) turnip, rapeseed and oriental vegetables.

It was recognized early on that amphidiploidy and aneuploidy have played important roles during the differentiation and evolution of *Brassica* species (Prakash and Hinata, 1980). U (1935) elucidated the genomic relationships among cultivated diploid and derived amphidiploid species. The three basic diploid cultivated species mentioned above engendered the three amphidiploids, *B. carinata* (2n=4x=34, genomes BC), *B. juncea* (2n=4x=36, genomes AB) and *B. napus* (2n=4x=n=38, genomes AC). After amphiploidy, the basic genomes have undergone further structural changes in the polyploid hybrid species (Cheung et al. 1997a, Song et al. 1995a). On the basis of limited cytological, biochemical and molecular data, it is believed that the B genome is more distantly related to both A and C genomes than the latter two are to each other (Vaughan, 1977; Attia and Robbelen, 1986; Song et al., 1988). The strongest evidence of closer homology between the A and C genomes derives from the chloroplast DNA studies of Erickson et al. (1983), Palmer et al. (1983) and Warwick and Black (1991). Warwick and Black (1991) resolved two *Brassica* lineages on the basis of chloroplast DNA RFLP's, one encompassing *B. oleracea* and *B. rapa*, and the second one including *B. nigra*. This is in agreement with the nuclear RFLP analysis of Song et al. (1988).

1.1. Ancient polyploidy of the *Brassica* genomes

Earlier research (Catcheside 1934, Robbelen, 1960, Prakash and Hinata, 1980) hinted that the cultivated diploids are ancient polyploids or paleopolyploids. This has been now widely demonstrated by RFLP mapping data from various laboratories (Quiros et al. 1994; Truco et al. 1996; Lagercrantz 1998, Lan 1999). An ancestral genome of five

(Sikka 1940) or six chromosomes (Robbelen 1960) presumably gave rise to the n=8, 9 and 10 chromosome genomes by aneuploidy. These were then further modified by hybridization and chromosomal rearrangements resulting in the present day genomes containing a wide range of chromosome numbers.

2. Mapping tools in *Brassica*

The development of genetic maps in *Brassica* will serve a double purpose: 1) understanding the origin and relationship among the genomes of the *Brassica* diploid cultivated species, and 2) utilization in applied genetics and breeding of the *Brassica* crops. The main mapping resources developed for *Brassica* crops are described below:

2.1. Genetic markers

Although morphological markers have been used for genetic analysis in some of the *Brassica* species (Sampson, 1966), they have had a minimal impact on gene mapping because of their small numbers. The study of Arus and Orton (1983) on the inheritance and linkage of isozyme loci in *B. oleracea* represented the first major attempt to develop genetic markers and their use to assemble linkage maps in *Brassica*. The nomenclature of isozyme loci in *Brassica* has been recently standardized by Chevre et al. (1995). Later, the advent of DNA based genetic markers (Quiros et al., 1994; Kresovich et al. 1995; Szewc-McFadden et al. 1996; Voorrips et al. 1997) provided sufficient markers to develop comprehensive maps for the *Brassica* genomes and related applications.

In the present chapter we will consider only DNA based markers, namely RFLPs and PCR based markers.

2.1.1. *RFLP markers*

RFLPs have been derived in *Brassica* from various sources. These include anonymous genomic clones, cDNA clones and gene specific probes (Quiros et al. 1994, Quiros 1999). Cheung et al. (1997a) generated RFLP probes by representational difference analysis. Cloned genes and ESTs from *Arabidopsis* now constitute a useful source of probes for mapping the *Brassica* genomes (Kowalski et al. 1994, Teutonico and Osborn 1994, Lagercrantz et al. 1996, Sadowski et al. 1996, Jourdren et al. 1996a, Sadowski and Quiros 1998, Lan 1999).

Extensive RFLP polymorphism for genomic clones has been detected in *Brassica* diploid cultivated species, which is in agreement with the extent of morphological variation observed in these species. This is explained in part by the almost obligate outcrossing of these species imposed by self-incompatibility. Figdore et al. (1988) in a comprehensive study concluded that the high level of polymorphism in *B. oleracea* and *B. rapa* warrant the use of only one restriction enzyme and many clones instead of several enzymes and few clones. Problems arise when doing comparative mapping due to the duplicated nature of the genomes. For example when comparing maps from all three genomes, A, B and C, the limiting factor is on the number of common polymorphic loci for meaningful comparisons among them (Lagercrantz and Lydiate, 1996, Truco et al. 1996, Cheung et al. 1997a).

RFLP polymorphism for the amphidiploid species is now available for *B. napus* and *B. juncea* (Cheung et al. 1997b). In general, polymorphism in amphidiploids is inferior to that observed in diploid species. For example, the level of polymorphism reported for *B. napus* is less than 45% (Ferreira et al. 1994, Uzunova et al. 1995, Cheung et al. 1997a & b), whereas in *B. oleracea* it can be higher than 80% (Cheung et al. 1997b). In *B. juncea* Cheung et al. (1997b) reported a polymorphism of approximately 60%. Lower levels of polymorphism in amphidiploids are expected since they have also a lower level of out-crossing due to a weak and often non-existing self-incompatibility system.

2.1.2. PCR Based markers

2.1.2.1. RAPD markers
These have been extensively used for map construction in *Brassica* species (Quiros et al. 1991, Jain et al. 1994, Truco and Quiros, 1994, Camargo et al. 1996, Lannér et al. 1996; Jørgensen et al. 1996; Mikkelsen et al. 1996; Chen et al. 1997, Cheung et al 1997a & b, Chevre et al. 1997). The levels of polymorphism of RAPDs and RFLPs are roughly similar in *Brassica* species. These two types of markers have been found equally useful to assess genetic relationships in several species (dos Santos et al. 1994; Halldén et al. 1994; Thormann et al. 1994). They have had multiple applications such as alien chromosomes identification in addition lines (Quiros et al. 1991; Struss et al. 1996; Chevre et al. 1997) and assessment of intra-specific variability for fingerprinting varieties of various crops and to follow their pedigrees (Hu and Quiros, 1991; Margalé et al. 1995). These markers are now extensively used to generate linkage maps in segregating progenies of practically all *Brassica* species (see for example Cheung et al. 1997a). However, the chance of finding the same marker segregating in two independent progenies is quite low (Quiros et al. 1994), therefore RAPDs cannot be used for comparative mapping. This makes difficult the wide application of pre-mapped RAPD markers on linkage maps to other breeding populations.

2.1.3. Microsatellite or Single Sequence Repeat (SSR) markers

This type of marker based on di- tri- and tetra-nucleotide tandem repeats was first developed in *Brassica napus* by Kresovich et al. (1995) and Bathia et al. (1995) in *B. juncea*. The use of these highly polymorphic co-dominant markers in map construction, is progressing in *B. napus* (Uzunova and Ecker 1999) and *B. oleracea* (Struss et al. 1999). The main limitation has been the lack of polymorphism across species for comparative mapping and the high cost associated to their development. They are as polymorphic as RFLP markers.

2.1.4. AFLP markers

The use of these markers is increasing in *Brassica*. Voorrips et al. (1997) used them in *B. oleracea* for mapping two clubroot resistance genes. Lim et al. (1998) constructed with these markers along with RAPDs a *B. rapa* linkage map. Recently AFLPs have been used to integrate two B. *oleracea* mapping populations (Sebastian et al. 2000). AFLPs markers are also highly polymorphic but are mostly dominant. Although bands of same migration in different populations correspond to the same alleles, it is not clear whether this is also the case across different species. This issue will have to be resolved before using AFLPs for comparative mapping.

2.1.5. *Other PCR-based markers*

The availability of sequencing information and gene characterization from A. *thaliana* make feasible to isolate without cloning homologous genes in *Brassica*. This has been accomplished by construction of degenerate primers aimed to amplify conserved coding sequences of specific genes. This type of marker is called Amplified Consensus Gene Markers (ACGM) (Fourmann et al. 1998, Brunel et al. 1999). A simplified PCR-based marker technique called SBAP (sequence-based amplified polymorphism) has been proposed by Li and Quiros (2000). This has the multiplexing ability and reproducibility of AFLPs but the simplicity of RAPDs.

2.2. Genetic markers

A new addition to the mapping toolbox of *Brassica* is the development of large size DNA clones. Basically there are two types of these developed for *Brassica*: cosmid clones with inserts of approximately 40 kb for *B. oleracea* (Acarkan et al. 1998; Wroblewski et al. 2000) and BAC clones for *B. oleracea* (Zhang et al. 1998) and *B. rapa* (Lim et al. 1998). These clones are aimed mostly to map-based gene cloning.

2.3. Cytogenetic stocks

In gene mapping projects, cytogenetic stocks provide the means to assign genes and linkage groups to specific chromosomes. Although early research in *Brassica* has involved extensively interploid crossing, only during the past 15 years there have been systematic attempts to develop cytogenetic stocks from the resulting aneuploid progenies. These stocks are primarily alien addition lines in both allotetraploid and diploid backgrounds, which are discussed in more, detain under the Synteny maps section below.

3. Maps

Two types of maps are available in *Brassica* - synteny maps based on alien addition lines, and F_2 linkage maps. The integration of these maps is an on-going effort by various laboratories.

3.1. Synteny maps

Since the last review, a few maps have been added to the existing ones for the B and C genomes (Quiros et al. 1994, Quiros 1999). No addition lines dissecting the A genome are yet available due to the difficulty encountered to develop these aneuploids (Quiros, unpublished).

3.1.1. *C genome maps*

Since the publication of the synteny maps for this genome developed by McGrath and Quiros (1990), McGrath et al. (1990) and Quiros et al. (1987) another partial map has been developed. This consists of a set of *B. campestris-alboglabra* for four C genome chromosomes (Cheng et al. 1994a & b, Cheng et al. 1995; Jørgensen et al. 1996, Chen et al. 1997). Genes for erucic acid content and a flower color gene were mapped along with RAPD markers in these lines.

Physical mapping by fluorescence *in situ* hybridization (FISH) is starting now in *Brassica*. Good examples of this activity is the mapping of DNA repetitive sequences in *B. oleracea* chromosomes by Armstrong et al. (1998) and rDNA genes by Schrader et al. (2000).

3.1.2. *B genome maps*

Since the four partial sets of addition lines mentioned in our earlier review for this genome (Quiros et al. 1994), a few more lines have been added to them. Struss et al. (1996) expanded the number of lines on *B. napus* background derived from three B genome sources, *B. nigra*, *B. carinata* and *B. juncea*. Each set now contains all or most of the B genome chromosomes. The B genome chromosomes carrying erucic acid, sinigrin and blackleg resistances were mapped along RAPD and isozyme markers. Chevre et al. (1997) developed a new set for eight B genome chromosomes in a series of *B. oleracea-nigra* addition lines. These were obtained by crossing a kale variety of *B. oleracea* by *B. nigra*. A similar series were previously developed from *B. carinata* (Quiros et al., 1986). This was a partial series including only four or five of the eight B genome chromosomes. Chevre et al. (1997) confirmed the synteny of two *6pdgh* loci in a single B genome chromosome previously reported by Quiros et al. (1986). Consensus maps for five B genome chromosomes was possible after partial alignment of synteny groups from the *B. napus-nigra* and *B. oleracea-nigra* addition lines.

3.1.3. *Determination of general stability of alien chromosomes*

With the availability of new sets of alien addition lines and additional synteny data, it is now possible to have a better idea on the stability level of alien chromosomes in *Brassica* addition lines. Basically two types of instability have been observed: Frequent terminal deletions as reported by Hu and Quiros, (1991) and Quiros et al. (1994) and intra-genomic recombination. The latter type of instability is quite frequent and makes difficult to obtain large synteny maps for the *Brassica* genomes. This has been observed for the C genome (Quiros et al. 1994) and more extensively for the B genome. For example, when Struss et al. (1996) compared maps from three different sources, only a small proportion of common markers conserved syntenic associations. A similar situation was observed by Chevre at al. (1997), where only some of the common markers across sets of lines conserved synteny for consensus map construction. Certainly, it could be argued that these structural changes are due to pre-existing rearrangements that may have occurred as part of the formation and evolution of natural amphidiploids. This would explain syntenic differences in lines extracted from natural amphidiploids and those extracted from diploids. Comparative mapping data indicates that indeed some of these changes could be explained by pre-existing rearrangements. For example, although Cheung et al. (1997a) found extensive linkage conservation for C genome chromosomes in *B. oleracea* and *B. napus,* rearrangements for some segments were detected. No comparative data of this sort for the B genome exists, however comparison of B genome synteny maps extracted from different *B. nigra* accessions revealed rearrangements for isozyme loci (This at al. 1990; Struss et al. 1996; Chevre et al. 1997). Therefore this indicates that most of the syntenic changes observed in different addition lines are likely due to intra-genomic recombination events taking place during addition line construction and little to pre-existing structural changes. This can be clearly seen in the *B. olerace-nigra* set of addition lines developed by Chevre et al. (1997), where blocks of five identical RAPD markers are present on two different synteny groups as "duplicated loci". The fact that the sizes of the fragments for these markers are identical makes unlikely the possibility that they represent truly duplicated loci. Extensive RFLP and isozyme loci data demonstrate that duplicated loci have diverged enough to display different size alleles. RAPD markers could hardly be an exception to this rule. Most likely these syntenic associations resulted by intra-genomic recombination.

3.2. F$_2$ linkage maps

Most of the mapping work in *Brassica* has taken place during the past 10 years. This activity has been focused mostly on rapeseed *B. napus* and on all three diploid cultivated species, *B. nigra, B. oleracea* and *B. rapa*. More recently, mapping has been expanded to include *B. juncea*. The maps produced in *Brassica* crops are based mainly on F$_2$ progenies developed independently by various laboratories, which will require in the future their integration for a more efficient use.

3.2.1. B. oleracea *maps*

Several maps have been developed independently for this species involving crosses between different crops. Slocum et al. (1990) reported an extensive RFLP map of 258 markers covering 820 recombination units in nine linkage groups, with average intervals of 3.5 units. This is a proprietary map developed from a broccoli x cabbage cross. Landry et al. (1992) constructed a map consisting of 201 RFLP markers distributed on nine major linkage groups covering 1112 cM. The F$_2$ progeny used to construct this map was developed by crossing a cabbage line resistant to clubroot to a rapid cycling stock. The cabbage line was derived from an interspecific cross involving *B. oleracea* and

B. napus, followed by a series of backcrosses *to B. oleracea* . According to the authors, it is likely that the resulting cabbage line had two chromosome segments of the A genome carrying the disease resistant genes. These had been introgressed or substituted in the C genome of the cabbage line. Recently, this map was expanded with additional RFLP and PCR-based markers (Cheung et al. 1997a). Kianian and Quiros (1992a) developed a map comprising 108 markers, spread in 11 linkage groups and covering 747 cM. This map was based on three intra-specific and one interspecific F_2 populations, namely: collard x cauliflower, collard x broccoli, kale x cauliflower and kohlrabi x *B. insularis*. The majority of the markers in the map are RFLP loci, with a few morphological and isozyme markers. Later, a few RAPD markers were added and the map was redrawn based only on the three intra-specific crosses using JOINMAP (Stam et al., 1993). It consists of 82 markers distributed on eight major linkage groups covering 431 cM (Quiros et al. 1994). Recombinant inbreds by single seed descent are under development for the collard by cauliflower and collard by broccoli progenies.

   Camargo et al. (1997) reported a map developed in an F_2 population of cabbage x broccoli. It includes 112 RFLP and 47 RAPD loci and a self-incompatibility locus on nine main linkage groups (Fig. 1). Kearsey et al. (1996) and Ramsay et al. (1996) constructed a linkage map based on backcross progenies obtained from doubled haploid lines (DH) of broccoli and *B. oleracea* var. *alboglabra*. Recombination of markers was higher in the BC_2 than in the BC_1 generation. These differences were attributed to differential chiasma frequency in female and male meiosis. Bouhon et al. (1996) used this map for aligning the C genome linkage groups of *B. oleracea* and *B. napus*. Lagercrantz and Lydiate (1995) also observed sex dependent recombination rates in *B. nigra*. In this species, however, recombination rate in each sex varied for some chromosome segments. If differential recombination rate in male and female gametes proves to be a widespread phenomenon, it posses another complexity to consider when applying linkage information to breeding problems. A subset of 169 DH lines from the population used Bouhon et al. (1996) was used by Sebastian et al. (2000) to developed and integrated map with a second DH mapping population resulting from crossing cauliflower by Brussels sprouts. This map produced by crossing cabbage by kale inbreds, consists of RFLP and AFLP markers. It has 547 loci on nine linkage groups covering 893 cM (average locus interval 2.6 cM). Moriguchi et al. (1999) produced another map based on RAPDs, RFLPs, isozymes, and morphological markers. It was used for the analysis of QTLs responsible for clubroot resistance. Most recently Lan (1999) developed an extensive *B. oleracea* map based on *A. thaliana* EST probes.

   We aligned four maps form different laboratories (Hu et al. 1998). For this purpose, a linkage map was constructed from an F_2 population of 69 individuals with sequences previously mapped independently in three linkage maps of this species. These were the maps published by Kianian and Quiros (1992a), Landry et al. (1992) and Camargo et al. (1997). The base map developed in this study consisted of 167 RFLP loci in nine linkage groups, plus eight markers in four linkage pairs, covering 1738 cM. Linkage group alignment was also possible with the map published by Ramsay et al. (1996), containing common loci with the map of Camargo et al. (1997) (Table 1). In general, consistent linear order among markers were maintained, although often, the distances between markers varied from map to map. A linkage group in Landry's map carrying

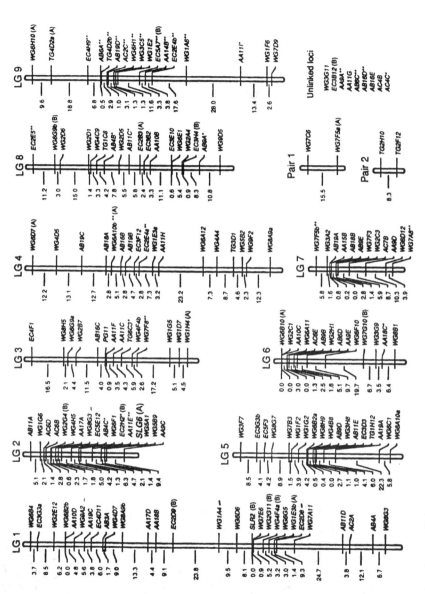

Figure 1.   Brassica oleracea linkage map produced by an F$_2$ population after crossing cabbage x broccoli. It includes 112 RFLP and 47 RAPD loci and a self-incompatibility locus on nine main linkage groups (Camargo et al. 1997).

a clubroot resistance QTL was rearranged, consisting of markers from two other linkage groups. This was not surprising considering that the resistance gene was introgressed from *B. napus*. The extensively duplicated nature of the C genome was revealed by 19 sequences detecting duplicated loci within chromosomes and 17 sequences detecting duplicated loci between chromosomes. The variation in mapping distances between linked loci pairs on different chromosomes demonstrated that sequence rearrangement is a distinct feature of the C genome. Although the consolidation of all linkage groups in the four *B. oleracea* maps compared was not possible, a considerable number of markers to corresponding linkage groups were added. Some of the chromosome segments in particular, were enriched with many markers, which may be useful for future gene tagging or cloning.

Table 1.   Correspondence among linkage groups of the four maps and common loci between the base map and two other maps unless specified. Tentative alignments in parenthesis.

Current Map	Camargo map	common loci	Landry's map	common loci	Ramsay's map	common loci[a]
C1	(LG3)	1	BL5	3	-	-
C2	LG2	3	-	-	R6	3
C3	LG8	3	BL2	3	-	-
C4	LG9	5	BL7	3	-	-
C5	(LG4)	2	BL1	3	R4	5
C6	(LG6)	0	-	(R8)	2	
C7	LG1	3	BL1	5	R3	4
C8	LG7	3	-	-	R2	3
C9	LG5	3	-	-	R9	6

(a) Common loci between Camargo et al.1997 and Ramsey et al. 1996 maps.

The assignment *B. oleracea* linkage groups to their respective chromosomes has been partially accomplished by alignment to synteny maps based on C-genome alien addition lines (Quiros et al. 1994, Quiros 1999). Although it was possible in most cases to physically assign linkage groups to chromosomes using these sets of lines, two major complications arising from this activity deserve comment. The first one is the frequent lack of polymorphism of interspecific markers on the alien chromosomes in the intra-specific crosses used to develop the linkage maps. Low coincidence of polymorphism between these two sets of materials makes chromosome assignment tedious and time consuming. The second complication is the instability of the alien chromosomes as explained above.

3.2.2. B. rapa *maps*

The two most extensive maps created for this species are proprietary. The first one, developed by Chyi et al. (1992) resulted by crossing sarson by canola. It includes 360 loci on 10 linkage groups covering 1876 recombination units. The average distance between markers is 5.2 map units. The second one developed by Slocum (1989) and Song et al. (1991) was obtained by crossing Chinese cabbage x "spring broccoli". This map includes 273 loci covering 1455 recombination units in 10 linkage groups with average intervals of 5 units. An updated version of this map was used to locate genes determining 28 phenotypic traits (Song et al. 1995b). Kole et al. (1996) created a map involving a white rust resistance line for the purpose of mapping the resistance gene. Other maps have been developed by Schilling and Bernatzky (Schilling 1991) including 58 RFLP loci covering approximately 700 cM after crossing the oilseed cultivar `Candle' by a rapid cycling strain. McGrath and Quiros (1991) developed another map from an F_2 of turnip x Pak Choi containing 49 markers in eight linkage groups and covering a total of 262 cM. These two maps have been consolidated by exchanging probes and DNA from the mapping populations (Hu, Bernatzky and Quiros, unpublished). More recently Lim et al. (1998) reported the construction of an AFLP and RAPD based linkage map map in doubled haploids of a hybrid between two Chinese cabbage varieties. Matsumoto et al. (1998) performed linkage analysis of RFLP markers for clubroot resistance and pigmentation in Chinese cabbage (*Brassica rapa* ssp. *pekinensis*). Ajisaka (1999) mapped a series of loci affecting the microspore culture efficiency in *B. rapa*.

3.2.3. B. nigra *maps*

Truco and Quiros (1994) developed a map for this species based on a single F_2 population of 83 plants, involving two parental individuals from geographically divergent populations. The map constructed with the program Mapmaker (Lander et al., 1987), has 67 markers arranged in eight major linkage groups which may correspond to the eight *B. nigra* chromosomes, plus two small groups (Fig. 2). The markers include RFLP's, RAPD's and a few isozymes. The RFLPs are based on *Eco*RI digestion. The map covers 561 cM with average intervals of 8.4 cM. Lagercrantz and Lydiate (1995) developed a RFLP map in a backcross population of *B. nigra*, consisting of 288 loci covering a length of 855 cM. A trend of higher recombination rates was observed in male gametes for the distal portions of the linkage groups. On the other hand, recombination rates tended to be higher in the proximal regions of some of the linkage groups.

It has been possible to assign only five linkage groups of the first map to their respective chromosomes by synteny mapping based on alien addition lines (Chevre et al. 1997).

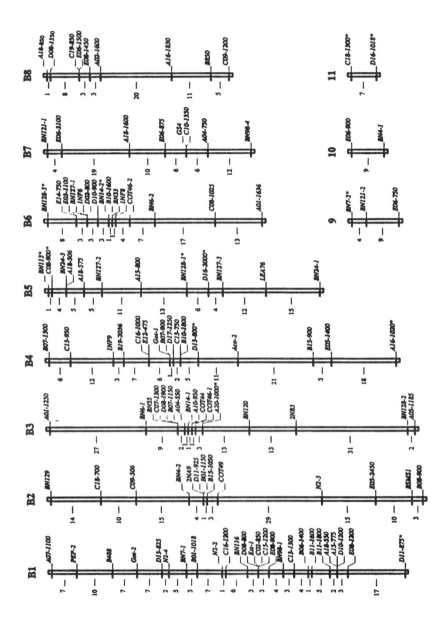

Figure 2.   Linkage map of *B. nigra* produced by a F$_2$ population after crossing two landraces of this species. (after Truco et al. 1994).

3.2.4. B. napus *maps*

Most of the mapping activity in *Brassica* has been focused to this species because of its economic importance. This has resulted in the construction of at least six independent maps. Several maps for this amphidiploid species have been reported by various laboratories. The first map was developed by Landry et al. (1991), and resulted from crossing two rapeseed cultivars, `Westar' and `Topaz', based on single-enzyme digestion by four enzymes, *Bam*HI, *Eco*RI, *Eco*RV and *Hin*dIII. It included 120 loci arranged in 19 linkage groups covering 1413 recombination units. More recently Landry's group produced a new *B. napus* map from a DH F_1 resulting from crossing two proprietary DH lines. It consists of 209 loci (RFLP and PCR-based) arranged in 19 linkage groups, covering 1955 cM. Hoenecke and Chyi, (1991) developed a proprietary map by crossing two breeding lines BN0011 x BN0031 based on *Eco*RI digestion. This map consisted of 125 markers arranged in 19 linkage groups covering 1350 map units. In the past few years, four other maps have been constructed based on F_2 progenies of doubled haploid lines. Ferreira et al. (1994) constructed a map by crossing the annual canola cultivar 'Stellar' by the biennial cultivar 'Major'. Doubled haploid lines generated from the resulting F_1 served to construct a map of 132 loci covering 1016 cM in 22 linkage groups. A partial map was also constructed from the F_2 progeny of the two parental lines. For the chromosome segments compared, no significant differences on linkage associations were observed between the two maps. Uzunova et al. (1995) constructed a map based on doubled haploid F_1 lines by crossing the two winter rapeseed varieties, 'Mansholt's Hamburger Raps' and 'Samourai'. It consisted in 204 RFLP and two RAPD markers distributed in 19 linkage groups covering 1441 cM (Fig. 3). Sharpe et al. (1995) reported an integrated map based on two populations, one of that included a resynthesized *B. napus* line crossed to a rapeseed cultivar. The chromosome instability of the resynthesized line added to the complexity of these maps. Parkin et al. (1995) published also a map based only on this synthetic population, consisting of 399 RFLP markers. The integration of the maps by Sharpe et al. (1995), based on the combination of the two populations resulted in a reduction of over 100 loci. Foisset et al. (1996) constructed a map based on RFLP, RAPD and isozyme markers developed from a doubled haploid progeny of the F_1 generated by crossing a dwarf rapeseed isogenic line 'Darmor-bzh' of cultivar 'Darmor' and 'Yudal', a spring Korean line. The map based on 153 doubled haploid lines had 254 markers on 19 linkage groups covering 1765 cM. Several genes of known function and economic importance were located on this map. (Cloutier et al. 1995).

3.2.5. B. juncea *maps*

The mapping activity in this species in fairly recent. Sharma et al. (1994) developed a small map based on a F_2 mapping population resulting of crossing 'Varuna', a brown seeded cultivar and BEC144, a yellow seeded accession. This map has only 25 markers on nine linkage groups covering 243 cM. A more extensive map has been developed by Cheung et al. (1997b), based on a doubled haploid progeny involving a canola quality line by a high oil content line. The map was developed with probes from *B. napus*, consisting of 343 loci on 18 main linkage groups covering 2073 cM (Fig. 4).

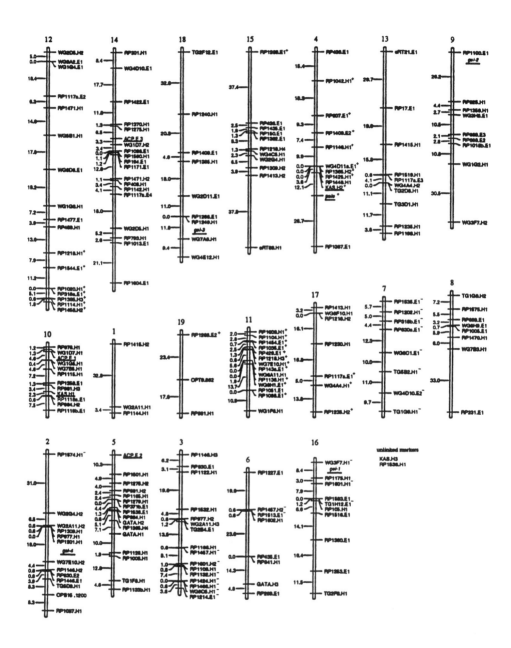

Figure 3. Linkage map of *B. napus* resulting from doubled haploid lines of the F_1 resulting from crossing the winter rapeseed cv "Mansholt' with cv 'Samourai". Mapped genes and QTLs are underlined: Pale (pale yellow flower), ACP (acyl-carrier-protein), KAS (beta-ketoacyl-ACP synthase) I, gls (see glucosinolate content QTLs) (after Uzunova et al. 1995).

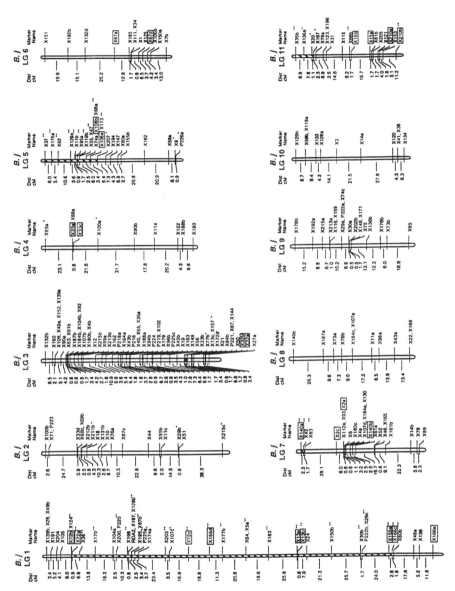

Figure 4. Linkage map of *B. juncea* (after Cheung et al. 1997b).

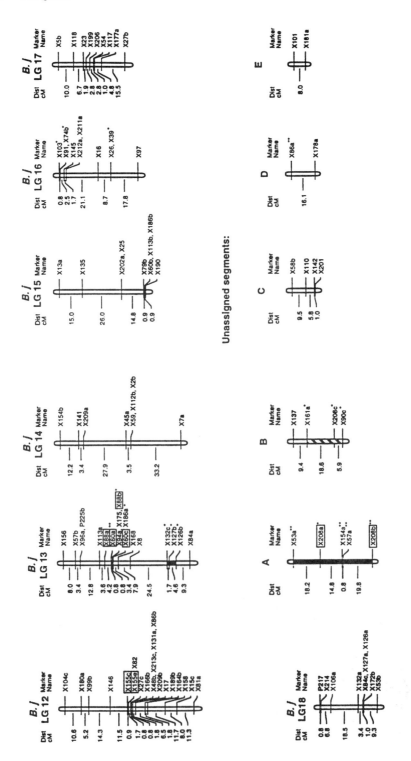

Figure 4. continued

4.   Structure of the *Brassica* genomes

4.1.   General attributes

A great deal of information has been gathered during the past few years on the genomic structure of all three cultivated genomes. Analyses of synteny maps for the B and C genomes (McGrath et al. 1990, Struss et al. 1996), and the linkage maps for the A, B and C genomes (Slocum et al. 1990, Song et al. 1991, McGrath and Quiros 1991, Kianian and Quiros, 1992a, Truco and Quiros 1994) reveal extensive sequence duplication. For example, Slocum et al. (1990) reported for *B. oleracea* that 35% of the genomic clones produced more than one locus, and 56% besides disclosing single locus segregations often produced other monomorphic fragments that may represent duplications. Kianian and Quiros (1992a) also in *B. oleracea* found 50% of their cDNA sequences mapping to more than one locus. McGrath et al. (1990) reported over 40% sequence duplications in *B. oleracea* chromosomes represented in a series of addition lines. Song et al. (1991) reported in *B. rapa* that 36% of their genomic clones produced segregating RFLPs at more than one locus and 41% detected sequences segregating as single locus but also revealed additional monomorphic fragments. Truco and Quiros (1994) found approximately 40% of RFLP loci duplicated in *B. nigra*. This finding was confirmed by Lagercrantz and Lydiate (1995 & 1996) who reported that "essentially every chromosomal region [is present] in three copies" in this species. Thus, these reports demonstrate that close to 50% of the loci in all three cultivated genomes are replicated, supporting the hypothesis that the *Brassica* diploid species are secondary polyploids (Robbelen, 1960, Prakash and Hinata, 1980, Quiros et al. 1994). In general, these duplications are distributed on more than one chromosome. Some of the linkage and synteny groups have sequences present in more than one group. When linkage arrangement is conserved, most commonly the distances are changed. The rearrangements of linkage groups may be explained by chromosomal translocations due in part to homoeologous recombination. Translocations are of common occurrence in *Brassica* and have been reported by independent investigators as a widespread event in various species (Snogerup 1980, Quiros et al. 1988, Kianian and Quiros (1992b). In addition to duplications and linkage rearrangements, deletions seem to be another important molding force of the *Brassica* genomes as explained before. The four independent F_2 linkage studies cited above have detected a large number of loci containing dominant markers, where one allele is apparently null, which may be due to deletions. This phenomenon, however, may be also due to the masking of common alleles in duplicated loci (Uzunova et al. 1995; Cloutier et al. 1997). Sometimes these loci are assembled in linkage blocks implying large deletions in some chromosomes (McGrath and Quiros 1991, Song et al. 1991). The occurrence of deletions has been demonstrated cytologically in the C genome using alien addition lines (Hu and Quiros 1991). Finally, Song et al. (1991) and Kianian and Quiros (1992a) have also observed inversions in the F_2 linkage maps of *B. oleracea*.

High levels of genome duplication have also been reported in *B. napus*. Landry et al. (1991) found that 88% of the probes disclosed more than one locus. *B. napus* being an amphidiploid, these duplications are expected to correspond to both intra-genomic and inter-genomic sequences. This indeed seems to be the case, because some of the probes disclosed three or four segregating loci located on different chromosomes (Landry et al. 1991; Hoenecke and Chyi 1991). The same trend was observed by Uzunova et al. (1995), detecting 10 different clusters of two to four duplicated loci in 11 linkage groups. Ferreira et al. (1994) also found a similar level of duplicated loci in *B. napus*.

Another interesting attribute of the *Brassica* maps in general is the relatively large number of loci deviating from Mendelian segregation, which often cluster in linkage blocks (Landry et al. 1991, McGrath and Quiros 1991, Chry et al. 1992, Kianian and Quiros 1992a, Figdore et al. 1993, Ferreira et al.1994, Uzunova et al. 1995, Foisset et al. 1996, Cheung et al. 1997a & b). The deviations in some cases may be due to genomic divergence between the parents involved in the crosses (McGrath and Quiros 1991). This conjecture is supported by the fact that the number of deviating loci increases with the divergence of the parents (Kianian and Quiros, 1992a). In androgenic plant material such as the doubled haploid lines, often used in *B. napus* for mapping purposes, loci with biased segregation ratios can reach up to 35% (Ferreira et al. 1994, Foisset et al. 1996). Inclusion of loci with distorted segregation ratios in linkage maps is risky. In such case, caution is necessary to avoid disclosing pseudolinkages due to biased statistical tests (Foisset et al. 1996; Cloutier et al. 1997).

4.2. Plasticity of the Brassica genomes

The highly duplicated nature of the *Brassica* genomes have important implications in structural changes of the chromosomes. The homoeologous regions arising by duplications under certain situations, such as those imposed by hybridization, will facilitate intra-genomic and inter-genomic recombination events. This is especially true in *Brassica*, where hybridization often results in aneuploidy and amphiploidy. Because of their plasticity, the *Brassica* genomes are prone to frequent structural changes.

4.2.1. *Intra-genomic homoeologous recombination*

We have discussed already the high frequency of intra-genomic recombination detected mostly in aneuploids, such as alien addition lines and newly synthesized amphidiploids. Undoubtedly natural amphiploids have gone thorough this same process as indicated by the comparative mapping data for the C-genome (Cheung et al. 1997a) and for the A-genome (Hoenecke and Chyi, 1991).

The structural changes observed in each of the three cultivated genomes indicate that this phenomenon is widespread and not exclusive of a single genome. Although amphiploidy and interspecific aneuploidy may serve to induce these genomic changes, some rearrangements might take place also in diploid species (Quiros et al. 1987, Kianian and Quiros 1992b).

4.2.2. *Inter-genomic homoeologous recombination*

Occasionally a few of the diploid individuals derived from alien addition lines will result from inter-genomic recombination. For example, *B. rapa-oleracea* monosomic addition plants were found to carry a few C-genome-specific markers present in the alien chromosomes of the parental plant, indicating that inter-genomic recombination had taken place. Earlier during the development of these lines, we detected inter-genomic recombination between the A and C genome chromosomes for the isozyme locus *Pgi-2* (Quiros et al. 1987). Recently we have observed inter-genomic recombinants for rDNA sequences, where *B. rapa* individuals display *Eco*RI fragments typical of *B. oleracea* (Hu et al., unpublished). Sharpe et al. (1995) who followed segregation of genome specific RFLP markers in *B. napus* progenies has obtained another line of evidence for this type of recombination. Parkin et al. (1995), detected non homoeologous recombination, resulting in non reciprocal homoeologous translocations in *B. napus* cultivars at approximately 0.3%. In resynthezided *B. napus* recombination of this type was estimated to be approximately 10%. Early generations of synthetic *Brassica* amphiploids after several cycles of selfing displayed genomic changes. Some of these changes could be attributed to inter-genomic recombination (Song et al. 1995a). Other instances of inter-genomic recombination have been reported by Struss et al. (1996) in *B. napus -nigra* addition lines. Euploids of constitution AACC derived from some of the lines carried genes for resistance to blackleg, erucic acid and sinigrin content from the B genome. It is unknown whether the B genome chromosome segments translocated to either the A or C genome chromosomes. However, the B genome segments from different origins always mapped on the same chromosomal location of the recipient genome, indicating that homoeologous recombination took place. Chevre et al. (1997) reported possible inter-genomic recombination for blackleg resistance from the B-genome of *B. juncea* to *B. napus*. A similar situation was reported by Landry et al. (1991), where resistance genes from *B. rapa* were transferred to the C genome by crossing and backcrossing *B. napus* to *B. oleracea*. More recently Osborn et al. (2000) reported a translocation between A and C genome chromosomes in *B. napus* probably resulting from homoeologous pairing and recombination.

4.3. Genomic relationships

Comparative mapping for all three basic genomes from the cultivated diploid species is at its infancy in *Brassica*. Previous work on this subject was limited to comparison of the A and C genomes (Slocum 1989; McGrath and Quiros 1991, Camargo 1994). Only recently comparative mapping for the A, B and C genomes has been reported, allowing to draw inferences on the origin of the three genomes based on their relationships (Lagercrantz and Lydiate 1996; and Truco et al. 1996). All three *Brassica* species share regions of homology in their genomes. Often a single linkage group showed regions of homology with more than one group of the other species. This is in agreement with a comparative study between maps of *B. rapa* and *B. oleracea* by Slocum (1989), who found that in some cases it was possible to align linkage groups from these two species. Some of the linkage groups, however, also shared homologous regions that were separated into more than one group in the other species. It is evident that extensive chromosome re-patterning has taken place during the evolution of *Brassica* species, even though there is considerable conservation among certain chromosome regions within and among the three genomes. This results in complex intra- and inter-genomic chromosomal relationships where gene and marker colinearity is maintained for some segments, but broken up for other chromosomal regions.

4.3.1. *Comparative mapping with Arabidopsis, used as a model for a simpler genome*

Although *A. thaliana* has a genome of only approximately 140-145 Mb (Arumuganathan and Earle 1991, Lin et al. 1999) and n=5 chromosomes. Sequencing data of two complete chromosomes, 2 (Lin et al. 1999) and 4 (Mayer et al. 1999), indicates extensive gene duplication and complex genome structure. This includes an insertion of 270 Kb corresponding to 75% of the *Arabidopsis* mitochondrial genome (Lin et al. 1999). The centromeres of these chromosomes are also quite complex, including large arrays of 180 bp repeats flanked by highly populated areas of retroelements and pseudogenes (Copenhaver et al. 1999). Taking advantage of the fact that *A. thaliana* (tribe *Arabideae*) is a related crucifer to *Brassica* species (tribe *Brassiceae*) (Price et al. 1994), comparative mapping between the two genera is taking place at an increasing pace (Paterson 1997, (Cavell et al. 1998). However, this comparative analysis has been limited mostly to RFLP maps constructed with common probes. For example, Kowalski at al. (1994) found islands of conserved marker content and order among extensive chromosomal rearrangements distinguishing the RFLP maps of *A. thaliana* and *B. oleracea*. More recently, Lagercrantz (1998) estimated that approximately 90 chromosomal rearrangements distinguish the genomes of *A. thaliana* and *B. nigra*. Sadowski et al. (1996) mapped in the three *Brassica* diploid cultivated species an *A. thaliana* gene complex carrying five genes within a 20 Kb span (Gaubier et al., 1993). This complex comprises a well characterized Em-like protein coding gene and other four flanking genes on chromosome 3. Although the five gene complex array from *A. thaliana* was conserved on a single chromosome of each *Brassica* genome, additional copies for most of the genes were found in one or two other chromosomes. A similar situation was observed for a six gene complex on *A. thaliana* chromosome 4, including the disease resistance gene *Rps2* (Sadowski and Quiros, 1998). In this case, besides the conserved array in one *Brassica* chromosome, four other chromosomes contained copies for some of the genes. The apparent conservation of the *A. thaliana* array detected by genetic mapping in one of the *Brassica* chromosomes of all three genomes indicated that this gene cluster arrays predates the separation of the genera. However, comparative sequencing of this segment revealed the presence of a *N-myrostyl transferase* gene in *B. oleracea* between the *Rps2* and *Ck1* genes (Quiros et al. 2000). Orthology of the segments could be still assigned based on the fact the *Rps2* is a single copy locus in both species (Wroblewski et al. 2000) and by the high similarity level for the rest of the compared chromosome segments (Quiros et al. 2000).

Lagercrantz et al. (1996) and Lagercrantz (1998) claimed the presence of general loci triplication in the *Brassica* genomes and in particular in the B genome. In contrast to this report, most other laboratories have found from two to six or more copies of replicated loci. For example, Lan (1999) estimated that 41% and 18% of the *B. oleracea* loci were duplicated and triplicated, respectively. This work was based on an extensive genetic map constructed using *Arabidopsis* ESTs as probes. These figures undoubtedly will be revised in the future, as additional sequencing data in *Brassica* becomes available, which will solve the present discrepancy based mostly on data interpretation. We can already appreciate the power of sequencing data by the recent finding that 60% of *Arabidopsis* protein coding genes on chromosome 2 are estimated to be duplicated somewhere else in the genome of this species. Before the overall estimate of duplicated genes in this species was only 12-17% (McGrath et al. 1993). So far, we know very little about gene spacer size and content in *Brassica* species, but there are instances in the latter species where spacers are larger for some genes, (Sadowski et al. 1996), smaller (Conner et al. 1998) or similar in size for others (Lagercrantz et al. 1996).

4.4. On the origin of the genomes

Based on the mapping data disclosing extensive loci duplication, it is clear that the three diploid *Brassica* cultivated species are indeed ancient polyploids as suggested earlier by Prakash and Hinata (1980). The reiteration of chromosomes within the genomes certainly agrees with the hypothesis that the existing *Brassica* genomes derive from a smaller ancestral genome. It is uncertain, however, what was the number of chromosomes of this ancestral genome. Truco et al. (1996) were able to narrow down this number to a range of 5 to 7 chromosomes but inspection of additional comparative mapping data (Lagercrantz et al. 1996) could drop this number to x=4. This is a viable number considering that the tribe *Lepidieae* in the Brassiceae contains species of x=4 chromosomes, which is the smallest chromosome number in the family (Mulligan 1968). Although this ancestral number is hypothetical, it serves to speculate on the origin of the *Brassica* genomes. The higher level of inter-genomic homology between the A and C genomes supports the conjecture that the former derived later on from an already established C genome (Song et al. 1990, Warwick and Black 1991, Pradhan et al. 1992). Mapping data from various laboratories indicate that it is unlikely that the *Brassica* genomes originated by polysomy or duplication of whole chromosomes. The complexity of the existing chromosomal relationships discards the possibility of autopolyploidy as an explanation for the higher chromosome numbers observed today in the cultivated genomes. Further, the genomes of *Brassica* cannot be simply described by formulae including a few founder chromosomes reiterated twice or trice (Robbelen 1960). Data gathered from various laboratories point to chromosomal rearrangements, and hybridization followed by aneuploidy as the main events involved in the origin of the *Brassica* genomes. This is essentially what Sikka (1940) postulated 60 years ago, namely that the two main forces molding the *Brassica* genomes departing from an ancestral genome of x=5 are hybridization and chromosomal structural changes. There is however, disagreement on whether the number founder species constituting each these genomes was two or three species. Lagercrantz and Lydiate (1996) and Lagercrantz (1998) postulate three donor species of x=5 consisting of genomes similar in structure and complexity to the *Arabidopsis* genome. By hybridization, an ancestral hexaploid was formed, which originated the A, B and C *Brassica* genomes by chromo-some number reduction caused by extensive chromosome fusion (Lagercrantz 1998). This conclusion was based on the observation that some of the loci detected by the RFLP probes used were triplicated. Another factor leading the authors to the hexaploid ancestor hypothesis is the fact that the *Brassica* genomes contain approximately three times the DNA found in the *Arabidopsis* genome (Arumuganathan and Earle 1991). Unfortunately the complexity of the *Brassica* genomes consisting a large proportion of heterochromatin and repetitive DNA (Gupta et al. 1990 & 1992, Iwabuchi et al. 1991) would argue against this possibility. Further, we know that *Arabidopsis* and *Brassica* are phylogenetically too distant to appeal to such close and direct relationship (Price 1994). Additionally, this hypothesis assumes that *Arabidopsis* is ancestral to *Brassica*, which has never been proved. An alternative hypothesis called "cyclic amphiploidy" (Fig. 5), was proposed by Truco et al. (1996) and Quiros (1998) and is based on the hybridization of different pairs of species containing x=4 and/or 5 chromosomes derived from an ancestral *Brassica* genome of similar chromosome number but unknown ancestry. Each pair of species originated the different *Brassica* A, B and C genomes. The ancestral cytotypes originating the cultivated genomes likely arose as a result of chromosomal structural modifications due to differential evolutionary forces caused by spatial isolation of the species containing the ancestral genome. Most likely reciprocal translocations, which are quite common in *Brassica* species (Quiros et al. 1988), have been responsible for these structural modifications. Further, translocations are known to generate not only duplications and deletions but also aneuploidy, opening also the possibility of changes in chromosome numbers previous to hybridization. Also deletions and rearrangements leading to duplications arose by intra-genomic recombination in hyperploids. The DNA content of the species possessing the A, B and C genomes have remained practically constant in spite of changes in chromosome number and structure (Arumuganathan and Earle 1991). Thus similar amounts of genetic information exist in each of these genomes but are packed in a different number of chromosomes. Because of their ancestral common origin, the three cultivated genomes have conserved chromosome segments and extensive duplications. After genomic stabilization, the species containing these genomes have generated by another cycle of hybridization the cultivated allotetraploid species we know today (Fig. 5).

5.   Applications of the maps in breeding

The *Brassica* linkage maps are now extensively applied to tag genes of interest, including quantitative trait loci (QTL) of economic importance.

5.1. Vernalization requirement and flowering time

Two genomic regions determining biennial habit in *B. rapa* have been identified, by crossing biennial to annual types. These regions associate to two linkage groups in *B. napus* carrying genes related to flowering time. (Ferreira et al.1995a, Teutonico and Osborn 1995). *co* gene homologs determining flowering time in *A. thaliana* were detected in *B. nigra* by Lagercrantz et al. (1996). Camargo and Osborn (1996) performed QTL analysis of flowering time in F_3 populations *B. oleracea* obtained by crossing cabbage and broccoli. A total of five QTLs were detected. One of these was

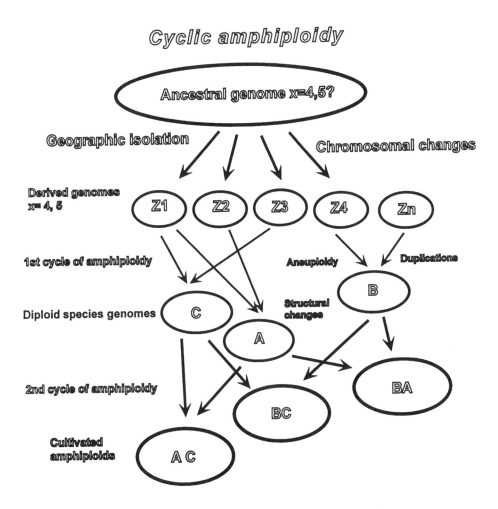

Figure 5. Cyclic amphiploidy and the origin of the *Brassica* diploid and allotetraploid species (from Quiros 1998).

linked to the S locus locus *slg6*. Another QTL was associated to petiole length. Bohuon et al. (1998) identified six QTLs for flowering time in *B. oleracea* . Three of these were on linkage group segments showing conservation to two linkage groups of *B. nigra* and to a segment on chromosome 5 of *A. thaliana*. Lack of common markers prevented the comparison of QTLs from the last two studies.

5.2. Freezing tolerance and winter

QTL analysis of these traits has been carried out in both species. In the former species, no QTLs were detected. In *B. rapa*, however, four QTLs were observed. Two of these linked to *A. thaliana* RFLPs for cold induced (COR6.6a) and stress related (DHS2) cDNAs (Teutonico et al. 1995).

5.3   Linolenic acid content in *B. napus*

Several laboratories have identified markers for this trait. One of the four markers detected by Tanhuanpaa et al. (1995a), corresponded to the dominant RAPD marker K01-1100 detected by Hu et al. (1995), which accounted for 37% of the variation for this trait. K01-1100 has now been converted to a SCAR co-dominant marker (Hu et al. 1999). Jourdren et al. (1996a & b) and Thormann et al. (1996) reported markers for this trait one of which was related the *fad3* (omega 3 desaturase) gene of *A. thaliana* (Arondel et al. 1992). Similarly, Sommers et al. (1998) identified three markers associated to linolenic acid content, one of which also corresponded to the *fad3* gene. Rajcan et al. (1999) reported two markers associated to linolenic acid. The equivalence of most markers disclosed from all these studies is yet to be done. In any case, a sufficient number of these seem to be now available for marker assisted selection of linolenic acid content in *B. napus*.

5.4.   Oleic acid content

A QTL in a linkage group of six markers associated to oleic acid content was detected by Tanhuanpaa et al. (1996) in *B. rapa*. OPH-17, a co-dominant RAPD marker was the best for selection of this trait. It was converted to a SCAR marker for better reproducibility. The oleic acid QTL also affected content of palmitic and linoleic acid content, which indicates that it controls either chain elongation or desaturation steps. Tanhuanpaa (1998) demonstrated that the identified QTL corresponds to the fatty acid desaturation (*fad2*) gene of *A. thaliana*. Hu et al (1999) identified two linked RAPD markers associated to oleic/linoleic acid content in *B. napus*.

5.5.   Palmitic acid content in *B. rapa*

Tanhuanpaa et al. (1995b) detected a RAPD marker associated to this trait in the same linkage group associated to oleic acid content.

5.6.   Erucic acid content in *B. napus*

Markers for loci determining this trait were reported by Jourdren et al. (1996c). All variation for this trait could be explained by genomic regions that correspond to two of three QTLs detected for seed oil content (Thormann et al. 1996, Ecke et al. 1996). Rajcan et al. (1999) reported a RAPD marker associated to erucic acid content and seed oil content. These studies demonstrate that erucic acid genes have a direct effect on seed oil content.

5.7. Glucosinolate content in *B. napus*

Most of our knowledge on glucosinolate synthesis in *Brassica* derives from *A. thaliana*. In this species several genes involved in the aliphatic glucosinolate pathway have been identified by genetic analysis, including Gls-*elong*, *Gls-alk*, and *Gls-ohp*. The *Gsl-elong* regulate aglycon side-chain length, *Gsl-alk* controls side-chain desaturation and *Gs-ohp* is responsible for side-chain hydroxylation (Magrath et al. 1993 & 1994). In rapeseed the genes *Gls-pro*, *Gls-elong-A*, *Gls-elong-C*, *Gls-oh-A*, *Gls-oh-C*, *Gls-alk-A* and *Gls-alk-C* regulating aglycon side-chain elongation and modification have also been reported (Parkin et al 1994, Halkier and Du 1997). The homology of these genes suggests that biosynthesis of glucosinolates is highly conserved in *Brassica* and *Arabidopsis*.

Regarding total glucosinolate content, four (Uzunova et al. 1995) to five QTLs Toroser et al.(1995) have been detected in *B. napus*. De Quiroz and Mithen (1996) integrated the two previous studies revealing RFLPs for low (GLS-1 marker) and high (GLS-2 marker) total glucosinolate content after screening varieties with contrasting amounts of these compounds. Varieties with intermediate content displayed both markers.

5.8. Clubroot (*Plasmodiophora brassicae*) resistance in *B. oleracea*

RAPD markers associated to this trait were detected by Grandclement et al. (1996). Previous work in *B. napus* by Figdore et al. (1993) had identified three QTLs for this trait. Landry et al. (1992) found two QTLs associated to resistance to this disease in *B. oleracea*. Voorips et al. (1997) found also two major QTLs conferring resistance to a Dutch field isolate of *Plasmodiphora brassiceae* explaining 68% of the variation for this trait. Moriguchi et al (1999) detected a single QTL for resistance to races 1 and 3 of this pathogen. It was located on linkage group 3. Kuginuki et al. (1997) reported RAPD markers associated to a clubroot resistance in *B. rapa*. These were converted to sequence tagged site (STS) markers by Kikuchi et al. (1999).

5.9. Blackleg resistance (*Leptosphaeria maculans*) in *B. napus*

For cotyledon and stem resistance, two to three common QTLs were detected. For field resistance two other unrelated QTLs were observed (Ferreira et al. 1995b). Dion et al. (1995) also discovered two QTLs for field resistance to this disease, one of which seems to correspond to a major resistance gene named *LmFr$_1$*. Two RFLP markers at 5% recombination on each side flanked this gene. Mayerhofer et al. (1997) mapped blackleg resistance genes in Australian rapeseed cultivars. Chevre et al. (1997) introgressed a chromosome segment from *B. juncea* into *B. napus* carrying a gene conferring complete resistance to this disease. The introgressed segment from the B genome could be identified by three specific RAPD markers. Plieske et al. (1998) reported also the transfer of resistance and associated markers from three B genome donor species to *B. napus* .

5.10. White rust (*Albugo candida* race 2) resistance in *B. napus*

A single gene responsible for resistance to this disease was mapped (Ferreira et al. 1995c). The locus responsible for the resistance, named *ACA1*, has been located on linkage group 4 of *B. rapa*, flanked by two markers at approximately 5 cM on each side (Kole et al. 1996). More recently, Cheung et al (1998) and Prabhu et al. (1998) have reported markers associated to white rust resistance in *B. juncea*.

5.11. Black rot (*Xanthomonas campestris*) resistance in *B. oleracea*

Two QTLs were found associated to this trait by Camargo et al. (1995). One of these was also associated to petiole length.

5.12. Markers for other disease resistance genes

Walsh et al. (1999) identified and mapped two genes associated to turnip mosaic virus resistance in *B. napus*. Joyeux et al. (1999) mapped a series of disease resistant homologs in *B. napus*.

5.13. Markers for fertility restorer gene for cytoplasmic male sterility in *B. napus*

An isozyme and a RAPD marker were found associated to the fertility restorer gene (*Rfo*) for the 'Ogura' radish system by Delourme et al. (1994 & 1998). Jean et al. (1997 & 1998) reported a series of DNA markers linked to the 'Polima' restorer gene (*Rpf1*) in canola, which is mapped on linkage group 18.

5.14. Markers for self-incompatibility (S)

This trait have been fully characterized genetically and most of the genes involved have been identified and cloned in *B. oleracea* (Boyes et al. 1997). Two of the main genes *SLG1* and *SRK* are highly polymorphic and multi-allelic serving to identify specific S lines in *B. oleracea* and *B. rapa* based on their DNA restriction or amplification profiles (Nishio et al. 1997, Okazaki et al. 1999).

5.15. Markers for morphological traits

Genes or markers for 28 traits, some of which were associated to as many as five QTLs were determined in a *B. rapa* progeny of Chinese cabbage 'Michihili x Spring broccoli (Song et al. 1995b). The same type of study was done by Kennard et al. (1994) for 22 traits in an F_2 progeny *B. oleracea,* resulting by crossing broccoli and cabbage. Upadhyay et al. (1996) found a marker for yellow seed coat color in *B. juncea*. Markers were also detected for six quantitative traits. An RFLP for seed coat content was detected by Teutonico and Osborn (1994) and van Deynze et al. (1995) in *B. napus*.

The DNA-based makers used for the creation of these maps have also been used extensively for variety identification and fingerprinting (see for example Dulson et al. 1998).

The increasing pace in sequencing activity for the *Arabidopsis* genome initiative will provide a very important resource to the *Brassica* geneticist and breeder in the very near future. The taxonomic proximity of *A. thaliana* to *Brassica* will male possible to transfer effectively this information for applied and basic research projects. This possibility will enhance our knowledge on the origin and evolution of the *Brassica* genomes.

Acknowledgments

Part of this work was funded by USDA Competitive Grant # 9600835.

6.   References

Acarkan, A., Koch, M., Rossberg, M. and Schmidt, R. (1998) Comparative genome analysis in cruciferous plants. Plant & Animal Genome VI, Abs P213.

Ajisaka, H., Kuginuki, Y., Shiratori, M., Ishiguro, K., Enomoto, S. and Hirai, M. (1999) Mapping of loci affecting the cultural efficiency of microspore culture of Brassica rapa L-syn. campestris L-using DNA polymorphism. Breeding Science 49: 187-192.

Armstrong, S.J., Fransz, P., Marshall, D.F. and Jones, G.H. (1998) Physical mapping of DNA repetitive sequences to mitotic and meiotic chromosomes of Brassica oleracea var. alboglabra by fluorescence in situ hybridization Heredity 81: 666-673.

Arondel, V., Lemieux, B., Hwang, T., Gibson, S., Goodman, H.M. and Somerville, C.R. (1992) Map-. based cloning of a gene controlling omega-3 fatty acid desaturation in *Arabidopsis*. Science 258: 1353-1355

Arumuganathan, K. and Earle, E.D. (1991) Nuclear DNA content of some important plant species. Plant Molec. Biol. Rept. 9: 208-219.

Arus, P. and Orton, T. (1983) Inheritance and linkage relationships of isozyme loci in *Brassica oleracea*. J. Hered. 74: 405-412.

Attia, T. and Robbelen, C. (1986) Cytogenetic relationship within cultivated *Brassica* analyzed in amphihaploids from the three diploid ancestors. Can. J. Genet. Cytol. 28: 323-329.

Bathia, S., Das, S., Jain, A. and Lakshmikumaran, M. (1995) DNA fingerprinting of *Brassica juncea* cultivars using microsatellite probes. Electrophoresis 16: 1750-1754.

Bouhon, E.J.R., Keith, D.J., Parkin, I.A.P., Sharpe, A.G. and Lydiate, D. (1996) Alignment of conserved C genomes of *Brassica oleracea* and B. napus. Theor. Appl. Genet. 93: 833-839.

Boyes, D.C., Nasrallah, M.E., Vrebalov, J. and Nasrallah, J.B. (1997) The self-incompatibility (S) haplotypes of Brassica contain highly divergent and rearranged sequences of ancient origin. Plant Cell 7: 1283-1294.

Brunel, D., Froger, N. and Pelletier, G. (1999) Development of amplified consensus genetic markers (ACGM) in Brassica napus from Arabidopsis thaliana sequences of known biological function Genome 42: 387-402

Camargo, L.E.A. (1994) Mapping RFLP and quantitative trait loci in *Brassica oleracea*. PhD thesis, Univ of Wisconsin, Madison WI., USA

Camargo, L.E.A., Williams, P.H. and Osborn, T.C. (1995) Mapping of quantitative trait loci controlling resistance of *Brassica oleracea* to *Xanthomonas campestris* pv *campestris* in the field and greenhouse. Phytopath 85: 1296-1300.

Camargo, L.E.A. and Osborn, T.C. (1996) Mapping loci controlling flowering time in *Brassica oleracea*. Theor. Appl. Genet. 92: 610-616.

Camargo, L.E.A., Savides, L., Jung, G., Nienhuis, J. and Osborn, T.C. (1997) Location of theself-incompatibility locus in an RFLP and RAPD map of *Brassica oleracea*. J. of Hered. 88: 57-59.

Carr, R.A. and McDonald, B.E. (1991) Rapeseed in a Changing World: Processing and utilization. GCIRC Eight Intl. Rapeseed Congress Saskatoon 1: 39-56.

Catcheside, D.G. (1934) The chromosomal relationship in the swede and turnip groups B*rassica*. Ann. Bot. Lond. 601: 33-56.

Cavell, A.C., Lydiate, D.J., Parkin, I.A.P., Dean, C. and Trick, M. (1998) Collinearity between a 30-centimorgan segment of Arabidopsis thaliana chromosome 4 and duplicated regions within the Brassica napus genome. Genome 41: 62-69.

Chen, B.Y., Jorgensen, R.B. Cheng, B.F. and Heneen, W.K. (1997) Identification and chromo-somal assignment of RAPD markers linked with a gene for seed colour in a Brassica campestris-alboglabra addition line. Hereditas 126: 133-138.

Cheng, B.F., Chen, B.Y. and Heneen, W.K. (1994a) Addition of *Brassica alboglabra* Bailey chromosomes to *B. campestris* L. with special emphasis to seed colour. Heredity 73: 185-189.

Cheng, B.F., Heneen, W.K. and Chen, B.Y. (1994b) Meiotic studies of a *Brassica campestris alboglabra* monosomic addition line and derived *B. campestris* primary trisomics . Genome 37: 584-589.

Cheng B.F., Heneen W.K. and Chen B.Y. (1995) Mitotic karyotypes of *Brassica campestris* and *B. alboglabra* and identification of the *B. alboglabra* chromosome in an addition line. Genome 38: 313-319.

Cheung W.Y., Champagne G., Hubert N. and Landry B.S. (1997a) Comparison of the genetic maps of Brassica napus and Brassica oleracea. Theor Appl Genet 94: 569-582.

Cheung W.Y., Friesen L., Rakow G.F.W., Seguin-Swartz G. and Landry B.S. (1997b) A RFLP-based linkage map of mustard [*Brassica juncea* (L) Czern and Coss] Theor Appl Genet 94: 569-582.

Cheung, W.Y., Gugel, R.K., Landry, B.S. (1998) Identification of RFLP markers linked to the white rust resistance gene (Acr) in mustard (Brassica juncea (L.] Czern. and Coss.) Genome 41: 626-628.

Chevre A.M., Delourme R., Eber F., Margale E., Quiros C.F. and Arus P (1995) Genetic analysis and nomenclature for seven isozyme systems in Brassica nigra, B. oleracea and B. campestris. Plant Breeding 114: 473-480.

Chevre A.M., Eber F., Barret P., Dupuy P. and Brace J (1997) Identification of the different *Brassica* nigra chromosomes from both sets of *B. oleracea-B.nigra* and *B. napus-B. nigra* addition lines with a special emphasis on chromosome transmission and self-incompatibility. Theor Appl Genet 94: 603-611.

Chyi Y.S., Hoenecke M. and Sernyk L (1992) A genetic linkage map of restriction fragment length polymorphism loci for *Brassica rapa* (syn. *campestris*). Genome 35: 746-757.

Cloutier S., Cappadocia M. and Landry B.S. (1995) Study of microspore culture responsiveness (*Brassica napus* L.) by comparative mapping of an F2 population and to microspore derived populations. Theor Appl Genet 91: 841-847.

Cloutier S., Cappadocia M. and Landry B.S. (1997) Analysis of RFLP mapping inaccuracy in Brassica napus L. Thoer Appl Genet 95: 83-91.

Copenhaver et al. (1999) Genetic definition and sequence analysis of Arabidopsis centromeres. Science 286: 2468-2474.

Delourme R., Bouchereau A., Hubert N., Renard M. and Landry B.S. (1994) Identification of RAPD markers linked to a fertility restorer gene for the Ogura radish cytoplasmic male sterility of rapeseed (*Brassica napus* L.) Theor Appl Genet 88: 741-748.

Delourme, R., Foisset, N., Horvais, R., Barret, P., Champagne, G., Cheung, W.Y., Landry, B.S., Renard, M (1998) Characterisation of the radish introgression carrying the Rfo restorer gene for the Ogu-INRA cytoplasmic male sterility in rapeseed (Brassica napus L.). Theor Appl Genet 97: 129-134.

De Quiroz H.C. and Mithen R (1996) Molecular markers for low-glucosinolate alleles in oilseed rape (*Brassica napus* L.). Mol Breed 2: 277-281.

Dion Y., Gugel R.K., Rakow G.F., Seguin-Swartz G. and Landry B.S. (1995) RFLP mapping of resistance to the blackleg disease [causal agent, *Leptosphaeria maculans*(Desm)]Cas et de Not] in canola (*Brassica napus* L.) Theor Appl Genet 91: 1190-1194.

dos Santos J.B., Nienhuis J., Skroch P., Tivang J. and Slocum M.K. (1994) Comparison of RAPD and RFLP genetic markers in determining genetic similarity among Brassica oleracea L. genotypes Theor Appl Genet 87: 909-915.

Dulson, J., Kott, L.S., Ripley, V.L. (1998) Efficacy of bulked DNA samples for RAPD DNA fingerprinting of genetically complex Brassica napus cultivars. Euphytica, 102: 65-70.

Ecke W., Uzunova M. and Weissleder K. (1995). Mapping thegenome of rapeseed (*Brassica napus* L.) 2. Localization of genes controlling erucic acid synthesis and seed oil content. Theor Appl Genet 91: 972-977.

Erickson L.R., Straus N.A. and Beversdorf W.D. (1983) Restriction patterns reveal origins of chloroplast genomes in *Brassica* amphiploids. Theor. Appl. Genet. 65: 201-206.

Fahey J.W., Zhang Y. and Talalay P. (1997) Broccoli sprouts: An exceptionally rich source of inducers of enzymes that protect against chemical carcinogens Proc Natl Acad Sci 94: 10367-10372.

Ferreira M.E., Williams P.H. and Osborn T.C. (1994) RFLP mapping of *Brassica* using doubled haploid lines. Theor Appl Genet 89: 615-621.

Ferreira M.E., Satagopan J., Yandell B.S., Williams P.H. and Osborn T.C. (1995a) Mapping loci controlling vernalization requirement and flowering time in *Brassica napus*. Theor Appl Genet 90: 727-732.

Ferreira M.E., Rimmer S.R., Williams P.H. and Osborn T.C. (1995b) Mapping loci controlling *Brassica napus* resistance to *Leptosphaeria maculans* under different screening conditions. Phytopath 85: 213-217.

Ferreira M.E., Williams P.H. and Osborn T.C. (1995c) Mapping of a locus controlling resistance to *Albugo candida* in *Brassica napus* using molecular markers. Phytopath 85: 218-220.

Figdore S.S., Kennard W.C., Song K.M., Slocum M.K. and Osborn T.C. (1988) Assessment of the degree of restriction fragment length polymorphism in *Brassica*. Theor. Appl. Genet. 75: 833-840.

Figdore S.S., Ferriera M.E., Slocum M.K. and Williams P.H. (1993) Association of RFLP markers with trait loci affecting clubroot resistance and morphological characters in *Brassica oleracea* L . Euphytica 69: 33-44.

Foisset, N., Delourme, R., Barret P., Hubert N., Landry B.S. and Renard M. (1996) Molecular mapping analysis of *Brassica napus* using isozyme, RAPD and RFLP markers on double haploid progeny. Theor. Appl. Genet. 93: 1017-1025.

Fourmann, M., Froger, N., Pelletier, G. and Brunel, D. (1998) Amplified consensus gene markers (ACGM): Tools designing for a genetic maps of known function genes. Plant & Animal Genome VI, San Diego, CA.Abs P215.

Gaubier, P., Raynal, M., Huestis, G., Hull, G., Grellet, F., Arenas, C., Pages, M. and Delseny, M. (1993) Two different Em-like genes expressed in *Arabidopsis thaliana* seeds during maturation. Mol Gen Genet 238: 409-418

Grandclement, C., Laurent, F. and Thomas, G. (1996) Detection and analysis of QTLs based on RAPD markers for polygenic resistance to *Plasmodiophora brassicae* Woron in *Brassica oleracea* L. Theor Appl Genet 93: 86-90.

Gupta, V., Jagannathan, and Lakshmikumaran, M.S. (1990) A novel AT-Rich tandem repeat in Brassica nigra. Plant Sci 68: 223-229.

Gupta, V., Lakshmisita, G., Shaila, M.S., Jagannathan, V. and Lakshmikumaran, M.S. (1992) Characterization of species-specific DNA sequences from *B. nigra* Theor Appl Genet 84: 397-402.

Halkier, B.A. and Du, L. (1997) The biosynthesis of glucosinolates. Trends in Plant Sci. 2: 425-431.

Harlan, J.R. and de Wet, J.M.J. (1975) On O Winge and a prayer: origins of polyploidy. The Botan Rev. 41: 361-390.

Hauge, B.M., Hanley, S.M., Cartinhour, S., Cherry, J.M., Goodman, H.M., Koorneef, M., Stam, P., Chang, C., Kempin, S., Medrano, L. and Meyerowitz, E.M. (1993) An integrated genetic/RFLP map of the *Arabidopsis thaliana* genome. The Plant Jour 3: 715-754.

Halldén, C., Nilsson, N.O., Rading, I.M. and Säll, T. (1994) Evaluation of RFLP and RAPD markers in a comparison of *Brassica napus* breeding lines. Theor Appl Genet 88: 123-128.

Hoenecke, M. and Chyi, Y.S. (1991) Comparison of *Brassica napus* and *B. rapa* genomes based on restriction fragment length polymorphism mapping. In Rapeseed in a Changing World, Proc. 8th International Rapeseed Cong. 4: 1102-1107. Saskatoon, Saskatchewan, Canada.

Hu, J. and Quiros, C.F. (1991) Molecular and cytological evidence of deletions in alien chromosomes for two monosomic addition lines of *Brassica campestris-oleracea*. Theor. Appl. Genet. 81: 221-226.

Hu, J., Quiros, C., Arus, P., Struss, D. and Robbelen, G. (1995) Mapping of a gene determining linolenic acid concentration in rapeseed with DNA-based markers. Theor Appl Genet 90: 258-262.

Hu, J., Sadowsky, J., Osborn, T.C., Landry, B.S. and Quiros, C.F. (1998). Linkage group alignment from four independent *Brassica oleracea* RFLP maps. Genome 41: 226-235.

Hu, J., Li, G., Struss, D., Quiros, C.F. (1999) SCAR and RAPD markers associated with 18-carbon fatty acids in rapeseed, Brassica napus. Plant Breeding 118: 145-150.

Iwabuchi, M., Itoh, K., Shimamoto, K. (1991) Molecular characterization of repetitive DNA sequences in Brassica. Theor Appl Genet 81: 349-355.

Jain, A., Bathia, S., Banga, S.S., Prakash, S. and Lakshmikumaran, M. (1994) Potential use of the random amplified polymorphic DNA (RAPD) to study the genetic diversity of Indian mustard (*Brassica juncea*) and its relationship to heterosis. Theor Appl Genet 88: 116-122

Jean M., Brown G.G. and Landry B.S. (1997) Genetic mapping of nuclear fertility restorer genes for the 'Polima cytoplasmic male sterility in canola (Brassica napus L.) using DNA markers. Theor Appl Genet 95: 321-328.

Jørgensen R.B., Chen B.Y., Cheng B.F., Heneen W.K. and Simonsen V. (1996) Random amplified polymorphic DNA markers of the *Brassica alboglabra* chromosome of a *B-campestris-alboglabra* addition line. Chromosome Res 4: 111-114.

Jourdren C., Barret P., Brunel D., Delourme R. and Renard M. (1996a) Specific molecular marker of genes controlling linolenic acid content in rapeseed. Theor Appl Genet 93: 512-518.

Jourdren C., Barret P., Horvais R., Delourme R. and Renard M. (1996b) Identification of RAPD markers linked to linolenic acid genes in rapeseed. Euphytica 90: 351-357.

Jourdren C., Barret P., Horvais R., Foisset N., Delourme R. and Renard M. (1996c) Identification of RAPD markers linked to the loci controlling erucic acid level in rapeseed. Mol Breed 2: 61-71.

Joyeux, A., Fortin, M.G., Mayerhofer, R., Good, A.G. (1999) Genetic mapping of plant disease resistance gene homologues using a minimal Brassica napus L. population. Genome 42: 735-743.

Kearsey M.J., Ramsay L.D., Jennings D.E., Lydiate D.J., Bohuon E.J.R. and Marshall D.F. (1996) Higher recombination frequencies in female compared to male meiosis in *Brassica oleracea*. Theor Appl Genet 92: 363-367.

Kennard W.C., Slocum M.K., Figdore S.S. and Osborn T.C. (1994) Genetic analysis of morphological variation in Brassica oleracea using molecular markers. Theor Appl Genet 87: 721-732.

Kianian S.F. and Quiros C.F. (1992a) Generation of a *Brasica oleracea* composite RFLP map: linkage arrangements among various populations and evolutionary implications. Theor. Appl. Genet. 84: 544-554.

Kianian S.F. and Quiros C.F. (1992b) Trait inheritance, fertility, and genomic relationships of some n=9 Brassica species. Gen Resources & Crop Evol 39: 165-175.

Kikuchi, M., Ajisaka, H., Kuginuki, Y., Hirai, M. (1999) Conversion of RAPD markers for a clubroot resistance gene of Brassica rapa into sequence - Tagged sites (STSs) Breeding Science 49: 83-88.

Kole C., Teutonico R., Mengistu A., Williams P.H. and Osborn T.C. (1996) Molecular mapping of a locus controlling resistance to *Albugo candida* in *Brassica rapa*. Phytopath 86: 367-369.

Kowalski S.P., Lan T.-H., Feldmann K.A. and Paterson A.H. (1994) Comparative mapping of *Arabidopsis thaliana* and *Brassica oleracea* chromosomes reveals islands of conserved organization. Genetics 138: 499-510.

Kresovich S., Szewc-McFadden A.K., Bliek S.M., NcFerson J.R. (1995) Abundance and characterization of simple-sequence repeats SSRS isolated from a size-fractionated genomic library of Brassica napus L. rapseed. Theor Appl Genet 91: 206-211.

Kuginuki, Y., Ajisaka, H., Yui, M., Yoshikawa, H., Hida, K., Hirai, M. (1997) RAPD markers linked to a clubroot-resistance locus in Brassica rapa L. Source: Euphytica 98: 149-154.

Lagercrantz U. and Lydiate D.J. (1995) RFLP mapping in *Brassica nigra* indicates differing recombination rates in male and female meiosis. Genome 38: 255-264.

Lagercrantz U., Putterill J., Coupland G. and Lydiate D. (1996) Comparative mapping of *Arabidopsis* and *Brassica*, fine scale genome collinearity and congruence of genes controlling flowering time. The Plant Jour 9: 13-20.

Lagercrantz U. and Lydiate D. (1996) Comparative Genome Mapping in Brassica. Genetics 144: 1903-1910.

Lagercrantz, U. (1998) Comparative mapping between Arabidopsis thaliana and Brassica nigra indicates that Brassica genomes have evolved through extensive genome replication accompanied by chromosome fusions and frequent rearrangements. Genetics 150: 1217-1228.

Lan T.-H. (1999) Molecular genetics of Brassica oleracea and Arabidopsis thaliana. PhD Dissertation Texas A&M University.

Lander E.S., Green P., Abrahamson J., Barlow A., Daly M.J., Lincoln S.E. and Newburg L. (1987) MAPMAKER: An interactive computer package for constructing primary genetic linkage maps of experimental and natural populations. Genomics 1: 174-181

Landry B.S., Hubert N., Etoh T., Harada J.J. and Lincoln S.E. (1991) A genetic map for *Brassica napus* based on restriction fragment length polymorphisms detected with expressed DNA sequences. Genome 34: 543-552.

Landry B.S., Hubert N., Crete R., Chang M.S., Lincoln S.E. and Etho T. (1992) A genetic map of *Brassica oleracea* based on RFLP markers detected with expressed DNA sequences and mapping of resistance genes to race 2 of P*lasmodiophora brassicae* (Woronin). Genome 35: 409-420.

Lannér C., Bryngelson T. and Gustafsson M. (1996) Genetic validity of RAPD markers at the intra- and inter-specific level in wild *Brassica* species with n=9. Theor Appl Genet 93: 9-14.

Lim Y.P., Kim J.H., Bang J.W., Nam H.G., Cho K.W. and Jang C.S. (1998) Genetic mapping of Chinese cabbage (Brassica rapa L. var. pekinensis) Plant & Animal Genome VI. Abs. P221.

Lin X. et al. (1999) Sequence and analysis of chromosome 2 of the plant Arabidopsis thaliana. Nature 402: 761-768.

Li G.Y. and Quiros C.F. (2000) Sequence-based amplified polymorphism as a marker source and its application to mapping genes detemining specific glucosinolates in Brassica oleracea Plant & Animal Genome VIII. P275.

Margalé E., Hervé and Quiros C.F. (1995) Determination of genetic variability by RAPD markers in cauliflower, cabbage and kale local cultivars from France. Genetic Resources & Crop Evol 42: 281-289.

Matsumoto, E., Yasui, C., Ohi, M., Tsukada, M. (1998) Linkage analysis of RFLP markers for clubroot resistance and pigmentation in Chinese cabbage (Brassica rapa ssp. pekinensis) Euphytica 104: 79-86.

Mayer K. et al. (1999) Sequence and analysis of chromosome 4 of the plant Arabidopsis thaliana. Nature 402: 769-777.

Mayerhofer, R., Bansal, V.K., Thiagarajah, M.R., Stringam, G.R., Good, A.G. (1987) Molecular mapping of resistance to Leptosphaeria maculans in Australian cultivars of Brassica napus. Genome 40: 294-301.

Magrath R., Herron C., Giamoustaris, A. and Mithen R. (1993) The inheritance of aliphatic glucosinolates in *Brassica napus*. Plant Breeding 111: 55-72

Magrath R., Bano F., Morgner M., Parkin I., Sharpe A., Lister C., Dean C., Turner J., Lydiate D. and Mithen R. (1994) Genetics of aliphatic glucosinolates. I. Side chain elongation in *Brassica napus* and *Arabidopsis thaliana*. Heredity 72: 290-299

McGrath J.M. and Quiros C.F. (1990) Generation of alien addition lines from synthetic *B. napus*: morphology, cytology, fertility, and chromosome transmission. Genome 33: 374-383.

McGrath J.M., Quiros C.F., Harada J.J. and Landry B.S. (1990) Identification of *Brassica oleracea* monosomic alien chromosome addition lines with molecular markers reveals extensive gene duplication. Molecular and General Genetics 223: 198-204.

McGrath, J.M. and Quiros C.F. (1991) Inheritance of isozyme and RFLP markers in *Brassica campestris* and comparison with *B. oleracea*. Theor. Appl. Genet. 82: 668-673.

McGrath J.M., Jancso M.M. and Pichersky E. (1993) Duplicate sequences with a similarity to expressed genes in the genome of *Arabidopsis thaliana*, Theor Appl Genet 86: 880- 888.

Mikkelsen T.R., Jensen J., Jørgensen R.B. (1996) Inheritance of oilseed rape (Brassica napus) RAPD markers in a backcross progeny with Brassica campestris. Theor Appl Genet 92: 492-497.

Moriguchi, K., Kimizuka-Takagi, C., Ishii, K., Nomura, K. (1999) A genetic map based on RAPD, RFLP, isozyme, morphological markers and QTL analysis for clubroot resistance in Brassica oleracea. Breeding Science 49: 257-265

Mulligan G.A. (1968) Physaria didymorcarpa, P. brassicoides and P. floribunda (Cruciferae) and their close relatives. Can J Bot. 46: 735-740.

Nishio T., Kusaba M., Sakamoto K. and Ockendon D.J. (1997) Polymorphism of the kinase domain of the S-locus receptor kinase gene (SRK) in Brassica oleracea L. Theor Appl Genet 95: 335-342.

Okasaki K., Kusuba M., Ockendon D.F. and Nishio T. (1999) Characterization of S tester lines and Brassica oleracea: polymorphism of restriction fragment length of SLG homologues and isoelectric point of S-glycoproteins. theor Appl Genet 98: 1329-1334

Osborn T.C., Butrille D., Sharpe A., Lydiate D., Hall, K. and Parker J. (2000) Translocations in linkage mapping and QTL analysis. Plant and Animal Genome VIII, W51.

Palmer J.D., Shield C.R., Cohen D.B. and Orton T.J. (1983) Chloroplast DNA evolution and the origin of amphidiploid *Brassica* species. Theor. Appl. Genet. 65: 181-189.

Parkin I.A.P., Sharpe A.G., Keith D.J. and Lydiate D.J. (1995) Identification of the A and C genomes of amphidiploid *Brassica napus* (oilseed rape). Genome 38: 1122-1131.

Paterson A. (1997) Comparative mapping of plant phenotypes. Plant Breed Rev 14: 13-38.

Plieske J., Knaack C and Struss D. (1998) Analysis in Brassica. Plant & Animal Genome VI, San Diego, CA. Abs P60.

Prabhu K., Somers D.J., Rakow G and Gugel R.K. (1998) Molecular markers linked to white rust resistance in mustard Brassica juncea. Theor Appl Genet 97: 865-870

Pradhan A.K., Prakash S., Mukhopadhyay A and Pental D. (1992) Phylogeny of *Brassica* and allied genera based on variation in chloroplast and mitochondrial DNA patterns: molecular and taxonomic classifications are incongruous. Theor Appl Genet 85: 331-340.

Price R.A., Palmer J.D. and Al-Shehbaz I.A. (1994) Systematic relationships of Arabidopsis: A molecular and morphological perspective. In Arabidopsis, Meyerowitz E.M. and Somerville C.R. (eds) CSH Lab Press, NY.

Prakash S. and Hinata K. (1980) Taxonomy, cytogenetics and origin of crop Brassicas, a review. Opera Bot. 55: 1-57.

Quiros C.F., Ochoa O., Kianian S.F. and Douches D. (1986) Evolutionary trends in *Brassica*: Gathering evidence from chromosome addition lines. Crucifer Newsletter 11: 22-23.

Quiros C.F., Ochoa O., Kianian S.F. and Douches D. (1987) Analysis of the *Brassica oleracea* genome by the generation of *B. rapa-oleracea* chromosome addition lines: Characterization by isozymes and rDNA genes. Theor. Appl. Genet. 74: 758-766.

Quiros C.F., Ochoa O. and Douches D.S. (1988) Exploring the role of x=7 species in *Brassica* evolution: hybridization with *B. nigra* and *B. oleracea*. J. Hered. 79: 351-358.

Quiros C.F., Hu J., This P., Chevre A.M. and Delseny M. (1991) Development and chromosomal localization of genome specific markers by polymerase chain reaction in *Brassica*. Theor. Appl. Genet. 82: 627-632.

Quiros C.F., Hu J. and Truco M.J. (1994) DNA-based marker *Brassica* maps. In: Advances in Cellular and Molecular Biology of Plants Vol. I: DNA based Markers in Plants. (Phillips R.L., Vasil I.K., eds) Kluwer Academic Publ, Dordrecht/Boston/London, pp 199- 222.

Quiros C.F. (1998) Molecular markers and their application to genetics, breeding and evolution of Brassica. J Japan Soc Hort Sci 67: 1180-1185.

Quiros C.F. (1999) Genome structure and mapping. In: Biology of Brassica coenospecies (Gomez-Campo ed). Elsevier Science B.V.

Quiros C.F., Grellet F., Sadowski J. and Wrobleski T. (2000). On exons, introns and spacers: complexity of comparing sequences of Arabidopsis thaliana and Brassica oleracea homologs. Plant & Animal Genome VIII. W56

Rajcan, I., Kasha, K.J., Kott, L.S., Beversdorf, W.D. (1999) Detection of molecular markers associated with linolenic and erucic acid levels in spring rapeseed (Brassica napus L.) Source: EUPHYTICA, 105: 173-181.

Ramsay L.D., Jennings D.E., Bohuon E.J.R., Arthur A.E., Lydiate D.J., Kearsey M.J. and Marshall D.F. (1996) The construction of a substitution library of recombinant backcross lines in *Brassica oleracea* for the precision mapping of quantitative loci. Genome 39: 558-567.

Robbelen G. (1960) Beitrage zur Analyse des *Brassica*-Genoms. Chromosoma 11: 205-228.

Sadowski J., Gaubier P., Delseny M. and Quiros C.F. (1996) Genetic and physical mapping in *Brassica* diploid species of a gene cluster defined in *Arabidopsis thaliana*. Mol Gen Genet 251: 298-306.

Sadowski J. and Quiros C.F. (1998) Organization of an *Arabidopsis thaliana* gene cluster on chromosome 4 including the RPS2 gene in the *Brassica nigra* genome. Theor Appl Genet 96: 468-474.

Sampson D.R. (1966) Genetic analysis of *Brassica oleracea* using nine genes from sprouting broccoli. Can J Gen and Cytol 8: 404-413.

Schilling A. (1991) Development of a molecular genetic linkage map in *Brassica rapa* (syn. *campestris)* L. MSc Thesis, U. of Massachusetts.

Schrader H., Budahn and Ahne R. (2000) Detection of 5S and 25S rRNA genes in Sinapis alba, Raphanus sativus and Brassica napus by double fluorescence in situ hybridization Theor Appl Genet 100: 665-669

Sebatian R.L., Howell E.C., King G.J., Marshall D.F. and Kearsey M.J. (2000) An integrated AFLP and RFLP Brassica oleracea linkage map from two morphologically distinct doubled-haploid mapping populations. Theor Appl Genet 100: 75-81

Sharma A., Mohopatra T. and Sharma R.P. (1994) Molecular mapping and character tagging in

Brassica juncea. 1 Degree, nature and linkage relationship of RFLPs and their association with quantitative traits. Jour Plant Biochem & Biotech 3: 85-89.

Sharpe A.G., Parkin I.A.P., Keith D.J. and Lydiate D.J. (1995) Frequent nonreciprocal translocations in the amphidiploid genome of oilseed rape (Brassica napus). Genome 38: 1112-1121.

Sikka S.M. (1940) Cytogenetics of *Brassica* hybrids and species. J. Genet. 40: 441-509.

Slocum M.K. (1989) Analyzing the genomic structure of *Brassica* species using RFLP analysis. In: Helentjaris, T., Burr, B. (eds). Development and application of molecular markers to problems in plant genetics. Cold Spring Harbor Lab. Press, NY.

Slocum M.K., Figdore S.S., Kennard W.C., Suzuki J.Y. and Osborn T.C. (1990) Linkage arrangement of restriction fragment length polymorphism loci in *Brasica oleracea*. Theor. Appl. Genet. 80: 57-64

Snogerup S. (1980) The wild forms of the *Brassica oleracea* group (2n=18) and their possible relations to the cultivated ones. In: *Brassica* crops and wild allies: biology and breeding. Tsonuda, S., Hinata, K. and Gomez-Campo, C. (eds). Japan Sci. Soc. Press.

Song K.M. *et al.* (1988) *Brassica* taxonomy based on nuclear restriction fragment length polymorphisms 1. Genome evolution of diploid and amphidiploid species. TAG 75: 784-794.

Song K., Osborn T.C. and Williams P.H. (1990) *Brassica* taxonomy based on nuclear restriction fragment length polymorphisms (RFLPs): 3. Genome relationships in *Brassica* and related genera and the origin of *B. oleracea* and *B. rapa* (syn. *campestris*). Theor. Appl. Genet. 79: 497-506.

Song K., Suzuki J.Y., Slocum M.K., Williams P.H. and Osborn T.C. (1991) A linkage map of *Brassica rapa* (syn. *campestris*) based on restriction fragment length polymorphism loci. Theor. Appl. Genet. 82: 296-304.

Song K., Lu P., Tang K., Osborn T.C. (1995a) Rapid genome change in synthetic polyploids of *Brassica* and its implications for polyploid eviolution. Proc Nat Acad Sci 92: 7719-7723.

Song K., Slocum M.K. and Osborn T.C. (1995b) Molecular marker analysis of genes controlling morphological variation in *Brassica rapa* (syn *campestris*). Theor Appl Genet 90: 1-10.

Stam P. (1993) Join Map Version 1.1. A computer program to generate genetic linkage maps. Center for Plant Breeding and Reproduction Research, CPRO-DLO. Wageningen, The Netherlands.

Struss D., Quiros C.F., Plieske J. and Robbelen G. (1996) Construction of *Brassica* B genome synteny groups based on chromosomes extracted from three different sources by phenotypic, isozyme and molecular markers. Theor Appl Genet 93: 1026-1032.

Struss D., Plieske J. and Saal B. (1999) Assignment of rapeseed microsatellite markers into the A and C genomes. Plant & Animal Genome VII, P140.

Szewc-McFadden A.K., Kresovich S., Bliek S.M., Mitchell S.E., McFerson J.R. (1996) Identification of polymorphic, conserved simple sequence repeats SSRS in cultivated Brassica species. Theor Appl Genet 93: 534-538.

Tanhuanpaa P.K., Vilkki J.P. and Vilkki H.J. (1995a) Association of a RAPD marker with linolenic acid concentration in the seed oil of rapeseed (*Brassica napus* L.) Genome 38: 414-416.

Tanhuanpaa P.K., Vilkki J.P. and Vilkki H.J. (1995b) Identification of a RAPD marker for palmitic-acid concentration in the seed oil of spring turnip rape (*Brassica rapa* ssp. *oleifera*).

Theor Appl Genet 91: 477-480.

Tanhuanpaa P.K., Vilkke J.P. and Vilkki H.J. (1996) Mapping a QTL for oleic acid concentration in spring turnip rape (*Brassica rapa* ssp *oleifera*). Theor Appl Genet 92: 952-956

Tanhuanpaa, P., Vilkki, J., Vihinen, M. (1998) Mapping and cloning of FAD2 gene to develop allele-specific PCR for oleic acid in spring turnip rape (Brassica rapa ssp. oleifera). Molecular Breeding 4: 543-550.

Teutonico R.A. and Osborn T.C. (1994) Mapping of RFLPs and quantitative traits loci in *Brassica* and comparison to the linkage maps of *B. napus, B. oleracea* and *Arabidopsis thaliana*. Theor Appl Genet 89: 885-894.

Teutonico R.A. and Osborn T.C. (1995) Mapping loci controlling vernalization requirement in *Brassica rapa*. Theor Appl Genet 91: 1279-1283.

Teutonico R.A., Yandell B., Satagopan J.M., Ferreira M.E., Palta J.P. and Osborn T.C. (1995) Genetic analysis and mapping of genes controlling freezing tolerance in oilseed *Brassica*. Mol Breed 1: 329-339.

This P., Ochoa O. and Quiros C.F. (1990) Dissection of the *Brassica nigra* genome by chromosome addition lines. Plant Breeding 105: 211-220.

Thormann C.E., Ferreira M.E., Camargo L.E.A., Tivang J.G. and Osborn T.C. (1994) Comparison of RFLP and RAPD markers to estimating genetic relationships within and among cruciferous species. Theor Appl Genet 88: 973-980.

Thormann C.E., Romero J., Mantet J. and Osborn T.C. (1996) Mapping loci controlling the concentrations of erucic and linolenic acids in seed oli of *Brassica napus* L. Theor Appl Genet 93: 282-286.

Toroser D., Thormann C.E., Osborn T.C. and Mithen R. (1995) RFLP mapping of quantitative trait loci controlling seed aliphatic-glucosinolate content in oliseed rape (*Brassica napus* L.). Theor Appl Genet 91: 802-808.

Truco M.J. and Quiros C.F. (1994) Structure and organization of the B genome based on a linkage map in *Brassica nigra*. Theor Appl Genet 89: 590-598.

Truco M.J., Hu J., Sadowski J. and Quiros C.F. (1996) Inter- and intra-genomic homology of the *Brassica* genomes: implications for their origin and evolution. Theor Appl Genet 93: 1225-1233

U N (1935) Genome-analysis in *Brassica* with special reference to the experimental formation of *B. napus* and peculiar mode of fertilization. Jpn. J. Genet. 7: 389-452.

Upadhayay A., Mohapatra T., Pai R.A. and Sharma R.P. (1996). Molecular mapping and character tagging in mustard *(Brassica juncea)* 2. Association of RFLP markers with seed coat colour and quantitative traits. Jour Plant Biochem & Biotech 5: 17-22.

Uzunova M., Ecke W., Weissleder K. and Robbelen G. (1995) Mapping the genome of rapeseed (*Brassica napus* L.) I. Construction of an RFLP linkage map and localization of QTLs for seed glucosinolate content. Theor Appl Genet 90: 194-204.

Uzunova, M.I., Ecke, W. (1999) Abundance, polymorphism and genetic mapping of microsatellites in oilseed rape (Brassica napus L.) Plant Breeding 118: 323-326.

Van Deynze A.E., Landry B.S. and Pauls K.P. (1995) The identification of restriction fragment length polymorphisms linked to seed colour genes in *Brassica napus*. Genome 38: 534-542.

Vaughn J.G. (1977) A multidisciplinary study of the taxonomy and origin of *Brassica* crops. BioSci 27: 35-40.

Voorrips R.E., Jongerius M.C. and Kanne H.J. (1997) Mapping of two genes for resistance to clubrot (Plasmodiophora brassicae) in a population of doubled haploid lines of Brassica oleracea by means of RFLP and AFLP markers. Theor Appl Genet 94: 75-82.

Walsh, J.A., Sharpe, A.G., Jenner, C.E. and Lydiate, D.J. (1999) Characterisation of resistance to turnip mosaic virus in oilseed rape (Brassica napus) and genetic mapping of TuRB01. Theor Appl Genet 99: 1149-1154.

Warwick S.I. and Black L.D. (1991) Molecular systematics of *Brassica* and allied genera (subtribe *Brassicinae, Brassiceae*)—chloroplast genome and cytodeme congruence. Theor. Appl. Genet. 82: 81-92.

Wroblewski T., Sadowski S and Quiros C.F. (1998) Positional and functional differences between the *ABI1-RPS2-CK1* gene complex from *Arabidopsis thaliana* and its homologues from *Brassica oleracea*. Plant and Animal Genome VI, San Diego CA. Abs P224

Wroblewski T., Coulibaly S., Sadowski J and Quiros C.F. (2000) Variation and phylogenetic utility of the Arabidopsis thaliana Rps2 homolog in various species of the tribe Brassiceae. Mol. Phylogenet & Evol (in press).

Zhang H.B. (1998) The Texas A&M BAC Center: Ten new libraries added in 1997. Plant & Animal Genome VI, San Diego, CA Abs P15.

14. Molecular genetic map of cotton

ANDREW H. PATERSON
Department of Crop and Soil Science, University of Georgia, Athens GA 30602
<paterson@uga.edu>

Contents

1. Introduction

World cotton commerce is largely derived from *Gossypium hirsutum* and *G. barbadense*, two tetraploid members of a diverse genus (Gossypium) comprised of about 50 diploid and tetraploid species distributed across America, Africa, Asia, and Australia. A central

R.L. Phillips and I.K. Vasil (eds.), DNA-Based Markers in Plants, 239 - 253.
© 2001 *Kluwer Academic Publishers. Printed in the Netherlands.*

issue in cotton molecular genetics is the genetic control of the development of spinnable fibers, the economic organ of cotton. The observation that tetraploid cottons transgress the levels of fiber yield and quality of diploid cottons has suggested that polyploidy has created novel opportunities for evolution. Development and application of a comprehensive set of DNA markers, and other molecular tools has created new avenues for investigating the genetic control of fiber development, and other questions in cotton genetics and evolution.

2.   Evolutionary relationships and biogeography of *Gossypium*

The genus *Gossypium* L. comprises about 50 diploid and tetraploid species indigenous to Africa, Central and South America, Asia, Australia, the Galapagos, and Hawaii (Fryxell 1979, 1992). Cultivated types derived from four species, namely *G. hirsutum* L. (n = 2x = 26), *G. barbadense* L. (n = 2x = 26), *G. arboreum* L. (n = x = 13), and *G. herbaceum* L. (n = x = 13), provide the world's leading natural fiber, cotton, and are also a major oilseed crop. Cotton was among the first species to which the Mendelian principles were applied (Balls 1906), and has a long history of improvement through breeding, with sustained long-term yield gains of 7-10 kg lint/ha/yr (Meredith and Bridge 1984). The annual world cotton crop of ca 65 million bales (of 218 kg/bale), has a value of ca US$15-20 billion/yr.

Diploid *Gossypium* species are all n = 13, and fall into 7 different "genome types", designated A-G based on chromosome pairing relationships (Beasley 1942; Endrizzi et al. 1984). A total of 5 tetraploid (n = 2x = 26) species are recognized, all exhibiting disomic chromosome pairing (Kimber 1961). The tetraploid sub-genomes resemble the extant A genome of *G. herbaceum* (n = 13) and D genome of *G. raimondii* Ulbrich (n = 13), respectively. The A- and D-genome species diverged from a common ancestor about 6-11 million years ago (Wendel 1989). The putative A x D polyploidization event occurred in the New World, about 1.1-1.9 million years ago, and required transoceanic migration of the maternal A-genome ancestor (Wendel 1989, Wendel and Albert 1992), which is indigenous to the Old World (Fryxell 1979). Polyploidization was followed by radiation and divergence, with distinct n = 26 AD genome species now indigenous to Central America (*G. hirsutum*), South America (*G. barbadense*, *G. mustelinum* Miers ex Watt), the Hawaiian Islands (*G. tomentosum* Nuttall ex Seemann), and the Galapagos Islands (*G. darwinii* Watt) (Fryxell 1979).

3.   Genome organization: repetitive DNA

Tetraploid cotton has a large and complex repetitive DNA fraction. DNA reassociation kinetics (Walbot and Dure 1976; Wilson et al. 1976; Geever et al. 1989) suggest that the tetraploid cotton genome is comprised of 50-65% low copy DNA and 35-50% repetitive DNA. Repetitive DNA elements have been isolated that are estimated to comprise 29-35% of the genome, and 60-70% of the repetitive fraction of tetraploid cotton (Zhao et al. 1995). Most repeat families are interspersed in the cotton genome and have copy numbers that range from 10^3-10^4 (except for three families, pXP004, pXP020, and pXP072 with a copy number of $\geq 10^5$), thus fall into the moderate-abundance class (10^2-10^5 copies, Bouchard 1982). Therefore, moderately-abundant interspersed repetitive DNA comprises the largest class of repetitive sequences in the cotton genome. Out of the cloned 103 repeat families, only 20 (19%) are tandem repeats. None of the tandem repeats meet the traditional threshold of 10^6 copies (Singer 1982) for classical "highly-repetitive" DNA. A paucity of high-copy tandem repeats in cotton is also supported by evidence from DNA/DNA reassociation kinetics, which showed the rapidly-reassociating fraction to be small (Wilson et al. 1976; Geever et al. 1989), and buoyant density centrifugation experiments which showed no detectable satellites (Walbot and Dure 1976). This feature makes the cotton genome different from many other higher plant genomes (Flavell 1982a; Lapitan 1992).

A total of 70 (68%) of the 103 families are methylated to some degree (Zhao et al. 1995). Cotton DNA has been estimated to have a G+C+5 MeC content of 36.1%, of which about 25% of the C (4.6% of the total bases) is the methylated derivative 5 MeC (Ergle and Katterman 1961; Walbot and Dure 1976). Since the majority of repeat families are methylated, methylated C residues may be concentrated in the repetitive fraction of the cotton genome. Genomic DNA digestion with P̲s̲t̲ I, a methylation-sensitive enzyme, is a highly effective means of enriching for 1-2 kb digestion products suitable as low-copy DNA probes (Reinisch et al. 1994).

3.1. Genome-specific (or enriched) families of dispersed repetitive elements account for much of the difference in DNA content between A and D genome diploid cottons

The genomic affinity of each repeat family was first evaluated by hybridization to stoichiometric quantities of DNA from the only two extant A-genome species (*G. arboreum*, *G. herbaceum*), and divergent representatives of the D-genome group (*G. trilobum*, *G. raimondii*), then further evaluated by quantitative slot-blot hybridization to genomic DNA from 20 different cotton species representing each of the 7 recognized genome types and also the tetraploids (Zhao et al. 1998). A-genome-specific or enriched clones were estimated to account for 0.41 pg more DNA than D-genome repeats, or about 48% of the 0.85 pg difference in total DNA content between the A and D genomes. This estimate must be considered only a first-order approximation, contingent on better delineating the precise boundaries of individual repetitive DNA elements within the clones.

With the exception of *G. gossypioides*, the phylogenetic distribution of repetitive element families is generally consistent with our present understanding of *Gossypium* phylogeny (cf. Wendel and Albert 1992). Families that were abundant in the D-genome, which is confined to the New World, were rare in the African/Arabian A, B, E, and F genomes. A-genome specific or enriched elements were found at moderate levels in the closely-related B, E, and F genomes, at low levels in the Australian C and G genomes, and were virtually absent from an outgroup, *Thespesia lampas*.

The set of dispersed repeat families found in *G. gossypioides* was incongruous with that of any other diploid *Gossypium* genome type. D-genome families occurred at similar levels in *G. gossypioides* and the other D-genome cottons. However, signals from A-genome specific repeats were found in *G. gossypioides* at about 36% of the level of A-genome diploids, and 600% higher than in other D-genome cottons.

3.2. Fluorescence in situ hybridization (FISH) reveals "spread" of some dispersed repetitive DNA families in polyploid cotton chromosomes

To further evaluate the physical distribution of dispersed repeats in tetraploid cotton, 20 of the 83 dispersed repeat probes have been applied to the chromosomes of four *Gossypium* species by FISH (Hanson et al. 1998). Relatively few dispersed repeat families clearly distinguish between the A- and D-subgenome chromosomes of tetraploid cotton. The majority of families show a continuous distribution of hybridization signal across the chromosomes of tetraploid cotton suggesting that they have "spread" to D-subgenome chromosomes. Further evidence in support of the "spread" comes from the FISH pattern for *G. gossypioides*, which contains otherwise A-genome-specific repeats at levels comparable to the D-subgenome chromosomes of *G. hirsutum*, that fall at the lower end of the continuum of signal intensities.

3.3. Most cotton nuclear repetitive elements do not correspond to previously-identified genes or DNA sequences

One-pass sequences revealed that 24 (23%) of the 103 repetitive DNA clones (including both tandem and dispersed repeats) showed significant correspondence (BLAST > 150) to previously-identified DNA sequences from a wide range of organisms (Zhao et al. 1998). Three of the clones show correspondence to parts of a transposable element from *Lilium henryi* (Smyth et al. 1989), that also corresponds to elements found in *Nicotiana* (Royo et al. 1996), *Brassica* (Elborough and Storey 1996), and *Arabidopsis* (Bevan et al. 1997). A fourth cotton element, pXP1-13 corresponds to part of a *Drosophila melanogaster* transposon (Biessmann et al. 1992).

3.4. Some cotton repetitive DNA clones hybridize to transcripts

Among the cotton repetitive DNA clones, 12 showed relative signal of >10, more than 100 times greater than the negative control. (Zhao et al. 1998). These include two of the three clones resembling the *Lilium* transposon, and clones that have DNA sequences similar to the *Arabidopsis* cyc2b gene, and *Plasmodium* DNA polymerase alpha gene.

3.5. Applications of dispersed repeats

The cotton dispersed repeats provide the means to quickly establish large numbers of landmarks throughout the genome, useful for integrating genetic and physical maps, fingerprinting individual BAC clones, and as sequence-tagged sites for future genomic sequencing. One impetus for this avenue of research was to develop "tags" to determine whether particular BACs from tetraploid cotton were from the A-subgenome or the D-subgenome, to expedite chromosome walking to agriculturally-important alleles that mapped to one specific subgenome of tetraploid cotton (most of the world's cultivated cotton is tetraploid). The ca 15-20% of dispersed repeat families that remain largely-restricted to their source genome in tetraploid cotton (Hanson et al. 1998), collectively provide about 280,000 such potential tags — however, a high degree of clustering observed within individual repeat families may impede some applications to plant genome analysis, such as BAC "tagging." To tag an A-subgenome-derived BAC clone of 100 kb, at least 7% (18,000) of individual family members would have to be distributed at average intervals of 100 kb along the ca 1800 Mb of the A-subgenome. By screening smaller (15 kb) lambda clones, we found a high degree of clustering among family members, with an individual family tagging an average of only 0.7% of lambda clones. Moderate-abundance families appear much less clustered than high-abundance families — pools of moderate-abundance families may be more suitable than individual high-abundance families as probes for comprehensive "fingerprinting" of large DNA clones (cf. Nelson et al. 1989).

Better understanding of cotton dispersed repeat families might yield practical strategies for insertional mutagenesis, analogous to those employed for multiple-copy transposons (cf. Chomet 1994). In total, the A-genome dispersed repeat families are estimated to include about 1.4 million individual elements, a potentially powerful mutagen even if only a small subset are still capable of spread. Naturally-occurring insertional mutagenesis might partly account for the widespread observation that inbred cottons remain far more variable in phenotype than would be expected after many generations of selfing.

4.   Status of the genetic map

Variation in ploidy among *Gossypium* spp., together with tolerance of aneuploidy in tetraploid species of *Gossypium*, has facilitated use of cytogenetic techniques to explore cotton genetics and evolution. Among 198 morphological mutants described in cotton, 61 mutant loci have been assembled into 16 linkage groups, through the collective results of many investigators. Using nullisomic, monosomic, and monotelodisomic stocks, 11 of these linkage groups have been associated with chromosomes (Endrizzi et al. 1984).

Reinisch et al. (1994) published a detailed molecular map of cotton, based on a cross between the two predominantly cultivated species, *Gossypium hirsutum*, and *G. barbadense*. The map is comprised of 705 RFLP loci in 41 linkage groups and 4675 cM, with average spacing between markers of about 7 cM. Since the gametes of these cotton species each contain 26 chromosomes, additional DNA markers are clearly needed, and much additional mapping is expected to result in a comprehensive revision of the map in the near future. Estimates of recombination in cotton based upon meiotic configuration analyses indicate a minimum overall map length of 4660 cM (Reyes-Valdes and Stelly 1995) suggesting that most regions of the genome are already covered by the present map, although gaps in some linkage groups remain to be filled. Given that the 26 chromosomes of cotton are presently represented by 41 linkage groups, and that we can detect linkage at up to 30 cM, filling the gaps between existing linkage groups should add at least 450 cM (= 30 cM x 15 gaps) to the map, for a minimum overall length of 5125 cM. Once the map reaches a density of about one marker per 5 cM, or a total of about 1025 markers, fewer than 1% of intervals between markers should measure >25 cM, and the map should "link up" into 26 linkage groups corresponding to the 26 gametic chromosomes of cotton.

4.1. Assignment of linkage groups to chromosomes

The polyploid nature of cotton renders it more tolerant of chromosomal deficiencies than many organisms. The ability of the cotton plant, and its megagametophytes, to survive even when deficient for entire chromosomes, has afforded a facile means to determine the correspondence between "linkage groups" and most chromosomes. A subset of mapped DNA probes were hybridized to genomic digests of a previously-developed series of monosomic and monotelodisomic interspecific substitution stocks, with a single *G. barbadense* chromosome substituted for one *G. hirsutum* chromosome, or chromosome arm. The hybrid nature of these stocks affords a modified form of deficiency mapping, in that high rates of interspecific DNA polymorphism obviate reliance upon dosage analysis. Loci in the affected chromosome are hemizygous for the *G. barbadense* allele, and lack the *G. hirsutum* allele. By contrast, loci on non-affected chromosomes exhibit the *G. hirsutum* allele. This method is conceptually similar to use of chromosome-deficient lines to determine chromosomal location of DNA probes (Galau et al. 1988; Helentjaris et al. 1988) except that assignment of probes to chromosomes is based on detection of an RFLP, thus is less subject to errors. Based on evidence from a minimum of three genetically-linked loci which correspond to the same aneuploid substitution stock, we have tentatively determined which linkage groups correspond to *chrs. 1, 2, 4, 6, 9, 10, 17, 22,* and *25*. The identity of *chrs. 5, 14, 15, 18,* and *20* is suggested by single loci, which are neither corroborated nor contradicted by any other locus on the linkage group (Reinisch et al. 1994). Identification of several additional chromosomes is in progress.

4.2. Determination of homoeology from map positions of duplicated loci

Mapping of multiple RFLPs derived from individual cotton DNA probes has revealed the consequences of genome-wide duplication in tetraploid (n = 2x = 26) cotton. Based on criteria described in detail (Reinisch et al. 1994), parts of 11 of the 13 expected homoeologous pairs in n = 26 cottons have been identified based on published data for 62 duplicated DNA probes, spanning 1668 cM or 35.6% of the genome. Among the homoeologous relationships found by RFLP mapping were three pairs of tentatively-identified chromosomes, chrs. 1 and 15, chrs. 5 and 20, and chrs. 6 and 25. Homoeology between chrs. 1 and 15 is supported by prior evidence that duplicated mutations in three morphological traits map to this pair of chromosomes (Endrizzi et al. 1984) with one inversion in order. Homoeology between chrs. 6 and 25 has been suggested previously (Endrizzi and Ramsay 1979) based upon similar phenotypes of plants monosomic for each of these chromosomes. Although several mutant phenotypes have been assigned to chr. 5, no evidence is available regarding homoeology of chromosomes 5 and 20.

Independent studies have demonstrated homoeology of chromosomes 7 and 16, and chromosomes 12 and 26 based on morphological markers (Endrizzi et al. 1984), and chromosomes 9 and 23 based on *in situ* hybridization (Crane et al. 1994)

Most of the homoeologous chromosomes in AD cottons are distinguished by one or more inversions. In only two cases are putative homoeologs homosequential over the entire region of the chromosome for which homoeology can be inferred, although large blocks of sequence conservation are evident in most cases. Four pairs clearly differ by at least one inversion, and two additional pairs differ by at least two inversions. At least one pair shows a translocation, while several additional pairs may represent either translocations or simply small linkage groups which have not yet linked up. New data (Brubaker et al., in press) will describe mapping of a subset of the tetraploid DNA probes in F_2 populations of *G. arboreum* x *G. herbaceum* (A-genome) and *G. raimondii* x *G. trilobum*. This should permit discrimination of structural mutations that occurred during divergence of the n = 13 A- and D-genomes from mutations that occurred after formation of the n = 26 AD-genome.

4.3.   Deducing the genomic origin of linkage groups in allotetraploid cotton

Some DNA probes detected genomic fragments in tetraploid cottons that were shared with either A- or D-genome ancestors, but not both. In a subset of these cases, polymorphism between *G. hirsutum* and *G. barbadense* permitted mapping of one or both of the homoeologous genomic fragments. Based on such "alloallelic" loci, the genomic origins of 33 of the present 41 linkage groups, including all of the known chromosomes, have been determined. One linkage group (L.G. U01) showed conflicting information (2 A, 2 D), and seven small linkage groups (totalling 224.8 cM and 34 loci) harbored no alloallelic loci. To evaluate the reliability of our subgenomic inferences, we calculated the frequency at which alloallelic information coincided with the classical assignment of cotton chromosomes to genomes based upon pairing relationships in diploid x tetraploid hybrids (A = *chrs. 1-13*; D = *chrs. 14-26*). Based on the 14 chromosomes we have identified, the majority of alloallelic data for a chromosome agreed with classical subgenomic assignment of chromosomes in all 14 (100%) cases. We conclude that alloallelic information is a reliable, although not infallible, indicator of genomic origin of a chromosome or linkage group.

Six of the 10 loci where alloallelic information *disagreed* with classical subgenomic assignment of chromosomes occurred as linked pairs on three linkage groups. In each of these three cases (Chr. 14, L.G. A03, L.G. U01), the two deviant loci are consecutive along the linkage group (with reference to alloallelic probes). This suggests the possibility that exchanges of chromatin between homoeologous chromosomes may have occurred during the evolution of tetraploid cottons.

4.4.   Identification of simple-sequence repeat polymorphisms

DNA microsatellites, or simple-sequence repeats, are short stretches of tandem repeats with a repeating unit of 1-5 nucleotides. Several features of microsatellites in eukaryotic genomes make them excellent molecular markers for many genetic investigations, including their abundance, high levels of allelic variation, amenability to PCR-based assay, and utility as sequence-tagged sites (STS) $(GA)_{15}$, $(CA)_{15}$, $(AT)_{15}$, and $(A)_{30}$ synthetic oligonucleotides, end-labeled with ^{32}P, were used to screen ca 20,400 lambda genomic clones (about 11.3% of the genome) from a partial Sau 3AI library of cotton (*G. barbadense* "K101"), of average length 15 kb (Zhao et al., in preparation). Assuming that each positive clone contained only one microsatellite array, the number of arrays for each of the four dinucleotides were estimated. Based upon a genome size of 2700 Mb, we calculate that one of these four arrays should occur an average of once every 170 kb. Considering only the GA and CA elements which are more reliable in PCR amplification, we find one array per 1155 kb, or about 2 centiMorgans in cotton. Comparison of 16 sequence-tagged microsatellites (STMs) and 39 low-copy DNA probes applied to 10 cotton cultivars indicated that microsatellites are more polymorphic than RFLP markers in cotton, but only moderately so. All 16 STMs amplified DNA fragments in both A and D genome diploid progenitors of cultivated (AD-genome) cotton, indicating the conservation of priming sequences over the ~6-11 million years since the divergence of A- and D-genome taxa from a common ancestor. Genetic mapping of 13 STMs revealed 20 polymorphic loci on 12 different linkage groups. Most of the duplicated loci mapped represented corresponding sites on homoeologous chromosomes, suggesting that not only priming sites, but also microsatellite arrays responsible for DNA polymorphism, have persisted for 6-11 million years. About 25% of the mapped loci were terminal in their respective linkage groups, possibly suggesting a telomeric bias in distribution of cotton STMs.

5.   Mapping of genes and QTLs

Application of the cotton map to the identification of genes and QTLs is progressing rapidly. Priorities have included genes associated with fiber quality and productivity, genes associated with disease reaction, and morphological features that affect cotton productivity, or confer botanical novelty. Ongoing priorities include these traits, plus additional complex traits such as response to water deficit.

5.1.   QTLs identified for fiber-related traits

A total of 14 QTLs have been reported to date, from a cross between *G. hirsutum* (GH) and *G. barbadense* (GB; Jiang et al. 1998). Mapping of several additional crosses is in progress. Allele effects for 12 of the 14 QTLs were consistent with the difference between parents, GB alleles being associated with long, strong, fine fibers and GH alleles associated with higher yield and early maturity. The two exceptions were each fiber strength QTLs, where the GB alleles reduced fiber strength. Individual QTLs detected were as follows:

5.1.1.   *Fiber strength (STR)*
Three QTLs collectively explained 30.9% of PVE. The GB allele increased fiber strength on LGD02, and decreased it on chromosomes 20 and 22.

5.1.2. *Fiber length*
One QTL explained 14.7% of PVE, with the GB allele on LGD03 increasing fiber length.

5.1.3. *Fiber thickness (Dn)*
One QTL explained 12.6% of PVE. The GB allele on chromosome 10 decreased fiber thickness.

5.1.4. *Coefficient of variation in mean length of fiber by number (LnCV)*
Two QTLs (Table 1) collectively explained 23.0% of PVE. The GB alleles on both chromosomes 10 and 15 increased the coefficient of variation.

5.1.5. *Fiber elongation (ELONG)*
Two QTLs collectively explained 21.1% of PVE. The GB alleles on both LGD03 and LGA02 increased fiber elongation.

5.1.6. *Fiber yield components*
(Log of seed cotton yield, *LogSDCT;* and number of locules per boll, *LB)*
Four QTLs were detected, two explaining 17.0% of PVE in LogSDCT and two explaining 58.5% of PVE in LB. The GB alleles on linkage groups A02 and D07 decreased seed cotton yield, and those on chromosomes 23 and 25 decreased locules per boll.

5.1.7. *Earliness*
Only one QTL was detected, on LGD04, explaining 8.1% of PVE. The GB allele delayed fiber harvest, increasing the percentage of fiber harvested after Sept. 12, the second of three weekly harvest dates.

A total of seven QTLs conferring bacterial blight resistance have been described (Wright et al. 1998), among four cotton populations. The inheritance of resistance in each population was as follows:

Empire B2 population.   A region near the DNA marker *G1219* on D-subgenome chromosome 20 explained 98.0% (LOD 103.1) of the phenotypic variation in reaction to *Xcm* Races 2 and 4. The GH (*G. hirsutum*) allele was dominant.

Empire B3 population.   A region near the DNA marker *pGH510a* on D-subgenome chromosome 20 accounted for 88.2% (LOD 23.2) of the phenotypic variation in reaction to Races 2 and 4. Although B_3 is more than 50 cM away from B_2, the finding that both are on the same chromosome supports classical data suggesting linkage between these genes (Knight 1944). B_3 showed a modest deviation from additivity toward dominance ($d/a = 0.41$), consistent with earlier reports (Innes 1983). Incompletely dominant gene action of B_3 has been reported (Innes 1983), and may account for the observed deviation from single-gene segregation in this population (see above).

Unexpectedly, B_3 also explains 53.4% (LOD 10.56) of phenotypic variation in reaction to *Xcm* Races 7 and 18. All individuals showed susceptible reaction to Races

7 and 18, however quantitative disease severity of individual plants ranged widely. The effect of B_3 on reaction to *Xcm* Races 7 and 18 was strictly additive ($d/a = 0.01$) differing from "partial dominance" of its effect on *Xcm* Races 2 and 4.

Empire B2b6 population.   The region near *G1219* detected in the Empire B2 population also explained 92.2% (LOD 53.36) of the phenotypic variation in reaction to *Xcm* Races 2 and 4 in Empire B2b6. The GH allele increased resistance and showed dominant gene action, as observed in the Empire B2 population. Scored as a discrete genetic locus, this trait mapped to an interval between *A1701b* and *pAR377*, again reinforcing the need for quantitative phenotypes to obtain reliable map positions.

The genetic basis of resistance to *Xcm* Races 7 and 18 was unexpectedly complex. Four QTLs (Qb_{6a}, Qb_{6b}, Qb_{6c} and Qb_{6d}) collectively explained 56.4% of the phenotypic variation in reaction to *Xcm* Races 7 and 18. Reduced models had LOD reductions of 2.0 or more, so the actions of these genes were considered largely independent. The unexpectedly high complexity of b_6 resistance presumably accounts for the deviation from simple segregation models (above). An initial scan of the genome detected two QTLs. A region (Qb_{6a}) near the DNA marker *A1666* on a linkage group of unknown subgenomic origin (U01) explained 23.5% (LOD 3.32) of the phenotypic variation in reaction to *Xcm* Races 7 and 18. The GH allele increases resistance in a manner that was partially recessive (dominant gene action could be ruled out, but additivity could not). The region (Qb_{6b}) near marker *pAR1-28* on the A-subgenome chromosome 5, mapped in a region that is homoeologous to the B_2 locus (on chromosome 20) and explained 22.4% (LOD 3.07) of the phenotypic variation in reaction to *Xcm* Races 7 and 18. This region may correspond to the bacterial blight resistance gene B_4, discovered in "A" genome species *G. arboreum* which has previously been assigned to chromosome 5 using cytological stocks (Endrizzi et al. *1984*). The GH allele decreases resistance in a manner that was dominant. Qb_{6a} and Qb_{6b} together explained 38.6% of phenotypic variation at a LOD score of 5.52.

Fixing the effect of Qb_{6a} uncovered two additional QTLs. A region (Qb_{6c}) near DNA marker *pAR827* on D-subgenome linkage group D04 explained 19.4% (LOD 3.53) of phenotypic variation in reaction to *Xcm* Races 7 and 18. The recessive GH allele increased plant resistance. The region (Qb_{6d}) near DNA marker *pAR723* on D-subgenome chromosome 14 explained 16.3% (LOD 3.01) of variation in reaction to *Xcm* Races 7 and 18. The dominant GH allele decreases plant resistance.

S295 population.   A region near the DNA marker *pAR043* on D-subgenome chromosome 14 accounted for 94.2% (LOD 50.46) of phenotypic variation in reaction to each of the four *Xcm* races. The locus corresponded closely to that of Qb_{6d} in the Empire B2b6 population. The GH allele was dominant.

5.2. Trichomes

Upland cotton (*Gossypium hirsutum* L.) genotypes have varying densities of trichomes on the leaves and stems of mature plants, hence their species name. Most modern cotton cultivars are "smooth", with few if any trichomes. Absence of trichomes reduces the attractiveness of the cotton plant to some major insect pests, reducing reliance on pesticides. Based on quantitative measures of young and mature leaves, four QTLs have been mapped (Wright et al. 1999). A QTL on chromosome 6 that imparts dense leaf pubescence is inferred to be the classical t_1 locus (Lee 1985). A second QTL on chromosome 25, which is homoeologous to chromosome 6, fits the description of the t_2 locus (Lee 1985). Two additional QTLs, QLP_1 and QLP_2, explained significant phenotypic variation in leaf pubescence — these may represent the t_3, t_4, or t_5 loci (Lee 1985). Some QTLs appeared to be specific to particular developmental stages; for example, QLP_1 reduced hairiness only in young leaves while QLP_2 increased hairiness in mature leaves. A single locus associated with variation in trichome density on the stem did not correspond to the genes/QTLs affecting leaf trichomes, suggesting that these traits may largely be controlled by different genes. A widely-used qualitative classification system for scoring trichome density (DTL) detected only the chromosome 6 locus, and was apparently not sensitive enough to detect alleles such as t_2 that had smaller phenotypic effects.

6.   Progress in physical mapping and prospects for positional cloning

The average physical size of a centiMorgan in cotton is about 400 kb (Reinisch et al. 1994), only moderately larger than that of *Arabidopsis* (ca 290 kb), and smaller than that of tomato (ca 750 kb), both species in which map-based gene cloning has been accomplished. However, the genetic map of >5000 cM will require ca 3000 DNA probes to map at average 1 cM density (assuming that the locations of duplicated loci can be inferred), and the physical genome of about 2500 Mb will require ca 75,000 YACs/BACs of average size 150 kb for 5x coverage. Techniques for isolation and cloning of megabase DNA in cotton have been established (Zhao et al. 1994). Two BAC libraries of cotton are known to be in progress — one is from the *G. hirsutum* cultivar Tamcot GCNH, and provides about 2 genome-equivalent-coverage, and the other is from the *G. barbadense* cultivar Pima S6, providing about one genome-equivalent coverage. Both libraries are being expanded.

Contig assembly in polyploids such as cotton introduces a new technical challenge not encountered in diploid (or highly diploidized) organisms, i.e. that virtually all "single-copy" DNA probes occur at two or more unlinked loci. This makes it difficult to assign megabase DNA clones to their site of origin. One possible approach to this problem is the utilization of diploids in physical mapping and map-based cloning; for example, exploiting the relatively small genome of D-genome diploid cottons. However, since D-genome cottons produce minimal fiber, and are not cultivated, they are of limited utility for dissecting the fundamental basis of cotton productivity. The relatively larger size of the A genome, despite low-copy sequence complexity similar to that of the D genome, suggests that the subgenome-specific repetitive elements may provide a means of characterizing the genomic identity of individual megabase DNA clones (YACs, BACs, P1, or others) from tetraploid cottons (Zhao et al. 1995).

7.   Interaction of genomes in the polyploid nucleus

The joining in a common nucleus of A- and D-genomes, with very different evolutionary histories, appears to have created novel avenues for response to selection in AD-tetraploid cottons. Tetraploid cottons are thought to have formed about 1-2 million years ago, in the New World, by hybridization between a maternal Old World "A" genome taxon resembling *G. herbaceum* (2n = 2x = 26), and paternal New World "D" genome taxon resembling *G. raimondii* (6) or *G. gossypioides* (7: both 2n = 2x = 26). Wild A-genome diploid and AD-tetraploid *Gossypium* taxa each produce spinnable fibers that were a likely impetus for domestication. Although the seeds of D-genome diploids are pubescent, none produce spinnable fibers.

Intense directional selection by humans has consistently produced AD-tetraploid cottons that have superior yield and/or quality characteristics than do A-genome diploid cultivars. Selective breeding of *G. hirsutum* (AADD) has emphasized maximum yield, while *G. barbadense* (AADD) is prized for its fibers of superior length, strength, and fineness. Side-by-side trials of 13 elite *G. hirsutum* genotypes and 21 *G. arboreum* diploids (AA) adapted to a common production region (India) show average seed cotton yield of 1135 (±90) kg/ha for the tetraploids, a 30% advantage over the 903 (±78) kg/ha of the diploids, at similar quality levels (Anonymous 1997). Such an equitable comparison cannot be made for *G. barbadense* and *G. arboreum*, as they are bred for adaptation to different production regions. However, the fiber of "extra-long-staple" *G. barbadense* tetraploids, representing ~5% of the world's cotton, commands a premium price due to ~40% higher fiber length (ca 35 mm), strength (ca 30 grams per tex or more), and fineness over leading A-genome cultivars (Anonymous 1997), at similar yield levels. Obsolete *G. barbadense* cultivars reportedly had up to 100% longer fibers (50.8 mm; Niles and Feaster 1984) than modern *G. arboreum* (25.5±1.6 mm; Anonymous 1997).

The D subgenome, from an ancestor that does not produce spinnable fiber, accounts for more genetic variation in fiber traits of modern *G. barbadense* and *G. hirsutum* than does the A subgenome, from a fiber-producing ancestor. Selection for new alleles at D-subgenome loci during domestication and scientific breeding is a likely basis for the observation that leading tetraploid cottons consistently exceed the yield and/or quality of comparable A-genome diploid cultivars. The presence of fiber on wild A-genome diploids suggests that when polyploid formation occurred, many A-genome loci relevant to fiber development may already have contained "favorable" alleles as a result of natural selection. By contrast, human selection for fiber attributes of tetraploid cotton may have conferred a new fitness advantage to mutations at D-subgenome loci. Ostensibly, the D-subgenome had rarely if ever been under selection for seed-borne fiber, as its diploid progenitors show inadequate promise to warrant domestication. Mutations that enhanced fiber development may have become favorable only after the D-genome was joined in the same nucleus with the fiber-producing A-(sub)genome. This suggests that the locations of D-subgenome QTLs in tetraploid cottons may guide us to the corresponding (homoeologous) locations of A-subgenome loci at which favorable alleles had already been fixed prior to domestication. Testing this hypothesis will require isolation of the underlying genes, and comparative analysis of alleles in both A- and D-genome diploid and tetraploid cottons, as well as other wild diploid *Gossypium*.

The joining of two genomes with divergent evolutionary histories into a common nucleus also appears to have had important consequences for interactions between the cotton plant and the *Xcm* pathogen (Wright et al. 1998). Molecular mapping has revealed that the genetic basis of this host-pathogen interaction is more complex than classical data had suggested, and also that the A and D-subgenomes have made very different contributions to the co-evolution of *Xcm* and cotton. Among the six resistance genes derived from tetraploid cottons, five (83%) mapped to D-subgenome chromosomes — if each subgenome were equally likely to evolve new R-gene alleles, this level of bias would occur in only about 1.6% of cases. The D-subgenome bias of R-gene alleles reinforces the suggestion based on the genetic control of cotton fibers that polyploid formation has offerred novel avenues for evolution in cotton.

8. References

Anonymous, (1997) Zonal Coordinators Annual Report of All-India Coordinated Cotton Improvement Project.

Balls, W.L. (1906) Studies in Egyptian cotton. Pp. 29-89 in *Yearbook Khediv. Agric. Soc. For 1906*. Cairo, Egypt.

Bevan, M., Stiekema, W., Murphy, G., Wambutt, R., Pohl, T., Terryn, N., Kreis, M., Kavanagh, T., Entian, K.D., Rieger, M., James, R., Puigdomenech, P., Hatzopoulos, P., Obermaier, B., Duesterhoft, A., Jones, J., Palme, K., Ansorge, W., Delseny, M., Bancroft, I., Mewes, H.W., Schueller, C. and Chalwatzis, N. (1997) Arabidopsis thaliana DNA chromosome 4, ESSA I contig fragment No. 7.

Beasley, J.O. (1942) Meiotic chromosome behavior in species hybrids, haploids and induced polyploids of *Gossypium*. Genetics 27: 25-54.

Biessmann, H., Valgeirsdottir, K., Lofsky, A., Chin, C. and Ginther, B. (1992) Het-A, a transposable element specifically involved in healing broken chromosome ends in *Drosophila melanogaster*. Mol. Cell. Biol. 12:3910-3918.

Bouchard, R.A. (1982) Moderately repetitive DNA in evolution. Int. Rev. Cytol. 76: 113-193.

Brubaker, C.L., Paterson, A.H. and Wendel, J.F. (1998) Comparative genetic mapping of allotetraploid cotton and its diploid progenitors. Genome, accepted.

Chomet, P. S. (1994) Transposon tagging with *Mutator*. In *The Maize Handbook*, eds. Freeling, M. and Walbot, pp. 243-249. Springer-Verlag, New York.

Crane, C.F., Price, H.J., Stelly, D.M. and Czeshin, D.G. (1994) Identification of a homoeologous chromosome pair by *in situ* hybridization to ribosomal RNA loci in meiotic chromosomes of cotton (*Gossypium hirsutum* L.). Genome 35: 1015-1022.

Elborough, K.M. and Storey, S.E.S. (1996) *B. Napus* DNA fragment with retrotransposon integrase motif. (Unpublished, genbank accession # x99804).

Endrizzi, J. E., and Ramsay, G. (1979) Monosomes and telosomes for 18 of the 26 chromosomes of *Gossypium hirsutum*. Can. J. Genet. Cytol. 21: 531-536.

Endrizzi, J.E., Turcotte, E.L. and Kohel, R.J. (1984) Qualitative genetics, cytology, and cytogenetics. pp. 81-129 in *Cotton*, edited by R. J. Kohel and C. F. Lewis. ASA/CSSA/SSSA

Publishers, Madison, WI.

Ergle, D.R. and Katterman, F.R.H. (1961) DNA of cotton. Plant Physiol. 36: 811-815.

Flavell, R.B. (1982a) Repetitive sequences and genome architecture. In Structure and Function of Plant Genomes. Edited by O. Ciferri and L. Dure. New York, Plenum Press, pp. 1-14.

Fryxell, P.A. (1979) The Natural History of the Cotton Tribe. Texas A&M University Press, College Station, TX.

Fryxell, P. A. (1992) A revised taxonomic interpretation of *Gossypium* L. (Malvaceae). Rheedea 2: 108-165.

Galau, G.A., Bass, H.W. and Hughes, D.W. (1988) Restriction fragment length polymorphisms in diploid and allotetraploid *Gossypium*: Assigning the late-embryogenesis-abundant (Lea) alloalleles in *G. hirsutum*. Mol. Gen. Genet. 211: 305-314.

Geever, R.F., Katterman, F. and J.E. Endrizzi, (1989 DNA hybridization analyses of a *Gossypium* allotetraploid and two closely related diploid species. Theor. Appl. Genet. 77: 553-559.

Hanson, R.E., Zhao, X., Paterson, A.H., Islam-Faridi, M.N., Zwick, M.S., Crane, C.F., McKnight, T.D., Stelly, D.M. and Price, H.J. (1998) Concerted evolution of 20 interspersed repetitive elements in a polyploid. Am. J. Bot, in press.

Helentjaris, T., Weber, D.F. and Wright, S. (1988) Use of monosomics to map cloned DNA fragments in maize. Proc. Natl. Acad. Sci. USA 83: 6035-6039.

Innes, N.L., (1983) Bacterial blight of cotton. Biol. Rev. 58: 157-176.

Jiang, C., Wright, R., El-Zik, K. and Paterson, A.H. (1998) Polyploid formation created unique avenues for response to selection in Gossypium (cotton). Proc. Natl. Acad. Sci. USA, 95: 4419-4424.

Kimber, G. (1961) Basis of the diploid-like meiotic behavior of polyploid cotton. Nature 191: 98-99.

Knight, R.L. (1944) The genetics of blackarm resistance. IV. *Gossypium punctatum* (SCH. & THON.) crosses. J. Genet. 46: 1-27.

Lapitan, N.L.V. (1992) Organization and evolution of higher plant nuclear genomes. Genome, 35: 171-181.

Lee, J.A. (1985) Revision of the genetics of the hairiness-smoothness system of *Gossypium*. J. Hered. 76:123-126.

Meredith, W.R. and Bridge, R.R. (1984) Genetic contributions to yield changes in upland cotton. pp. 75-86 in Genetic Contributions to Yield Gains of Five Major Crop Plants, edited by W.R. Fehr. Crop Science Society of America, Madison WI.

Nelson, D.L., Ledbetter, S., Corbo, L., Victoria, D.H. and Caskey, C.T. (1989) Alu polymerase chain reaction: A method for rapid isolation of human-specific sequences from complex DNA sources. Proc. Natl. Acad. Sci. U.S.A. 86: 6686-6690.

Niles, G.A. and Feaster, C.V. (1984) in Cotton, eds. Kohel, R. J. & Lewis, C. F. (ASA/CSSA/SSSA Publishers, Madison, WI), pp. 202-229.

Reinisch, A., Dong, J.-M., Brubaker, C.L., Stelly, D.M., Wendel, J.F. and Paterson, A.H. (1994) A detailed RFLP map of cotton, Gossypium hirsutum x Gossypium barbadense: chromosome organization and evolution in a disomic polyploid genome. Genetics, 138: 829-847.

Reyes-Valdes, M. and Stelly, D.M. (1995) A maximum likelihood algorithm for genome mapping of cytogenetic loci from meiotic configuration data. Proc. Natl. Acad. Sci. USA; 9824-9828.

Royo, J., Nass, N., Matton, D.P., Okamoto, S., Clarke, A.E. and Newbigin, E. (1996) A retrotransposon-like sequence linked to the S-locus of *Nicotiana alata* is expressed in styles in response to touch. Mol. Gen. Genet. 250, 180-188.

Singer, M.F. (1982) Highly repetitive sequences in mammalian genomes. Int. Rev. Cytol., 76: 67-112.

Smyth, D.R., Kalitsis, P., Joseph, J.L. and Sentry, J.W. (1989) Plant retrotransposon from *Lilium henryi* related to Ty3 of yeast and the gypsy group of Drosophila. Proc. Natl. Acad. Sci. U.S.A. 86, 5013-5019.

Walbot, V. and Dure, L.S. (1976) Developmental biochemistry of cotton seed embryogenesis and germination. VII. Characterization of the cotton genome. J. Mol. Biol. 101: 503-536.

Wendel, J.F. (1989) New World cottons contain Old World cytoplasm. Proc. Natl. Acad. Sci. USA 86: 4132-4136.

Wendel, J.F. and V.A. Albert, (1992) Phylogenetics of the Cotton genus (*Gossypium*): Character-state weighted parsimony analysis of chloroplast-DNA restriction site data and its systematic and biogeographic implications. Syst. Bot. 17: 115-143.

Wilson, J.T., Katterman, F.R.H. and Endrizzi, J.E. (1976) Analysis of repetitive DNA in three species of Gossypium. Biochem. Genet. 14: 1071-1075.

Wright, R., Thaxton, P., Paterson, A.H. and El-Zik, K. (1998) Polyploid formation in *Gossypium* has created novel avenues for response to selection for disease resistance. Genetics 149: 1987-1996.

Wright, R., Thaxton, P., Paterson, A.H. and El-Zik, K. (1999) Molecular mapping of genes affecting pubescence of cotton. J. Heredity, in press.

Zhao, X., Zhang, H., Wing, R.A. and Paterson, A.H. (1994) An efficient and simple method for isolation of intact megabase-sized DNA from cotton. Plt. Mol. Biol. Rptr. 12:126-131.

Zhao, X., Si., Y., Hanson, R., Price, H.J., Stelly, D., Wendel, J. and Paterson, A.H. (1998) Dispersed repetitive DNA has spread to new genomes since polyploid formation in cotton. Genome Res. 8: 479-492.

15. Maize molecular maps: Markers, bins, and database

EDWARD H. COE, MARY L. POLACCO,
GEORGIA DAVIS, *and* MICHAEL D. McMULLEN
Plant Genetics Research Unit, ARS-USDA, and Department of Agronomy,
University of Missouri, Columbia, MO 65211, U.S.A. <ed@teosinte.agron.missouri.edu>

Contents

1. Introduction

Following several years of vigorous communication with co-workers in maize, Emerson et al. (1935) presented the first comprehensive maps, linkage data, and genetic descriptions for maize. Sharing of data and stocks, i.e., maize genetics cooperation, enabled M.M. Rhoades to construct the first linkage maps (Emerson et al., p. 71). These maps included phenotypic variants and a few reciprocal translocations. One pest resistance gene (rp1, resistance to Puccinia sorghi) had been placed to chromosome arm by deletion analysis, but no biochemically defined loci were yet identified. Accretion of sixty years of data on the 1935 foundation is reflected in the genetic linkage maps in Mutants of Maize (Neuffer et al. 1997).

2. Initial molecular maps

Burr et al. (1983), Burr and Burr (1985), and Helentjaris et al. (1985) projected the use of RFLP markers, combined with isozymes and visible genes, to map quantitative traits and contribute to plant breeding. T. Helentjaris presented, at the 1985 Maize Genetics Conference, maps under development at Native Plants Incorporated (NPI), which were updated in the 1986 and 1987 Maize Genetics Cooperation Newsletters. D. Hoisington and B. Burr presented University of Missouri (UMC) and Brookhaven

R.L. Phillips and I.K. Vasil (eds.), DNA-Based Markers in Plants, 255 - 284.

National Laboratory (BNL) maps in the 1988 and 1989 Newsletters et seq. Working copies of NPI, BNL, and UMC maps, among others, have been distributed at meetings and conferences as they developed. Recent versions of maps can be found in the Maize Genetics Cooperation Newsletter, but are kept most current in the Maize Genome Database, MaizeDB (http://www.agron.missouri.edu). The molecular-marker maps were definitively oriented relative to chromosome arms and to established gene maps by monosomy, and with B-A translocations, by Helentjaris et al. (1986) and Weber and Helentjaris (1989). By merging data for recombinant inbreds from two hybrid populations, Burr et al. (1988) added efficiently to the BNL maps by use of the fact that polymorphism is substantially more frequent in joint populations. The numbers of markers now placed on the BNL maps (1,699) and on the UMC maps (1,858) have reached the point of diminishing returns for resolution of the order of loci.

3.   Bin maps

Gardiner et al. (1993) proposed designating 'Bins' for intervals along the maps, to systematize placement of the increasing numbers of markers and to simplify defining locations of genes and markers on the increasingly complex maps. Bins are segments of the map, around 20cM in length, delimited by markers (Core Markers) chosen on the basis of spacing, polymorphism, clarity of band pattern, and minimal multiplicity of bands. A map showing the current Bins and Core Markers is presented in Fig. 1. The coodinates for bins are numbers, styled chr#.bin#, with the first bin# assigned 00. A bin is an interval between two fixed Core Marker loci and includes the beginning (leftmost or top) Marker. It is statistically defined (rather than absolute) in the sense that chance can lead to inaccurate left-right placement of loci. As in the classical mapping order problem, some placement to bins is dependent only on approximated relationship to the Core Markers, and order will be determined only by 3-point data.

4.   Genetic and cytological maps

Combined maps, in parallel presenting the UMC molecular-marker map and mapped genes on the same coordinate scale, are given in Mutants of Maize (Neuffer et al. 1997). Essentially the same maps were presented in the 1995 Newsletter (No. 69, pp. 247-256). Updates of the molecular-marker map, incorporating large numbers of ESTs, are presented in MaizeDB and in the Maize Genetics Cooperation Newsletter. Updates of the genetic map will be made available as soon as they are revised. Cytological maps are also presented in Mutants of Maize and MaizeDB - cytological coordinates are represented from -1.0 (short arm) to 1.0 (long arm) and represent proportional physical distances along a chromosome arm.

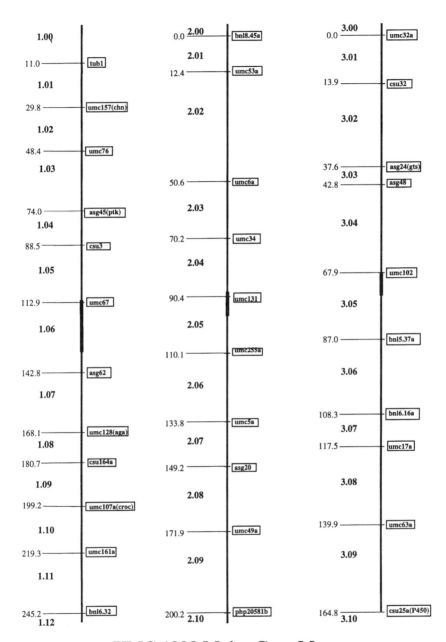

UMC 1998 Maize Core Map

Figure 1.   Map of Bins and Core Markers for Maize. Bin numbers are shown between the bin boundaries, and coordinates are shown at the bin boundaries. Core Markers are shown at the right of the vertical line of the map.

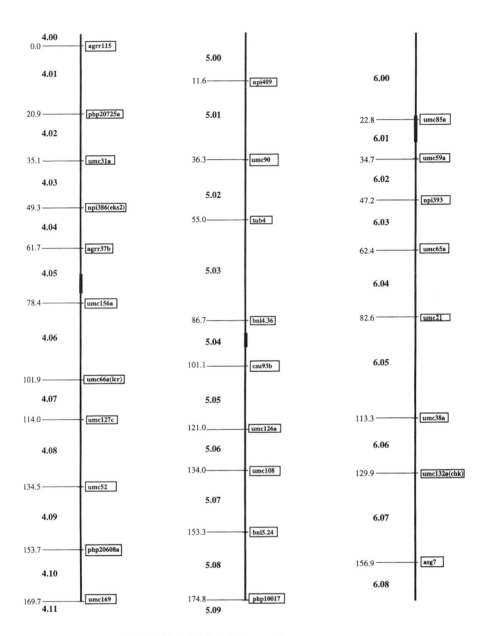

UMC 1998 Maize Core Map

Figure 1. Continued.

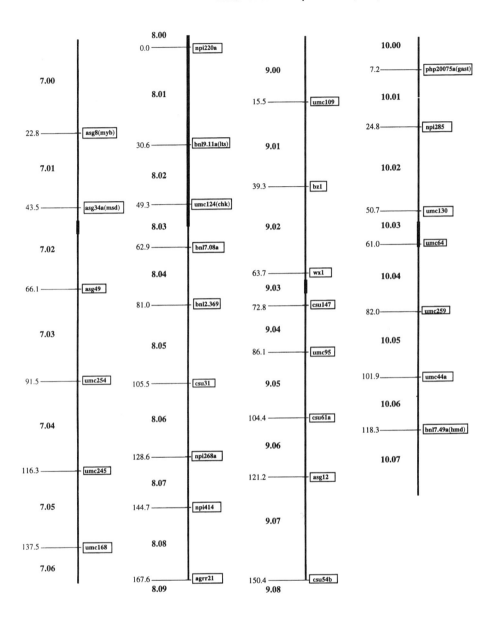

UMC 1998 Maize Core Map

Figure 1. Continued.

5.   Maize genome database

MaizeDB currently presents 16,105 loci, including quantitative trait loci and gene candidates; 16,156 references; 4,385 colleague addresses; 4,943 traits; 2,788 genetic stocks; 5,449 elite germplasm pedigrees; 27,154 variations with over 3,000 images and 801 phenotypes; 70,186 links to external databases from 47,575 records. SwissProt, GenBank (Entrez format), and GRIN link reciprocally to MaizeDB pages. The WWW page 'Previous What's New' provides an historical listing of data and design changes. Since January 1995 there have been 3,688,593 accesses, with 256,043 in January 1998. A more detailed usage analysis is accessible from the database homepage.

6.   Numbers of loci in different maps

Loci have been initially placed onto the classical genetic map (845 loci), the cytogenetic map (2,307 loci), or maps derived with molecular markers applied to specific mapping populations. The most comprehensive of the latter are the UMC and BNL mapping populations, which share markers to varying degrees with the other maps (Table 1). This table reflects that a total of 6,521 sites have been mapped with sufficient accuracy to place them on a chromosome and in a bin or, when appropriate, in a range of bins. The number of loci shared between maps varies widely — for example, only 72 cytological sites are placed in bins. Of 3,267 loci placed on the BNL or UMC map populations, only 390 are on both maps, i.e., these maps are anchored to each other by that many common sites. Well over 145 loci on the genetic map are on either a BNL or a UMC map, i.e., the genetic maps are anchored by that many sites in common with molecular maps.

7.   Distributions of loci

The distributions by bin of loci for which information is documented in MaizeDB are presented in Table 2. Of 1,518 designated genes, 785 have been placed to bin, 540 of which have described phenotypes, 459 of which have defined gene products, and 214 of which have clone sequences, including a number of genes probed by EST's and by simple sequence repeats (SSR). Genes mapped to chromosome or better total 1,277, and 4,486 probed sites have been mapped. Public EST's (partially sequenced cDNA's or expressed sequence tags) total 2,511, of which 1,192 have been placed onto public maps as one or more probed sites and/or genes. Of the probed sites that are not designated genes, 1,559 have probes with sequence information.

Table 1.   Numbers of loci shared and unique, among several maps of maize.

	Bin	Cyt	Gen	UMC	BNL	CU	SSR	BxM	INRA	MxH	Myc	NPI	Pio	Uniq
Bin[1]	6521													—
Cyt	72	2307												2202
Gen	770	105	845											533
UMC	1856	1	145	1858										1209
BNL	1699	2	109	390	1699									1026
CU	670	1	51	237	285	670								311
SSR	864	2	46	250	336	132	864							390
BxM	203	0	24	124	128	50	183	203						4
INRA	279	1	37	126	111	47	82	47	279					134
MxH	126	0	19	69	74	36	107	33	25	128				13
Myc	368	1	22	133	108	73	83	37	40	28	368			218
NPI	143	0	2	4	25	4	5	1	1	3	1	143		117
Pio	220	0	9	127	126	67	142	61	49	34	49	0	220	47

[1] Bin: Bins map constructed by evaluation of compiled data.
Cyt: Cytological map based on microscopic measurements of chromosomal aberration breakpoints.
Gen: Genetic map constructed by evaluation of compiled data.
UMC: University of Missouri 1998, Tx303/CO159 Immortalized F2 panel; Davis et al. (submitted); map score data in MaizeDB.
BNL: Brookhaven National Laboratory 1996, CO159/Tx303 and T232/CM37 Recombinant Inbred (RI) panels; Burr et al. (1994); map score data in MaizeDB.
CU: Cornell University 1997, with the BNL T232/CM37 RI panel; Van Deynze et al. (1995a, b); map score data in MaizeDB.
SSR: Composite map prepared by J. Romero-Severson from 5 panels containing RFLP and SSR markers; personal communication to MaizeDB (1998).
BxM: ARS-USDA/NC State Univ. map 1997, B73/Mo17 RI panel; Senior et al. (1996); map score data in MaizeDB.
INRA: Inst. Nat. Rech. Agron. composite map 1996, from 4 panels; Causse et al. (1996).
MxH: ARS-USDA/NC State Univ. map 1997, Mo17/H99 RI panel; Senior et al. (1996); map score data in MaizeDB.
Myc: Mycogen composite map 1992, from 3 panels; Shoemaker et al. (1992).
NPI: Native Plants Incorporated 1990; unpublished.
Pio: Pioneer Hi-Bred Intl. Inc. composite map 1993, from several panels; Grant et al. (1993).
Uniq: Loci uniquely on that map and on no other in the table.

Table 2.   Distributions of loci by type and bin. SSR, simple sequence repeat variations identified by primer amplification; RFLP, restriction fragment length polymorphisms identified by hybridizations with probes; EST, sites identified by hybridizations with cDNA probes; Pheno, sites identified with a phenotype; Gene Prod, sites identified with a gene product; Pheno Gene Prod Clone, sites identified with a phenotype, gene product, and DNA clone. The first three types are counted based on locations in one bin or a range of bins, while the last three types are counted only if placed definitively in one bin.

Bin	SSR	RFLP	EST	Pheno	Gene Prod	Pheno Gene Prod Clone
1.00	1	9	1	1	0	0
1.01	6	31	10	1	2	2
1.02	9	37	9	3	3	3
1.03	13	68	20	6	8	9
1.04	3	43	21	3	2	2
1.05	5	62	24	4	2	2
1.06	9	64	21	3	1	1
1.07	4	51	13	6	1	0
1.08	9	50	20	7	4	5
1.09	10	50	20	7	4	6
1.10	2	50	13	6	4	4
1.11	8	41	16	5	6	3
1.12	0	14	8	0	1	0
2.00	0	20	2	1	0	0
2.01	3	7	3	2	0	0
2.02	9	36	18	5	2	1
2.03	6	41	10	3	1	1
2.04	8	52	16	6	1	1
2.05	16	45	23	3	2	2
2.06	7	44	12	3	5	2
2.07	3	39	8	1	3	3
2.08	23	69	18	0	3	3
2.09	1	32	13	3	3	3
2.10	0	11	1	2	0	0
3.00	0	6	0	1	0	0
3.01	1	8	1	2	1	0
3.02	0	14	6	4	2	1
3.03	5	11	3	1	2	2
3.04	14	82	25	13	3	4
3.05	12	83	27	7	9	9
3.06	8	51	18	9	1	1
3.07	6	34	8	2	1	1

Table 2.   Continued.

Bin	SSR	RFLP	EST	Pheno	Gene Prod	Pheno Gene Prod Clone
3.08	5	32	12	2	3	0
3.09	5	59	23	4	3	3
3.10	1	13	7	0	0	0
4.00	2	12	5	1	0	0
4.01	5	44	7	5	7	4
4.02	0	25	9	4	4	0
4.03	4	18	7	2	2	2
4.04	3	24	11	4	5	5
4.05	12	131	39	10	9	7
4.06	4	29	15	4	1	1
4.07	6	26	10	3	1	1
4.08	9	62	18	6	2	2
4.09	3	46	26	1	1	1
4.10	2	19	8	0	1	1
4.11	7	11	4	1	1	1
5.00	2	24	6	1	0	0
5.01	4	45	19	1	2	3
5.02	5	31	11	3	3	3
5.03	10	94	31	6	7	7
5.04	6	70	25	10	7	5
5.05	6	47	17	7	4	5
5.06	4	41	17	2	0	0
5.07	12	30	7	1	2	2
5.08	0	15	4	2	1	0
5.09	2	2	0	0	0	0
6.00	7	20	5	0	2	2
6.01	19	68	24	7	6	6
6.02	3	35	13	6	6	5
6.03	0	5	3	1	0	0
6.04	5	28	4	2	3	2
6.05	14	81	34	2	4	3
6.06	3	27	7	1	1	1
6.07	5	23	9	3	4	2
6.08	1	17	4	0	0	0
7.00	4	17	2	1	0	1
7.01	5	32	12	3	2	2
7.02	17	85	20	12	7	7
7.03	7	42	9	6	2	1
7.04	7	104	37	1	5	4

Table 2.   Continued.

Bin	SSR	RFLP	EST	Pheno	Gene Prod	Pheno Gene Prod Clone
7.05	4	34	13	0	2	2
7.06	4	11	1	2	0	0
8.00	1	17	8	1	0	0
8.01	3	14	5	0	1	1
8.02	4	25	4	0	3	3
8.03	8	72	24	1	5	4
8.04	5	47	19	4	3	3
8.05	12	53	17	2	1	1
8.06	4	57	13	2	3	2
8.07	3	27	12	1	1	1
8.08	2	25	11	1	0	0
8.09	3	12	1	3	2	2
9.00	2	7	0	4	1	0
9.01	9	24	4	2	2	2
9.02	8	48	16	9	4	4
9.03	11	74	14	12	8	9
9.04	6	41	20	3	2	3
9.05	8	28	7	1	1	1
9.06	1	28	13	0	1	1
9.07	9	26	11	1	0	0
9.08	0	5	2	1	0	0
9.09	0	1	0	0	0	0
10.00	3	20	5	1	0	0
10.01	0	21	7	5	0	0
10.02	4	18	7	2	3	2
10.03	12	60	20	1	3	1
10.04	7	68	27	8	6	4
10.05	3	20	6	2	1	1
10.06	7	26	4	9	5	5
10.07	3	37	12	3	2	2
Total	563	3735	1232	326	245	214

8.   Binned loci

The bin maps, first presented in MNL 66:127 (1992), currently represent all loci that
have been mapped, to whatever precision. Within a bin, order can often be ascertained

by examining other maps, although a large number of loci have not been ordered on more than one map, or with more than one other marker or cytological point. Table 3 presents a list of bins, and all genes and molecular markers in each bin listed in alphabetical order after the Core Marker that sets off the start of that bin. This table contains all known loci, extracted from MaizeDB and marked to identify those for which there is a phenotypically known expression.

Table 3.   All genes and molecular markers in maize, listed in bins. The Core Marker, which is the fixed defining point for the beginning of that bin, is listed first, followed by all other loci in alphabetical order.
- signifies that a visible or known-biochemical phenotype exists.
= signifies that both a visible or known-biochemical phenotype and a molecular probe exist.
~ signifies that the bin location is in a range, extending beyond the bin listed.

1.00 ~-ad*-N605B ~-blh*-N487C ~-blh*-N495B ~-bnk*-N1519C ~-car1 ~-cp*-N1078B ~-cp*-N1393A ~-cp*-N1399A ~-cp*-N628 ~-cp*-N991 ~-dcr*-N1176B ~-de*-N1142 ~-de*-N1162 ~-de*-N1345B ~-de*-N1390A ~-dek32 ~-emb*-8517 ~-emb*-8549 ~-emb1 ~-emb10 ~-emb4 ~-emb7 ~-et*-N617 ~-et*-N745 ~-fl*-N1208A ~-gm*-N1390C ~-gt1 ~-hcf31 ~-hcf6 ~-hmp1 ~-l16 ~-ms*-6034 ~-ms12 ~-ms28 ~-pg*-N484A ~-pg*-N484B ~-pg*-N526C ~-pg*-N619 ~-pg15 ~-ptd*-N923 ~-sen3 ~-smk*-N1057A ~-v*-5688 ~-vp*-N1136B ~-w*-4791 ~-w*-8345 ~-wt*-N650A ~-ys2 ~cdo63 ~npi322 ~npi327 ~npi593b ~opn11-710 ~ucsd61g ~umc166b bnl(tas1c) bnl(tas1h) bnlg149 csu804b(dnp) gsy266c(ptkr) php20603 rgpc654

1.01 = tub1 = cat2 ~-ct2 ~-emp1 ~-rab30 ~=lls1 ~cdo20a ~cdo353c ~uaz146b ~umc194c(gpr) ~umc8f agrr152 agrr22 agrr294a asg31 asg59a bmc1014 bmc1112 bmc1124 bmc1179 bnl5.62a bnl8.05a cdo1081a cdo507a(ant) csu454(gst) csu589 csu680a csu738 fus6 knox1 npi97a pbs11 php20537b php20689 rgpc385a(rpL5) std20b tda47 uaz104 uaz182 uaz260b(rpL5) umc164c umc266b(ptk) umc94a umn857b

1.02 umc157(chn) -les2 -srl = gsr1 = pds1 = vp5 ~-blh1 ~-cfr1 ~-rth3 ~cdo860a ~csu145c(pck) ~npi106 ~rz103b ~rz742a agrc467 agrr77 bmc1007 bmc1083 bmc1178 bmc1429 bmc1614 bmc1627 bmc1803 bnl9.13 bnlg109 bnlg147 csic(rab30) csu1171 csu1190 csu320 csu680c csu691 csu860b dnap9705(Ac) dpg12b ifbf33 ltk1 npi109a npi209b npi403b npi406 npi411b npi415 npi423 npi579b php20640 rgpc1122c(rpL15) std2c(dba) tda50 uaz1 uaz2a umc115 umc194a(gpr) umc37b

1.03 umc76(gne) -dek1 -ms17 -zb4 = gln6 = hcf3 = hsp26 = ibp2 = ncr(sod4a) = npi(sod4) = p1 = pdc3 = ts2 ~-ht4 ~cdo64b ~cdo689a ~gsy297a(emb) ~npi(G10C-11) ~npi293b ~npi589 ~rz574a(cwp) ~rz585a ~uaz146a ~uaz3 ~umc6b agrr117 agrr92a asg26 asg35b bmc1203 bmc1458 bmc1484 bmc1866 bmc1953 bmc2180 bmc2204 bnl10.38a bnl159 bnl35d(blr) bnl5.21c bnl8.35b bnlg176 bnlg182 bnlg424 bnlg432 bnlg439 cdo1387b(emp70) cdo189a cdo38a(ntp) cdo541 cdo586 cdo938a chs5054 csu179a(hsp70) csu181a csu214b(grp) csu215b(grp) csu238a(apx) csu254b csu315c csu392b csu59b csu710a(apx) csu745d(rpPo) csu753 csu75b csu814b csu859(gol) dgg9b dpg10 dpg11 dpg4 dpg8 dpg9 gsy59c(sh2) ias8 ifbf91 isu2117c mpik36 npi234 npi241b npi242b npi286 npi354 npi425c npi427b npi439a npi448 opn11-1070 opz13-900 pbs16b

Table 3.   Continued.

pge(phyB1) php20006b uaz120 uaz139 uaz264b uaz266a(nad) uaz267 umc11a umc13 umc137c umc162b umc230 umc243b umc266a(ptk) umc8a wusl1032 ynh(ts2) ynh21

1.04 asg45(ptk) -ms9 -msv1 -nec2 = rz672a(cgs) = sod4 ~-les5 ~npi(C3D-03) ~npi(G2B-03d) ~npi13 ~uaz10a ~uaz12 agrc634 agrr153a agrr88a asg30 asg69 asg75 bcd450e bmc1016 bmc1811 bmc2238 bnl12.06a bnl2.323 bnl9.11b(lts) csu1082 csu207 csu323 csu389 csu452 csu632a csu633 csu649(scp) csu737(npc) csu887 csu924(wsi) csu941 dpg12a fmi1(pki1) isu61 les29 npi262 npi453 psl25 rgpc198a(sik) rgpc361(ppi) rgpr44a rz251a uaz11 uaz198d(rpL10) uaz248a(his3) ufg9 umc227

1.05 csu3 -as1 -bsd2 -ms14 -rs2 = eno2 = obf1 ~-cp3 ~-les20 ~bnl7.21a ~isu81 ~npi(G12C-02) ~npi20 ~npi205 ~npi224f ~npi40 ~npi429 ~npi482(a1) ~npi84 ~npi96 agrc516 agrr197 agrx23 asg3 bcd1092b bcd734a bmc1832 bmc1886 bmc2086 bmc2295 bnl1.556 bnlg652 cdo103 cdo344c(rga) cdo373 cdo749 csu1041a(ptk) csu1138 csu263b csu653(fbn) csu694b(uce) csu710f(apx) csu781b csu781c csu793 csu822 csu95c mpik41d(mem1) npi214 npi304 npi401 npi598 opn11-920 opt01-900 pbs6d pbs9b php20682 pop1 rgpc316 rz296a rz323a rz421 rz500(stp) rz892a(alt) uaz13 uaz198c(rpL10) uaz203 uaz253 uaz4 uaz5 uaz6 uaz7a uaz8a(spr1) uaz9 umc167a umc260 umc26b umc29a umc31c

1.06 umc67 -br2 -ws4 = hm1 ~-ad*-N613B ~-amp2 ~-bl*-N43 ~-cl*-N801 ~-cp*-N1311C ~-cp*-N918A ~-cps1 ~-d*-N1352B ~-d*-N454A ~-de*-N1310B ~-de*-N1420 ~-de*-N978 ~-dek2 ~-dek22 ~-dnt*-N1185A ~-et*-N1001A ~-gm*-N1303 ~-hcf12 ~-hcf2 ~-hcf4 ~-hcf41 ~-hcf44 ~-hcf50 ~-hcf7 ~-l*-N129 ~-l17 ~-msc1 ~-olc1 ~-pg*-N1822A ~-pg16 ~-smp*-N706A ~-spc2 ~-tlr1 ~-v*-N1806 ~-v*-N245 ~-v*-N55 ~-w*-6577 ~-w*-8054 ~-w*-N1890 ~-w*-N547A ~-wl*-N1831 ~-wl*-N47 ~-wl*-N56 ~-wl*-N60 ~-wl*-N709B ~agrc175 ~opn09-960 ~opt07-800 ~php20044 ~psl6 agrc512 agrc569 agrc584 agrc587 agrr193 asg11 asg58 bmc1023b bmc1041 bmc1057 bmc1273 bmc1556 bmc1598 bmc1908b bmc2057 bnl1.326a bnl23b bnl34 bnl5.59 bnlg421 cdo1091b cdo116a cdo1395f cdo464a cdo475b cdo57c cdo595 cdo678b cdo786c cent1 csu1132 csu1150 csu194(met) csu256(hsp90) csu503(met) csu505(rpL7) csu574b(eif2B) csu590(rpL17) csu60b csu61b csu675a(prh) csu805 csu881(cys) csu899b(ant) csu91b csu92 dup382 isu2191i mpik34 npi258 npi272 npi279 ntf1 opy20-550 php20644 psl18 ptk3 rgpc356 rz28a uaz14a uaz15 uaz17a uaz249a(ubf9) ucsd61a ucsd72a umc104c umc105b umc119 umc177a umc196 umc276c(hm1) umc53c umc58 umc59c umc64b umc82b uwo2 zmm6

1.07 asg62 -amp1 -br1 -f1 -hcf13 -lpe1 -vg1 -wlu5 ~cdo598 ~cdo94b ~npi19 ~npi54 ~npi82 ~tum4 ~tum5 ~umc33a ~umc37a agrc489 agrp83b agrr175 agrr185 agrr200 agrr250 agrr291 bcd450d bcd98a bcd98g bmc1025 bmc1564 bnl17.15b(bt2) bnl5.02b bnl7.08b bnlg257 bnlg615 btl1 cdo99 csh11 csh12 csu542 csu614a csu660a csu921b(ppp) csucmt11b dpg7c gsy243(ndnr) gsy60b(bt2) isu2117a isu2117h med63b ncr(nrA) npi224g npi236 npi566 npi605 opn09-980 opx03-1290 opy20-1550 phi102 php20527b php20661 php20668a php20855 rz698a(ppy) std1b(his2B1) uaz151(sar) uaz18d uaz19c uaz205b(hsp70) uaz20a uaz228d(his2b) uaz2b umc16d umc23a umc89b wsu(nia4)

Table 3. Continued.

1.08 umc128 -ad1 -mmm1 -v22 = an1 = bz2 = id1 = mdh4 = uce1 ~npi10 ~npi340 ~opv01-920 ~umc49c ~umc50b agrr110 agrr299 agrr71 agrx1176 an2.6 bcd207a bcd386a bcd98m bmc1044 bmc1629 bmc1643 bmc2228 bnl17.06 bnl29d(pds) cdo680a cdo98b csh4(id1) csu1007(eif4F) csu12b(cin4) csu20a(lhcb) csu531 csu580a(mdh) csu66a(lhcb) csu780b csu889b(lhcb) csu982(goa) dup103 dup135a dupssr12 gsy282a(cab) mpik37 mwg645f npi120 npi255 npi441 npi447 npi569a npi573 npi614 phi039 rgps10558a rny(pcr)a rz561a rz583a(msb) rz630e(sat) ucsd61e ufg(vp2274a) ufg4 umc174b umc83a

1.09 csu164a -ptd1 -ts3 -zb7 = glb1 = pgm1 = rth1 = tb1 = tbp1 = umc24b(lhcb) ~-tls1 ~=adh1 ~=phy1 ~bmc1331 ~bmc1502 ~bmc1597a ~bmc1720 ~bnl17.04(tua) ~bnlg100 ~ope01-1000 ~opi13-1700 ~php20518 agrc362b agrr235a agrr238a agrr246 asg63b bcd1072c(hsp70) bmc1268 bnl8.10a bnlg400 cdo669 cdo795a csu1097c csu110b csu1174 csu200b csu21a(ago) csu222a(wsi) csu511a csu554a(rnh) csu696 csu745e(rpPo) csu921a(ppp) dup218b ias7 msu2(iaglu) Mu4-1 npi615 rgpc746(rnp) rgpr250 rpa5b rpa6b rz403 rz474b(dnaj) uat1(lox) uaz268c umc129(geb) umc140a umc197a(rip) umc252b umc27b umc66b(lcr) ynh20

1.10 umc107a(croc) -d8 -gs1 -lw1 -mpl1 -w18 = gln2 = kn1 = tua1 = tua2 ~npi(C4F-03b) ~pge5b agrr278a agrr290 agrr34a asg54a bcd265b bcd450b bcd808a bmc1347 bmc1671 bnl10.13c bnl15.18 bnl17.18b bnl17.21(tua) bnl6.29b bnl7.25a cdo122a(nad) cdo246 cdo353b cdo465 csu137b(ap) csu248 csu261 csu272a(tua) csu947 csu954 gsy52(root) gsy56(tua1) isu18 knox3 knox8 mpik22a(zmm4) mta1 npi(adh1) npi282b npi407 npi581a npi75b npi98a ope13-1900 opp07-1230 pge19 phyA1 psl24 rz630a(sat) rz632c rz912a(phy) u22 uat4a uaz166a uaz167a uaz208(mta) uaz21a ucsd104b(zag6) ucsd106d ucsd64c umc106a umc147b umc39e umc57b umc72b

1.11 umc161a -alh1 -bm2 -dia2 -gdh1 -rd1 -ts6 -vp8 = ccr1 = chi1 = phi1 ~-ij2 ~-py2 ~npi24 agrc259a agrc669b agrc707 agrr103b agrr88b asg68b bmc1055 bmc2123 bmc2331 bnl8.08a bnl8.29a bnlg131 bnlg504 cdo457b cdo87b(ptk) csu1169b csu134a(thf) csu175e(eif5A) csu266 csu33b csu381 csu536(ccr) csu570b(mtl) csu604a(trh) csu63a(cdj) csu663b(psaD) csu755 csu868(trp) dpg1a isu6 mps2.11 npi238 npi241a npi97c pge18 phi064 phi120 php15058 php20557 psl44 rpa7a uaz240c ucsd44a umc257 umc264 umc42d umc84a umc86a usu1a(fnr)

1.12 bnl6.32 -acp4 csu1084 csu1089 csu1114 csu1146 csu1154 csu1193 csu865(phb) fd3 mpik9 npi294i psl13 psl33 rgpr3239a rz614(fdx) uaz22

2.00 ~-cp*-N1076A ~-cp*-N1319A ~-d*-3685 ~-d*-N155B ~-d*-N208B ~-dcr*-N1233A ~-de*-N1122A ~-dek3 ~-et2 ~-fl*-N1426 ~-gm*-N1312 ~-nec*-N1119B ~-o*-N1189A ~-os1 ~-ptd*-N901A ~-sen5 ~-v26 ~-wt*-N136A ~agrc805 ~bcd512 ~bnl10.38b ~bnl12.06d ~cdo36b ~csu1192(apx) ~csu326 ~npi105b ~npi249 ~npi329 ~npi365 ~npi367c ~rz569b ~umc120b ~umc2Stelo-1 ~umc2Stelo-2 ~umc87b agrr13 bnl(tas1a) bnl(tas4k) npi239 npi417a pgs1 rny(pcr)c uaz21b

Table 3.   Continued.

2.01 bnl8.45a -al1 -ws3 ~bcd135a ~fht1 ~npi321b agrc938 bmc1092 bmc1338 csu29a csu300a mps1.02 mps1.05

2.02 umc53a -d5 -dks8 -gl2 -les11 -Mut = lg1 ~-nec4 ~-tr1 ~bcd131b ~bnl5.62d ~cdo1417a(ptk) ~cdo375a ~cdo539b ~npi(G1B09) ~opb01-400 agrc539a agrp168a bcd98x bmc1017 bmc1297 bmc1302 bmc1327 bmc2042 bmc2277 bnl7.49c(hmd) bnlg125 bnlg469b cdo244b(crp) cdo524 cdo93a csu1053 csu1091 csu1113 csu1148 csu12d(cin4) csu348a csu425(gct) csu552 csu642 eks1 gsy54b(rpl7) mir3b(thp) npi254a npi290b npi421a npi577a pbs5 phi098 rz387a rz590a uaz24b uaz251b(rpS11) uaz25b uaz26a uaz288b(ppi) ucsd(lfyB)

2.03 umc6a -abph1 -gl11 = b1 ~-gs2 ~-les18 ~-sk1 ~bnl12.36 ~npi(C6A-06) agrc321 agrr113a agrr167a bcd1086d bcd855a(ext) bmc1064 bmc1537 bmc1621b bmc2248 bmc469b bnl10.42a bnl12.06c bnl17.23b(pal) bnlg381 cdo684 csh6 csh7 csu1058 csu1167 csu176 csu498 csu571a(ipp) csu761 csu821 csu861 inra1(tmp) mpik4b ncr(ars1) ncsu1 npi208c npi269a npi287a npi340c npi402 npi57 npi583 npi587 oec2 opn09-850 uaz27b ufg3a(ivr) umc11b umc44b umc61 umc78 umc8b/c

2.04 umc34 -fl1 -les1 -les15 -mn1 -o8 -wt1 = hrg1 ~-ba2 ~-gl14 ~-sks1 ~-ts1 ~-wrp1 ~=ole1 ~cdo456b ~cdo79 ~cent2 ~opf16-610 ~opv07-1110 ~opx4-1420 ~psl11 agrc593 agrp54a bcd1302b bcd348b bcd450f bmc1018 bmc1175 bmc1613 bmc1818 bnl(plB) bnl1.45b bnl12.09 bnl35e(blr) bnl7.25b bnl8.04 bnl8.35c bnlg108 bnlg166 cdo1328b cdo202b(mcf) cdo544b cdo650 cdo667b cdo677 cdo941 csu110d csu1117b csu148b(clx) csu255a csu334 csu348b csu350(gpdh) csu393(fbn) csu40(grx) csu56c(ohp) csu6a(sam) csu735(geb) csu762 dpg6a isu2117b isu7 mps2.02 npi220d npi242c npi248 npi607 opf16-590 opr03-500 pbs12 php10012 pic3 prp2 rgpg271 rz103a rz585b tug4 ufg6(incw) umc134b umc135 umc184b(glb) umc234 umc8b

2.05 umc131 -v4 -wlv1 = hcf106 = pmg1 ~-de*-N1175 ~-de*-N660C ~-dek16 ~-dek23 ~-dek4 ~-et*-N1078A ~-fl*-N1287 ~-hcf1 ~-hcf15 ~-les10 ~-les4 ~-mn*-N1120A ~-o*-N1195A ~-px1 ~-spt*-N579B ~-spt1 ~-v24 ~-w*-N332 ~-w*-N346 ~-w*-N77 ~cdo783b ~csu46b ~isu4 ~mpik33i ~mwg645i ~npi11 ~npi565b ~opg09-960 ~opu20-740 ~psl31 ~rz206b ~uaz265a(sbe1) accA agrp173 agrr216a agrr267b asg29b bmc1036 bmc1047c bmc1063b bmc1831 bmc1893 bmc1909 bmc1914 bmc2039 bmc2328a bnl17.25 bnl20 bnl29(pds3) bnlg180 bnlg371 csu1059 csu1060 csu1066 csu1073a csu1163 csu143 csu2a csu337 csu4a csu50c csu671a csu833 csu842 csu850 csu851a dupssr21 isu89 mwg6451 npi123c npi242a npi271a npi273 opg06-900 pbs15 psr109a ssu2 uaz135 uaz179 uaz181 uaz18b uaz198b(rpL10) uaz234 uaz235(px) uaz236b(ser) uaz25c uaz262 uaz28b uaz29 ucsd1.8c umc16c umc8c umc8d umc8g umn1(acc) umng1 umng2 uox(ssu1b)

2.06 umc255a -ask2 -dia1 -emp2 -tpi2 -w3 = agp1 = akh2 ~-emb5 ~-l18 ~-les19 ~npi(C1E-05) ~npi277a ~npi347(EMu) ~npi348(alr) ~npi4 ~npi45a ~npi468(al) ~uaz30c ~umc139a ~umc14b agrp58 agrp62 agrr239 agrr265b agrr85b bcd1087a bcd249a bmc1138 bmc1184 bmc1225 bmc1396 bmc1887 bnl17.24 bnlg121 cdo59c(gos2) csu1051 csu270 csu281b csu747a(arf) gsy199(ptk) ici99 klp1d mpik331 nc003 npi221b npi297 npi356 npi456 npi565a opg09-920 opr03-1110 opz07-1200

Table 3.   Continued.

---

opz10-790 pbs13b rz509b(mip) rz698c(ppy) tda217b tug3 uaz194b(ugu) uaz228a(his2b) uaz352b
ucsd1.8b ucsd141b ufg2(agp2) uky1(P450) umc112a umc176 umc267(kapp) umc2b umc55a

2.07 umc5a -ht1 = amy3 = psl32 = ugp1 ~bnl6.29c ~csh1b(chi) ~csu37a ~gsy266a(ptkr) ~npi216a
~npi392 ~npi465 ~npi59 ~umc22a agrc333a agrc479 agrc939 agrr111b asg72 asg84a bmc1045
bmc1413 bmc1633 bnl27 bnl8.21b cdo412a csu1103 csu54a csu635 csu75c gsy252a(zn14) isu117
mpik27a(zmm7) npi47a npi49 npi613 ope16-1410 php20005 php20569b psl1 rgpc1122a(rpL15)
tda66b uaz140 uaz194a(ugu) uaz269b(kri) ucsd641 ucsd64i umc110b umc29b umc36b umc88(P450)
umc98a

2.08 asg20 = ck2 = rDNA5S = tua5 ~-d10 ~-MuDR-1(p1) ~npi32 ~umc12b agrr85a apd2 asg23
asg28c asg56 ast(amyBS2)b bcd1088b bcd249h bcd808c bcd926b bmc1140 bmc1141 bmc1169
bmc1233 bmc1258 bmc1267 bmc1316 bmc1329 bmc1335 bmc1606 bmc1662 bmc1721 bmc1746
bmc1767 bmc1908a bmc1940 bmc2077 bmc2144 bnl17.30b bnl5.21b bnl5.61b bnl6.20 bnl8.44b
bnlg198 cdo337 cdo375b cdo385 cdo533a cdo57a cdo786a cdo938c csu1097b csu1151b
csu154a(eif5A) csu175a(eif5A) csu17b(rnp) csu203b(eif5A) csu657(atpd) csu658(mam) csu749b
csu800(lhca) csu847a(lhcb) csu894a csu909 csu920a dpg6d dupssr24 dupssr25 gsy282b(cab)
ias4a mgs1.41 npi113b npi210 npi274 npi278a npi298 npi413a npi41a npi452 npi46 npi576b
npi591 npi61 npi610 opb18-780 opp01-960 opx20-430 phi127 php20668b pur1 rgpc643c rgpc74c
rgpg99 rz530a uaz239b uaz23b uaz241b uaz31b uaz32 ucsd106f umc10b umc116b umc122
umc125a umc137a umc150b umc31b umc4a umc80b

2.09 umc49a -ch1 -rf3 = betl1 = csu64a(grf) = whp1 ~-ms*-6019 ~opb10-800 ~opv03-940
~opv07-960 agrc39b agrx825 bcd855c(ext) bcd98k bcd98l bmc1520 csu109a csu200a csu304a
csu315d csu611a(grp) csu622 csu665a(adt) csu728b csu810a csu9c(trf) fco1a(pex) fco1b(pex)
kev2 mha1 mpik(chs1b) mpik26 npi294a npi294d npi47c ope08-1200 pbs10 srk1 uaz33a
umc171b(oec) umc32c umc35b umc36a umc93c

2.10 php20581b -gn1 -se1 bcd98n bnl(tas1g) bnl(tas1p) bnl17.14 bnl17.19b csu251b knox4
npi400b php20622b rgpc12b ucsd106a ucsd113c ucsd61c umc2Ltelo

3.00 -g2 ~-dcr*-N1053A ~-dek24 ~-dek5 ~-gl19 ~-hcf*-N1257B ~-les14 ~-les17 ~-nl*-1517
~-ns1 ~-rgh*-N1112 ~-sdw2 ~-sen1 ~-si*-N1323 ~-wi2 ~mpik30b ~npi(G2D-12) ~umc3Stelo
bnl(tas1b) bnl(tas4l) bnl8.15 npi71 uaz109

3.01 umc32a -brn1 -e8 ~psl5 ~umc121 ~umc229c agrr209a asg64 bcd130 csu628 php20905 umc249

3.02 csu32 -cg1 -cg2 -cr1 -d1 = me3 ~-h1 ~-hex1 ~-ra2 ~bcd734b ~cdo511 ~psl47 asg16
csu1062 csu199b csu230 csu75a mpik24b(zmm2) php20042a zag4

3.03 asg24(gts) -tp3 = hsp18f = me1 ~npi218c ~npi219 ~php20006a ~uaz164d agrr116a bmc1144
bmc1325 bmc1523 csu56b(ohp) csu56e(ohp) csu728c gsy252b(zn14) mps1.03 mps1.04 mps2.15
php12006 uaz210(hsp18) umc29d

Table 3.   Continued.

3.04 asg48 -Ac(CO159) -cl1 -e4 -gl6 -lg3 -pg2 -ref1 -rf1 -rg1 -rp3 -rt1 -wrk1 = bet1 =
rps25 = tha1 = tpi4 ~-tru1 ~-ts4 ~=phys1 ~=phys2 ~cdo226 ~cdo419a ~csu912b(hsp70)
~npi(G2B-06) ~npi(G4B-11) ~npi109c ~opj05-1970 ~php20797 ~umc51c Ac(PvuII) Ac340
asg46a bcd98j bmc1019a bmc1447 bmc1452 bmc1628 bmc1638 bmc1647 bmc1904 bmc2047
bmc2136 bnl13.05b bnl5.33e bnl8.35a bnlg602 cdo1160b(kri) cdo459 chs566 csh10b(cycII)
csu1070 csu212b csu242 csu290 csu29b csu2b csu404b csu408(grp) csu60d csu621 csu795
csu851b csu949a dup104 dup162 dup183b dup287a dup53 gsy298b(pmg) gsy60g(bt2) ici273c
ici286b isu1719b isu1719h isu2117i isu2191c klp1a mpik32e(zag2) mpik35d mpik35e mps2.04
mps2.10 mps2.12 npi(C6A-11) npi(tpi) npi114b npi220b npi247 npi249a npi276a npi398a npi446
npi83 npi89 opv12-1300a opx04-1190 pbs14a pbs14e phi036 php20017b php20509 php20576
php20753b psl10 rgpc131a rgpc601b rz244b(dia) rz382a rz543a rz995b(fbp) std1d(his2B1)
ttu1(hsp18) uaz19a uaz249b(ubf9) uaz255 uaz34a uaz34b uaz35 ucla(obf6) umc10a umc154
umc161b umc42b umc50a umc92a umc94c umc97

3.05 umc102 -mv1 -pm1 -wsm2 -ys3 = abp1 = atp1 = cyp7 = gst4 = myb2 = pgd2 = sps2 = te1 =
vp1 ~-bif2 ~-cp*-N1379A ~-cp*-N1436A ~-crp*-N2207 ~-d*-N282 ~-de*-N1126A ~-de*-N1166
~-de*-N932 ~-dek17 ~-e3 ~-emb*-8512 ~-emb*-8515 ~-emb*-8532 ~-emb9 ~-et*-N1322C
~-gl*-N352A ~-gl9 ~-gm*-N1311B ~-hcf19 ~-hcf46 ~-ms23 ~-nec*-N720C ~-rgh*-N1060
~-rgh*-N802 ~-smk*-N1168A ~-smp*-N1324B ~-spc3 ~-su*-N748A ~-v*-N1886 ~-wl*-N4
~-wlu1 ~npi(G4D-02) ~npi70 ~npi90 ~ope16-880 ~opn20-675 agrc332 agrc476a agrc514a agrc923
agrp50 agrp91 agrp97 agrr179 agrr19 agrr206 asg1b asg52b asg67b bmc1022 bmc1035 bmc1113
bmc1117 bmc1246b bmc1399 bmc1456 bmc1505 bmc1957 bnl(tas11) bnl31b bnl6.06a bnl8.08g
bnlg420 cdo105 cdo109 cdo1528 cdo250 cdo344a(rga) cdo473 cdo689b cent3 csu134d(thf)
csu154b(eif5A) csu229b(oec) csu234a(gbp) csu237b(psaN) csu268 csu362 csu382c(cld) csu439(trm)
csu44(gst) csu561a csu636 csu961(fnr) klp3 ldp1 mpik2 mpik30a mwg645c npi609 npi611b npi612
phi053 php20501 php20508 php20558a psl28 psl4 rgpc385b(rpL5) rgpc529 rgpc6(rpS9) rz14
rz141a(emp70) rz261b(sad) rz296b rz390c(cyb5) tda30 tda64 uaz189(rpL5) uaz288a(ppi) uaz36
uiu8(geb) umc18a(psaN) umc252a umc268 umc26a umn41 umn857a zag2

3.06 bnl5.37a -ba1 -ig1 -lg2 -lxm1 -na1 -rd4 -rea1 -spc1 -yd2 = tub6 ~-ms3 ~gsy252c(zn14)
~npi52 ~npi78 ~npi94 ~tda217u ~umc47b agrc435 agrp40 agrr184b agrr271 agrr274a asg15
asg34b(msd) asg39 asg61a bmc1047a bmc1063a bmc1449 bmc1601 bmc1796 bmc1798 bmc2241
bnl10.24a bnl5.14 bnl8.01 cdo251a csu1029 csu1183 csu180 csu191 csu215a(grp) csu223b(psei)
csu264 csu351 csu38a(taf) csu690 csu776b csu96a(psei) csucmt165 dupssr23 gsy298c(pmg) ici98
ksu1a npi108a npi268b npi296 npi328b opb08-1300 ope13-700 php20802 psl16 rz444b rz538b
rz630d(sat) uaz260a(rpL5) uaz37 uaz8b(spr1) ucsd72d umc164b umc165a umc208(cppgk)
umc252c umc39a umc60 umc82c

3.07 bnl6.16a -dek6 -y10 = hox3 ~npi88 ~ope16-1610 ~umc175 agrc461 agrr144a agrr50a asg10
asg4 bcd738b(pgk) bcd805 bmc1160 bmc1605 bmc1779 bmc1931 bmc1951 bnl1.297 bnl1.326b
bnl15.20 bnl3.18 bnl5.33b bnlg197 cdo1160c(kri) cdo1395d cdo241a csu1130 csu567(ces)
csu680b csu706 ici273a npi212b npi291k npi42 php15033 php20026 php20521 rgpc643b uaz38a
ucla(obf3A) ufg21 umc3b

Table 3.   Continued.

3.08 umc17a -a3 -got1 -mdh3 bcd1127a bcd828a(atpb) bmc1108 bmc1350a bmc2243 bnl17.27
bnl24b cdo1174 cdo118 cdo345b csu1117a csu189(thr) csu240 csu456(uce) csu703 csu744
csu772a dgc13 dpg1 isu1 npi(C9G-06) npi201a npi215b npi257 npi432 npi565c npi91a php10080
rny(atpb) rz527a uaz164c uaz176b uaz18c uaz243a(atpb) uaz251e(rpS11) umc103b umc15b
umc16a umc226a umc228b umc231 umc72c umc999a

3.09 umc63a -et1 -ga7 = a1 = lhcb1 = sh2 ~cdo1007a ~cdo1338a ~opg17-1330 agrc568a
bcd134b bmc1182 bmc1257 bmc1496 bmc1536 bmc1754 bnl1.123 bnl1.67 bnl12.30b bnl7.26
cdo1053a cdo455b cdo665a cdo920b(egl) cdo962b csu1086 csu1142 csu125a(cah) csu21b(ago)
csu289 csu303 csu305b csu36c(rpL19) csu397(cah) csu58a csu768 csu780a csu845 csu869(cah)
csu899a(ant) csu919a dup214 dup216 H3.119B#2 ici94 ILS-1 isu1410h med68 mwg645j npi420
npi425a npi457 php20726 pic6a rgps10558b rz776a uaz10b uaz110 uaz133 uaz213b uaz39
uiu1a(pog) umc126c umc174c umc184d(glb) umc187 umc199(a1) umc96

3.10 csu25a(P450) agrc638 agrr43b bmc1098 csu1061 csu71(cab) csu728a pbs13f rpa2 tda117
uaz114 uaz117a uaz198a(rpL10) uaz5c05d03(rpl12) umc255b umc2a

4.00 ~-cb*-N719A ~-cp*-N1313 ~-de*-N929 ~-dek11 ~-dsc1 ~-Dt4 ~-gl7 ~-hcf23 ~-lp1
~-nec*-N562 ~-nec*-N673B ~-o*-N1119A ~-o*-N1228 ~-o*-N1244A ~-pg*-N1881 ~-pg*-
N673A ~-ra3 ~-sh*-N1105B ~-sh*-N1324A ~-sh*-N1519B ~-smp*-N156A ~-spt2 ~-v17
~-wst*-N413A ~-wt2 ~=rca1 ~adp4 ~csu221 ~csu606 ~msf1 ~opr05-1180 ~opr05-1390 ~pep7
~uaz59 ~umc4Stelo ~zrp4 bmc1370 bnl(tas1e) bnlg372 cdo344b(rga) rz536

4.01 agrr115 -asr1 -bx1 -bx2 -bx3 -bx5 -ph1 -ri1 -rp4 = bx4 = cyp3 = cyp4 = cyp5 ~-zpl1a ~-zpl1e
~-gsy260(zn22) ~-gsy4a(zpl1) ~-opr05-1220 ~uaz52a ~uaz60 ~uaz61a ~uaz62a bmc1241 bmc1318
bmc1434 cdo442 csu618(P450) cyp1710 cyp2707 dnap3 mpik5a mpik5b mtl1 npi294j npi604a
php20713 rz329b(bga) uaz103a uaz14b uaz26b uaz41d uaz42b uaz43b uaz43d uaz44b(zp19)
uaz45b uaz46a uaz47b uaz48a uaz49b uaz49d(zp19) uaz50 uaz51 uaz52a uaz53a uaz54 uaz55
uaz57a uaz58a umc123

4.02 php20725a -dek25 -dzr1 -ga1 -sos1 -zpl1b -zpl1c -zpl1d -zpl1f bnl17.13b cdo520(ser) inra2(prp)
uat3b(fl2) uaz103b uaz129 uaz149(zp19) uaz17b uaz184(hfi) uaz185(zp22) uaz30a uaz38b uaz41c
uaz63a uaz64b uaz65b uaz66a uaz67 uaz68b(zp19) uaz69a uaz70c umc171a(oec) umc277

4.03 umc31a -ts5 = adh2 = zp19/22*-L34340 ~-la1 ~=psl45 ~opv03-970 ~psl35 ~uaz30b
agrr109 bmc1126 bmc1162 bnl5.46 chs556 csu235 csu583 csu585 csu63b(cdj) dpg2 nk1(ck)
rgpc496b(adh) rz53b rz630b(sat) uaz180 uaz239a uaz298(PDsI) umc200a(adh2) umc55b umc87a

4.04 npi386(eks) -cp2 -st1 -ysk1 = fl2 = gpc3 = pgd3 = zp1 = zp22.1 ~-aco1 ~-zpl2a ~-zpl3a
~-bnl8.37b ~cdo244c(crp) ~cdo551a ~csu12c(cin4) ~csu19(colp) ~gsy292(rab28) ~gsy4b(zpl1)
~opr05-1430 ~opu16-850 ~umc193a(orp1) ~umc226b agrc39a csu1135 csu1185 csu298a csu449
csu855 med63c mps1.14 npi(C5A-09a) psb3 rgps2470 rz900(ahh) uaz145(ahh) uaz53b uaz57b
uaz62b uaz63b

Table 3.   Continued.

4.05 agrr37b -als1 -Dt6 -dup(als1) -gl5 -su3 -tga1 -v23 = akh1 = betl2 = bm3 = bt2 = gpc1 = orp1 = su1 ~bmc1937 ~cdo400b ~cdo772a ~npi250b ~php20597a ~rz630c(sat) ~ucsd61i(zag4) ~zag3 agrc4 agrc567 agrp54b agrp67 agrp83a agrr190 agrr286 agrr301 agrr321 agrr62b agrr89 bcd421b bet2 bmc1159 bmc1168 bmc1217 bmc1265 bmc1729 bmc1755 bnl(tas3a) bnl12.06b bnl15.27a bnl15.45 bnl17.09 bnl17.10 bnl17.13c bnl17.23c(pal) bnl35b(blr) bnl5.33a bnl5.71b bnl7.20 bnlg490 bnlg667 cdo116b cdo1380b cdo350 cdo395a(ypt) cdo475c cdo497 cdo588 cdo59b(gos2) cent4 csu100(ptk) csu1026 csu1063 csu1098 csu1125 csu26b(ant) csu294 csu358b(pal) csu474(rpS14) csu509 csu565(rpPo) csu599a csu693(lrr) csu716 csu74(fdx) csu742b(rpS7) csu81b(ank) csu84 csu902 csu93c dnap4 dpg7a gsy54c(rpl7) med63a mpik11d mpik11e mpik11f mpik15a mpik15b mpik16e mpik19b ncr(nrB) npi18 npi259a npi267 npi284 npi289 npi367b npi383 npi395 npi574 npi594b npi6 npi73 npi77 npi95a ope10-1000 ope10-450 pbs13a pic1b rz143b(gpc) rz28b rz467b rz66a rz86a std16a std1a(his2B1) std1c(his2B1) tda62b uaz130b(tlk) uaz157(rpL19) uaz195(ms) uaz212 uaz216 uaz218a(gss) uaz230a uaz246a(mbf) uaz261b uaz265c(sbe) uaz41b uaz42a uaz42c uaz43a uaz44a(zp19) uaz45a uaz46b uaz46c uaz48b uaz48c uaz49A uaz49c uaz56 uaz61b uaz69b uaz71a uaz72 uaz73 ucsd62j(zag4) ucsd64f ucsd72g ucsd72l umc(orp1) umc191(gpc1) umc193d(orp) umc201(nr) umc23b umc242 umc256b umc263 umc273b umc33b umc42a umc47a umc49d umc999b wsu(nia2)

4.06 umc156a -gl4 -lw4 -trg1 -zb6 = gln5 ~-cl*-N795 ~-dek10 ~-dek8 ~-emb*-8501 ~-emb*-8509 ~-emb*-8534 ~-emb*-8537 ~-emb*-8538 ~-emb11 ~-emb3 ~-emb6 ~-emb8 ~-mdr1 ~-nec*-N1487 ~-nec*-N193 ~-nec5 ~-ns2 ~-rgh*-N1105A ~-v*-N378A ~-v8 ~-wl*-N311B agrc563b agrr27 bmc1023a bmc1621a bmc1741 bnl8.08h bnlg252 cdo507c(ant) csu428(cyb561) csu587a csu638 csu640 csu643b csu661 csu720b csu816 csu907a dge18 isu136a mpik3 npi340b npi396 npi584 rgpc601a rgpr663a rz273a(ant) rz567b(klc) uaz144a uaz170 uaz228c(his2b) uaz257 uaz47a ucsd64g umc126b umc14a umc65c

4.07 umc66a(lcr) -dek31 -o1 -tu1 = prh1 ~asg84c ~umc12c ~umc63d agrp168c asg33 asg9a bmc1137 bmc1189 bmc1784 bmc1927 bmc2291 bnl5.67a bnl8.08e bnl8.08i csu525(rpL17) csu597d(dah) csu672b dupssr34 klp1e npi292 opu16-810 pbs13c php20595b rz446a uaz222 uaz263 uaz66b uaz74 umc104a umc19 umc229a umc244a wsu(nia3)

4.08 umc127c -gl3 -j2 -ms*-LI89 -ms41 -ms44 = c2 = fer1 ~cdo57b ~cdo786b ~csu9b(trf) ~npi(C4H-05a) ~npi17a ~npi36 ~npi371a ~npi390 ~npi450 ~npi56 ~pep3 ~wsu(nia1) agrc303a agrp166 agrr324 asg27a asg74b asg85a bmc1444 bmc2162 bmc2244 bnl10.05 bnl10.17b bnl17.05(ssu) bnl22 bnl23a bnl29(pds2) bnl7.65 bnl8.45b bnlg57 cdo127a(pyk) cdo365(pet) csu1038b csu1073b csu166a csu178a csu202(rpL7) csu20b(lhcb) csu313a(fer) csu597a(dah) csu704 csu91a dupssr28 gol1 gsy298d(pmg) gsy34a(pep) mpik(chs1a) npi12 npi208a npi253b npi270 npi300c npi410 npi444 npi570 npi910 phi066 phi086 php10025 php20071 php20562 psr109b rgpg111 rgpg124a rgpg24 rgpl102 ssu1 tda44 uaz122 uaz137 uaz142 uaz171 uaz252a(ptk) umc133a umc158 umc15a umn433 uox(ssu1a)

4.09 umc52 -dp1 = ris2 ~bmc1019b ~bmc1565 ~cdo270 ~knox7 ~npi104b ~npi317 ~npi333 ~psl75 ~uaz123b agrc300 agrr273 asg22 bnlg292b cdo1395c cdo370 cdo534a(cts) csu1107

Table 3. Continued.

csu201 csu241b csu304b csu324a(cts) csu34b(rpS8) csu39 csu50a csu631 csu674(gts) csu719(lox)
csu745b(rpPo) csu848a(vpp) csu862b(rpL11) csucmt11a cuny9 mgs2 mpik39(myb) mwg645e
npi116a npi294g npi449b opp01-1110 opt01-750 pge14.1 rgpc643a rgpr3235b rz103c rz476a
uaz115 uaz118 uaz279(cbp) uaz33b uaz41a uaz65a

4.10 php20608a = cas1 ~bnl15.07a agr(ubi)1002a agrr248a asg41 bmc1917 csh2a(cdc2) csu283a
csu330(ubi) csu36a(rpL19) csu377a(ubi) csu758 dba1 gsy249a(b70) gsy257(ubi) ici281 isu2191a
isu77 mps2.09 ncr(b70b) ncr(cat3) npi593a ufg(ivr2a) uwo3

4.11 umc169 = cat3 agrc445a bmc1337 bmc1890 bmc2186 bnl(tas1o) bnl32 bnl8.23a bnlg589
csu315b csu380 csu710b(apx) npi451 pge17 php20563b psl26 uaz247(ubi) umc10c umc111a(psy)
umc112c uwo8

5.00 ~-ad*-N664 ~-br3 ~-cl*-N818A ~-cp*-N1430 ~-crp2 ~-d*-6 ~-dcr*-N925A ~-de*-N1002A
~-dek18 ~-emb*-8504 ~-fl*-N1333B ~-ga10 ~-gl*-N681A ~-gl25 ~-l*-N1838 ~-ms13 ~-ms42
~-msc2 ~-sen6 ~-smk*-N1529 ~-wl*-N44 ~cdo1007b ~ope11-520 ~php1163 ~tum3 ~umc149b
~umc59d agrc259b agrc66 agrc669a agrc926 agrp53 agrr103a asg60 bmc1006 bnl(tas1n)
bnl(tas2b) bnl(tas2g) bnl(tas4j) bnl8.33 cdo457a cdo484 csh1c(chi) csu1087 csu134b(thf) csu277
csu527(crm) isu62 npi890 php20045b rgpg164 uaz214 uaz259 uaz75 uaz76a umc84b umc86b

5.01 npi409 = hcf108 = ohp2 = tua3 ~bcd265a ~csu272b(tua) ~rz53a ~rz630f(sat) agrc23 asg54b
bcd450a bmc1382 bmc1836 bnl17.18a bnl6.25a bnl7.24b bnl8.29b bnlg219 cdo1395h cdo345a
cdo87a(ptk) csu1169a csu137a(ap) csu318 csu33a csu570a(mtl) csu604b(trh) csu663a(psaD)
csu707 cuny7 mpik22b(zmm4) npi282a npi305a npi579a npi581b npi75a rgpc975(rpS27) rpa7b
sca1 uat4b uaz163 uaz166b uaz201(tua) ucsd104a(zag6) ucsd106c ucsd44b ucsd64a ucsd72j
umc103c umc132b(chk) umc144a umc147a umc240 umc72a umccmt146

5.02 umc90 -d9 = ole2 = pgm2 = tua4 ~-wi4 ~knox10 ~npi474(oec) agrc329 agrc362a agrr238b
agrr278b agrr34b agrx43 asg73 bcd1072a(hsp70) bnlg105 bnlg143 bnlg565 cdo122b(nad) cdo542
csu108(gbp) csu10a csu554b(rnh) dupssr1 phyA2 rpa5a rpa6a rz632a rz912b(phy) uaz134
uaz167b uaz211 uaz215b(odo) uaz219(hsp) umc106b umc107b(croc) umc144b umc197b(rip)
uwm2(rnp)

5.03 = tub4 -am1 -anl1 -na2 -nl2 = cat1 = cpn1 = ivr2 = mdh5 = rab15 = tbp2 ~-als2 ~-got3 ~-lu1
~=amy2 ~=ole3 ~asg29a ~npi(G20B-08) ~npi116c ~npi213 ~npi233 ~npi384 ~npi593c ~opf15-620
~oph13-1050 ~opk19-1290 ~opx06-1290 ~opx09-1000 agrc614 agrc637 agrc814 agrp52 agrr142
agrr199 agrr235b agrr248b agrr70 agrx1128 bcd207b bmc1046 bmc1063c bmc1208 bmc1660
bmc1700 bmc1879 bmc1902 bnl1.380 bnl10.06 bnl5.02a bnl5.27 bnl6.10 bnl6.16b bnl6.22a
bnl7.43 bnl7.56 bnlg557 cdo1173b cdo32b cdo347 cdo431a cdo456c cdo475d cdo795b cdo94a
cdo98a csic(mah9) csu150b csu164b csu168a csu175c(eif5A) csu222b(wsi) csu252a(cdc2) csu338
csu340 csu419 csu511b csu574a(eif2B) csu580b(mdh) csu60c csu652(rpL27) csu68d(mcf) csu720c
dnap2 gsy249b(b70) ici287 ici97 isu2192a jc162 mpik33e mps1.12 ncr(b70a) ncr(cat1) ncr200b(rip)
niu2::Bs1 npi256 npi275 npi41b npi434 npi601 opg02-680 ops14-880 php20597b php20622a

Table 3. Continued.

---

php20872 psl20 psl43 psl7 rab28 rgpc1122e(rpL15) rgpc643d rgpr440a(gap) rny(pcr)b rz17b rz20b rz474a(dnaj) rz508 rz561b rz583b(msb) rz892b(alt) std2b(dba) tda37a tda66a uaz111 uaz158(alt) uaz159 uaz205a(hsp70) uaz25d uaz25e uaz77 ucr1b(eif) uky2(P450) umc1 umc166a umc186b(Bs1) umc27a umc43 umc50c umc83b xet1

5.04 bnl4.36 -amp3 -bm1 -dup(als2) -nec3 -nec6 -ps1 -ris1 -td1 -v3 -vp2 = a2 = bt1 = gl17 = incw1 = pep6*- ~-bv1 ~-dek33 ~-ms5 ~bcd1302a ~cdo718 ~npi(G2D-01) ~npi60 ~opb10-1350 ~oph13-640 ~opo20-720 ~psl39 ~psl8 ~rz2c ~rz476b ~rz599 ~ucsd64h agr(ubi)1002b agrr106 agrr127 agrr298 asg43 asg51 bcd1086c bcd855e(ext) bmc1287 bmc2323 bnl17.30a bnl31a bnl35a(blr) bnl7.71 bnlg150 bnlg603 bnlg653 cdo375c cdo534b(cts) cent5 chs572 csu241a csu283b csu302 csu305a csu308 csu315a csu36b(rpL19) csu377b(ubi) csu562b(ubi) csu660b csu670 csu765 csu774(lhcb) csu862a(rpL11) dupssr10 gsy34b(pep) gsy60e(bt2) ici273b isu2191j knox6 koln2a(hox) mip1 npi(pmr15) npi104 npi295 npi408 npi424 npi449a npi53b npi569b npi571 pep2 pge10 php06012 php10014 php15018 php20589 rz87(clp) uaz131 uaz132a(dts) uaz186 uaz213a uaz238(ppi) uaz275 uaz70a umc138b umc250 umc40 umc66c(lcr) umn388 uwo4 uwo6 uwo7

5.05 csu93b -ga2 -lw2 -mep1 -pr1 -sh4 = ae1 = gl8 = gpc4 = nbp35 = pal1 ~-cp*-N1275A ~-cp*-N1369 ~-cp*-N1385 ~-cp*-N863A ~-cp*-N935 ~-de*-N1196 ~-dek26 ~-dek27 ~-dek9 ~-dnj*-N1534 ~-fl*-N1145A ~-grt1 ~-hcf18 ~-hcf21 ~-hcf38 ~-hcf43 ~-mn*-N1536 ~-nec7 ~-o*-N1065A ~-pg*-N408C ~-ppg1 ~-pr*-N850 ~-prg1 ~-psb2 ~-rth2 ~-smk*-N1160 ~-sms*-N146C ~-v*-N26 ~-v*-N735 ~-v12 ~-w*-N1126B ~-w*-N21A ~-wgs1 ~-ys1 ~-zn*-N571D ~bcd421a ~bmc1237 ~bnlg278 ~cdo202c(mcf) ~cdo400a ~cdo474 ~npi294 ~npi313 ~npi346(tpi5) ~npi362 ~rz86b ~uaz164a agrr215 asg71 bmc1246a bnl10.12 bnl15.27b bnl5.71a bnlg161a bnlg609 cdo1091a cdo1380a csu1080 csu1105 csu173 csu28b(rps22) csu550 csu600 csu713 csu95b ias3 mpik12c mps1.08 mps1.10 mps1.13 mps1.17 mps1.18 mps2.05 npi237 nrz5 psl21 rgpc174a rpl19 rz166(nac) rz467a rz585c std16b tda62a uaz168 uaz190(gpc) uaz226(cat1) uaz248b(his3) uaz261a uaz78 uaz79 ucsd106e umc211 umc48b umc54

5.06 umc126a -hsf1 -lw3 ~-Cy ~-eg1 ~-ren1 ~cdo348b ~cdo686 ~cdo772c ~npi426b ~umc133d agrp90 agrr252 agrx701 asg81 bmc1847 bnl5.40 cdo395b(ypt) cdo507b(ant) csh10a(cycII) csu1164 csu26a(ant) csu434 csu440 csu587b csu615a csu643a csu777 csu907b ici229 npi458a npi562 phi087 phi101 phi107 php20566 rgpg57 rz273b(ant) rz567a(klc) uaz138c uaz144b uaz204 uaz215a(odo) uaz254a umc141 umc14c umc156b umc253a umc262 umc26c umc51a zag5

5.07 umc108 -yg1 = gln4 = lhcb4 ~bcd269 ~npi(G3F-11) ~npi419b ~npi469(a1) ~umc127b ~umc229b agrc563a agrr288 asg74a asg84b asg85b asg9b bmc1118 bmc1306 bmc1346 bmc1695 bmc1711 bmc1885 bmc2305 bnl9.07b bnlg118 cdo516a csu1074 csu288 csu672a klp5 mpik10 mpik40(myb) npi253c npi442 pbs6a phi048 ppp1 psr3b umc139b umc241 umc39c umc68 wsu(nia5)

5.08 bnl5.24 -dap1 -got2 ~-v2 ~-zb1 ~npi363 ~npi74 ~rz20a agrr211 agrr45a cdo836 csu695(rpL9) csu799(rpCL9) csu834(mss) npi288a php20523b php20909a rpa3 rz446b uaz240a uaz71b umc104b umc228a umc57d

---

Table 3.   Continued.

5.09 php10017 bnlg386 bnlg389 gsy60f(bt2) php20042b umc63c

6.00 = adk1 = fdx1 ~-dek28 ~-dek34 ~-dep1 ~-hcf26 ~-hcf323 ~-hcf5 ~-les13 ~-ln1 ~-oro1
~-rd2 ~-rDNA18S ~-rhm1 ~-rhm2 ~-sr4 ~=rDNA25S ~=rDNA5.8S ~bmc1597b ~npi(C5A-09b)
~npi100 ~npi302 ~npi305 ~npi418 ~ubc281-900 ~ubc425-700 agrc67 asg79 bmc1043 bmc1600
bnl161 bnlg161b bnlg238 cdo358 cdo772b csu150a csu178b csu710d(apx) csu926(frk) jc1270
npi2 npi340a npicmt386 phi106 phi126 rgpc174b rz143a(gpc) uaz18a

6.01 umc85a -mdm1 -mn3 -po1 -rgd1 -w15 -wsm1 = gpc2 = nor = pgd1 = uaz237b(prc) = uck1
= zp15 ~-afd1 ~-cps2 ~-d*-9 ~-dek19 ~-hcf34 ~-hcf36 ~-hcf408 ~-hcf48 ~-l10 ~-l12 ~-psb1
~-sbd1 ~-wi1 ~cdo420 ~cdo545 ~npi387 ~php20719b agrc12 agrc3 agrc611 agrp144 agrr221
agrr47 asg40 bcd98f bmc1047b bmc1139 bmc1165 bmc1188 bmc1246d bmc1422 bmc1432
bmc1433 bmc1538 bmc1641 bmc1753 bmc1867 bmc2097 bmc2191 bnl17.28 bnl6.29a bnl7.28
bnlg107 bnlg249 bnlg391 bnlg426 cdo580b(ivd) cent6 csu1120 csu1187 csu1196 csu243 csu680e
csu699 csu70 csu700 csu71a csu809 csu94a gsy224c(sps) gsy261a(be) isu1410b isu1410j
isu1774a isu2232h mpik(DH7) mpik11b mpik33d mwg645b npi101c npi235a npi285b(cac)
npi594a npi606 npi7 pbs8 pge23 phi077 php20045a php20528 php20854 tug2 uaz102 uaz150
uaz169 uaz197a(cdpk) uaz197b(cdpk) uaz227(end) uaz233b(act) uaz233d(act) uaz23a uaz258a
uaz269c(kri) uaz80(iron) uiu1b(pog) umc159a umc163b umc36c umc44c

6.02 umc59a -enp1 -l15 -ms1 -pg11 -si1 = cdc48 = cyc3 = mir1 = oec33 = y1 ~-de*-N1400
~-gs3 ~-l*-N113 ~-l*-N612B ~-l*-N62 ~-o*-N1320A ~-o*-N1368 ~-o*-N1384A ~-ol4
~-pg*-N1885 ~-ptd*-N1425A ~-sh*-N1320B ~-smp*-N272A ~-v*-N634A ~-v*-N69A ~-w*-
N278A ~-w*-N335 ~-wl*-N217A ~-wl*-N358A ~-wl*-N362B ~csu116b(elf1) ~opg06-1170
~opv12-1300b ~psl15 agrr189 agrr87a bcd221d bmc1371 bmc2151 bnl107 bnl28(sbe1)
bnl6.22b csu158b(eno) csu183a(cdc48) csu309(atpc) csu395a csu548 csu56a(ohp) csu605
csu747b(arf) mir2(thp) mir4(thp) mpik1 mpik18 npi373 npi377 opg06-740 php06007 psu1a(spe1)
rz242 rz2b rz698b(ppy) tda204 uaz162 ucr1a(eif) uiu5(chn) uiu6(chn) umc39d umc51b
umn361

6.03 npi393 -l11 csu199a csu226b(elf) csu923(sec61) npi98b php20856 uaz106a

6.04 umc65a -Dt2 -hex2 = dzs23 = pl1 ~-sm1 ~-su2 ~npi580b ~psl29 agrr118a agrr37a bnlg480
csu95d ici96 mps1.11 npi223a npi224i npi253d npi293c npi617 opg05-800 phi124 pic2a psr108
rgpc643e rgpc74b rz144b rz588b std6b(dba) tda51 tug6 tug7 tug8 uat2(noi) uaz160 uaz161a(elf)
uaz244a(prh) umc113b umc180(pep) umc204 umc248b

6.05 umc21 -npi330(me) -w1 -w14 = dhn1 = ncr(sod3a) = pdk1 ~cdo113c ~cdo312b ~npi(G2B-
03a) ~rz67a agrr261 asg52c bcd221a bcd454a bcd855b(ext) bmc1154 bmc1443 bmc1617
bmc1702 bmc1732 bmc1922 bmc2249 bnl15.37a bnl17.22 bnl17.26 bnl3.03 bnl5.47a bnl8.06b
bnl8.23b bnlcmt9.08 cdo771 chs562 csu1065 csu1083a csu1095 csu1101a csu116a(elf1) csu1189
csu16b csu225 csu236(eif2) csu259 csu310(ptk) csu360(elf) csu382a(cld) csu46c csu481 csu578a
csu581b(tua) csu60a csu666(his2A1) csu760a csu782 csu807b csu812 csu835 dup400(pac)

Table 3.   Continued.

gsy298e(pmg) isu1410i nc013 npi102 npi252 npi265 npi294c npi38 npi560 npi608 npi616 npi63b npi67 opw18-720 pge20 phi129 php10016 php20608b rgpc43b rgpc74a rz455b std7a(dba) uaz121a uaz209 uaz220(elf) ucsd78a(zag1) uky3b(P450) umc137b umc152c umc265(ptk) umc46 ynh(me2) zag1

6.06 umc38a -pt1 = mlg3 ~npi260 ~npi9 ~umc71b ~umc85b asg50a asg6a bcd738a(pgk) bnl17.12 bnl5.62e bnl8.05b bnl8.08c bnl8.08j bnlg345 cdo348a cdo548 cdo64c cdo89(aat) csu727(trh) csu841a dupssr15 gsy299(abasi) npi280 npi419a php20904 rz206a uaz19d uaz243b(atpb) uaz251d(rpS11) uaz256 uaz43e uaz81 ufg(vp2274b) ufr1(cal) umc138a umc140c umc160a umc32d(cgn)

6.07 umc132a(chk) -idh2 -mdh2 -tan1 = agp2 = hox2 ~bmc1136 ~bmc1521 ~gsy59b(sh2) ~npi(5-05a) ~php20569c ~uaz123c ~umc32e asg18 asg47 bcd828b(atpb) bmc1740 bmc1759a csu238b(apx) csu291 csu293 csu928 npi597a phi123 uaz269d(kri) umc237 umc238a umc246 umc266c(ptk) umc63b

6.08 asg7 agrr213 cdo1160d(kri) cdo202a(mcf) cdo346b cdo393 csu68a(mcf) npi561 phi089 php20599 uaz229 uaz240b ufg(agp1) umc133b umc134a umc167b umc28 umc62 uor1a(rpS12)

7.00 = rs1 ~-cp*-N1294 ~-de*-N1136A ~-hcf101 ~-hcf103 ~-hcf104 ~-hs1 ~-mn2 ~-sen2 ~-sh6 ~-w*-N42 ~bmc1642 ~bmc2132 ~isu86 ~npi109b ~npi278b ~npi332 ~npi353 ~npi399 ~opc06-1010 ~opn09-1200 ~umc49e ~zp50 agrc261 agrc36 agrr128 bmc1367 bmc1686 bnl25 csu582 csu60e knox8b npi567 npi576A opm08-2000 pge5c rgpg124b ucsd106b umc7Stelo

7.01 asg8(myb) -w17 -y8 = mdh6 = o2 ~-ask1 ~-de*-B30 ~bcd855d(ext) ~bnl17.13a ~npi(C5D-06) ~npi277b ~npi28 ~npi313 ~npi361b/e ~npi388 ~npi430 ~npi44 ~npi470(a1) ~npi48 ~npi500 ~npi564 ~php20690b bmc1200 bmc1292 bmc2160 cdo353a csu129 csu13 csu251a csu486b csu611b(grp) csu794 csu810b csu93d cuny12 gsy297c(emb) isu145a lc12 npi400a php20558b php20581a rgpc1122b(rpL15) uaz20b uaz7b uaz83 umc235 usu1b(fnr)

7.02 asg34a(msd) -cp1 -les9 -ms7 -o5 -ra1 -rcm1 -rs4 -v5 -vp9 -w16 -zpl2b = crt2 = cyp6 = gl1 = in1 = nbp1 = pep4 = zpb36 ~-cp*-N1104B ~-cp*-N1417 ~-crp1 ~-de*-N1177A ~-et*-N1332 ~-gl*-N1845 ~-ms*-6004 ~-o*-N1298 ~-o*-N1310A ~-sh*-N1341 ~-smp*-N586B ~-spc*-N357A ~-v27 ~-wl*-N629A ~-wlu2 ~dupssr11 ~dupssr9 ~npi112 ~npi23 ~npi375 ~npi475(oec) ~opy10-1110 ~tum2 ~umc116a agrc203 agrc914 agrr168 agrr241 agrr265a agrr267a agrr49 ast(dcm1) bcd1087b bcd450c bcd98i bmc1003 bmc1094 bmc1164 bmc1247 bmc1380 bmc1759b bmc1792 bmc1808 bmc2203 bmc2233 bnl15.40 bnl5.33g bnlg398 bnlg657 cdo218b cdo407 cdo412 cdo412b cdo544a cdo662b cdo678a cdo680b cent7 chs606 ciw(S10) csu11 csu233(psaN) csu241c csu281a csu34a(rpS8) csu4b csu7a csu81a(ank) csu848b(vpp) csu919b csu936 ias9a isu1410c mpik4a mps1.06 mps1.15 mps2.06 npi111 npi216d npi221a npi224a npi294b npi294e npi367a npi391 npi47b npi568 npi596 npi600 oec17*-Z26824 opv08-800 psr3a psu2(bZip) rny(pcr)d rz413 rz422 rz509a(mip) rz617 tda45 tug5 tug9 uaz123a uaz143 uaz173 uaz187 uaz19b uaz224(eif2) uaz268b uaz351a(rpS12) uaz352a uaz64a

Table 3. Continued.

uaz68a(zp19) uaz84 uaz85 uaz86 uaz87 uaz88 uaz89 ucsd107b ucsd141a ucsd81c(zag2)
ufg(inv1A) ufg1 umc(nabp1) umc112b umc113c umc136 umc193c(orp) umc258 umc270 umc39f
umc5b umc98b uor1c(rpS12) zds1

7.03 asg49 -bn1 -sl1 -tp1 -tpi1 -va1 -wyg1 = ij1 ~npi(7-317b) ~npi(G2A-09) ~npi440 ~npi459
~npi58 ~psl27 ~rz596 ~tyk30 agrc333b agrr111a agrr174 agrr73 bcd1088a bcd926a bmc1070
bmc1305 bmc1579 bnl15.21 bnl15.37b bnl6.27 bnl6.29d bnlg339 bnlg434 bnlg572 csh2b(cdc2)
csu1124 csu253 csu274(hsp90) csu296 csu395c csu820 mpik27b(zmm7) npi122 npi349(alr)
npi389 npi394 npi455a opu10-800 php15037 php20569a php20708 psl23 tda37b uaz123d uaz136
uaz200 uaz205c(hsp70) uaz221(his2a) uaz233c(act) uaz82 uaz92 uaz922 ucsd106g umc110a
umc111b(psy) umc149a umc222(fgh) umc56 umc57c

7.04 umc254 -e1 -ren2 = lhcb2 = ndk1 = rip2 = tua6 ~npi335 ~npi35 ~npi392b ~npi59b ~umc59b
agrc542 agrc701 agrr101 agrr131 agrr132 agrr186 agrr207 agrr44 asg14a asg32 asg36a asg5
ast(amyBS2)a bas1 bcd221c bcd249b bcd249i bcd349 bcd808b bcd9 bmc1161 bmc1666 bmc1805
bmc2259 bmc2271 bnl13.24 bnl14.07 bnl14.34 bnl15.07b bnl15.27b bnl4.24 bnl5.21a bnl5.61a
bnl6.06b bnl7.61 bnl8.21a bnl8.32 bnl8.37a bnl8.39 bnlg155 cdo59a(gos2) csh14 csh3(tyk30)
csu1055 csu154c(eif5A) csu16c csu175d(eif5A) csu17a(rnp) csu209 csu213a csu21d csu229a(oec)
csu36d csu37(atp54) csu37b csu5 csu597c(dah) csu749a csu8 csu818b(lhca) csu847b(lhcb) csu904
csu906 csu996 dupssr13 gsy250(amy) ias4b ias5 isc(b32b) mus1 ncr(b32c3b) npi217 npi235b
npi240 npi263 npi283 npi300b npi352 npi385 npi398b npi413b npi433 npi435 pge3 php20563a
php20746 rgpc12a rgpg20 rgpr440b(gap) rgpr663b rz395 rz404(ccp) rz530b rz753(cdpk) tda66c
uaz117c uaz119b(rpS6) uaz199 uaz207 uaz225(lox) uaz241a uaz245(gbp) uaz28a uaz292(gdh)
uaz31c uaz90 uaz91(ndk) ukd(hotr) umc125b umc137d umc80a uor2(crp)

7.05 umc245 = ncr(sod2) = sod2 ~-Dt3 ~-gzr1 ~-o15 ~-ptd2 ~-px3 ~cdo533b ~npi29 ~npi465b
~phi051 agrc6 agrr202 agrr55 asg28a asg28b bmc2328b bnl16.06 bnl8.44a bnlg469c cdo38b(ntp)
cdo405 csu1097a csu1106 csu1151a csu163b csu27 csu578b csu632b csu814a csu894b csu920b
npi113a npi380 phi043 phi069 phi082 php20020 php20523a php20593 php20690a php20890
php20909b rgpr44c std16c umc151 umc251 umc45 umc91a

7.06 umc168 -bd1 -pn1 abg373 agrc525 bmc469c bnl(tas1j) bnlcmt7.13 csu705 npi45b npi611a
pbs7 phi116 php20728 umc35a umc7Ltelo

8.00 -rf4 ~-ats1 ~-clm1 ~bmc1252 ~bnl10.17c ~cdo293 ~cdo406 ~cdo681 ~cent8 ~csu1076
~csu312 ~csu319 ~csu368(phr) ~csu597b(dah) ~csu675b(prh) ~mpik41b(mem1) ~npi253e
~npi301 ~npi315 ~npi398c ~npi476(oec) ~npi91b ~rz382b ~rz995a(fbp) cuny19 npi114a pbs6c

8.01 npi220a = ncr(sod3b) ~-crp*-N1429A ~-ct1 ~-fl*-N1163 ~-v*-N29 ~bmc1194 ~bnl6.25b
~cdo480a ~npi(G1E-05) ~npi110 ~npi79 agrr169 bmc1073 bmc2037 bnl13.05a bnl8.08k cdo64d
csu29c csu332 csu891(rpL30) npi222b psl42 rnp2 rz323b

Table 3.   Continued.

8.02 bnl9.11a(lts) = hsp18c = pdc2 = tpi3 ~-bif1 ~cdo328 ~cdo32a ~csu229c(oec) ~npi(5-05b) ~npi376 ~npi585 ~php20727 ~umc65d bcd98b bmc1352 bmc2235 bmc2289 bnl21 cdo460 csu949b isu1410a mpik12b mpik15d mpik15e mpik17c mpik17d mpik35f npi206 npi218a npi276b npi64 opc06-800 pbs13d phi119 php10040 psl38 rgpc131b rz543b ucsd64b umc103a umc91b umc92b wusl1042 zmm2

8.03 umc124(chk) -mdh1 = gpa1 = lhcb3 = ncr(sod3c) = tub2 ~bmc1460 ~bnl9.08 ~csu116c(elf1) ~csu75d ~opf06-1100 ~opu10-990 act1 agrc20 agrc514b agrc747 agrr116b agrr209b agrx975 bcd1127b bcd355 bcd828c(atpb) bmc1067 bmc1834 bmc2082 bnl1.45a bnl10.39 bnl13.05c bnl17.16(bt2) bnl17.20 bnl8.06a bnl9.44 bnlg669 cdo113b cdo1160a(kri) cdo346a cdo480b csu1175 csu244(imp) csu275a(mtl) csu329 csu620 csu760b csu849(atpb) csu910 dupssr3 ici277 ici286a isu1719c isu2191b isu2191h klp1b ksu1c niu1::Bs1 npi(C9A-04) npi260b npi37 npi618 opd13-1210 pbs4 pge2 pge21 phi125 rgpc161 rpa5c rz244a(dia) rz28c rz556b stp1 tda164 tda217e tda52 tug1 uaz121b uaz243c(atpb) uaz244b(prh) uaz249c(ubf9) uaz251a(rpS11) uaz252b(ptk) uaz25a uaz269a(kri) uaz290(SDAg) ucsd61f uky3a(P450) umc120a umc152b umc206(hsp70) umc236 umc238b umc32b uor1b(rpS12)

8.04 bnl7.08a -clt1 -fl3 -pro1 -sdw1 = pdc1 = pdk2 = rip1 ~-blh*-N2359 ~-dek20 ~-dek29 ~-des17 ~-el1 ~-gl18 ~-hcf102 ~-ms43 ~-nec1 ~-pet1 ~-v*-N779A ~-v*-N826 ~-wlu3 ~ope01-1160 ~opf01-1200 ~psl19 agrc1 agrc478 agrr222 bcd454b bmc1863 bnl24a bnlg119 caat1 cdo1395e cdo202e(mcf) csh9(cyc1) csu1101b csu179d(hsp70) csu226a(elf) csu254d csu66b(lhcb) csu68c(mcf) csu720a csu807a gsy60d(bt2) isc(b32a) koln2c mwg645k ncr(b32c3a) npi(pdk2) npi224c npi224h npi294f phi121 rz455a rz67b sb32 std7b(dba) uaz147 uaz165 uaz193(rip) uaz202 ucsd106h umc160b umc186a(Bs1) umc209(prk) uwm1a(uce) wusl(pdc1)

8.05 bnl2.369 -ht2 -lg4 = hox1 ~bmc1152 ~bmc1782 ~bnl5.62c ~cdo1056 ~opb18-920 ~opr03-750 ~opr03-760 ~rz527b ~uaz164b agrc568b bcd134a bcd738c(pgk) bmc1176 bmc1246c bmc1446 bmc1599 bmc1651 bmc1812 bmc2046 bmc2181 bnl12.30a bnl162 bnl8.26 bnlg162 bnlg666 cdo1081b cdo1173a cdo312a cdo455a cdo464b cdo580a(ivd) cdo64a cdo708 cdo920a(egl) cdo962a csu1023 csu1041b(ptk) csu125b(cah) csu292 csu742a(rpS7) csu829 csu841b dba2 dgg9a dgg9h gstIIB ici222 knox11 knox5 mwg645a npi101b npi371b npi595 opg05-200 opp08-1120 pic6b rgpc597(prs) rgpg81 rz206c rz390a(cyb5) rz390b(cyb5) rz556c rz66b rz776b scri1(msf) uaz233a(act) ucb(anp1) uiu1c(pog1) umc12a umc184c(glb) umc189(a1) umc2c umc38b umc89a umc93a umn430

8.06 csu31 -htn1 -idh1 = ald1 = sps1 ~-v21 ~npi(C4D-02) ~npi(C9B-07) ~npi(G2B-06a) ~npi1 ~npi101a ~npi3 ~npi33 ~npi39 ~npi425b ~npi43 ~npi50 ~npi69 ~npi72 ~umc17b ~umc32g ~umc36d ~umc85c aba2 agrr144b agrr50b agrr51 asg17 asg1a asg52a asg53 asg61b bcd1086b bcd134c bmc1031 bmc1065 bmc1607 bnl10.24b bnl17.01 bnl17.17 bnl5.33d bnlg240 cdo470 cdo54 csu110a csu2c csu382b(cld) csu384 csu597e(dah) csu685 csu772b ici95 isu1774b ksu1b ksu1d npi108b npi201b npi299 npi590 npi599 pbs6b pbs9a rgpc112 rgpc198b(sik) rgpc949 tum1 uaz119a(rpS6) uaz138b uaz176a uaz291(F-bA) uaz94 uaz95 umc117 umc150a umc16b umc271 umc30a umc48a umc53b umc71a umc84c

Table 3. Continued.

8.07 npi268a -v16 = psy2 ~-ms8 ~-tpi5 ~cdo113a ~cdo251b ~cdo662a ~csu163b ~npi328a ~rz444a agrr184a agrr269 agrr274b agrr322 bmc1350b bmc1823 bmc1828 bnl10.11 bnl10.38c csu110c csu179c(hsp70) csu254c csu38b(taf) csu776a rgpc86(ptk) rz538a uaz174 umc164a umc165b umc266d(ptk) umc66d(lcr) uwm1c(uce)

8.08 npi414 -j1 ~cdo187 ~cdo430 ~csu64b(grf) ~csu9a(trf) ~php892 ~rz460 agrr262 bmc1056 cdo241b csh8a(cyc4) csu1155a csu165a csu223a(psei) csu591(uce) csu786(uce) csu922(arf) csu96b(psei) dupssr14 npi224b npi438b sb21 uaz128 ucla(obf3B) umc7 umc82d uwo1

8.09 agrr21 -emp3 -rgh1 = gst1 = hox4 asg50b bcd98e bmc1131 bnl(tas1m) csu103b(aba) csu1155b npi107 npi119(hsp70) npi212a sb2.1 ucsd113a umc39b umc3a umc4b

9.00 -Dt1 -Dt5 -v28 -yg2 ~-cp*-N1092A ~-da1 ~-dek12 ~-emb2 ~-hcf113 ~-ms45 ~-sem1 ~-sh*-N399A ~-skb1 ~-v*-N1893 ~-v*-N829A ~-w*-N1830 ~-w*-N1854 ~-w*-N1865 ~-w*-N627B ~-w1*-N1803 ~-wl*-N1857 ~npi(G2C-10) ~npi211 ~npi463(sod) ~npi86 ~umc70 agrr118b bmc1272 bmc1724 bnl9.07a gsy261b(be)

9.01 umc109 = c1 = sh1 ~-g6 ~-Ins1 ~-Ins2 ~-zb8 ~cdo1053b ~cdo665b ~isu136b ~rz588a agrr147 agrr41 bmc1288 bmc1583 bmc1810 bmc2122 bnl17.11 csu250b(aba) csu95a isu1146 koln2b(hox) mir3a(thp) mpik19a npi116b npi253a opw13-1110 phi067 phi068 phi122 php10005 rz144a rz144c ucsd72f umc113a umc148 umc248a

9.02 = bz1 -baf1 -ga8 -l6 -l7 -lo2 -Mr -v31 -w11 = eno1 = prc1 = uaz237a(prc) ~bcd93 ~dpg1b ~mgs3 ~opr11-880 ~opx04-1230 agrc255b agrc273 agrr58 agrr87b asg19 asg82 bmc1082 bmc1372 bmc1401 bmc1913 bnl213 bnl3.06 bnl5.67b bnlg244 cdo1395g cdo475a csu1077 csu1083b csu228(pfk) csu466(lhcb) csu471 csu486a csu651(rpL39) csu665b(adt) csu733(rpL39) csu94b dupssr19 dupssr6 gsy266b(ptkr) gsy89 isu124 klp1c mpik11c mpik25(zmm3) npi266 npi300a phi022(wx) phi061(wx) phi130 psl3 rz2a rz516 ucsd62k(zag4) umc105a umc256a umc82a = fdx3

9.03 = wx1 -acp1 -ar1 -d*-3010 -les8 -ms2 -pg12 -v1 = d3 = dzs10 = gl15 = hsk1 = obf2 = pep1 = rf2 ~-cp*-N1381 ~-dcr*-N1409 ~-dek13 ~-dek30 ~-dsc*-N749 ~-et*-N357C ~-gm*-N1319B ~-hcf42 ~-ms*-6011 ~-pg*-N660A ~-trn1 ~-v*-N1871 ~-v*-N53A ~-v*-N806C ~-wlu4 ~bmc1626 ~bmc1688 ~cdo775 ~npi34 ~npi454 ~opb17-1 ~opx04-1240 ~opy13-730 ~opz10-1 ~umc112d agrc445b agrr125 agrr205 agrr64 asg37 asg63a asg65 asg66 asg67a asg68a bcd221b bmc1687 bmc1730 bmc469a bmcp21.2 bnl26 bnl5.10 bnl5.21d bnl5.33c bnl7.24a bnlg127 bnlg430 bnlg469a cdo17 cdo319 cdo590(ppr) cdo673 cdo689c cdo78 cent9 csu193 csu321 csu616 csu623 csu680d csu857 gsy54d(rpl7) isu2191d isu2191k knox2 npi215a npi222a npi25 npi416 pbs14b pbs14d php1460 php20052 php20075b(ext) rgpr1908a(acb) rgpr3235a rz476c rz612 rz953 std6a(dba) tda66d uaz161b(elf) uaz166c uaz223(vpp) uaz246b(mbf) uaz284 ucsd1.8a umc127a umc153 umc194b(gpr) umc20 umc247 umc253b umc273a umc81 uwm1b(uce)

Table 3.   Continued.

9.04 csu147 -Mrh = hm2 = sus1 = tub7 ~-bk2 ~-ms*-6006 ~-ms*-6021 ~bnl5.71c ~bnl56.2d ~gsy297b(emb) ~isu98a ~npi80 ~php20554 agrr153b agrr90 bmc1714 bnl5.04 bnl7.13 bnlcmt6.06a csu110e csu179b(hsp70) csu181b csu183b(cdc48) csu212a csu214a(grp) csu252b(cdc2) csu254a csu263a csu404a csu43 csu557 csu56d(ohp) csu694a(uce) csu778(lhcb) gsy264(zn10) ici266 mps1.07 mps1.16 mps2.08 mps2.14 mwg645g pbs14c pic1a psl22 rgpc524 rgpr3239b rz251b rz672b(cgs) rz682 uaz112 uaz119c(rpS6) uaz236a(ser) uaz266b umc114 umc140b umc38c wsu1(ptk)

9.05 umc95 -v30 = hsp18a ~-wc1 ~bmc1091 ~bmc1129 ~bmc1270 ~bnl10.13b ~cdo189b ~chs5046 ~npi17b ~npi404c agrc595 agrp1000 agrr171 agrr92b bmc1012 bmc1209 bmc1884 bnl7.50 bnl8.05c bnl8.08d bnl8.17 cdo260 cdo419b cdo770 cdo938b csu219(tgd) csu355(ext) csu392a csu395b csu58b csu710e(apx) dpg6c mps2.07 mps2.13 ncr(sod4b) npi293a npi427a npi443 npi580 opd02-920 pge(phyB2) phi040 rgpr44b rpa8 uaz125 uaz264a umc29c

9.06 csu61a = ibp1 ~bnl14.28a ~bnl5.09a asg44 bcd1086a bcd131a bcd855f(ext) bnl7.57 bnlg292a cdo1387a(emp70) cdo1395a chs504 csu1004 csu109b csu145a(pck) csu28a(rpS22) csu59a csu634 csu877 csu93a dba4 isu3 mpik28a(zmm8) mpik28b(zmm8) npi403a npi425d npi439b opu19-1080 rz574b(cwp) rz742b uaz148 uaz96a uom1(hb)

9.07 asg12 -bf1 ~-bm4 ~bmc1506 ~bnlg619 ~cdo20b ~csu145b(pck) ~npi98d ~rz329a(bga) ~uaz231(zag) agrr294b asg59b bmc1191 bmc1375 bmc1525 bmc1588 bnlg128 bnlg279 csu1005 csu1118 csu12a(cin4) csu285(his2B) csu860a csu870 csu883(rpL21) dupssr29 gsy266d(ptkr) klp6 npi209a npi291 npi97b opg05-600 phi108 psl46 rz561c rz632b std20a std2a(dba) ucsd61b umc272(vfa)

9.08 csu54b -rld1 csh2c(cdc2) csu50b csu804a(dnp) dpg12c uaz31a ucsd107a umc94b npi208d

10.00 ~-dek14 ~-gl21 ~-hcf316 ~-hcf47 ~-ij*-N504A ~-l19 ~-les*-NA7145 ~-les12 ~-les16 ~-les3 ~-les6 ~-mac1 ~-o*-N1046 ~-rgh*-N1524 ~-ufo1 ~adp0 ~npi216b ~npi253f ~npi366 ~npi371c ~npi593d ~opo20-760 agrr43a bnl10.17a cdo1338b csu25b(P450) csu306(fer) mpik12a mpik13a mpik15f mpik17f npi208b phi041 phi117 phi118 php20626 php20753a ucsd64d ucsd72b

10.01 php20075a(gast) -rp1 -rp5 -rp6 -rpp9 -sr3 ~bcd1072b(hsp70) ~cdo218a ~rz141b(emp70) agrc561 bnl3.04 cdo127b(pyk) csu1042 csu136(plt) csu359(alp) csu577 ksu1e ksu1f ksu2 ksu3 ksu3/4 mpik33::cin4 uaz21c

10.02 npi285 -oy1 -sad1 = cr4 = gdcp1 ~-og1 ~-rlc1 ~csu103a(aba) ~opw02-530 agrc255a agrc528 agrc690 agrc714 bmc1451 bnl5.62b csu1054 csu250a(aba) csu30b(atp) csu561b csu825 isu167 ksu5 phi059 phi063 rz400(gbp) tda217a tda217c tda217d umc152a

10.03 umc130 -gdh2 -php1 = glu1 ~-ad*-N377B ~-rgh*-N799A ~-y9 ~bcd348a ~bnl5.09b ~bnlcmt6.06b ~cdo456a ~cent10 ~csu110f ~fgp1 ~npi(C6F-03) ~npi(G5A-06) ~npi92 ~oph12-780

Table 3.   Continued.

~opw06-680 ~opy10-1100 ~rz556a agrr104 agrr18 agrr216b agrr232 agrr255a agrr295 agrr57
asg76 bcd147(gbp) bcd98c bmc1037 bmc1079 bmc1547 bmc1655 bmc1712 bmc1716 bmc1762
bnlg210 bnlg640 cdo504 cdo551b cdo783a chs5008 csu1050 csu213b csu234b(gbp) csu237a(psaN)
csu625 csu745c(rpPo) dpg3 dpg5 eoh1 gcsh1 ias13c mpik20b mpik41c(mem1) mps2.03 npi105a
npi223b npi250a npi327b npi378 npi417b npi445 npi597b npi85 npi98c ov23 phi050 phi054
php06005 php10033 php20646 psl9 rgpc1122d(rpL15) rgpc496c(adh) rgpr440c(gap) rz261a(sad)
uaz100(prl) uaz116 uaz153 uaz178 uaz17c uaz242(clp) uaz24a uaz97 uaz98 ucsd72k ucsd72m
umc(orp2) umc155 umc18b(psaN) zmm1

10.04 umc64a -acc1 -bf2 -cx1 -du1 -li1 -ms11 -nl1 -zn1 = gpa2 = grf2 = mgs1 = orp2
~-dek15 ~-gs4 ~-l*-N1879 ~-l*-N1908 ~-l*-N195 ~-l*-N31 ~-l*-N392A ~-ms10 ~-o*-N1422
~-tp2 ~-v*-N114A ~-v*-N354B ~-v*-N470A ~-v18 ~-v29 ~-vp13 ~-vsr1 ~-w*-N24 ~bcd304
~cdo1328a ~cdo259 ~cdo366 ~isu5 ~npi22 ~npi582 ~npi602 ~php20719a ~ufg1433 agrc459
agrr113b agrr62a amo1 asg2 bcd1302c bmc1074 bmc1518 bmc1526 bmc2336 bnl7.49b(hmd)
cdo1395b cdo482 cdo667a csh::stAc csu276 csu298b csu333 csu46a csu599b csu613(acb)
csu671b csu797(uce) csu815 csu86 csu864 csu893(isp) csu898 csu913 csu929(his3) csu948
csu951(eno) csu981(eif5A) gstIIA gsy87 hsp90* isu1719a nac1 ncsu2 npi264 npi294h npi303
npi305b npi327a php15013 psu1b(spe1) rgpr1908b(acb) rz69 stAc tda205 uaz117b uaz175a
uaz76b uaz99 ufg8(grf) umc146 umc159b umc161c umc243a umc261

10.05 umc259 -g1 -wsm3 = sam1 ~-hcf28 ~bmc1028 ~npi269b agrp168b bcd386b bmc1185 bnlg137
csu745a(rpPo) dpg6b gf14-12B npi232a npi563 npi578 rz740(sam) std4(dba) uaz228b(his2b)
ufg3b(ivr) ufg7B umc162a umc163 umc182(r1)

10.06 umc44a -cm1 -I-R -isr1 -l1 -mst1 -o7 -w2 = lc1 = r1 = rps11 = S = sn1 ~bmc1677 ~bmc2190
~bnlg153 ~cdo539a ~csu148a(clx) ~gsy242b(hox1) ~isu2192b ~rz387b bmc1250 bnl10.13a
bnl17.02 bnl17.07 bnl17.08 bnlg236 bnlg594 cdo1417b(ptk) cdo662c csu615b gsy54a(rpl7) isu12
klp1f npi287b npi290a npi306 npi461 phi035 uaz251c(rpS11) ucsd(lfyA) umc57a

10.07 bnl7.49a(hmd) -l13 -ren3 -sr2 = gln1 = uaz294(rpl12) agrr167b asg50d bcd1092a bcd135b
bmc1360 bmc1450 bmc1839 cdo244a(crp) cdo36a cdo431b cdo93b csu1028(lhcb) csu1039
csu300b csu48 csu571b(ipp) csu781a csu844 dba3 gsy53(rps11) mir3c(thp) mwg645l npi254b
npi321a npi421b npi577b npi604b psl48 rgpc285 rz17a rz569a rz590b umc232 umc269(ptk)
umc49b

---

9.   Other maps

Other maps in MaizeDB include (1) mitochondrial; (2) maps for tropical or mixed
subspecies germplasm; (3) prior releases of maps (e.g., BNL, UMC, genetic) for
historical and data tracking; and (4) maps of a few related species. A full list of

the maps in MaizeDB can be viewed by typing % in the Map Name field and performing Retrieve.

10. Mapping data

Complete sets of map scores for a given population ("Panel of Stocks") may be retrieved using the WWW query form, "Map Score Tables". Subsets, constrained by a range of bins, may be retrieved from "Map Scores By Bins" queries. Alternatively, individual locus pages link to all the map scores for a locus, and to the recombination data used for the genetic map.

11. Clone sets

DNA clones in the public bank and distribution center at the UMC Maize RFLP Lab now total over 6400, of which more than 4700 are publicly distributed. This includes the sets of sequenced cDNAs generated by Chris Baysdorfer, California State University - Hayward (designated csu), and by Tim Helentjaris (designated uaz, or 5C, 6C, etc.), which represent the main collections of candidate genes publicly available for maize to date. Descriptive information and request procedures are detailed in MaizeDB.

12. Developing markers

Current research efforts will have a major impact on probe resources for maize. The first is the development of SSR probes. The total of publicly available SSR primer pairs for maize as of June 1998 was 195. These were derived by Lynn Senior, USDA-ARS Raleigh; Emily Chin, Pioneer; and Ben Burr, Brookhaven Natl. Lab. Programs in progress to derive SSR primer pairs at other loci include work by Ben Burr, Brookhaven Natl. Lab. (with a consortium organized by Linkage Genetics); Mike McMullen, USDA-ARS Columbia; Keith Edwards, Long Ashton, and others. As this number expands, coverage of the genome will permit many of the applications currently conducted with RFLPs to be replaced with SSRs. The second development is the expanded use of AFLP technology. For many private and large scale public applications the cost effectiveness of AFLPs is found attractive. The major drawbacks for mapping applications in maize are the lack of codominant information for most AFLP alleles, and the non-transferability of mapping information for bands from one pedigree to the next. The third development is the initiation of large-scale EST sequencing projects for maize. Three or more private concerns are sequencing ESTs in projects targeting 100,000 or more, and are seeking to derive assemblies from them that approach the total number of functional genes in the maize genome. Public EST efforts also are under development.

13. Requirements for the next generation

What are the requirements that might be identified for the next generation of mapping efficiency? Greatly improved biotechnical methods, automatable, and more rapid, efficient, accurate, and economical than the methods now in use, are essential. Higher resolution is needed, yet the limit of resolution has been reached with the conventional types of mapping panels that are the sources for today's RFLP maps (F_2, Recombinant Inbred, and Immortal F_2). In fact, expansion to larger families, or pooling of families to make composite maps, is subject to compounding and confounding of rare events or errors, and often proves to be an illusory exercise. Finally, larger numbers of markers, more discriminatable with small changes in alleles, as small as single base pair changes, are needed. Ultimately, physical mapping must contribute to meeting these requirements. The question may be more immediate, whether mapping by genetic or cytogenetic approaches will precede, will be conjoint with, or will be preceded by, physical mapping. On the other hand, were there today no genetic or cytogenetic maps for maize, mapping would begin with physical mapping of random, large fragments and would be a very long time to resolution. The genetic and cytogenetic maps are thus preparatory for the next phase of maize genome analysis, physical mapping, which is in essence an elevation of cytogenetic mapping conjoined with genetic mapping.

14. References

Burr, B. and Burr, F.A. (1985) Toward a molecular characterization of multiple factor inheritance. In: Zaitlin, M. et al., Biotechnology in Plant Science, pp.277-284. Academic Press, New York.

Burr, B., Burr, F.A. and Matz, E.C. (1994) Mapping genes with recombinant inbreds. In Freeling, M. and Walbot, V. (eds.), The Maize Handbook, pp.249-254. Springer-Verlag, New York.

Burr, B., Burr, F.A., Thompson, K.H., Albertsen, M.C. and Stuber, C.W. (1988) Gene mapping with recombinant inbreds in maize. Genetics 118:519-526.

Burr, B., Evola, S.V., Burr, F.A. and Beckman, J. (1983) The application of restriction fragment length polymorphism to plant breeding. In: Setlow, J.K. and Hollaender, A. Genetic Engineering Principles and Methods, pp.45-58. Plenum Press, New York.

Causse, M., Santoni, S., Damerval, C., Maurice, A., Charcosset, A., Deatrick, J. and de Vienne, D. (1996) A composite map of expressed sequences in maize. Genome 39:418-432.

Emerson, R.A., Beadle, G.W. and Fraser, A.C. (1935) A summary of linkage studies in maize. Cornell Univ. Agric. Exp. Stn. Memoir 180:1-83.

Gardiner, J., Coe, E., Melia-Hancock, S., Hoisington, D.A. and Chao, S. (1993) Development of a core RFLP map in maize using an Immortalized-F_2 population. Genetics 134:917-930.

Grant, D., Blair, D.L., Owens, T., Katt, M. and Beavis, W.D. (1993) Updated Pioneer Hi-Bred maize RFLP linkage map. Maize Genetics Cooperation Newsletter 67:55-61.

Helentjaris, T., King, G., Slocum, M., Siedenstrang, C. and Wegman, S. (1985) Restriction fragment polymorphisms as probes for plant diversity and their development as tools for applied plant breeding. Plant Mol. Biol. 5:109-118.

Helentjaris, T., Weber, D.F. and Wright, S. (1986) Use of monosomics to map cloned DNA fragments in maize. Proc. Natl. Acad. Sci., U.S.A. 83:6035-6039.

Neuffer, M.G., Coe, E. and Wessler, S. (1997) Mutants of Maize. Cold Spring Harbor Laboratory, New York.

Senior, L., Chin, E., Lee, M., Smith, J.S.C. and Stuber, C.W. (1996) Simple sequence repeat markers developed from maize sequences found in the GENBANK database: Map construction. Crop Sci. 36:1676-1683.

Shoemaker, J., Zaitlin, D., Horn, J., DeMars, S., Kirschman, J. and Pitas, J.M. (1992) A comparison of three Agrigenetics maize RFLP linkage maps. Maize Genetics Cooperation Newsletter 66:65-69.

Van Deynze, A., Nelson, J., Yglesias, E., Harrington, S., Braga, D., McCouch, S. and Sorrells, M. (1995a) Comparative mapping in grasses. Wheat relationships. Mol. Gen. Genet. 248:744-754.

Van Deynze, A.E., Nelson, J.C., O'Donoughue, L.S., Ahn, S.N., Siripoonwiwat, W., Harrington, S.E., Yglesias, E.S., Braga, D.P., McCouch, S.R., and Sorrells, M.E. (1995b) Comparative mapping in grasses. Oat relationships. Mol. Gen. Genet. 249:349-356.

Weber, D.F. and Helentjaris, T. (1989) Mapping RFLP loci in maize using B-A translocations. Genetics 121:583-590.

16.   RFLP map of peanut

H. THOMAS STALKER[1], TRACY HALWARD[2], *and* GARY KOCHERT[3]
[1] Department of Crop Science, North Carolina State University,
Raleigh, NC 27695-7629, U.S.A.
[2] Department of Biology, Colorado State University, Fort Collins, CO 80523
[3] Department of Botany, University of Georgia, Athens, GA 30602 7271, U.S.A.
<kochert@dogwood.botany.uga.edu>

Contents

1.   Introduction

Cultivated peanut (*Arachis hypogaea* L.) provides a significant source of oil and protein for large segments of the population, particularly in the less developed regions of Asia, Africa, and South America. In the United States, peanut is considered a high-value cash crop of regional importance, with production concentrated in the Southeast region of the country along with parts of Texas, Oklahoma, North Carolina and Virginia. Domestically, peanuts are grown primarily for use in the snack-food, peanut butter and confection industries, but also serve as an excellent source of mono-unsaturated cooking oil, as well as a source of meal for livestock.

The genus *Arachis*, a member of the family Leguminosae, is native to South America, and central Brazil is believed to be the center or origin. Wild species of *Arachis* are widely distributed from the Atlantic Ocean to the foothills of the Andes Mountains, and from the mouth of the Amazon in the north to Uruguay in the south (Stalker et al. 1998). *Arachis* species have evolved in such diverse environments as rock outcroppings, heavy to sandy soils, marshy areas and streams, forest-grassland margins, and from sea level to an elevation of 1600 meters (Valls et al. 1985). A taxonomic treatise was published by Krapovickas and Gregory (1994) who divided the genus into nine sections based on morphology, geographic distribution, and cross compatibilities. Cultivated peanut, an allotetraploid ($2n = 4x = 40$), has been assigned to section *Arachis* along with one other allotetraploid species, *A. monticola* Krapov. et Rig., and 25 diploid species ($2n = 2x = 20$). Cultivated peanut has been further subdivided into two subspecies and six varieties. Subspecies *hypogaea* contains var. *hypogaea* (Virginia and runner market types) and var. *hirsuta* (Peruvian runner); and subspecies *fastigiata* consists of var. *fastigiata* (Valencia market type), var.

R.L. Phillips and I.K. Vasil (eds.), DNA-Based Markers in Plants, 285 - 299.

vulgaris (Spanish market type), var. *aequatoriana*, and var. p*eruviana* (Krapovickas Gregory 1994). Cultivated · peanut likely originated in Bolivia at the base or in the foothills of the Andes mountains. This region is an important source of variability for the subspecies *hypogaea* and is the only region where *A. monticola* is known to occur (Krapovickas 1969; Kochert et al. 1996a).

Abundant germplasm resources of both the cultivated and related wild species are available to peanut breeders. However, a large portion of these potential genetic resources have been inadequately evaluated for useful traits, and most peanut breeding programs have traditionally relied on the crossing of elite breeding lines for developing improved cultivars. As a result, the germplasm base of domesticated peanut is extremely narrow. Although considerable levels of morphological variability have been observed among the germplasm resources of cultivated peanut (Wynne and Halward 1989; Knauft and Ozias-Atkins 1995), very little genetic polymorphism has been detected with molecular markers. Grieshammer and Wynne (1990) evaluated a broad range of *A. hypogaea* genotypes for isozyme variation and observed very little polymorphism. The researchers concluded that isozyme analysis would not be useful for characterizing genetic diversity in cultivated peanut. Using RFLPs, only a few polymorphisms have been observed among cultivars (Kochert et al. 1991) or exotic germplasm lines (Halward et al. 1991) of *A. hypogaea*. Similarly, studies using RAPD marker analysis indicated that relatively little genetic polymorphism exists within *A. hypogaea* (Halward et al. 1992). He and Prakash (1997) reported that only 3% of the primers used for DNA amplification fingerprints (DAF) were polymorphic. Forty-three of the primer combinations using amplified fragment length polymorphisms had at least one polymorphic band but, considering that each primer set amplifies many fragments, this is a very low level of polymorphism indeed. Conversely, large amounts of polymorphism have been detected among related wild species of peanut with isozymes (Lu and Pickersgill 1993; Stalker et al. 1994), RFLPs (Kochert et al. 1991; Paik-Ro et al. 1992), and RAPD markers (Halward et al. 1992; Hilu Stalker 1995). Further, Stalker et al. (1995) reports significant variation within the species *A. duranensis* and related the variation to geographical origin of accessions.

The apparent contraction between the abundance of morphological variability observed and the lack of detectable genetic polymorphisms within the cultivated germplasm is not uncommon. Similar results have been observed in tomato (*Lycopersicon* sp.) (Miller and Tanksley 1990), melons (*Cumunis* sp.) (Shattuck-Eidens et al. 1990), soybean (*Glycine* sp.) (Keim et al. 1990), and common bean (*Phaseolus* sp.) (Gepts 1991). Morphological traits are often controlled by a few major genes and may be subjected to intense selection pressure. As a result, morphological variation is likely to increase during domestication. On the other hand, biochemical and molecular markers, which are not subject to direct selection, often decrease during domestication of a species (Gepts 1991). A review of genetic studies indicates that a large number of phenotypic traits in cultivated peanut are controlled by a few major genes with expression influenced by the action of modifier genes and epistatic interactions among loci (Wynne and Halward 1989). Intense selection for a few major genes affecting obvious morphological traits during domestication of *A. hypogaea* which, after polyploidization, had been cut off from introgression and gene exchange with its wild ancestors could explain the lack of variability observed at the molecular level in cultivated peanut (Williams 1996). The highly self-pollinating nature of cultivated peanut would serve to enhance genetic isolation. A similar situation has been observed in domesticated bean (*Phaseolus vulgaris*) (Gepts 1991). Results from the isozyme, RFLP, and RAPD marker studies mentioned above support the hypothesis of Smartt and Stalker (1982) that *A. hypogaea* originated from a single polyploidization event. Further, Kochert et al. (1996b) concluded that *A. duranensis* was the female parent and *A. ipaensis* was the male parent of the original hybrid. Thus, including exotic germplasm lines of *A. hypogaea* in a breeding program will result in the addition of only limited genetic variability due to the severe narrowing of the germplasm base that occurred during evolution of the cultivated species. The reduction in genetic diversity that accompanied the evolution of cultivated peanut represents a genetic bottleneck for peanut improvement.

In light of the relatively narrow germplasm base of cultivated peanut, a greater emphasis should be placed on the evaluation and utilization of related wild species to enhance the available genetic variability for development of improved cultivars. The relative levels of variability observed at the molecular level for wild and cultivated peanut species is consistent with that observed for disease and insect resistance and for tolerance to a number of environmental stresses. Extensive screening of wild *Arachis* species has revealed these potential genetic resources to be valuable as sources of disease resistance (see Stalker 1992), insect resistance (see Lynch and Mack 1995), tolerance to environmental stresses (ICRISAT 1982), and variation for protein and oil quality (Young et al. 1973; Amaya et al. 1977; Cherry 1977; Stalker et al. 1989). The exploitation of these valuable genetic resources would be greatly enhanced through use of molecular markers to tag and follow the introgression of chromosome segments containing desirable traits from the wild species into cultivated peanut and through the development of a genetic linkage map in peanut to expedite the location and transfer of these chromosomal regions.

Basic genetic research on peanut has not proceeded as rapidly as it has in many other species of agricultural importance. This is at least partially due to the limited acreage devoted to domestic peanut production, as compared to other major agronomic crops, and the relative importance of peanut as a staple crop only in less developed regions of the world. As a result, little information is known about the molecular biology or evolutionary history of the genus *Arachis*. Although there have been a few reports of linkage between various morphological traits (Badami 1928; Patel et al. 1936; Patil 1965; Coffelt and Hammons 1973; Stalker et al. 1979; Murthy et al. 1988; Halward et al. 1991; Knauft et al. 1991), there is no genetic linkage map available for peanut. However, an RFLP map was developed as a cooperative effort by the Department of Botany at the University of Georgia and North Carolina State University. The mapping populations are also being evaluated for a number of agronomic traits by researchers in the Department of Plant Pathology at the University of Georgia and the Departments of Crop Science and Plant Pathology at North Carolina State University, with the goal of eventually producing a genetic map containing both conventional and molecular markers. The present status of the genetic map in peanut will be presented here.

2.   Libraries and probe sources

Both random genomic peanut clones and cDNA clones have been used in the development of the linkage map. A genomic DNA library was constructed using the peanut cultivar 'GK-7' (*A. hypogaea* subsp. *Hypogaea*) as the genomic DNA source. The library was constructed by cloning gel-isolated *Pst*I fragments (1.0-2.0 kb in length) into pUC8 plasmids and transforming the recombinant plasmids into DH5*a* strains of *E. coli*, according to the procedures of Sambrook et al. (1989). Recombinant clones were selected on IPTG-Xgal plates, and plasmids were isolated by a miniprep procedure (Wilimzig 1985). Recombinant plasmids were digested with *Pst*I and subjected to electrophoresis on 0.8% agarose gels. The molecular weight of each insert was determined by comparison to molecular weight standards, and blots were prepared on nylon filters (Gene Screen Plus, Du Pont) following the method of Southern (1975). Total peanut DNA was labeled with 32P-dCTP by nick-translation (Rigby et al. 1977) and hybridized to the filters to detect inserts containing repeated sequences. Cotton chloroplast DNA (compliments of G. Galau, Dept. of Botany, Univ. of Georgia) was used to screen the library for members containing chloroplast DNA inserts. Those inserts that showed no signal with either chloroplast or total peanut DNA were assumed to represent low-copy number nuclear sequences and were selected for RFLP analysis.

Two separate cDNA libraries were prepared, one from root tissue and one from shoot tissue, using the Stratagene ZAP-cDNA GigapackII Gold Cloning Kit according to the manufacturer's instructions. Approximately 200 seeds of the peanut cultivar 'GK-7' were germinated in moist paper towels to yield a total of 10-15 grams each of root and shoot tissue for the two libraries. Poly A* RNA was isolated independently from young shoots and young roots according to the procedures described in Hong et al. (1987) and used to construct the two cDNA libraries.

Clones in the random genomic library are being maintained as glycerol cultures and as mini-preparations of plasmids as described in Sambrook et al. (1989). The cDNA libraries are being maintained as whole phage for current use in addition to being cloned into Bluescript plasmids for long-term storage of clones. All clones from both the random genomic and cDNA libraries are available for distribution to researchers who are interested in using them for genetic analyses. Genomic clones can be distributed as plasmid minipreps or as PCR products following amplification with M13 forward and reverse primers: clones from either cDNA library are available as small quantities of PCR product which can subsequently be amplified for use in random primer labeling reactions using forward and reverse M13 primers and the following amplification conditions: 94°C for 1 min. 20 sec. (denaturation); 37°C for 2 min. (annealing); and 72°C for 3 min. (extension) for 30 cycles.

3.   Mapping population

A number of factors were taken into consideration when selecting an appropriate population for developing an RFLP map in peanut. Insufficient variability for RFLPs or

RAPD markers exists within cultivated peanut to allow construction of a genetic linkage map directly in *A. hypogaea*. In addition, since cultivated peanut is an allotetraploid, segregation of traits is inherently complex. The genetic basis for inheritance of most agronomically important traits in peanut is not completely understood. Although most traits in cultivated peanut follow a diploid pattern of inheritance, many appear to be under the influence of duplicate loci (Wynne and Halward 1989). As a result, segregation analysis is more difficult in the allotetraploid cultivated species than in the related diploid species. A complete series of aneuploids, often used in the construction of RFLP maps (Beckmann and Soller 1986), is not available in peanut. For these reasons, the diploid species of *Arachis* were considered the best choice for a mapping population. The main advantage to constructing an RFLP map in diploid peanut is the abundance of RFLP variability observed among the wild species. The relative ease of analyzing segregation data in a diploid cross versus a cross between allotetraploids provides an additional advantage. RFLP maps that have been developed using populations derived from interspecific crosses between a cultivated species and a related wild species have proven useful for cultivar improvement in such crops as tomato (Miller and Tanksley 1990) and soybean (Keim et al. 1990). Although the peanut RFLP map was developed using populations derived from crosses among wild *Arachis* species, it should also be useful for cultivar improvement programs, especially when applied to following the introgression of chromosome segments from wild species into cultivated peanut.

In peanut, we were fortunate to have available several F_2 populations derived from interspecific crosses between various diploid species of *Arachis*. Two of these populations were chosen for map construction. Simultaneously mapping in two independent populations has allowed us to evaluate the homosequentiality of chromosomes and to verify linkage relationships among populations. The two mapping populations evaluated [(*A. stenosperma* x *A. cardenasii*) and (*A. duranesis* x *A. diogoi*)] were developed at North Carolina State University. Analyses of a large number of clones have revealed more polymorphisms between *A. stenosperma* and *A. cardenasii* than between *A. duranensis* and *A. diogoi*. Therefore, we concentrated mapping efforts in the F_2 population resulting from the *A. stenosperma* x *A. cardenasii* cross. The *A. duranensis* x *A. diogoi* F_2 population is being evaluated primarily to compare mapping in different peanut populations and for verifying linkage relationships. A total of 87 F_2 individuals were evaluated in the *A. stenosperma* x *A. cardenasii* cross, and 83 F_2 individuals were evaluated in the *A. duranensis* x *A. diogoi* cross.

The mapping populations were not originally developed for the purpose of RFLP analysis, therefore our F_2 populations consist of bulked progeny from several F_1 plants. Given the highly self-pollinating nature of peanut, the original parents of the crosses should have been essentially homozygous and the F1s from which our mapping populations were developed should have been genetically similar. The original crosses from which the populations were developed have been remade such that a single F_1 gave rise to each of the F_2 populations. The original F1s are being maintained as cuttings, and seeds of the new F_2s have been planted to produce new mapping populations. These new populations, derived from single F_1 plants, will be evaluated using the RFLP markers on the existing map to confirm the accuracy of the linkage groups developed in the original mapping populations. Once linkage relationships are confirmed we will begin mapping in the new F_2 populations. These new F_2 populations and the availability of the F1s will also enable us to begin mapping RAPD markers with a higher degree of certainty.

4.   The map

Genomic DNA was isolated from the parents, F_1s and F_2s using a crude nuclear preparation as described in Kochert et al. (1991). The parents used to generate the two mapping populations were screened for clones that revealed polymorphisms between them in order to identify clones that would be segregating in the F_2 generation. Surrey filters consisted of genomic DNA from each of the parents (A. stenosperma, A. cardenasii, A. duranensis, and A. diogoi) that is digested with seven restriction enzymes (BamHI, DraI, EcoRI, EcoRV, HaeIII, HindIII, RsaI). Several enzymes were used in the initial screening of the parents because previous results (unpub. data) indicated that a given clone often revealed polymorphism with some enzymes but not others. By simultaneously screening the parents with a number of probe-enzyme combinations, a greater number of polymorphic clones could be detected. Clones that revealed polymorphisms between the parents were used to evaluate the F_2 mapping populations for segregation. Initially, only those clones that revealed polymorphism at a single locus were used in construction of the RFLP linkage map. A total of 100 random genomic clones and 300 cDNA clones have been evaluated for polymorphism to date. Of these, 15 (15%) of the genomic clones and 190 (63%) of the cDNA clones were polymorphic with one or more enzyme. Of those clones that were polymorphic, 7 (47%) genomic and 92 (48%) cDNA clones, respectively, detected a single polymorphic locus; while 8 (53%) and 98 (52%) detected polymorphisms at multiple loci.

A large number of probes detected polymorphisms with more than one restriction enzyme, suggesting that many of the RFLP markers identified in peanut resulted from DNA rearrangements. Similar results have been observed in a number of other species including soybean (Apuya et al. 1988) and rice (McCouch et al. 1988). Of the cDNA clones evaluated, 38 produced ladders with one or more restriction enzymes suggesting the presence of a number of repeat sequences within the peanut genome. Unfortunately, no variability was observed among the parents of either mapping population for these clones and therefore they could not be mapped. Mapping these complex probes would allow us to determine whether the repeat sequences detected are scattered among linkage groups throughout the genome or are clustered within particular linkage groups. The information obtained from such analyses would provide insight into the evolution of the peanut genome. Therefore, we will continue to look for clones which produce ladders that reveal polymorphisms between the parents so that these can then be mapped in the future.

Several probes hybridized to more than one polymorphic locus. While the majority of these appeared to be co-segregating, a few were found to be segregating independently, indicating the presence of duplicated loci within the peanut genome. Unfortunately, the majority of the clones detecting duplicated loci produced very complex banding patterns which were difficult to score and could not be mapped with accuracy. We chose not to apply these markers to the map. As the map develops and becomes saturated with markers, it would be desirable to attempt to map some of these more complex probes in order to evaluate the extent to which duplicated loci have evolved within diploid *Arachis* species. Only two clones detecting duplication could be mapped with confidence: both of these clones hybridized to loci that were repeated on different linkage groups. The level of duplication observed within the peanut genome may have important implications regarding genome evolution in the genus. High levels of sequence duplication, similar to those detected in peanut, have been observed in *Brassica* spp. (Slocum et al. 1990; Song et al. 1991). These levels of duplication are much greater than those reported for tomato (Bernatsky and Tanksley 1986), potato (*Solanum* spp.) (Bonierbale et al. 1988), or rice (*Oryza* spp.) (McCouch et al. 1988).

To analyze segregation in the mapping populations, the F_2 progeny were scored as either 'A' (homozygous like parent A), 'B' (homozygous like parent B), or 'H' (heterozygous). Chi-square analyses were conducted to determine goodness-of-fit of the segregation data to the expected 1:2:1 Mendelian ratio. Thirty three of the 103 mapped loci showed deviation from the expected ration (P = 0.05). Most of these loci had an excess of one or the other parental type, while four had an excess number of heterozygotes. The segregation data obtained were analyzed using the MAPMAKER computer package which is specifically designed for the construction of primary genetic linkage maps (Lander et al. 1987). The locus arrangements and map distances for each linkage group were determined based on the output from the MAPMAKER program using the Kosambi mapping function and the constraints of minimum LOD score of 3.0 and a maximum recombination frequency (theta) of 0.25 RFLP loci detected by different probes were assigned different numbers using the following convention: Xugo.gk (or cr, or cs) -#, where gk indicates and genomic clone and cr and cs indicate cDNA clones from the root and shoot libraries, respectively. Multiple segregating loci detected by a single probe were assigned the same number followed by a letter (a, b, etc.) to indicate each duplicated locus. In addition to random genomic and cDNA clones, one cDNA clone (Xuga.cc 315 obtained from Dr A. Abbott, Clemson University) representing the gene for stearoyl ACP desaturase has been mapped in peanut. The segregating loci are presently distributed among 11 linkage groups (Fig. 1). However, two of the linkage groups contain only two markers each and are expected to link up with one of the larger linkage blocks as additional markers are added to the map. Eventually, the map should consist of 10 linkage groups corresponding to the haploid chromosome number of diploid *Arachis* spp. To date, a total map distance of 1400 cM has been covered to 20 cM resolution. This is estimated to represent approximately 80% coverage of the peanut genome.

Both conventional RFLP markers and RAPD markers can be useful for expanding the existing peanut map, each having advantages and disadvantages as molecular markers. The main advantages to using RAPD markers are the speed with which large numbers of markers can be added to the map, the relatively small amounts of DNA required for analysis, and the elimination of the need for radioisotopes, restriction enzymes, and the laborious procedure of Southern blotting. Disadvantages of RAPD markers include the production of complex banding patterns with most primers, making

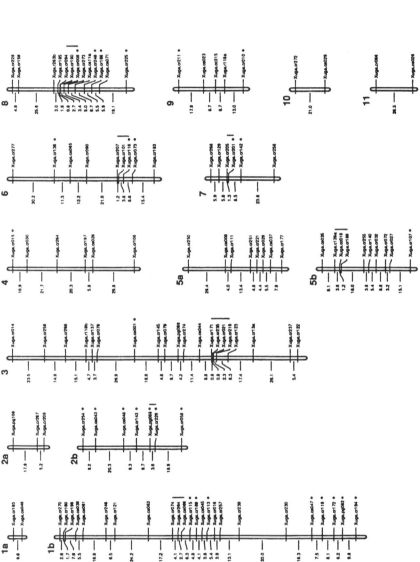

Figure 1. A peanut RFLP map developed using an F₂ population derived from the cross *Arachis stenosperma* x *A. cardenasii*. Loci were ordered using the MAPMAKER program (minimum LOD 3.0, maximum recombination 0.25). RFLP markers were either random genomic clones or cDNA clones. Random genomic clones have 'pg' in the clone designation, cDNAs from a root library have 'cr' and cDNAs from a shoot library have 'cs'. Areas of closely linked markers where the order is uncertain are indicated by a vertical bar next to the affected markers. Markers which segregated with ratios significantly different from the expected 1:2:1 are denoted by a dot (•).

comparisons of mapped markers among populations or laboratories difficult; the degree of reproducibility among different DNA extraction preparations and different researchers is still being debated: and the dominant-recessive (presence or absence of bands) nature of the majority of RAPD markers restricts the information that can be derived from linkage analysis. Conversely, RFLPs behave as co-dominant markers giving the maximum amount of information obtained from a mapping population. Other advantages of the RFLPs as molecular markers are the high degree of reproducibility and the relatively large number of simple banding patterns produced for most organisms evaluated, particularly if libraries have been screened for single copy clones. However, larger amounts of DNA are required for RFLP analysis relative to RAPD marker analysis. The use of radioisotopes, large quantities of restriction enzymes, and Southern blotting procedures are all disadvantages associated with using RFLPs as molecular markers.

One approach to mapping in peanut, for expanding the map as quickly as possible would be to use RFLPs to construct a solid framework for the map and to subsequently fill it in with RAPD markers. Garcia (1995) used an *A. stenosperma* x (*A. stenosperma* x *A. cardenasii*) backcross population and 39 shared RFLPs to place 167 RAPD loci onto the RFLP map. The RAPDs covered a total genetic length of 800 cM and mapped to 11 linkage groups, with all common markers mapping to the expected map location. However, a reduction in the recombination fraction was observed in the RAPD backcross map as compared to the RFLP map which was constructed from an F_2 population. Techniques developed recently for selecting markers in specific chromosomal regions using RAPD markers and pooled DNA samples from segregating populations (Giovannoni et al. 1991; Michelmore et al. 1991) could also be useful for obtaining additional markers in specific areas of interest, such as in the bracketing of QTLs. However, such approaches will not be effective in peanut until a more extensive basic mapping framework is in place.

We are also investigating the use of microsatellite markers for mapping in peanut. Southern blots of peanut DNA digested with restriction enzymes were probed with a number of synthetic oligomers composed of simple nucleotide repeats such as (GT)n or (CAC)n. The initial survey indicated that such microsatellites are present in peanut. When genomic DNA from a diverse group of exotic germplasm lines of *A. hypogaea* was amplified using PCR with microsatellite primers isolated from rice, slight differences in banding patterns were observed among the genotypes evaluated. The random genomic peanut library is being probed with a number of short oligomers to isolate clones containing microsatellites. After sequencing the region found to contain a microsatellite, PCR primers ranking the microsatellite can then be used to convert the markers to sequence tagged sites (STSs) as was successfully done in rice (*Oryza sativa*) (Zhao and Kochert 1992). Microsatellites have potential advantages as molecular markers in that they can be analyzed with simple procedures utilizing PCR amplification with flanking markers, and because they are often more polymorphic than conventional RFLPs or RAPD markers (Edwards et al. 1991; Nanda et al. 1991; Stallings et al. 1991). This latter advantage will be especially important in transferring the presently developing peanut map to cultivated peanut since no other molecular markers evaluated to date have revealed sufficient variability within *A. hypogaea* for mapping applications.

5.   Applications of the map to peanut breeding and genetics

Peanut chromosomes are small with few distinctive cytogenetic markers, making identification of individual chromosomes tedious (Stalker and Dalmacio 1986). In addition, no classical genetic map exists for peanut, and no trait has yet been mapped to a specific peanut chromosome (Stalker 1991). However, Garcia et al. (1996) associated with RAPD marker with two genes conferring root-knot nematode resistance. The marker was subsequently mapped to linkage group 1 of the RFLP map. Burow et al. (1996) identified a second RAPD marker with root-knot nematode resistance derived from another *Arachis* species.

As the peanut RFLP map continues to develop, a number of potential advantages will be realized in peanut improvement programs. Our understanding of genetic segregation analysis and linkage relationships will be greatly enhanced as molecular markers accumulate in peanut. The chromosomal location and distribution of duplicated regions within the peanut genome will allow for a better understanding of the underlying basis for the complex patterns of segregation often observed in peanut, particularly among the progeny of intersubspecific crosses (Wynne and Coffelt 1980; Wynne and Halward 1989).

Given the relatively narrow germplasm base available in *A. hypogaea*, and the abundance of polymorphism observed for both morphological traits and molecular markers in wild species of *Arachis*, increased emphasis should be placed on the utilization of wild relatives in peanut improvement programs. This is perhaps the area for which a comprehensive RFLP map will be of most benefit to peanut breeders. A number of methods have been investigated for their potential utilization in the introgression of traits from related diploid wild species into cultivated peanut (Simpson 1991; Stalker 1992). One such method, the 'hexaploid route', involves making crosses between diploid wild species and the tetraploid cultivated species to produce triploid hybrids. The chromosome number of these sterile hybrids are then doubled with colchicines to form hexaploids, which are subsequently backcrossed to cultivated peanut to produce pentaploid hybrids. Repeated selfing of these pentaploids often results in the spontaneous loss of chromosomes with eventual stabilization of progeny at the tetraploid level (Stalker 1992). In addition to investigations involving the above route of introgression, both amphidiploids and autotetraploids have been constructed among several of the wild diploid species, and these have then been crossed directly with *A. hypogaea* (Singh 1986a, b; Simpson et al. 1993). Some success has been achieved using any of the above methods of introgression as part of a conventional peanut breeding program. In early generations, the progeny resulting from interspecific hybridizations between wild and cultivated peanut species usually appear very different from *A. hypogaea* and retain many of the undesirable traits of the wild parent(s). Subsequent selection for agronomic traits during selfing generations typically results in the loss of desirable characters the breeder was attempting to introgress from the wild parent. The net result is an enormous waste of time and resources. The development of a genetic linkage map in peanut will greatly enhance the ability of breeders to tag and follow the introgression of specific chromosome segments linked to desirable traits from wild species into breeding lines of cultivated peanut.

The developing peanut RFLP map will be useful for monitoring introgression following the mapping of valuable agronomic traits in wild species.

An example of the application of introgression analysis to peanut can be illustrated using an interspecific population which was derived from a cross between *A. hypogaea* x *A cardenasii* that contains highly variable hybrid derivatives which are stable at the tetraploid level (Stalker et al. 1979). Many of the lines appear to have traits that were introgressed from *A. cardenasii* and which are not found in the cultivated parent used for the cross, including large seed size, high yields, and resistance to rust (*Puccinia arachidis* Speg.), early and late leafspot [*Cercospora arachidicola* Hori and *Cercosporidium personatum* (Berk. And Curt.) Deighton, respectively], root-knot nematodes, and a number of insect pests (Stalker 1991). Breeding lines from this population were analyzed for molecular markers, and one to four introgressed DNA segments from *A. cardenasii* were found in each of 46 lines. The smallest segments were detected by a single RFLP marker whereas the largest ones were detected by three or four adjacent markers and represented segments up to 40 cM long. Introgressed segments were detected on 10 of the 11 linkage groups in the RFLP map. The ability to identify and follow specific chromosome segments from wild *Arachis* species through three generations of backcrossing following interspecific hybridization with cultivated peanut has already been demonstrated (Garcia 1995). In addition, Kochert et al. (1996a) proposed a breeding scheme involving the use of molecular marker-assisted interspecific hybridization aimed at the development of multiple populations, each with a number of desirable characters, for general use in peanut breeding programs. The proposed method would involve the development of several breeding populations derived from interspecific hybridizations among various wild *Arachis* species. The chromosome number of F_1 interspecific hybrids would be doubled and then be backcrossed to cultivated peanut. Plants in the BC2 or BC3 generation would be selected and families would subsequently be developed by a combination of selfing and single seed descent for several generations to produce lines that are homozygous for specific introgressed segments which could be detected by molecular marker analysis. Introgressed segments would then be correlated with specific agronomic traits and eventually introgressed into cultivated breeding lines. By providing a means for breeders to screen interspecific populations in the seedling stage, rather than evaluating large segregating populations in the field for the trait of interest, molecular markers will become a valuable tool for enhancing the efficiency of peanut breeding programs involving the introgression of genes from related wild species.

Protocols for transformation in peanut are being investigated in a number of laboratories and it appears that a satisfactory system for peanut transformation and regeneration is now available (Ozias-Akins et al. 1993; Schnall and Weisinger 1996). The potential exists for protocols to be combined with map-based cloning procedures which are being rapidly developed and should become widely applicable (Young 1990; Collins 1991). As these procedures become routinely available, knowing the map location of desirable genes will greatly enhance their utilization in peanut breeding programs, making the peanut RFLP map a valuable resource for use in cultivar improvement programs.

6.   References

Amaya, F., Young, C.T., Hammons, R.O. and Martin, G. (1977) The tryptophan content of the U.S. commercial and some South American genotypes of the genus *Arachis*: a survey. Oleagineux 32:225-229.

Apuya, N., Frazier, B.L., Keim, P., Roth, E.J. and Lark, K.G. (1988) Restriction length polymorphisms as genetic markers in soybean, *Glycine max* (L.) Merr. Theor. Appl. Genet. 75: 889-901.

Badami, V.K. (1928) *Arachis hypogaea* (the groundnut). PhD thesis, Cambridge England University Library Inheritance Studies.

Beckmann, J.S. and Soller, M. (1986) Restriction fragment length polymorphisms and genetic improvement of agricultural species. Euphytica 35: 111-124.

Bernatzky, R. and Tanksley, S.D. (1986) Toward a saturated link-see map in tomato based on isozymes and random cDNA sequences. Genetics 112: 887-898.

Bonierbale, M.W., Plaisted, R.L. and Tanksky, S.D. (1988) RFLP maps based on a common set of clones reveal modes of chromosomal evolution in potato and tomato. Genetics 120: 1095-1103.

Burow, M.D., Simpson, C.E., Paterson, A.H. and Starr, J.L. (1996) Identification of peanut (*Arachis hypogaea* L.) RAPD markers diagnostic of root-knot nematode (*Meloidogyne arenaria* (Neal) Chitwood) resistance. Molecular Breeding 2: 369-379.

Cherry, J.P. (1977) Potential sources of peanut seed proteins and oil in the genus *Arachis*. J. Agric. Food Chem. 25: 186-193.

Coffelt, T.A. and Hammons, R.O. (1973) Influence of sizing peanut seed on two phenotypic ratios. J. Hered. 64: 39-42.

Collins, F.S. (1991) Of needles and haystacks — finding human disease genes by positional cloning. Clin. Res. 39: 615-623.

Edwards, A., Civitello, A., Hammond, H.A. and Caskey, C.T. (1991) DNA typing and genetic mapping with trimeric and tetrameric tandem repeats. Am. J. Hum. Genet. 49: 746-756.

Garcia, G.M. (1995) Evaluating efficiency of germplasm introgression from *Arachis* species to *A. hypogaea* L. PhD Dissertation, N.C. State University, Raleigh.

Garcia, G.M., Stalker, H.T. and Kochert, G.D. (1995) Introgression analysis of an interspecific hybrid population in peanuts (*Arachis hypogaea* L.) using RFLP and RAPD markers. Genome 38: 166-176.

Garcia, G.M., Stalker, H.T., Shroeder, E. and Kochert, G. (1996) Identification of RAPD, SCAR and RFLP markers tightly linked to nematode resistance genes introgressed from *Arachis cardenasii* to *A. hypogaea*. Genome 39: 836-845.

Gepts, P. (1991) Biotechnology sheds light on bean domestication in Latin America. Diversity 7: 49-50.

Giovannoni, J.J., Wing, R.A., Ganal, M.W. and Tanksky, S.D. (1991) Isolation of molecular markers from specific chromosomal intervals using DNA pools from existing mapping populations. Nucl. Acids Res. 19: 6553-6558.

Grieshammer, U. and Wynne, J.C. (1990) Mendelian and non-Mendelian inheritance of three isozymes in peanut (*Arachis hypogaea* L.). peanut Sci. 17: 101-105.

Halward, T.M., Stalker, H.T., LaRue, E.A. and Kochert, G. (1991) Genetic variation detectable with molecular markers among unadapted germplasm resources of cultivated peanut and related wild species. Genome 34: 1013-1020.

Halward, T.M., Stalker, H.T., LaRue, E.A. and Kochert, G. (1992) Use of single-primer DNA amplications in genetic studies of peanut. Plant Mol. Biol. 18: 315-325.

He, G. and Prakash, C.S. (1997) Identification of polymorphic DNA markers in cultivated peanut (*Arachis hypogaea* L.). Euphytica 97: 143-149.

Hong, J.C., Nagao, R.T. and Key, J.L. (1987) Characterization and sequence analysis of a developmentally regulated putative cell wall protein gene isolated from soybean. J. Biol. Chem. 17: 8367-8376.

Hilu, K.W. and Stalker, H.T. (1995) Genetic relationships between peanut and wild species of *Arachis* sect. *Arachis (Fabacaea)*: Evidence from RAPDs. Plant Syst. Evol. 198: 167-178.

Icrisat (1982) Annual Report. Patancheru, A.P., India

Keim, P., Dier, B.W., Olson, T.C. and Shoemaker, R.C. (1990) RFLP mapping in soybean: Association between marker loci and variation in quantitative traits. Genetics 12: 735-742.

Knauft, D.A., Branch, W.D. and Gorbet, D.W. (1991) Two dominant genes for white testa color in peanut. J. Hered. 81: 73-75.

Knauft, D.a. and Ozias-Akins, P. (1995) Recent methodologies for germplasm enhancement and breeding. In: H.E. Pattee and H.T. Stalker (eds.), Advances in Peanut Science, pp. 54-94. Am. Peanut Res. And Educ. Soc., Stillwater OK.

Kochert, G., Halward, T.M., Branch, W.D. and Simpson, C.E. (1991) RFLP variability in peanut cultivars and wild species. Theor. Appl. Genet. 81: 565-570.

Kochert, G.D., Halward, T. and Stalker, H.T. (1996a) Genetic variation in peanut and its implications in plant breeding. In: B. Pickersgill and J.M. Lock (eds.), Advances in Legume Science 8: Legumes of Economic Importance, pp. 19-30. Royal Botanic Gardens, Kew, UK.

Kochert, G., Stalker, H.T., Gimenes, M., Galgaro, L., Lopes, C.R. and Moore, K. (1996b) RFLP and cytogenetic evidence on the origin and evolution of allotetraploid domesticated peanut, *Arachis hypogaea* (Leguminosae). Am. J. Bot. 83: 1282-1291.

Krapovickas, A. (1969) Evolution of the genus *Arachis*. Seminario Advanzado de Genetica Agricola para America Latina, Maracay, Venezuela SAGA/B(d): 1-4.

Krapovickas, A. and Gregor, W.C. (1994) Taxonomia del genero *Arachis* (Leguminosae). Bonplandia 8: 1-186.

Lander, E.S., Green, P., Abrahamson, J., Barlow, A., Daly, M.J., Lincoln, S.E. and Newburg, L. (1978) MAPMAKER: An interactive computer package for constructing primary genetic linkage maps of experimental and natural populations. Genomics 1: 176-181.

Lu, J. and Pickersgill, B. (1993) Isozyme variation and species relationships in peanut and its wild relatives (*Arachis* L. — Leguminosae). Theor. Appl. Genet. 85: 550-560.

Lynch, R.E. and Mack, T.P. (1995) Biological and biotechnical advances for insect management in peanut. In: H.E. Pattee and H.T. Stalker (eds.), Advances in Peanut Science, pp. 95-159. Am. Peanut Res. And Educ. Soc., Stillwater, OK.

McCouch, S.R., Kochert, G., Yu, Z.G., Wang, Z.Y., Khush, G.S., Coffman, W.R. and Tanksley, S.D. (1988) Molecular mapping of rice chromosomes. Theor. Appl. Genet. 76: 815-829.

Michelmore, R.W., Paran, I. And Kesseli, R.V. (1991) Identification of markers linked to disease resistance genes by bulked segregant analysis — a rapid method to detect markers in specific genomic regions by suing segregating populations. Proc. Natl. Acad. Sci. USA, 88: 9828-9832.

Miller, J.c. and Tanksley, S.D. (1990) RFLP analysis of phylogenetic relationships and genetic variation in the genus *Lycopersicon*. Theor. Appl. Genet. 10-437-448.

Murthy, T.G.K., Tiwari, S.P. and Reddy, P.S. (1988) A linkage group for genes governing pod characters in peanut. Euphytica 39: 43-46.

Nanda, I., Zischler, H., Epplen, C., Guttenbach, M. and Schmid, M. (1991) Chromosomal orga-nization of simple repeated DNA sequences used for DNA fingerprinting. Electrophoresis 12; 193-203.

Ozias-Akins, P., Schnall, J.A., Anderson, W.F., Singsit, C., Clemente, T.E., Adang, M.J. and Weissinger, A.K. (1993) Regeneration of transgenic peanut plants from stably transformed embryogenic callus. Plant Science 93: 185-194.

Paik-Ro, O.G., Smith, R.A. and Knauft, D.A. (1992) Restriction fragment length polymorphism evaluation of six peanut species within the *Arachis* section. Theor. Appl. Genet. 84: 201-208.

Patel, J.S., John, C.M. and Seshadri, C.R. (1936) The inheritance of characters in the groundnut. Proc. Indian Acad. Sci. 3: 214-233.

Patil, V.H. (1965) Genetic studies in groundnut (*Arachis hypogaea* L.). MSc thesis, Poona University, India.

Rigby, P., Dieckmann, M., Rhodes, C. and Berg, P. (1977) Labeling deoxyribonucleic acid to high specific activity in vitro by nick-translation with DNA polymerase I. J. Mol. Biol. 113: 236-351.

Sambrook, J., Fritsch, E.F. and Maniatis, T. (1989) Molecular Cloning: A Laboratory Manual. Cold Spring Harbor Laboratory Press, Cold Spring Harbor, NY.

Schnall, J.A. and Weissinger, A.K. (1996) Genetic transformation in *Arachis hypogaea* L. (Peanut). Biotechnology in Agriculture and Forestry, vol. 34.

Shattuck-Eidens, D.M., Bell, R.N., Neuhausen, S.L. and Helentjaris, T. (1990) DNA sequence variation within maize and melon: Observations from polymerase chain reaction amplifica-tion and direct sequencing. Genetics 126: 207-217.

Simpson, C.E. (1991) Global collaborations find and conserve the irreplaceable genetic resources of wild peanut in South America. Diversity 7: 59-61.

Simpson, C.E., Starr, J.L., Nelson, S.C., Woodard, K.E. and Smith, O.D. (1993) Registration of TxAG-6 and TxAG-7 peanut germplasm. Crop Sci. 33: 1418.

Singh, A.K. (1986a) Utilization of wild relatives in the genetic improvement of *Arachis hypogaea* L. 7. Autotetraploid production and prospects in interspecific breeding. Theor. Appl. Genet 72: 164-169.

Singh, A.K. (1986b) Utilization of wild relatives in the genetic improvement of *Arachis hypogaea* L. 8. Synthetic amphidiploids and their importance in interspecific breeding. Theor. Appl. Genet. 72: 433-439.

Slocum, M.K, Figdor, S.S., Kennard, W.C., Suzuki, J.Y. and Osborn, T.C. (1990) Linkage arrangement of restriction fragment length polymorphism loci in *Brassica oleracea*. Theor. Appl. Genet. 80: 56-64.

Smartt, J. and Stalker, H.T. (1982) Speciation and cytogenetics in *Arachis*. In: H.E. Pattee and C.T. Young (eds.), Peanut Science and Technology, pp. 21-49. Am. Peanut Res. Educ. Soc., Yoakum, TX.

Song, K.M., Suzuki, J.Y., Slocum, M.K., Williams, P.H. and Osborn, T.C. (1991) A linkage map of *Brassica rapa* (syn. *Campestris*) based on restriction fragment length polymorphism loci. Theor. Appl. Genet. 82: 296-304.

Southern, E.M. (1975) Detection of specific enzyme sequences among DNA fragments separated by gel electrophoresis. J. Mol. Biol. 98: 503-517.

Stalker, H.T. (1991) A morphological appraisal of wild species in section *Arachis* of peanuts. Peanut Sci. 17: 117-122.

Stalker, H.T. (1992) Utilizing *Arachis* germplasm resources. In: Proc. 2nd Intl. Workshop on Groundnut, pp. 24-29. ICRISAT, Patancheru, A.P. India.

Stalker, H.T. and Dalmacio, E.D. (1986) Karyotype analysis and relationships among varieties of *Arachis hypogaea* L. Cytologia 51: 617-629.

Stalker, H.T., Dhesi, J.S. and Kochert, G. (1995) Genetic diversity within the species *Arachis duranensis* Krapov. and W.C. Gregory, a possible progenitor of cultivated peanut. Genome 38: 1201-1212.

Stalker, H.T. and Moss, J.P. (1987) Speciation, cytogenetics, and utilization of *Arachis* species. Adv. Agron. 41: 1-40.

Stalker, H.T., Phillips, T.G., Murphy, J.P. and Jones, T.M. (1994) Diversity of isozyme patterns in *Arachis species*. Theor. Appl. Genet. 87: 746-755.

Stalker, H.T., Valls, J.F.J., Pittman, R.W., Simpson, C.E. and Bramel-Cox, P. (1998) Germplasm Catalog of *Arachis* Species. International Crops Research Institute Semi-Arid Tropics, Hyderabad, India.

Stalker, H.T., Wynne, J.C. and Company, M. (1979) Variation in progenies of an *Arachis hypogaea* (diploid wild species hybrid). Euphytica 28: 675-684.

Stalker, H.T., Young, C.T. and Jones, T.M. (1989) A survey of the fatty acids of peanut species. Oleagineux 44: 419-424.

Stallings, R.L., Ford, A.F., Nelson, D., Torney, D.C., Hildebrand, C.E. and Moyzis, R.K. (1991) Evolution and distribution of (GT)n repetitive sequences in mammalian genomes. Genomics 19: 807-815.

Valls. J.F.M., Rao, V.R., Simpson, C.E. and Krapovickas, A. (1985) Current status of collection and conservation of South American groundnut germplasm with emphasis on wild species of *Arachis*. In: Proc. Intl. Workshop Cytogenetics of *Arachis*, pp. 15-33. ICRISAT, Patancheru, A.P., India.

Weissinger, A.K. (1991) Biotechnology for improvement of peanut (*Arachis hypogaea* L.). In: Proc. Intl. Workshop Cytogenetics. *Arachis* ICRISAT, Patancheru, A.P., India

Wilimzig, R. (1985) LiCl method for plasmid minipreps. Trends Genet. 1: 158.

Williams, D.E. (1996) Aboriginal farming system provides clues to groundnut evolution. In: B. Pickersgill and J.M. Lock (eds.), Advances in Legume Systematics 8: Legumes of Economic Importance, pp. 11-17. Royal Botanic Gardens, Kew.

Wynne, J.C. and Coffelt, T.A. (1982) Genetics of *Arachis hypogaea*. In: H.E. Pattee and C.T. Young (eds.), Peanut Science and Technology, pp. 50-94. Am. Peanut Res. Educ. Soc., Yoakum, TX.

Wynne, J.C. and Halward, T.M. (1989) Cytogenetics and genetics of *Arachis*. Crit. Rev. Plant Sci. 8: 189-220.

Young, C.T., Waller, G.R. and Hammons, R.O. (1973) Variations in total amino acid content of peanut meal. J. Am. Oil Chem. Soc. 50: 521-523.

Young, N.D. (1990) Potential applications of map-based cloning to plant pathology. Physiol. Mol. Plant Pathol. 37: 81-94.

Zhao, X.P. and Kochert, G. (1992) Characterization and genetic mapping of a short, highly repeated, interspersed DNA sequence from rice (*Oryza sativa*). Mol. Gen. Genet. 231(3): 353-359.

17. *Phaseolus vulgaris* — The common bean integration of RFLP and RAPD-based linkage maps

C. EDUARDO VALLEJOS[1],
PAUL W. SKROCH[2], *and* JAMES NIENHUIS[2]
*[1]Department of Horticultural Sciences, and Graduate Program in Plant Molecular and Cellular Biology. 1143 Fifield Hall. University of Florida. Gainesville, FL 32611. U.S.A.
<vallejos@ufl.edu>
[2]Department of Horticulture. 1575 Linden Drive. University of Wisconsin. Madison, WI 53706. U.S.A.*

Contents

1. Introduction

Gregor Mendel (1866) conducted the first genetic analysis of common beans. Mendel studied the inheritance of growth habit, and pod color and shape in a progeny between *P. vulgaris* and *P. nanus* (= *P. vulgaris*, bush type) in order to confirm his findings with peas. Unfortunately, further studies on the inheritance of flower and seed coat color were hampered by his use of interspecific hybrids between *P. nanus* and *P. multiflorus* (= *P. coccineus*), which are now known to yield aberrant ratios. Later, Shaw and Norton (1918) used intraspecific crosses and determined that pigmentation and pigmentation patterns of the seed coat are controlled by multiple independent factors. A few years later Sax (1923) began to identify the multiple components that determine the inheritance of these traits. A single factor was identified as responsible for pigmentation, while two linked factors were identified to control mottling; this appears to be the first report of linkage in beans. Furthermore, Sax (1923) was the first to report linkage between a Mendelian character (seed coat pigmentation) and a QTL (for seed size). Although the common bean was used as experimental material at the inception of genetics, its genetic characterization has lagged behind that of many other crop species.

The common bean is a diploid organism (n = 11) with relatively small chromosomes (Zheng et al. 1991) and a small genome estimated by flow cytometry to be 637 Mbp or 0.66 pg/1C (Arumuganathan and Earle 1991). It has also been estimated, via DNA

R.L. Phillips and I.K. Vasil (eds.), DNA-Based Markers in Plants, 301 - 317.
© 2001 *Kluwer Academic Publishers. Printed in the Netherlands.*

reassociation kinetics, that 60% of the genome is comprised of single copy sequences (Talbot et al. 1984). The chromosome number and genome size of *P. vulgaris* are very similar to those of *P. acutifolius* and *P. coccineus* (Arumuganathan and Earle 1991), both of which are partially compatible with the common bean and represent an important source of germplasm for plant improvement (Hucl and Scoles 1985). A rudimentary linkage map has been developed through the years with mostly morphological markers and a few isozymes (Bassett 1991; Vallejos and Chase 1991a, b). Electrophoretic analysis of the major seed storage protein (phaseolin) and a group of isozymes has led to the identification of a Mesoamerican and an Andean gene pool (Gepts et al. 1986; Koenig and Gepts 1989). Moreover, preliminary survey of DNA restriction fragment length polymorphisms showed that DNA probes can be used to differentiate the two groups because low levels of polymorphism were detected within each gene pool, but moderate levels were found between the gene pools (Chase et al. 1991).

The first RFLP-based linkage map of the common bean was constructed with 250 markers that were distributed among 11 linkage groups and covered 960 cM of the genome (Vallejos et al. 1992). Two smaller maps were reported later, one with 152 markers that assorted into 15 linkage groups spanning over 827 cM (Nodari et al. 1993), and another with 157 markers (51 RFLPs and 100 RAPDs, and 6 other markers) distributed among 12 linkage groups covering 568 cM (Adam-Blondon et al. 1994). The mapping populations for the construction of these maps were obtained from crosses between representatives of the Andean and the Mesoamerican gene pools. The choice of parental genotypes for these mapping populations was dictated by the low level of restriction fragment length polymorphisms present within each gene pool (Chase et al. 1991).

Among PCR-based markers (Chapter 3), RAPD markers (Williams et al. 1990) offer the advantage of technical simplicity and lower cost over conventional RFLP technology. RAPD markers can not only easily distinguish Andean from Measoamerican accessions (Johns et al. 1997), but can also effectively detect polymorphisms within a gene pool (Beebe et al. 1995; Skroch and Nienhuis 1995). These studies have demonstrated that RAPD markers have a greater diversity index than RFLPs in beans, a similar situation has been shown in other species (Milbourne et al. 1997, Williams and St. Clair 1993). Further demonstration of the higher diversity index of RAPD markers is given by the recent construction of two small RAPD-based linkage maps of the common bean. The first of these maps was constructed with a recombinant inbred family obtained between two Mesoamerican genotypes and contained 75 RAPD markers distributed among 9 linkage groups covering 545 cM (Jung et al. 1996). The second map was constructed with an RI family obtained between two Andean genotypes, and comprised of 168 RAPD markers distributed among 10 linkage groups covering 426 cM (Jung et al. 1997). A larger RAPD-based linkage map has recently been constructed with an RI family obtained between representatives of the two gene pools. This map comprises 361 markers distributed in 11 linkage groups covering 825 cM (Skroch and Nienhuis, in preparation). We present here an alignment of this RAPD-based linkage map with the previously constructed RFLP-based linkage map. The alignment of these maps was obtained by constructing a new linkage map using previously mapped RAPD and RFLP markers. Each of these maps is described in the following section.

2.   Construction of three linkage maps

2.1.   The RFLP-based XC map

The RFLP-based linkage map was constructed using a backcross progeny between XR-235-1-1 and Calima, a Mesoamerican and an Andean breeding line, respectively (Vallejos et al. 1992). A genomic library of size-selected (500-4000 bp) *Pst*I fragments was the main source of probes (Chase et al. 1991). This library was enriched for single copy fragments, 95% of the 362 clones tested yielded hybridization patterns typical of single copy sequences. Sixty percent of the clones tested revealed polymorphisms between the parental genotypes with at least one of four restriction enzymes (*Dra*I, *Eco*RI, *Eco*RV, and *Hin*dIII). Subsequently, 28 clones that had not detected RFLPs with any of the previously used enzymes were tested with four new enzymes: *Bam*HI, *Bgl*II, *Kpn*I, and *Xba*I. About half of these clones revealed polymorphisms with at least one of the new enzymes. These results bring the estimated polymorphism between the parental genotypes at 80% with at least one of the 8 restriction enzymes. These enzymes differed in their ability to detect polymorphisms between the parental genotypes: *Kpn*I (40%); *Dra*I (42%); *Bam*HI, *Bgl*II and *Hin*dIII (53%); *Eco*RI (62%); and *Eco*RV (64%).

Segregation data obtained from a backcross between the Mesoamerican breeding line 'XR-235-1-1' and the Andean cultivar 'Calima' were used to assemble the first linkage map of the common bean (Vallejos et al. 1992). After the publication of this map, 36 loci detected with mungbean clones (Young et al. 1992; Boutin et al. 1995), and nine additional loci of known sequences have been analyzed in the same population. Linkage analysis was performed using **Mapmaker/EXP 3.0** (Lander et al. 1987; Lincoln et al. 1992). This map was constructed by first sorting the markers into linkage groups using the "group" command with stringent linkage criteria (LOD 4.0 and maximum distance of 40 cM). The markers were ordered with the "order" command after setting the multipoint criteria with a LOD threshold of 3.0, a strict threshold of 4, and a window size of 7 markers. Finally, additional markers were incorporated into the map using the "try" command with an exclusion LOD threshold of 2.0. Markers that could not be uniquely placed on the map with a log-likelihood greater than 2.0 were drawn by the closest "framework" marker (Fig. 1).

The linkage between *Bng205a* and *Bng7* in group *F* has a LOD score of 2.22; although weak, this linkage is supported by data from two different progenies (Vallejos et al. 1992). This map comprises 294 marker loci that are distributed among 11 linkage groups (n = 11) and covers 900 cM. This is less than the total length reported previously because this time markers that could not be uniquely placed on the map (LOD > 2.0) were not inserted. The linkage groups range in length from 64 to 109 cM (Kosambi). Of these markers, 230 were detected with 221 genomic clones (Bng's) (Table 1). Nine of them hybridized to homologous sequences located in different linkage groups. The development of the bean genomic clones at the University of Florida was financed in part by a grant from the Agency for International Development (AID). These clones have been transferred to the International Center for Tropical Agriculture (CIAT) in Cali, Colombia, and can be requested from its Biotechnology Research Unit.

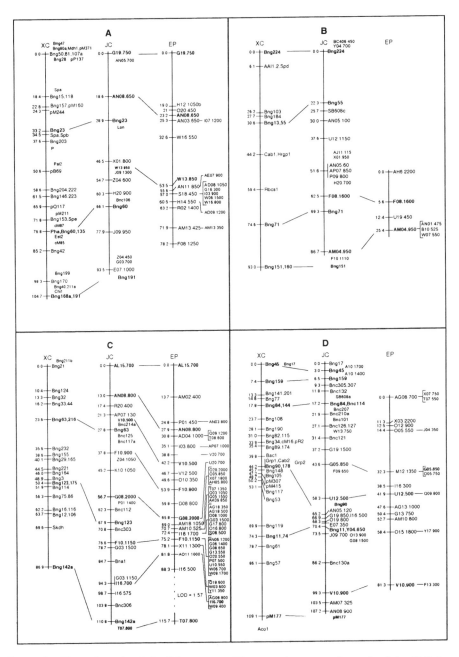

Figure 1. *Phaseolus vulgaris* L. linkage maps. There are three maps for each of the 11 linkage groups (*A-K*). The XC map was constructed with a BC$_1$ progeny from the XR-235-1-1 x Calima cross, the JC map was constructed with an RI family from the Jamapa x Calima cross, and the EP map with an RI family from the Eagle x Puebla 152 cross. Framework markers have been included in each of these maps; markers that could not be uniquely place with a LOD value of

Continued.

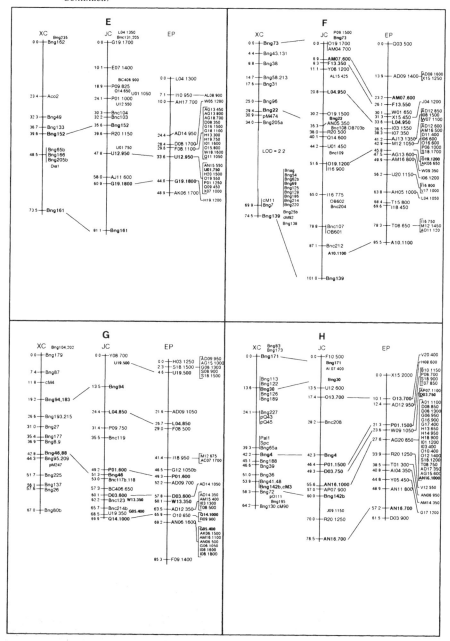

2.0 or higher were drawn next to the closest marker. Bng markers in the JC map were used to align this map with the RFLP-based XC map, these markers are written in bold typeface and lines between the maps indicate the alignment. Likewise, RAPD markers in the JC map that had been mapped in the EP map are written in bold typeface in both maps, and the alignments are indicated by lines between the maps.

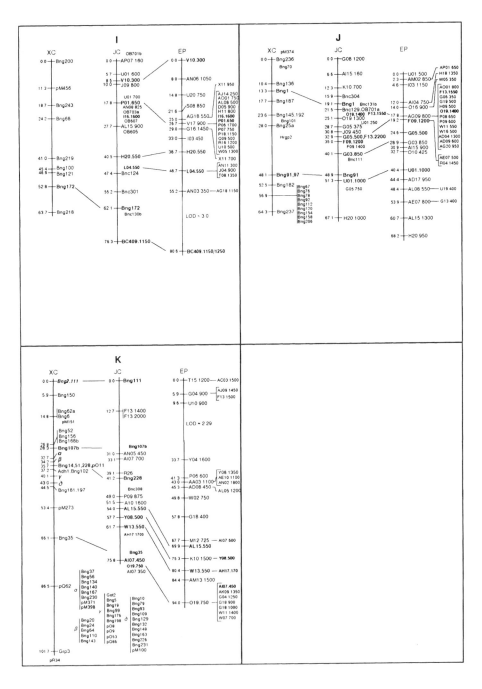

Figure 1.   Continued.

Table 1. List of loci identified with genomic clones (pBngN), their corresponding linkage group association, and relative position. The insert size of each clone is also included.

Locus	Linkage Group	Relative Position	Insert size (bp)	Locus	Linkage Group	Relative Position	Insert size (bp)
Bng1	*J*	3	1790	*Bng14*	*K*	9	3370
Bng2	*K*	1	1800	*Bng15*	*A*	6	3220
Bng3	*C*	12	1200	*Bng16*	*C*	16	2040
Bng4	*H*	7	1270	*Bng17*	*D*	2	1600
Bng5	*K*	12	1470	*Bng18*	*F*	10	1100
Bng6	*K*	3	1400	*Bng19*	*K*	12	2370
Bng7	*F*	9	2800	*Bng20*	*K*	7	1430
Bng8	*G*	8	800	*Bng21*	*C*	2	1980
Bng9	*G*	8	4510	*Bng22*	*F*	7	1780
Bng10	*K*	13	4190	*Bng23*	*A*	8	620
Bng11	*D*	19	3180	*Bng24*	*K*	7	1070
Bng12	*C*	17	2700	*Bng25*	*J*	7	1100
Bng13	*B*	5	1560	*Bng26*	*G*	13.	1960
Bng27	*G*	6	1260	*Bng43*	*F*	2	1060
Bng28	*A*	4	2110	*Bng44*	*C*	5	970
Bng29	*C*	9	1120	*Bng45*	*D*	1	890
Bng30	*H*	4	1350	*Bng46*	*G*	9	2370
Bng31	*F*	5	1130	*Bng47*	*A*	2	2480
Bng32	*C*	4	950	*Bng48*	*H*	11	1790
Bng33	*C*	5	1100	*Bng49*	*E*	4	1320
Bng34	*D*	11	4330	*Bng50*	*A*	3	850
Bng35	*K*	16	1020	*Bng51*	*K*	10	1150
Bng36	*H*	10	2500	*Bng52*	*K*	4	1140
Bng37	*K*	6	1250	*Bng54*	*F*	11	1170
Bng38	*F*	3	1800	*Bng55*	*B*	5	1650
Bng39	*H*	9	2300	*Bng56*	*K*	6	2480
Bng40	*A*	21	800	*Bng57*	*D*	21	2090
Bng41	*H*	11	2780	*Bng58*	*F*	4	2400
Bng42	*A*	18	1500	*Bng60*	*A*	16	2290
Bng61	*D*	20	1580	*Bng76*	*J*	10	3120
Bng62a	*K*	3	2310	*Bng77*	*D*	5	2500
Bng62b	*F*	11	2310	*Bng78*	*J*	10	2270
Bng63	*C*	6	1560	*Bng79*	*K*	13	2610
Bng64	*K*	7	1620	*Bng80a*	*A*	1	1300
Bng65a	*H*	6	2080	*Bng80b*	*G*	14	1300
Bng65b	*E*	7	2080	*Bng81*	*A*	3	1090
Bng67	*J*	10	2170	*Bng82*	*D*	10	1480
Bng68	*I*	3	2530	*Bng83*	*H*	1	870
Bng69	*F*	11	1000	*Bng84*	*D*	6	2190

Table 1. Continued.

Locus	Linkage Group	Relative Position	Insert size(bp)	Locus	Linkage Group	Relative Position	Insert size (bp)
Bng70	J	1	1970	Bng86	C	15	1650
Bng71	B	7	2240	Bng87	G	3	3350
Bng72	H	12	3880	Bng88	G	9	2510
Bng73	F	1	2410	Bng89	D	12	840
Bng74	D	19	650	Bng90	D	13	1850
Bng75	C	15	1600	Bng91	J	8	1470
Bng92	J	10	1130	Bng107b	K	5	2560
Bng93	K	13	800	Bng108	D	8	1540
Bng94	G	4	1460	Bng109	K	13	1930
Bng95	G	10	1580	Bng110	K	7	1270
Bng96	F	6	1140	Bng111	K	1	1970
Bng97	J	8	1380	Bng112	J	10	820
Bng98	D	17	1750	Bng113	H	4	660
Bng99	K	12	1260	Bng114	C	14	1540
Bng100	I	5	2160	Bng115	D	10	1930
Bng101	J	6	670	Bng116	C	16	1930
Bng102	K	11	1400	Bng117	D	16	1430
Bng103	B	3	1410	Bng118	A	6	1760
Bng104	G	1	1830	Bng119	D	18	2220
Bng105	D	15	1260	Bng120	J	10	1520
Bng106	C	17	2680	Bng121	I	6	1710
Bng107a	A	3	2560	Bng122	H	4	1330
Bng123	C	13	1180	Bng140	K	6	1780
Bng124	C	3	1270	Bng141	D	4	1340
Bng125	F	11	1140	Bng142a	C	19	3470
Bng126	H	4	2510	Bng142b	H	11	3470
Bng128	F	11	2340	Bng143	K	8	950
Bng129	K	13	1710	Bng144	D	6	2220
Bng130	H	14	950	Bng145	J	5	1930
Bng131	F	2	1190	Bng146	A	14	1550
Bng132	K	13	1120	Bng148	D	14	2220
Bng133	E	5	3120	Bng149	K	13	820
Bng134	K	6	3830	Bng150	K	2	1940
Bng135	A	16	3000	Bng151	B	8	1030
Bng136	J	2	1190	Bng152	E	6	1070
Bng137	G	12	1710	Bng153	A	15	1900
Bng138	F	9	1320	Bng154	J	10	1050
Bng139	F	12	2540	Bng155	C	8	2810
Bng156	K	4	2500	Bng172	I	7	3120
Bng157	A	7	1310	Bng173	H	2	2050

Table 1. Continued.

Locus	Linkage Group	Relative Position	Insert size (bp)	Locus	Linkage Group	Relative Position	Insert size (bp)
Bng158	J	10	2310	*Bng174*	D	12	2600
Bng159	D	3	3210	*Bng175*	C	13	880
Bng160	B	8	2280	*Bng176*	K	12	3120
Bng161	E	9	1290	*Bng177*	G	7	930
Bng162	E	2	590	*Bng178*	D	13	1310
Bng163	K	13	2180	*Bng179*	G	2	2240
Bng164	C	11	1480	*Bng180*	D	17	2550
Bng165	C	9	2260	*Bng181*	K	15	1770
Bng166	E	7	2170	*Bng182*	J	9	1180
Bng167	K	6	520	*Bng183*	G	4	1210
Bng168a	A	22	2210	*Bng184*	B	4	530
Bng168b	K	4	2210	*Bng186*	F	11	1350
Bng170	A	20	1170	*Bng187*	J	4	1860
Bng171	H	3	1900	*Bng188*	H	8	1960
Bng189	H	4	2030	*Bng206*	J	10	1290
Bng190	D	9	1330	*Bng209*	G	10	1020
Bng191	A	22	2030	*Bng211a*	A	21	1810
Bng192	J	5	1300	*Bng211b*	C	1	1810
Bng193	G	5	2620	*Bng212*	D	16	1380
Bng195	H	13	2700	*Bng213*	F	4	2180
Bng197	K	14	2560	*Bng214*	F	11	2330
Bng198	K	12	2140	*Bng215*	G	5	1350
Bng199	A	19	1230	*Bng216*	C	6	2040
Bng200	I	1	2340	*Bng218*	I	8	2020
Bng201	D	4	1320	*Bng219*	I	4	580
Bng202	G	1	1620	*Bng220*	F	11	4250
Bng203	A	10	2990	*Bng221*	C	10	2100
Bng204	A	12	1420	*Bng222*	A	12	2480
Bng205a	F	8	1350	*Bng223*	A	13	2350
Bng205b	E	7	1350	*Bng224*	B	1	880
Bng225	G	11	1030	*Bng231*	K	13	2100
Bng226	K	13	2220	*Bng232*	C	7	1770
Bng227	H	5	700	*Bng234*	I	2	900
Bng228	K	9	1840	*Bng235*	E	1	2030
Bng230	K	6	500	*Bng236*	J	1	nd

An additional set of 34 loci were identified with the same number of mungbean clones obtained from Nevin Young, University of Minnesota (Young et al. 1992). These clones (Table 2) were used as part of a project designed to compare the genome structures of mungbean, the common bean, and soybean (Boutin et al. 1995). Twelve loci of known sequences have also been mapped in this population and are listed in Table 3. Also included in this map are: one phenotypically identified pigmentation locus (*P*), nine isozymes, and nine seed proteins (Table 4).

Table 2. List of loci detected with mungbean clones.

Locus	Linkage Group	Locus	Linkage Group	Locus	Linkage Group	Locus	Linkage Group
pM371	A	*pR2*	D	*pQ45*	H	*pR26*	K
pP137	A	*pM307*	D	*cM3*	H	*pO8*	K
pM160	A	*pM415*	D	*pO111*	H	*pO9*	K
pM244	A	*pM177*	D	*cM90*	H	*pO53*	K
pB69	A	*pM474*	F	*pM456*	I	*pQ86*	K
pQ117	A	*cM11*	F	*pM374*	J	*pM100*	K
pM211	A	*cM92*	F	*pM151*	K	*pR34*	K
cM87	A	*cM4*	G	*pM371*	K		
cM85	A	*pM247*	G	*pM398*	K		
cM16	D	*pQ43*	H	*pO11*	K		

Table 3.   List of known sequences.

Sequence Name	Probe	Locus	Group	Reference
Bean abscission cellulase	pBAC1	*Bac1*	D	Tucker et al., 1988
Chalcone isomearse	pCHI	*Chi1*	A	Mehdy and Lamb, 1987
Chlorophyll a/b binding protein	pMB123	*Cab1*	B	Thompson et al., 1983
Chlorophyll a/b binding protein	pMB123	*Cab2*	D	Thompson et al., 1983
Glycine-rich CW proteins	GRPp211	*Grp1*	D	Keller et al., 1988
Glycine-rich CW proteins	GRPp211	*Grp2*	D	Keller et al., 1988
Glycine-rich CW proteins	GRPp211	*Grp3*	K	Keller et al., 1988
Hydroxyproline-rich glycoprotein	HRGP4.1	*Hrgp1*	B	Sauer et al., 1990
Hydroxyproline-rich glycoprotein	HRGP4.1	*Hrgp2*	J	Sauer et al., 1990
Lon protease	pAtLon	*Lon*	A	Sarria et al., 1988
Phenylalanine Ammonia Lyase	pPAL1	*Pal1*	H	Edwards et al., 1985
Phenylalanine Ammonia Lyase	pPAL1	*Pal2*	A	Edwards et al., 1985
Ribulose 1,5 bis P carboxylase	pPvSS191	*Rbcs1*	B	Knight and Jenkins, 1992

Table 4. List of Protein Marker Loci: Isozymes and Seed Storage Proteins.

Isozyme Locus	Linkage Group	Relative Position	References[1]	Seed Protein Locus	Linkage Group	Relative Position	References
Aco1	D	-	1,3	*AAI1*	B	2	3
Aco2	E	3	1,3	*AAI2*	B	2	3
Adh1	K	9	1,3	*Pha*	A	15	2, 3
Bnag	F	11	1,3	*Spa*	A	9	2, 3
Dia1	E	8	1,3	*Spb*	A	9	2, 3
Est2	A	16	1,3	*Spba*	A	5	2, 3
Got2	K	10	1,3	*Spc*	H	6	2, 3
Mdh1	A	1	1,3	*Spd*	B	2	2, 3
Skdh	C	18	1,3	*Spe*	A	14	2, 3

[1] 1, Vallejos and Chase, 1991a; 2, Vallejos and Chase, 1991b; 3, Vallejos et al., 1992.

The seed proteins are: phaseolin, the α-amylase inhibitor proteins (identified by western blots with antibody provided by M. Chrispeels (Moreno and Chrispeels 1989), and other globulins (Vallejos and Chase 1991b). Electrophoretic variation at the protein lvel could be due to either variation at the DNA level and/or variation in post-translational events-protease processing and/or glycosylation. Nevertheless, Southern analysis of genomic blots with a phaseolin clone (Sun et al. 1981) has shown perfect co-segregation between a restriction fragment identified by this clone and phaseolin protein bands identified by SDS-PAGE. Thus, these results strongly suggest that variation at the protein level in phaseolin is due to variation at the DNA level and that the locus mapped with protein data corresponds to the structural gene of phaseolin.

2.2. The RAPD-based EP map

This map was constructed with a recombinant inbred population of 72 lines (Burr et al. 1988) at the F_7 level, and generated from the cross Eagle (Andean) x Puebla 152 (Mesoamerican). RAPD analysis was conducted essentially as described by Skroch and Nienhuis (1995), except that the concentration of xylene cyanole in the reaction was lowered from 0.02 to 0.01%. RAPD primer kits A-Z and AA-AP were purchased from Operon Technologies (Alameda, CA.). The sequences for the primers can be obtained at the following address: http://www.operon.com. PCR reactions were carried out in a 10 μl volume dispensed in 96 well V bottom plates sealed with Microseal 'A' film in an MJ Research PTC 100 thermocycler, fitted with a Hot Bonnet lid (MJ Research, Inc. Watertown, MA). Cycling temperature settings were 91°C for denaturation, 42°C for primer annealing and 72°C for elongation. For the first two cycles denaturation was 60 sec, annealing for 15 sec, and elongation for 70 sec. For the subsequent 38 cycles, denaturation was 15 sec, annealing 15 sec, and elongation 70 sec. RAPD products were separated on agarose gels and data scored as described by Skroch and Nienhuis (1995). Markers that could not be verified in RAPD data replicated for parental genotypes were discarded from all subsequent analyses. Markers were assigned names based on the single or double letter and number identifier from the Operon primer name followed by a period and the estimated size of the amplicon. A 100 bp ladder (Life Technologies, Gaithersburg, MD) was used to size amplicons in the gels to the nearest 50 base pairs.

Linkage analysis was carried out with Mapmaker 2.0 (Lander et al. 1987) for the Macintosh. Linkage groups were established with a minimum LOD for linkage of 3.0 and a maximum theta of 0.12 using the 'Group' command. For each linkage group, map construction was initiated with up to 6 markers that could be ordered with a log-likelihood difference, between the most probable and second most probable order, of at least 3.0. Markers were then added using Mapmaker's 'Try' command. All markers were placed with a minimum LOD threshold of 2. A marker had to have an estimated map distance of at least 1 cM from a previously mapped marker to be considered a separate locus. Map distances were based on the Kosambi mapping function.

A total of 361 RAPD markers have been assigned to 11 linkage groups (n = 11) and cover a span of 825 cM, a larger genome coverage than that provided by previous RAPD-based linkage maps. About 150 markers were assigned to a framework; the rest of the markers could not be assigned a unique map position with an LOD greater than 2.0 and were drawn on a column to the right of the framework markers (Fig. 1). The association (<1cM) of these markers to the closest framework marker is indicated by a line. An interesting feature of this map is the clustering of markers in eight of the linkage groups.

2.3.   The JC map for integration of the RFLP and RAPD maps

This map was constructed with a RI family (Burr et al. 1988) at the F_8 level, that was generated from the cross between the Mesoamerican genotype Jamapa and the Andean genotype Calima. The construction of this map followed the same strategy described for the RFLP-based linkage map (see Section 2.1). This map contains a total of 243 markers: 34 Bng loci, 2 mungbean loci, 42 bean cDNAs (Bnc loci), 10 resistance gene analogs (RGA) (S/OB600s, Rivkin et al. 1998; S/OB700s, Rivkin, McClean and Vallejos, unpublished results), 1 known sequence, and 155 RAPD markers (Fig. 1). These markers were assigned to 11 linkage groups and cover 950 cM of the bean genome.

The JC map has been used to align and integrate the RFLP and RAPD-based linkage maps. This has been accomplished by including in each JC linkage group an average of three BNG markers from the XC map, and four to five RAPD markers placed from the EP map. Marker loci placed in two adjacent maps have been indicated in bold typeface and are joined by lines (Fig. 1). The JC map has helped solve the problem of gaps or weak linkages in the RFLP and RAPD maps indicated by thin or broken lines. These linkages were found in linkage group *F* of the XC map and in linkage groups *C*, *I*, and *K* of the EP map. Taking the three maps into consideration, a total of 814 markers have been placed in the common bean genome; these include nine seed proteins, nine isozymes, 230 Bng markers, 37 mungbean markers, 42 Bnc (cDNA) markers, ten RGA markers, 13 known sequences, and 464 RAPD markers. The identity of RAPD markers present in both the JC and EP maps was established by the specificity of the primer, the size of the amplicon, and the relative map position in each of the maps. Further progress on the integrated map should include the development of sequence tagged sites (STS) (Olson et al. 1989) from both RFLP and RAPD markers. One of the problems with RAPD markers is that it is difficult to predict the fragments that will amplify in any given genotype. The generation of STSs from the mapped RAPD markers can overcome this problem. An extensive on-line *P. vulgaris* STS database can provide researchers with the ability to develop PCR markers for any genome region without the need to access any biological material.

3.   Some applications of molecular markers in common bean

Although the current and potential uses of molecular markers have been treated in detail in the first chapters of this volume, we would like to point out three targets of interest in common beans: identification of genetic factors that affect gene flow between gene pools, analysis of QTLs that affect yield, and tagging disease resistance genes for plant breeding purposes and direct genomic cloning.

Mesoamerica and the northern Andean region of South America have been identified as the two major centers of diversity (Gepts et al. 1986; Koenig and Gepts 1989). In addition to the group of landraces and modern cultivars that constitute the primary gene pool of common beans, additional genetic variation can be found in the secondary gene pool that comprises the wild forms of *P. vulgaris*, the tertiary gene pool that corresponds to the *P. coccineus* complex, and finally in the quaternary gene pool that includes *P. acutifolius* and other species (Hidalgo 1991). These pools represent a valuable source of genes of economic importance. Unfortunately, gene transfer between the pools can be hampered by intra- and interspecific genetic barriers. For instance, differential photoperiodic responses within the andean gene pool alone can present some difficulties for within-pool gene transfer (Brücher 1988). Even more dramatic is the dwarf-lethal two gene system (DL_1 and DL_2) that restricts gene flow between some mesoamerican and andean accessions (Gepts and Bliss 1985; Shii et al. 1980; Singh and Gutierrez 1984). The existence of some genetic factors that affect compatibility in interspecific crosses has been suggested and it is supported by the identification of certain genotype combinations with increased compatibility (Hucl and Scoles 1985; Parker and Michaels 1986). The extent of structural similarities in the chromosomes of different gene pool members can also affect the effectiveness of gene transfer. Only a modest characterization of the cytogenetics of the *Phaseolus* group has been achieved due to the small size of the chromosomes. It is not known for instance whether the chromosomes of closely related species are homosequential. For example, Cheng et al. (1981) reported two chromosome inversions on different chromosomes in an interspecific hybrid between *P. vulgaris* and *P. coccineus*; however these findings have been contested by Shii et al. (1982). Identification and molecular tagging of genetic factors that restrict gene flow between the gene pools of beans will permit the design of strategies to facilitate an effective gene transfer. The availability of the RFLP map in beans has permitted the comparion of the genomes of mungbean, common bean, and soybean (Boutin et al. 1995). Comparisons of this kind have also been performed in the Solanaceae (Bonierbale et al. 1988; Tanksley et al. 1988), and in the cereals (Ahn and Tanksley 1993; Melake Berhan et al. 1993).

Common beans lend themselves as a good model system to study QTLs that affect yield. Wallace and Masaya (1988) have developed a "yield system analysis" to investigate the genetic components of yield and their interactions with the environment. In addition, Hoogenboon et al. (1988) have developed a computer simulation model for common beans — BEANGRO. The application of molecular markers to recombinant inbred lines (Burr et al. 1989), generated from suitable contrasting genotypes, will be useful in the identification of both genetic factors that affect yield and the responses these factors have to different environments. Molecular markers have been used to detect a number of QTLs in tomato (Paterson et al. 1988). This information can in turn be used to refine computer simulation programs that would take into account specific genetic factors. For instance, a negative correlation between seed size and yield has been reported for beans (Coyne 1968). A genetic factor that affects seed size has been identified via isozyme linkage analysis (Vallejos and Chase 1991a). Thus, tagging genes that affect the different components of yield can lead to the construction of specific genetic stocks carrying one or multiple combinations of these genes. These stocks can then be used to ask specific questions about the role certain genes play in different physiological processes such as sink-source relationships or photosynthate partitioning.

Finally, a large number of genes involved in disease resistance is available for molecular tagging in beans. There are at least 21 genetically characterized monogenic virus resistances (Provvidenti 1987), and a few other resistances to bacterial and fungal pathogens (Bassett 1989). Molecular tags can facilitate the efficient pyramiding of appropriate resistances into single breeding lines tailored for specific environments. The relatively small size of the bean genome opens the possibility for molecular cloning of any of these resistances via chromosome walking techniques (Rommens et al. 1989). The average ratio of physical distance to map distance has been estimated at 530 Kb/cM in beans (Vallejos et al. 1992). The possibility of positional cloning in the common bean has come close to reality by the recent development of a *P. vulgaris* BAC library with a five genome redundancy in the lab of Sally Mackenzie at Purdue University (personal comm.).

4.   References

Adam-Blondon A., Sévignac M. and Dron M. (1994) A. genetic map of common bean to localize specific resistance genes against anthracnose. Genome 37: 915-924.

Ahn S. and Tanksley S.D. (1993) Comparative linkage maps of the rice and maize genomes. Proc. Natl. Acad. Sci. USA 90: 7980-7984

Arumuganatham K. and Earle E.D. (1991) Nuclear DNA content of some important plant species. Plant Molec. Biol. Rept. 9: 208-218.

Bassett M.J. (1989) List of genes-*Phaseolus vulgaris* L. Annu. Rept. Bean Improv. Coop. 32: 1-7.

Bassett M.J. (1991) A revised linkage map of common bean. HortScience 26: 834-836.

Beebe S.E., Ochoa I., Skroch P.W., Nienhuis J. and Tivang J. (1995) Genetic diversity among common bean breeding lines developed for Central America. Crop Sci. 35: 1178-1183.

Bonierbale M., Plaisted R. and Tanksley S.D. (1988) Construction of comparative genetic maps of potato and tomato based on a common set of cloned sequences. Genetics 120: 1095-1103.

Boutin S.R., Young N.D., Olson T.C., Yu Z.H., Shoemaker R.C., Vallejos C.E. (1995) Genome conservation among three legume genera detected with DNA markers. Genetics 38: 928-937.

Brücher H. (1988) The wild ancestor of *Phaseolus vulgaris* in South America. In: Gepts P (ed.) Genetic resources of *Phaseolus* beans (pp. 185-214) Kluwer Academic Publishers, Dordrecht.

Burr B., Burr F.A., Thompson K.H., Albertson M.C. and Stuber C.W. (1988) Gene mapping with recombinant inbreds in maize. Genetics 118: 519-526.

Cheng S.S., Bassett M.J. and Quesenberry K.H. (1981) Cytogenetic analysis of interspecific hybrids between common bean and scarlet runner bean. Crop Sci. 21: 75-79.

Chase C.D., Ortega V.M. and Vallejos C.E. (1991) DNA restriction fragment length polymorphisms correlate with isozyme diversity in *Phaseolus vulgaris* L. Theor. Appl. Genet. 81: 806-811.

Coyne D.P. (1968) Correlation, heritability, and selection of yield components in field beans, *Phaseolus vulgaris* L. Proc. Amer. Soc. Hort. Sci. 93: 388-396.

Edwards K., Cramer C.L., Bolwell G.P., Dixon R.A., Schuch W. and Lamb C.J. (1985) Rapid transient induction of phenylalanine ammonia-lyase mRNA in elicitor-treated bean cells. Proc. Natl. Acad. Sci. USA 82: 6731-6735.

Gepts P. and Bliss F.A. (1985) F₁ hybrid weakness in the common bean: Differential geographic origin suggests two gene pools in cultivated germplasm. J. Hered. 76: 447-450.

Gepts P., Osborn T.C., Rashka K. and Bliss F.A. (1986) Phaseolin protein variability in wild forms and land races of the common bean (*Phaseolus vulgaris*): evidence for multiple centers of domestication. Econ. Bot. 40: 451- 468.

Hidalgo R. (1991) CIAT's world *Phaseolus* collection. In: van Schoonhoven A. and Voysest O (eds.) Common beans: research for crop improvement (pp. 163-197) C. A.B. International, Wallingford, UK.

Hoogenboon G., Jones J.W., White J.W. and Boote K.J. (1988) BEANGRO V 1.0: Dry bean crop growth simulation model: user's guide. Department of Agricultural Engineering. University of Florida.

Hucl P. and Scoles G.J. (1985) Interspecific hybridizations in the common bean: a review. HortSci. 20: 352-357.

Johns M.A., Skroch P.W., Nienhuis J., Hinrichsen P., Bascur G., Muñoz-Schick C. (1997) Gene pool classification of common bean landraces from Chile based on RAPD and morphological data. Crop Sci 37: 605-613.

Jung G.W., Coyne D.P., Skroch P.W., Nienhuis J., Arnaud-Santana E., Bokosi J., Ariyarathne H.M., Steadman J.R., Beaver J.S., and Kaeppler S.M. (1996) Molecular markers associated with plant architecture and resistance to common blight, web blight, and rust in common beans. J. Am. Soc. Hort. Sci. 121: 794-803.

Jung G.W., Skroch P.W., Coyne D.P., Nienhuis J., Arnaud-Santana E., Ariyarathne H.M., Kaeppler S.M., and Bassett M.J. (1997) Molecular-marker- based genetic analysis of tepary bean-derived common bacterial blight resistance in different developmental stages of common bean. J. Am. Soc. Hort. Sci.122: 329-337.

Keller B., Sauer N. andLamb C.J. (1988). Glycine-rich cell wall proteins in bean: gene structure and association of the protein with the vascular system. EMBO 12: 3625-3633.

Knight M.R. and Jenkins G.I. (1992) Genes encoding the small subunit of ribulose 1,5-bisphosphate carboxylase/oxygenase in *Phaseolus vulgaris* L.: nucleotide sequence of cDNA clones and initial studies of expression. Plant Mol. Biol. 18: 567-579.

Koenig R. and Gepts P. (1989) Allozyme diversity in wild *Phaseolus vulgaris*: further evidence for two major centers of genetic diversity. Theor. Appl. Gent. 78: 809-817.

Lander E.S., Green P., Abrahamson J., Baarlow A., Daly M.J., Lincoln SE and Newburg L. (1987) MAPMAKER: An interactive computer package for constructing primary genetic linkage maps of experimental and natural populations. Genomics 1: 174-181.

Lincoln S.E., Daly M. and Lander E.S. (1992) Constructing genetic maps with MAPAMKER/EXP 3.0. Whitehead Institute Technical Report. 3rd edition.

Mehdy M.C. and Lamb C.J. (1987) Chalcone isomerase cDNA cloning and mRNA induction by fungal elicitor, wounding and infection. EMBO 6: 1527-1533.

Melake Berhan A., Hulbert S.H., Butler L.G. and Bennetzen J.L. (1993) Structure and evolution of the genomes of *Sorghum bicolor* and *Zea mays*. Theor. Appl. Genet. 86: 598-604.

Mendel G. (1866) Experiments in plant hybridisation. Translated by the Royal Horticultural Society of London in 1938. Harvard University Press, Cambridge, MA.

Milbourne D., Meyer R., Bradshaw J.E., Baird E., Bonar N., Provan J., Powell W. and Waugh R. (1997) Comparison of PCR-based marker systems for the analysis of genetic relationships in cultivated potato. Molec. Breed. 3: 127-136.

Moreno J. and Chrispeels M.J. (1989) A lectin gene encodes the -amylase inhibitor of the common bean. Proc. Natl. Acad. Sci. USA 86: 7885-7889.

Nodari R.O., Tsai S.M., Gilbertson R.L., Gepts P. (1993) Towards an integrated linkage map of common bean. II. Development of an RFLP-based linkage map. Theor. Appl. Genet. 85: 513-520.

Olson M., Hood L., Cantor C., and Botstein D. (1989) A common language for physical mapping of the human genome. Science 245: 1434-1435.

Parker J.P. and Michaels T.E. (1986) Simple genetic control of hybrid plant development in interspecific crosses between *Phaseolus vulgaris* L. and *P. acutifolius* A. Gray. Plant Breeding 97: 315-323.

Paterson A.H, Lander E.S., Hewit J.D., Peterson T.S., Lincoln S.E. and Tanksley S.D. (1988) Resolution of quantitative traits into Mendelian factors by using a complete linkage map of restriction fragment length polymorphisms. Nature 335: 721-726.

Provvidenti R. (1987) List of genes in *Phaseolus vulgaris* for resistance to viruses. Annu. Rept. Bean Improv. Coop. 30: 1-4.

Rivkin M.I., Vallejos C.E. and McClean P.E. (1998) Disease-resistance related sequences in common bean. Genome (In press).

Rommens J.M., Iannuzzi M.C., Kerem B.-S., Drumm M.L., Melmer G., Dean M., Rozmahel R., Cole J.L., Kennedy D., Hidaka N., Zsiga M., Buchwald M., Riordan J.R., Tsui L.-C. and Collins F.S. (1989) Identification of the cystic fibrosis gene: chromosome walking and jumping. Science 245: 1059-265.

Sarria R., Lyznik A., Vallejos C.E. and Mackenzie S.A. (1998) A cytoplamic male sterility-associated mitochondrial peptide in common bean is post-translationally regulated. Plant

Cell 10:1217-1228.

Sauer N., Corbin D.R., Keller Band Lamb C.J. (1990) Cloning and characterization of a wound-specific hydroxyproline-rich glycoprotein in *Phaseolus vulgaris*. Plant Cell Environ. 13, 257-266.

Shaw J.K. and Norton J.B. (1918) The inheritance of seed-coat color in garden beans. Massachusetts Agric. Exp. Sta. Bull. 185: 58-104.

Shii C.T., Rabakoarihanta A., Mok M.C., and Mok D.W.S. (1982) Embryo development in reciprocal crosses of *Phaseolus vulgaris* L. and *P. coccineus* Lam. Theor. Appl. Genet. 62: 59-64.

Skroch P.W., Nienhuis J. (1995) Qualitative and quantitative characterization of RAPD variation among snap bean (*Phaseolus vulgaris*) genotypes. Theor. Appl. Genet. 91: 1078-1085.

Sun S.M., Slightom J.L. and Hall T.C. (1981) Intervening sequence in a plant gene - comparison of the partial sequence of cDNA and genomic DNA of French bean phaseolin. Nature 289: 37-41.

Talbot, D.R., AdangM.J., Slighton J.L. and Hall T.C. (1984) Size and organization of a multigene family encoding phaseolin, the major seed storage protein of *Phaseolus vulgaris* L. Molec. Gen. Genet. 198: 42-49.

Tanksley S.D., Bernatzky R., Lapitan L.L., and Prince J.P. (1988) Conservation of gene repertoire but not gene order in pepper and tomato. Proc. Natl. Acad. Sci. USA 85: 6419-6423.

Thompson W.F., Everett M., Polans N.O. and Jorgensen R.A. (1983) Phytochrome control of RNA levels in developing pea and mungbean leaves. Planta 158: 487-500.

Tucker M.L., Sexto R., Del Campillo E. and Lewis L.N. (1988) Bean abscission cellulase. Characterization of a cDNA clone and regulation of gene expression by ethylene and auxin. Plant Physiol. 88: 1257-1262.

Vallejos C.E. and Chase C.D. (1991a) Linkage between isozyme markers and a locus affecting seed size in *Phaseolus vulgaris* L. Theor. Appl. Genet. 81: 413-419.

Vallejos C.E. and Chase C.D. (1991b) Extended linkage map for the phaseolin linkage group of *Phaseolus vulgaris* L. Theor. Appl. Genet. 82: 353-357.

Vallejos C.E., Sakiyama N.S. and Chase C.D. (1992) A molecular marker-based linkage map of *Phaseolus vulgaris* L. Genetics 131: 733-740.

Wallace D.H. and Masaya P.N. (1988) Using yield trial data to analyze the physiological genetics of yield accumulation and the genotype X environment interaction effects on yield. Annu. Rept. Bean Improv. Coop. 31: vii-xxiv.

Williams C.E. and St. Clair D.A. (1993) Phenetic relationships and levels of variability detected by restriction fragment length polymorphism and random amplified polymorphic DNA analysis of cultivated and wild accessions of *Lycopersicon esculentum*. Genome 36: 619-630.

Williams J.G.K., Kubelik A.R., Livak K.J., Rafalski J.A. and Tingey S.V. (1990) DNA polymorphisms amplified by arbitrary primers are useful as genetic markers. Nucl. Acids Res. 18: 6531-6535.

Young N.D., Kumar L., Menacio-Hautea D., Danesh D., Talekar N.S., Shanmugasundarum N.S. and Kim D.H. (1992) RFLP mapping of a major bruchid resistance gene in mungbean (*Vigna radiata* L. Wilczek). Theor. Appl. Genet. 84:839-844.

Zheng J., Nakata M., Uchiyama H., Morikawa H. and Tanaka R. (1991) Giemsa C-banding pattern in several species of *Phaseolus* L. and *Vigna* Savi, fabacea. Cytologia 56: 459-466.

18.   RFLP map of the potato

CHRISTIANE GEBHARDT,
ENRIQUE RITTER[1], and FRANCESCO SALAMINI
Max-Planck-Institut für Züchtungsforschung, Carl-von-Linné-Weg, D-50829 Köln, Germany
<gebhardt@mpiz-koeln.mpg.de>
[1] *NEIKER, Instituto Vasco de Investigacion y Desarrollo Agraria, Apdo 46,*
E-01080 Victoria, Spain

Contents

1.   Introduction

The potato, *Solanum tuberosum*, a species of the family Solanaceae, is cultivated in most temperate and subtropical zones of the world. After wheat, rice and corn it occupies the fourth position in terms of world production per year (FAO Yearbook 1988).

The history of the potato and its impact during the development of the industrial society has been described in detail by Salaman (1985). Most likely, the potato originated in the Andean region of South-America where it has been cultivated for at least 2000 years. After the Spanish conquest, the potato was introduced into Europe via Spain and England in the 16th century. Historical evidence indicates that the first potato tubers reaching the European continent were botanically *Solanum tuberosum* subsp. *andigena*, producing tubers only under short-day photoperiod conditions. The conversion from an exotic ornamental plant being of interest only to few botanists into a major food crop supporting the growing population in the industrial society of the 18th and 19th centuries, was made possible by natural or unintentional selection of genotypes producing tubers under the long-day photoperiod of Europe. The cultivated potato of today has the botanical name *Solanum tuberosum* subsp. *tuberosum*. It is a tetraploid plant with 48 chromosomes and tetrasomic inheritance. One genome complement has, therefore, twelve chromosomes.

As potatoes are vegetatively propagated via tubers, there was no need to select for highly fertile genotypes during potato breeding. This fact, together with the tetrasomic inheritance of the crop, prevented the development of classical genetic linkage maps.

R.L. Phillips and I.K. Vasil (eds.), DNA-Based Markers in Plants, 319 - 336.

Only a few linkages were reported, for example among anthocyanin pigmentation genes and between pigmentation and tuber shape (reviewed by De Jong 1991). Centromere map distances of isozyme loci were determined by Douches and Quiros (1987). Reduction of the ploidy from the tetraploid to the diploid level is possible either by pollination of tetraploid genotypes with certain diploid strains of *Solanum phureja* which induce the parthenogenetic development of diploid gametes into plants (Hougas et al. 1964; Hermsen and Verdenius 1973), or by regenerating plants from diploid male gametes via anther or microspore culture of tetraploid parents (Dunwell and Sunderland 1973; Powell and Uhrig 1987). Diploid potatoes are, however, largely self-incompatible. This fact, and the high genetic load present in the species make the construction of pure lines in most cases impractical. The RFLP map of potato is based, therefore, on segregating progeny of highly heterozygous diploid parents. This linkage map is, of course, useful for understanding the genome of tetraploid potatoes.

2.   The mapping population

Two diploid, heterozygous *S. tuberosum* subsp. *tuberosum* breeding lines were crossed to give a F_1. The heterozygosity of the parents was estimated to be 57% and 59%, respectively, based on RFLP alleles. The parents exhibited 82% informative polymorphisms among each other when compared with 147 DNA probes and three restriction enzymes tested per probe. Under the given experimental conditions (Gebhardt et al. 1989) this value was close to the mean value (80%) found in a gene pool of 38 diploid potato genotypes (Gebhardt et al. 1989). Pollinating an individual F_1 plant with one of the parents yielded a backcross progeny of 67 lines (Gebhardt et al. 1989, 1991). Parents and progeny were clonally propagated via tubers and are available from the BGRC *in vitro* collection (Bundesforschungsanstalt für Landwirtschaft, Braunschweig-Völkenrode, FAL, Bundesallee 50, D-38116 Braunschweig, Germany).

3.   The experimental system

Details of the experimental system by which the segregation data for the potato RFLP map were obtained are described by Gebhardt et al. (1989). In short, total genomic DNA was isolated from freeze-dried leaves and shoots of plants grown in the greenhouse under normal daylight conditions. The DNA was digested with the four-base-cutter restriction enzymes *Taq*I, *Rsa*I and *Alu*I, respectively. Approximately 55% of the segregation data were obtained with *Taq*I-, 35% with *Rsa*I- and only 10% with *Alu*I-restricted DNA. Restriction fragments were separated on 4% polyacrylamide gels under denaturing conditions and transferred to nylon membranes by electroblotting. The separation range was between 250 and 2000 bases resolving minimum length differences of ca 5 bases. The membranes were hybridized to ^{32}P labelled probes and washed at a moderate stringency.

4. Map construction

The principles and algorithms on which the RFLP segregation data were converted into linkage groups have been described by Ritter et al. (1990). Whereas a single genetic model applies to segregating alleles of F_2 or backcross mapping populations derived from pure lines, in populations derived from partially heterozygous parents, RFLP alleles segregate according to several different genetic models. In our backcross-type mapping population, between one and three RFLP alleles segregated per locus. Autoradiographs were evaluated by scoring presence versus absence of segregating individual restriction fragments without taking into consideration possible allelism among different fragments. No distinction was made between a fragment A in the homozygous (*AA*) or heterozygous (*AO*) state. In scoring RFLPs phenotypically, three types of segregation patterns were distinguished: fragments being present in the heterozygous state either in the female or in the male parent expected to segregate with a 1:1 ratio (presence versus absence, testcross type segregation); fragments being present in the heterozygous state in both parents expected to segregate with a 3:1 ratio (presence versus absence, F_2 type segregation). Recombination frequencies among all fragments scored over all markers tested were estimated using algorithms proper for the genetic models fitting the observed mode of segregation (Ritter et al. 1990). Linkage subgroups were constructed based on the recombination frequencies found among fragments with the same mode of segregation, and, in the case of the 1:1 segregation, separately for the female and male parent. The linkage subgroups were connected and oriented relative to each other considering the fraction of RFLP loci with allelic fragments belonging to at least two linkage subgroups. Allelism among RFLP fragments detected by the same probe was assumed when they were found linked with zero percent recombination (in coupling or repulsion phase). Following this strategy, three linkage groups were obtained for each chromosome: one for the female and one for the male parent, and the third resulting from both parents.

5. Nomenclature and origin of RFLP markers

5.1. CP markers

The RFLP map contains 84 marker loci based on random cDNA sequences of potato cloned in the Bluescript vector (Stratagene). The source tissues for the cDNA library were young leaves and shoots harvested from several diploid potato genotypes. Selection criteria were - besides polymorphism - the insert size (> 200 bp) and pattern complexity (low repetitiveness) (Gebhardt et al. 1989). Each cDNA marker locus is identified by the letters CP followed by an identification number. Multiple loci detected with the same marker probe are indicated by small letters a, b, c, ... in parentheses behind the identification number. Of 65 cDNA probes mapped, 52 (80%) detected a single locus, 11 (17%) detected two loci, one detected 4 and one 6 loci. Most of the cDNA markers have been sequenced from both ends.

5.2. GP markers

The majority of marker loci is based on random genomic sequences of potato. *Pst*I digested total genomic DNA was cloned in the Bluescript vector and selected for size (500-2000 bp) and low copy number (Gebhardt et al. 1989). This marker class is indicated by the letters GP. Otherwise the nomenclature is the same as for CP markers. Seventy nine percent of the genomic probes detected a single locus in the mapping population used, 16 percent detected two loci and five percent detected more than two loci. Most of the GP markers have been sequenced from both ends.

5.3. PSTR markers

A repetitive sequence isolated from the wild potato *S. spegazzinii* was also mapped (Gebhardt et al. 1995). This sequence corresponds to a subtelomeric repeat and detects seven most distal RFLP loci, which indicate positions of telomeric regions on the linkage groups.

5.4. Markers encoding functional genes

Map positions of a number of genes of known or operationally defined function are included in the map. Identification of genes and corresponding loci, map positions, sources and references are listed in Table 1. Besides cloned genes of potato, one cloned gene of maize (ShM) and seven tobacco genes (*Nt* markers) were also used as markers. Seven marker genes, *Sr1, AmyZ1, TPT, PHA1, Nt ChtA-Q, Cyt-c red.53 kD and Cyt-c red.25 kD*, were mapped using a population resulting from a different cross (Gebhardt et al. 1991), because they did not segregate in the progeny on which the map shown in Fig. 1 is based. The loci identified by those markers are included at approximate positions in the linkage groups. Several loci harbour clusters of members of multigene families: acidic and basic glucanases, actin and patatin genes, chitinases and members of the *prp1* family.

5.5. Tomato markers

Twenty-seven reference RFLP markers of the twelve largely colinear tomato chromo-somes (Bonierbale et al. 1988) were included in the RFLP map of potato, therewith aligning the genetic maps of potato and tomato (Gebhardt et al. 1991). The markers are indicated by TG or CT followed by an identification number. The markers were obtained from S.D. Tanksley (Cornell University, Ithaca, New York, USA).

5.6. Morphological markers

One morphological trait segregated in the mapping population with a 1:1 ratio. This was the tuber skin color (purple versus colorless). The *PSC* (purple skin color) locus is situated on chromosome X.

Table 1. Functional gene markers.

Specification	Marker name	No. of loci	Map position	Sources and references
4-Coumarate: CoA ligase	*4CL*	2	III(a), VI(b)	(3), (1), (2)
Phenylalanine ammonia-lyase	*PAL*	5	IX(a,b,c), X(d,e)	(4), (1), (2)
Glutathione S-transferase (pathogenesis related)	*prp1*	1	IX	(5), (6), (1), (2)
Induced upon infection of potato leaves with *P. infestans*	*PC116*	1	IV	(5), (1), (2)
Lipoxygenase	*Lox*	2	VIII(a), I(b)	(7), this chapter
Chitinase, class I, basic	*Cht B*	2	X(a,b)	(8), (9)
Chitinase, class III, acidic	*Nt Cht A III*	1	V	(10), this chapter
Chitinase, acidic, PR-Q	*Nt Cht A-Q*	1	II	(10), this chapter
1,3 β-glucanase, basic	*Glu B*	1	I	(8), (2), (36)
1,3-β-glucanase, acidic	*Glu A*	1	I	(35), (36)
1,3-β-glucanase, acidic, extracellular, PR-Q'	*Nt Glu A-Q'*	1	X	(10), this chapter
PR protein, acidic, extracellular, PR-1	*Nt PR-1*	1	I	(10), this chapter
PR protein, PR-1a	*Nt PR-1a*	1	X	(11), this chapter
PR protein, PR-4	*Nt PR-4*	1	X	(10), this chapter
PR protein, PR-5	*Nt PR-5*	1	XII	(10), this chapter
WUN-1 (wound induced)	*WUN1*	1	VI	(12), (2)
WUN-2 (wound-induced)	*WUN2*	1	VI	(12), (2)
Patatin	*PAT*	2	VIII(a), IV(b)	(13), (1), (2)
Actin	*Actin*	1	V	(14), (1), (2)
S-locus, selfincompatibility	*Sr1*	1	I	(15), (2)
Stylar endochitinase	*SK2*	2	II(a), I(b)	(16), this chapter
Ribulose bisphosphate carboxylase, small subunit	*rbcS*	3	II(rbcS-c,rbcS-2), III(rbcS-1)	(17), (18), (1), (2)
Granule bound starch synthase (GBSS)	*wx*	1	VIII	(19), (1), (2)
Sucrose synthase	*ShM*	1	XII	(20), (2), (36)

Table 1.   Continued.

Specification	Marker name	No. of loci	Map position	Sources and references
α-Amylase	*AmyZ3/4*	1	IV	(21), (9), (36)
	AmyZ1	1	IV	
Branching enzyme	*BE*	1	IV	(22), (9), (36)
ADP glucose pyrophos-phorylase B	*AGPase B*	2	VII(a), XII(b)	(23), (9), (36)
ADP glucose pyrophos-phorylase S	*AGPase S*	2	I(a), VIII(b)	(23), (9), (36)
Invertase, apoplastic	*Inv-apo*	2	X(a), IX(b)	(24), (36)
Plasma membrane H$^+$-ATPase	*PHA1*	2	III(a), VI(b)	(25), (9), (36)
Plasma membrane H$^+$-ATPase	*PHA2*	1	VII	
Triosephosphate 3-phos-phoglycerate translocator	*TPT*	1	X	(26), (9), (36)
Sucrose transporter	*StSUT1*	1	XI	(27), (36)
Inorganic pyrophosphatase	*Ppa1*	2	XII(a), VIII(b)	(28), (36)
Phosphoenolpyruvate carboxylase	*Pepc*	2	X(a), XII(b)	(29), (36)
Cytochrome-c reductase: 10 kD subunit	*cyt-c red, 10 kD*	2	VI(a, b)	(30), this chapter
12 kD subunit	*cyt-c red, 12 kD*	1	IX	(30), this chapter
14 kD subunit	*cyt-c red, 14 kD*	2	VIII(a), I(b)	(30), this chapter
25 kD subunit	*cyt-c red, 25 kD*	1	XI	(30), this chapter
30 kD subunit	*cyt-c red, 30 kD*	1	XI	(30), this chapter
33 kD subunit	*cyt-c red, 33 kD*	1	VI	(30), this chapter
51 kD subunit	*cyt-c red, 51 kD*	2	V(a), XII(b)	(30), this chapter
53 kD subunit	*cyt-c red, 53 kD*	1	II	(30), this chapter
55 kD subunit	*cyt-c red, 55 kD*	1	V	(30), this chapter
Cold regulated genes	*ci7*	1	IV	(31)
	ci13	1	V	(31)
	ci19	1	XI	(31)

Table 1. Continued.

Specification	Marker name	No. of loci	Map position	Sources and references
Cold regulated genes	*ci21*	2	IV(a, b)	(31)
mas-binding factor	*MBF*	1	IV	(32), this chapter
Potato protein kinase	*potkin*	1	XII	(33), this chapter
Pto resistance gene homologues	*StPto*	2	V(a, b)	(34)

(1) Gebhardt et al. (1989), (2) Gebhardt et al. (1991), (3) Becker-André et al. (1991), (4) Fritzemeier et al. (1987), (5) Taylor et al. (1990), (6) Hahn and Strittmatter (1994), (7) Royo et al. (1996), (8) Beerhues and Kombrink (1994), (9) Gebhardt et al. (1994), (10) Ward et al. (1991), (11) Uknes et al. (1993), (12) Logemann et al. (1988), (13) Rosahl et al. (1986), (14) Thangavelu (pers. comm.) (15) Kaufmann et al. (1991), (16) Wemmer et al. (1994), (17) Eckes et al. (1985), (18) Wolter et al. (1988) (19) Hergersberg (1988), (20) Werr et al. (1985), (21) J. Kreiberg, Danisco, Copenhagen, Denmark (pers. comm.), (22) Koßmann et al. (1991), (23) Müller-Rber et al. (1990), (24) Hedley et al. (1994), (25) Harms et al. (1994), (26) Schulz et al. (1993), (27) Riesmeier et al. (1993), (28) Du Jardin et al. (1995), (29) Merkelbach et al. (1993), (30) Braun et al. (1994), (31) Van Berkel et al. (1994), (32) Feltkamp et al. (1994), (33) R. Thompson, MPIZ, Cologne, Germany (pers. comm.), (34) Leister et al. (1996), (35) E. Kombrink, MPIZ, Cologne, Germany (pers. comm.), (36) Chen et al. (2000).

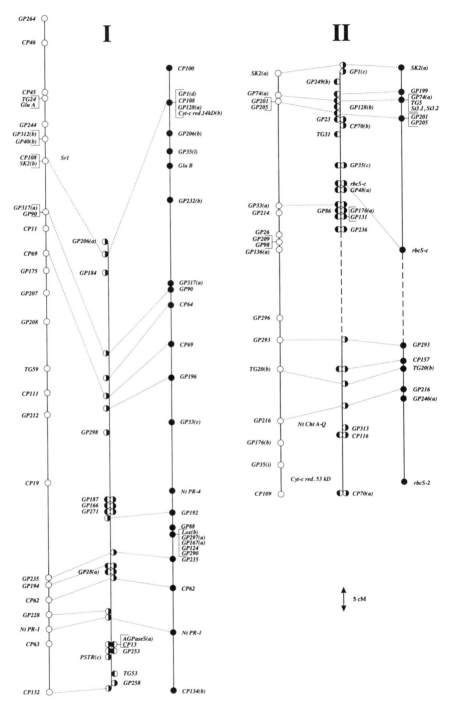

Figure 1.   RFLP map of the potato. For each chromosome there are three linkage groups. The linkage group on the left was derived from restriction fragments descending from the male

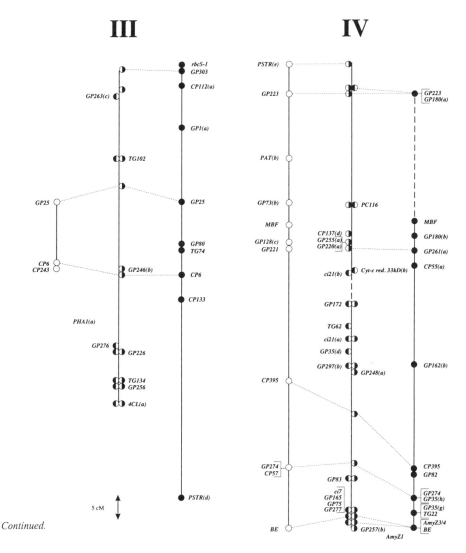

Continued.

parent (open circles). The linkage group on the right was derived from restriction fragments descending from the female parent (solid circles). The central linkage group was obtained with fragments descending from both parents (half open, half solid circles). Pairs of circles in the central group indicate allelism between two common fragments. Restriction fragments in the central group which are linked with each other in coupling are shown as half circles in the same orientation. The half circles are shown inverted relative to each other when the fragments were in repulsion or coupling/repulsion phase (see Ritter et al. 1990). Dotted lines connecting the linkage groups indicate that those restriction fragments are alleles of the same RFLP locus (allelic bridges). The orientation of the linkage subgroups relative to each other is determined by these allelic bridges. Marker fragments not showing significant linkage, but nevertheless on the same chromosome, are connected by a broken line. Map distances are given in Centimorgans

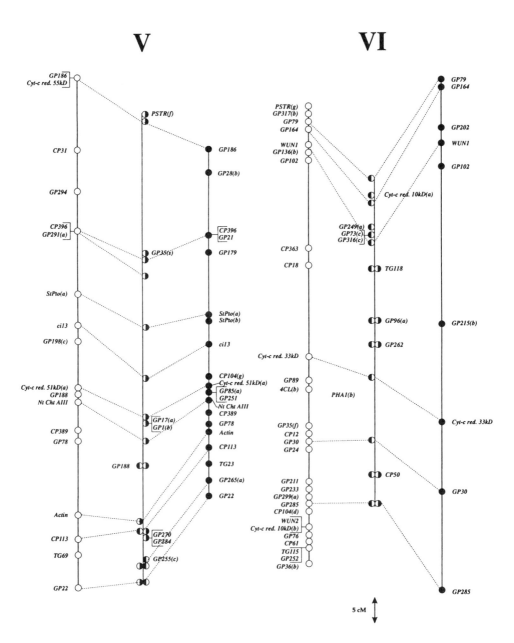

Continued.

(Kosambi 1944). Marker loci resulting from hybridization with characterized sequences are shown with their specific names (Table 1). Small letters in parentheses indicate that multiple loci were detected by the same probe.

VII

VIII

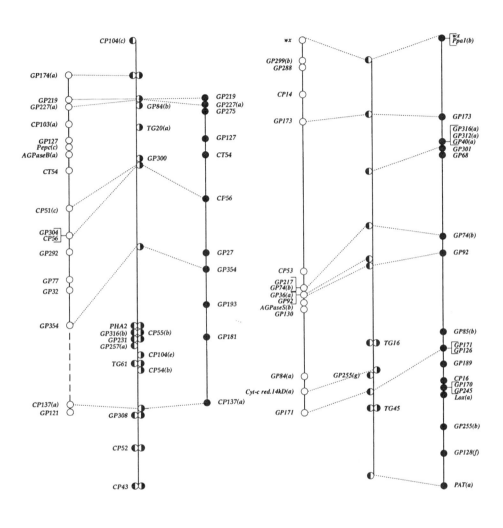

Figure 1. Continued.

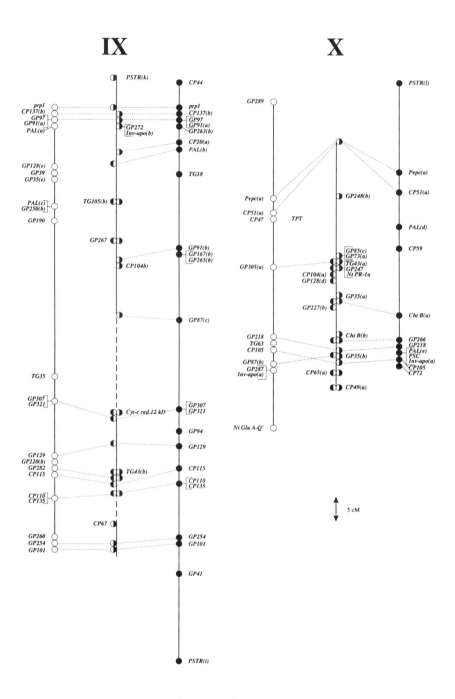

Figure 1. Continued.

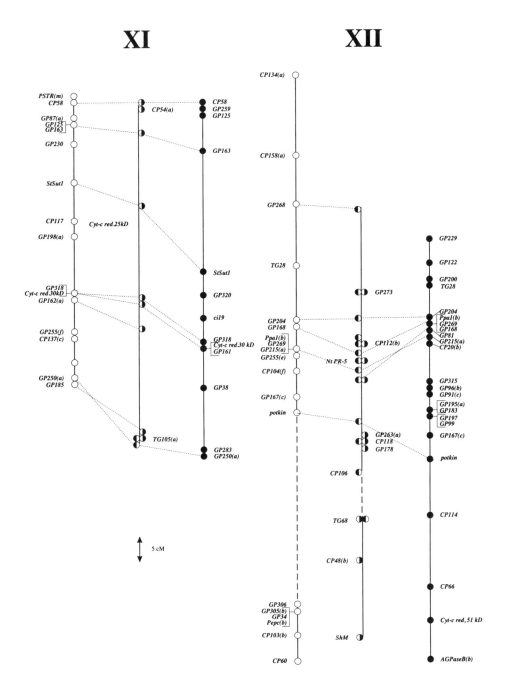

Figure 1. Continued.

6.   The RFLP map of potato

The molecular map of potato as based on ca 350 marker probes is shown in Fig. 1. Each chromosome is split into three linkage groups, two for the male and female parents, being equivalent to backcross type segregation analysis, and the third one based on F_2 type segregation data. Chromosome nomenclature is the same as in the syntenic tomato genome (see Chapter 20). Twenty-one percent of the markers detect two or more loci. Only two duplicated linkage blocks involving at least three markers have been identified so far: *CP70(b) rbcS-c*, and *GP176(a)* on chromosome II are duplicated on the same chromosome *(GP176(b), rbcS-2, CP70(a))*. A second duplicated linkage block involving four markers is detectable on chromosomes IX and XII *(GP91(a)*, *GP263(b), CP20(a), GP167(b)* on IX and *CP20(b), GP263(a), GP91(c), GP167(c)* on XII). Unlike maize (Helentjaris et al. 1988) or *Brassica* (Slocum et al. 1990) it seems that there was no extensive duplication of chromosomes or chromosomal segments during the evolution of the potato genome.

The potato map, as derived from the intraspecific mapping population used, is ca 1100 cM long (map units defined as by Kosambi 1944). Genome coverage by the map is estimated to be at least 90% based on mapping of subtelomeric repeats (Gebhardt et al. 1995). This is also supported by the observation that only three marker probes from more than 350 tested did not show significant linkage to any of the twelve established linkage groups.

Linkage maps derived from different genetic backgrounds in potato vary in map length (Bonierbale et al. 1989; Gebhardt et al. 1991). The order of loci, however, is conserved within the limits of the standard error associated with each map position (Gebhardt et al. 1991).

Segregation ratios deviating significantly from the expected 1:1 and 3:1 ratios, respectively, were found for 27% of the segregating restriction fragments. Several chromosomal segments with distorted segregation ratios were identified by linked distorted marker loci. Some of these segments were detectable in different genetic backgrounds, while others were specific for certain parents in particular cross combinations. The strongest distortion of segregation ratios was found on linkage group I and is assumed to be caused by the activity of the self-incompatibility locus (Gebhardt et al. 1991).

The RFLP map of potato is informative for other intra- and interspecific crosses. Mapped probes have been used in different genetic backgrounds to locate dominant resistance loci acting against the root cyst nematode *Globodera rostochiensis* (Barone et al. 1990; Gebhardt et al. 1993), potato virus X (Ritter et al. 1991), and the oomycete *Phytophthora infestans* (Leonards-Schippers et al. 1992, El-Kharbotly et al. 1994, 1996). The markers are also suitable for mapping quantitative trait loci (QTL) in potato (Leonards-Schippers et al. 1994, Schäfer-Pregl et al. 1998, Oberhagemann et al. 1999).

7.   References

Barone, A., Ritter, E., Schachtschabel, U., Debener, T., Salamini, F. and Gebhardt, C. (1990). Localization by restriction fragment length polymorphism mapping in potato of a major dominant gene conferring resistance to the potato cyst nematode *Globodera rostochiensis*. Mol. Gen. Genet. 224: 177-182.

Becker-André, M., Schulze-Lefert, P., Hahlbrock, K. (1991) Structural comparison, mode of expression, and putative *cis*-acting elements of the two 4-coumarate:CoA ligase genes in potato. The J. of Biol. Chem. 266: 8551-8559.

Beerhues, L., Kombrink, E. (1994) Primary structure and expression of mRNAs encoding basic chitinase and 1,3-ß-glucanase in potato. Plant Mol. Biol. 24: 353-367.

Bonierbale, M.W., Plaisted, R.L. and Tanksley, S.D. (1988) RFLP maps based on a common set of clones reveal modes of chromosomal evolution in potato and tomato. Genetics 120: 1095-1103.

Braun, H.-P., Kruft, V., Schmitz, U.K. (1994) Molecular identification of the ten subunits of cytochrome-c reductase from potato mitochondria. Planta 193: 99-106.

Chen, X, Salamini, F, Gebhardt, C. (2000) A potato molecular function map for carbohydrate metabolism and transport. Theor Appl Genet, in press.

De Jong, H. (1991) Inheritance of anthocyanin pigmentation in the cultivated potato. A critical review. Am. Potato J. 68: 585-593.

Douches, D.S. and Quiros, C.F. (1987) Use of 4x-2x crosses to determine gene-centromere map distance of isozyme loci in *Solanum* species. Genome 29: 519-527.

Du Jardin, P., Rojas-Beltran, J., Gebhardt, C., Brasseur, R. (1995) Molecular cloning and characterization of a soluble inorganic pyrophosphatase in potato. Plant Physiol. 109: 853-860.

Dunwell, J.M. and Sunderland, N. (1973) Anther culture of *Solanum tuberosum* L. Euphytica 22: 317-323.

Eckes, P., Schell, J. and Willmitzer, L. (1985) Organ-specific expression of three leaf/stem specific cDNAs from potato is regulated by light and correlated with chloroplast development. Mol. Gen. Genet. 199: 216-224.

FAO Yearbook (1988) Production Vol. 42, FAO Statistics Series No. 88, Rome.

Feltkamp, D., Masterson, R., Starke, J., Rosahl, S. (1994) Analysis of the involvement of *ocs*-like bZip-binding elements in the differential strength of the bidirektional *mas*1'2' promoter. Plant Physiol. 105: 259-268.

Fritzemeier, K.H., Cretin, C., Kombrink, E., Rohwer, F., Taylor, J., Scheel, D. and Hahlbrock, K. (1987) Transient induction of phenylalanine ammonia-lyase and 4-coumarate: CoA ligase mRNAs in potato leaves infected with virulent or avirulent races of *Phytophthora infestans*. Plant Physiol. 85: 34-41.

Gebhardt, C., Ritter, E., Debener, T., Schachtschabel, U., Walkemeier, B., Uhrig, H. and Salamini, F. (1989). RFLP analysis and linkage mapping in *Solanum tuberosum*. Theor. Appl. Genet. 78: 65-75.

Gebhardt, C., Ritter, E., Barone, A., Debener, T., Walkemeier, B., Schachtschabel, U., Kaufmann, H., Thompson, R.D., Bonierbale, M.W., Ganal, M.W., Tanksley, S.D. and Salamini, F. (1991). RFLP maps of potato and their alignment with the homoeologous tomato genome. Theor. App. Genet. 83: 49-57.

Gebhardt, C., Mugniery, D., Ritter, E., Salamini, F. and Bonnel, E. (1993). Identification of RFLP markers closely linked to the *H1* gene conferring resistance to *Globodera rostochiensis* in potato. Theor. Appl. Genet. 85: 541-544.

Gebhardt, C., Ritter, E., Salamini, F. (1994) RFLP map of the potato. In: DNA-based markers in pants. Vasil, I.K., Philipps, R.L. (eds.) Vol. I: Kluwer Academic Publ., Dordrecht, The Netherlands, pp. 271-285.

Gebhardt, C., Eberle, B., Leonards-Schippers, C., Walkemeier B., Salamini, F. (1995) Isolation, characterization and RFLP linkage mapping of a DNA repeat family of *Solanum spegazzinii* by which chromosome ends can be localized on the genetic map of potato. Genet. Res., Camb. 65: 1-10.

Hahn, K., Strittmatter, G. (1994) Pathogen defence gene *prp1-1* from potato encodes an auxin-responsive glutathione S-transferase. Eur. J. Biochem. 226: 619-626.

Harms, K., Wöhner, R., Schulz, B., Frommer, W.B. (1994) Expression of plasmamembrane H^+-ATPase genes in potato. Plant Mol. Biol. 26: 979-988.

Hedley, P.E., Machray, G.C., Davies, H.V., Burch, L., Waugh, R. (1994) Potato (*Solanum tuberosum*) invertase-encoding cDNAs and their differential expression. Gene 145: 211-214.

Helentjaris, T., Weber, D. and Wright, S. (1988) Identification of the genomic locations of duplicate nucleotide sequences in maize by analysis of restriction fragment length polymorphisms. Genetics 118: 353-363.

Hergersberg, M. (1988) Molekulare Analyse des waxy Gens aus *Solanum tuberosum* und Expression von antisense RNA in transgenen Kartoffeln. Ph.D. thesis, University of Cologne.

Hermsen, J.G.Th. and Verdenius, J. (1973) Selection from *Solanum tuberosum* group phureja of genotypes combining high-frequency haploid induction with homozygosity for embryo spot. Euphytica 22: 244-259.

Hougas, R.W., Peloquin, S.J. and Gabert, A.C. (1964) Effect of seed parent and pollinator on the frequency of haploids in *Solanum tuberosum*. Crop Sci. 4: 593-595.

Kaufmann, H., Salamini, F. and Thompson, R. (1991) Sequence variability and gene structure at the self-incompatibility locus of *Solanum tuberosum*. Mol. Gen. Genet. 226: 457-466.

Kosambi, D.D. (1944) The estimation of map distances from recombination values. Ann. Eugen. 12: 172-175.

Koßmann, J., Visser, R.G.F., Müller-Röber, B.T., Willmitzer, L. and Sonnewald, U. (1991) Cloning and expression analysis of potato cDNA that encodes branching enzyme: evidence for co-expression of starch biosynthetic genes. Mol. Gen. Genet. 230: 39-44.

Leister, D., Ballvora, A., Salamini, F., Gebhardt, C. (1996) A PCR-based approach for isolating pathogen resistance genes from potato with potential for wide application in plants. Nature Genetics 14: 421-429.

Leonards-Schippers, C., Gieffers, W., Salamini, F. and Gebhardt, C. (1992). The *R1* gene conferring race-specific resistance to *Phytophthora infestans* in potato is located on potato chromosome V. Mol. Gen. Genet. 233: 278-283.

Logemann, J., Mayer, J.E. Schell, J. and Willmitzer, L. (1988) Differential expression of genes in potato tubers after wounding. Proc. Natl. Acad. Sci. USA 85: 1136-1140.

Merkelbach, S., Gehlen, J., Denecke, M., Hirsch, J., Kreuzaler, F. (1993) Cloning, sequence analysis and expression of a cDNA encoding active phosphoenolpyruvate carboxylase of the C3 plant *Solanum tuberosum*. Plant Mol. Biol. 23: 881-888.

Müller-Röber, B.T., Koßmann, J., Hannah, L.C., Willmitzer, L. and Sonnewald, U. (1990) One of two different ADP-glucose pyrophosphorylase genes from potato responds strongly to elevated levels of sucrose. Mol. Gen. Genet. 224: 136-146.

Oberhagemann, P, Chatot-Balandras, C, Bonnel, E, Schäfer-Pregl, R, Wegener, D, Palomino, C, Salamini, F, Gebhardt, C. (1999) A genetic analysis of quantitative resistance to late blight in potato: Towards marker assisted selection. Mol Breeding 5: 399-415.

Powell, W. and Uhrig, H. (1987) Anther culture of *Solanum* genotypes. Plant Cell Tissue and Organ Culture 11: 13-24.

Riesmeier, J., Hirner, B., Frommer, W.B. (1993) Potato sucrose transporter expression in minor veins indicates a role in phloem loading. The Plant Cell 5: 1591-1598.

Ritter, E., Gebhardt, C. and Salamini, F. (1990). Estimation of recombination frequencies and construction of RFLP linkage maps in plants from crosses between heterozygous parents. Genetics 125: 645-654.

Ritter, E., Debener, T., Barone, A., Salamini, F. and Gebhardt, C. (1991) RFLP mapping on potato chromsomes of two genes controlling extreme resistance to potato virus X (PVX). Mol. Gen. Genet. 227: 81-85.

Rojo, J., Vancameyt, G., Perez, A.G., Sanz, C., Stormann, K., Rosahl, S., Sanchez-Serrano, J.J. (1996) Characterization of three potato lipoxygenases with distinct enzymatic activities and different organ-specific and wound-regulated expression patterns. J. Biol. Chem. 271: 21012-21019.

Rosahl, S., Schmidt, R., Schell, J. and Willmitzer, L. (1986) Isolation and characterization of a gene from *Solanum tuberosum* encoding patatin, the major storage protein of potato tubers. Mol. Gen. Genet. 203: 214-220.

Salaman, R. (1985) The history and social influence of the potato, revised impression (Hawkes, J.G., ed.). Cambridge University Press, Cambridge, New York, New Rochelle, Melbourne, Sydney.

Schulz, B., Frommer, W.B., Flügge, U.I., Hummel, S., Fischer, K., Willmitzer, L. (1993) Expression of the triose phosphate translocator gene from potato is light dependent and restricted to green tissues. Mol. Gen. Genet. 238: 357-361.

Slocum, M.K., Figdore, S.S., Kennard, W.C., Suzuki, J.Y. and Osborn, T.C. (1990) Linkage arrangement of restriction fragment length polymorphism loci in *Brassica oliracea*. Theor. Appl. Genet. 80: 57-67.

Taylor, J.L., Fritzemeier, K.H., Häuser, I., Kombrink, E., Rohwer, F., Schröder, M., Strittmatter, G. and Hahlbrock, K. (1990) Structural analysis and activation by fungal infection of a gene encoding a pathogenesis-related protein in potato. Mol. Plant-Microbe Interactions 3: 72-77.

Uknes, S., Dincher, S., Friedrich, K., Negrotto, D., Williams, S., Thompson-Taylor, H., Potter, S., Ward, W., Ryals, J. (1993) Regulation of pathogenesis-related protein-1a gene expression in tobacco. The Plant Cell 5: 159-169.

Van Berkel, J. Salamini, F., Gebhardt, C. (1994). Transcripts accumulating during cold storage of potato (*Solanum tuberosum* L.) tubers are sequence related to stress-responsive genes. Plant Physiol. 104: 445-452.

Ward, E.R., Uknes, S.J., Williams, S.C., Dincher, S.S., Wiederhold, D.L., Alexander, D.C., Ahl-Goy, P., Métreaux, J.-P., Ryals, J.A. (1991) Coordinate gene activity in response to agents that induce systemic acquired resistance. The Plant Cell 3: 1085-1094.

Wemmer, T., Kaufmann, H., Kirch, H.-H., Schneider, K., Lottspeich, F., Thompson, R.D. (1994) The most abundant soluble basic protein of the stylar transmitting tract in potato (*Solanum tuberosum* L.) is an endochitinase. Planta 194: 264-273.

Werr, W., Frommer, W.B., Maas, C. and Starlinger, P. (1985) Structure of the sucrose synthase gene on chromosome 9 of *Zea mays* L. EMBO J. 4: 1373-1380.

Wolter, F.P., Fritz, C.C., Willmitzer, L. and Schell, J. (1988) rbcS genes in *Solanum tuberosum*: conservation of transit peptide and exon shuffling during evolution. Proc. Natl. Acad. Sci. USA 85: 846-850.

19.   Rice molecular map

S. R. McCOUCH
Department of Plant Breeding, Cornell University, Ithaca, NY 14853-1901, U.S.A.
<srm4@cornell.edu>

Contents

1.   Introduction

Rice is ideally suited to genetic and molecular studies. Some of its positive features include a small nuclear genome (450 Mb), diploidy (2n = 2x = 24) and the ability to be transformed by exogenous DNA (Uchimaya et al. 1986; Arumuganathan and Earle 1991; Kothari et al. 1993). Moreover, because of its importance as a crop species, scientists have been identifying and studying the genetic basis of morphological and physiological mutants for nearly a century.

2.   History of genetic mapping in rice

As with most plant species, the first genetic linkage maps of rice were constructed with morphological markers (Kinoshita 1986 and references therein). By the mid 1980's isozymes were being added to the genetic maps of rice (Ranjhan et al. 1988; Wu et al. 1988). Mapping of both morphological and isozyme markers was greatly aided by the development of a series of primary trisomics which could be used to associate genetic loci (and thus linkage groups) with each of the 12 chromosomes (Khush et al. 1984). In 1988, the first rice RFLP linkage map was reported (McCouch et al. 1988). This map was developed at Cornell University and contained 135 loci corresponding to single or low copy *Pst*I-generated genomic clones. Since that time, a large number of additional morphological and molecular marker loci have been mapped in rice. Using a set of telotrisomic stocks, the centromeres of all 12 chromosomes have been located in relation to mapped markers (Singh et al. 1996). The main molecular mapping efforts on rice have been at Cornell University (sponsored by the Rockefeller Foundation; Causse et al. 1994; Chen et al. 1997), at the Ministry of Fishery and Agriculture in Japan

R.L. Phillips and I.K. Vasil (eds.), DNA-Based Markers in Plants, 337 - 345.

(Kurata et al. 1994; Harushima et al. 1998), in China (Xiong et al. 1997) and Korea (Cho et al. 1998). RFLP/SSLP maps developed by these groups are now being integrated and together comprise more than 3,000 loci corresponding to genomic and cDNA clones (RFLP), AFLP and microsatellite markers (SSLP). The map developed at Cornell currently contains 630 loci and is the one presented in this chapter.

3.   RFLP/SSLP map

The RFLP/SSLP map presented in Fig. 1 represents a composite picture of marker order based on information derived from two mapping populations; a doubled haploid population derived from an inter-subspecific cross (*indica* X *japonica*), and a backcross population derived from the interspecific cross *Oryza sativa* X *O. longistaminata*. *O. longistaminata* is a wild rice from Africa which possesses the same genome (AA) as *Oryza sativa*. The genetic divergence between the two species and between the *indica* and *japonica* subspecies allows for great efficiency in genetic mapping. A total of 1,491 map units was measured in the interspecific backcross; however, the overall rate of recombination appears to be only 70-80% of that found in the doubled haploid population, which was approximately 1900 cM. On average, this corresponds to a DNA:cM ratio of approximately 250 kb/cM which is considerably less than for most crop species, and very close to the value for *Arabidopsis thaliana*. The abundance of mapped markers in rice provides a genetic landmark approximately every 0.5 cM, or every 125 kb. This paves the way for the development of a complete physical map of the rice genome (Kurata et al. 1997; Zhang and Wing 1997). End-sequence information is available for virtually all of the clones on both the Cornell and the Japanese maps, making it possible to move seamlessly between genetic and physical maps via PCR.

Of highest priority for rice breeders has been the mapping of genes and QTLs important in rice production - especially for disease and insect resistance, abiotic stress tolerance, yield and yield components, maturity and plant architecture (summarized in the RiceGenes database, http://probe.nalusda.gov). Rice blast (caused by the fungal pathogen *Pyricularia oryzae*) and bacterial leaf blight (caused by *Xanthomonas oryzae* pv. *oryzae*) represent two of the most serious disease problems in rice worldwide and many major genes and QTLs conferring resistance to these pathogens have been mapped with respect to molecular markers. Several have now been cloned using a combination of positional, candidate gene and genomic sequencing approaches (Song et al. 1995; Yoshimura et al. 1998; B. Valent, DuPont, pers. comm.; M. Blair, Cornell University, pers. comm.). The toolkit available for rice, including high resolution maps, sequence information, an efficient transformation system, an extensive germplasm collection, mutant stocks and a long history of crop improvement, makes this species an excellent target for full scaled structural (sequencing) and functional genomic analysis.

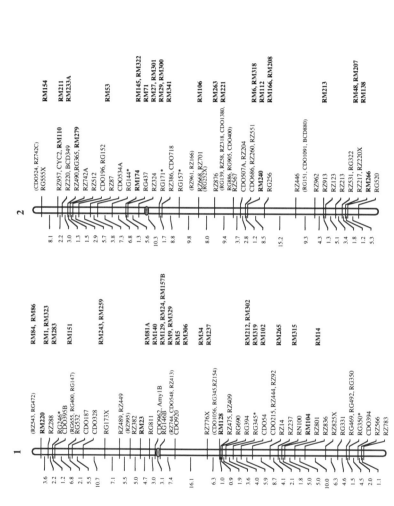

Figure 1. Rice linkage map: on the left is the molecular map based on the interspecific backcross population (*O.sativa/O.longistaminata//O.sativa*); on the right is the list of microsatellite markers mapped onto itersubspecific cross IR64 (*indica*) x Azucena (*japonica*) with the map positions determined based on linkage to common markers. cDNA markers are designated as RZ (from rice), CDO (from oat), BCD (from barley), rice genomic markers are designated as RG, microsatellite markers are RM and shown in bold. Map distances are presented in centimorgans (Kosambi function) to the left of the chromosome bars. Markers located to intervals with a LOD < 2.0 are represented in parentheses in the appropriate interval. Approximate positions of centomeres are indicated by gray ovals.

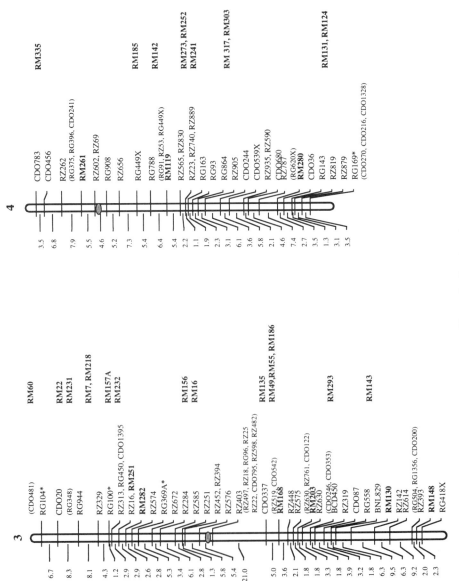

Figure 1. Continued.

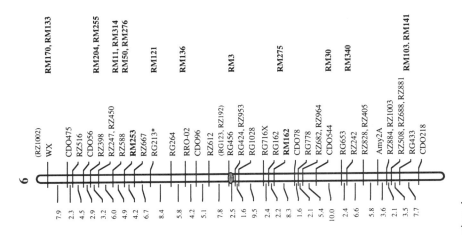

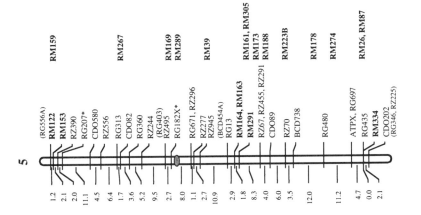

Figure 1. Continued.

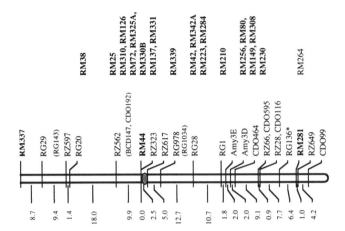

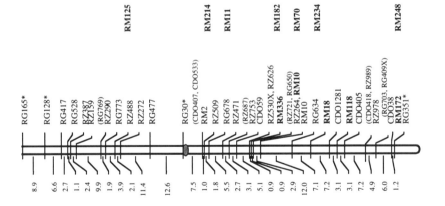

Figure 1. Continued.

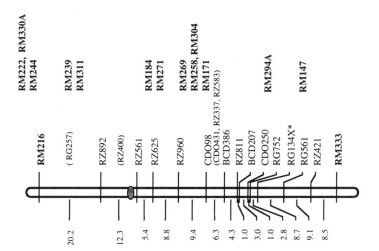

Figure 1. Continued.

4.   Future work

Because of its great economic importance and favorable genomic properties, rice is poised to become a model system for plant genome research. Comparative mapping among the grasses (Devos and Gale 1997; Van Deynze et al. 1998; Wilson et al. 1998) and between monocots and dicots has demonstrated that the rice genome provides a useful model for examining synteny within and between families of plants. Database tools are being developed to facilitate comparative genomic analysis aimed at understanding not only how genes are organized along the chromosomes in a wide array of species, but to define the entire genetic repertoire of rice and to use that information to study how gene structure relates to gene function. Ultimately, genomic science aims to provide a roadmap that will allow breeders and geneticists to use information about the structure and function of individual genes to predict how allelic variation in those genes leads to new functionality, and to understand how genes interact with each other and with the environment in the expression of the phenotype.

As a target of intensive genomic research today, rice provides a valuable template for developing new strategies for harnessing basic insights from genomic science to drive the development of novel crop varieties for the future and to pave the way for the investigation of larger, more complex biological questions in the years to come.

5.   References

Arumunagathan, K. and Earle, E.D. (1991) Nuclear DNA content of some important plant species. Plant Mol. Bio. Reporter 9: 208-218.

Causse, M., Fulton, T.M., Cho, Y.G., Ahn, S.N., Chunwongse, J., Wu, K., Xiao, J., Yu, Z., Ronald, P.C., Harrington, S.B., Second, G.A., McCouch, S.R. and Tanksley, S.D. (1994) Saturated molecular map of the rice genome based on an interspecific backcross population. Genetics 138: 1251-1274.

Chen, X., Temnykh, S., Xu, Y., Cho, Y.G. and McCouch, S.R. (1997) Development of a microsatellite framework map providing genome-wide coverage in rice (*Oryza sativa* L.). Theor. Appl. Genet. 95: 553-567.

Cho, Y.G., McCouch, S.R., Kuiper, M., Kang, M.R., Pot, J., Groenen, J.T.M. and Eun, M.Y. (1998) Integrated map of AFLP, SSLP and RFLP markers using a recombinant inbred population of rice (*Oryza sativa* L.). Theor. Appl. Genet. 97: 370-380.

Devos, K.M. and Gale, M.D. (1997) Comparative genetics in the grasses. Pl. Mol. Biol. 35: 3-15.

Harushima, Y., Yano, M., Shomura, A., Sato, M., Shimano, T., Kuboki, Y., Yamamoto, T., Yang Lin, S., Antonio, B., Parco, A., Kajiya, H., Huang, N., Yamamoto, K., Nagamura, Y., Kurata, N., Khush, G. and Sasaki, T. (1998) A high density rice genetic linkage map with 2275 markers using a single F_2 population. Genetics 148: 479-494.

Khush, G.S., Singh, R.J., Sur, S.C. and Librojo, A.L. (1984) Primary trisomics of rice: origin, morphology, cytology, and use in linkage mapping. Genetics 107: 141-163.

Kinoshita, T. (1986) Standardization of gene symbols and linkage maps in rice. In: Rice Genetics, pp. 215-228. Proc. Int. Rice Gen. Symp., May 1985, IRRI, Los Banos, Philippines.

Kothari, S.L., Davey, M.R., Lynch, P.T., Finch, R.P. and Cocking, E.C. (1993) Transgenic Rice.

In: Transgenic Plants Vol. 2 (Kung, S.D. and Wu, R., eds.), pp. 3-20. Academic Press Inc., NY.

Kurata, N., Nagamura, Y., Yamamoto, K., Harushima, Y., Sue, N., Wu, J., Antonio, B.A., Shomura, A., Shimizu, T., Lin, S.Y., Inoue, T., Fukuda, A., Shimano, T., Kuboki, Y., Toyama, T., Miyamoto, Y., Kirihara, T., Hayasaka, K., Miyao, A., Monna, L., Zhong, H.S., Tamura, Y., Wang, Z.X., Momma, Y., Umehara, Y., Yano, M., Sasaki, T. and Minobe, Y. (1994) A 300 kilobase interval genetic map of rice including 883 expressed sequences. Nature Genetics 8: 365-372.

Kurata, N., Umehara, Y., Tanoue, H. and Sasaki, T. (1997) Physical mapping of the rice genome with YAC clones. PMB 35: 101-113.

McCouch, S.R., Kochert, G., Yu, Z.H., Khush, G.S., Coffman, W.R. and Tanksley, S.D. (1988) Molecular mapping of rice chromosomes. Theor. Appl. Genet. 76: 815-829.

Ranjhan, S., Glazmann, J.L., Ramirez, D.A. and Khush, G.S. (1988) Chromosomal location of four isozyme loci by trisomic analysis in rice (*Oryza sativa* L.). Theor. Appl. Genet. 75: 541-545.

Singh, K., Ishii, T., Parco, A., Huang, N., Brar, D.S. and Khush, G.S. (1996) Centromere mapping and orientation of the molecular linkage map of rice (*Oryza sativa* L.). PNAS 93: 6163-6168.

Song, W.-Y., Wang, G.-L., Chen, L., Kim, H.-S., Wang, B.,Holsten, T., Zhai, W.-X., Zhu, L.-H., Fauquet, C. and Ronald, P. (1995) The rice disease resistance gene *Xa21* encodes a receptor kinase-like protein. Science 270: 1804-1806.

Uchimiya, H., Fushimi, T., Hashimoto, H., Harada, H., Syono, K. and Sugawara, Y. (1986) Expression of a foreign gene in callus derived from DNA-treated protoplasts of rice (*Oryza sativa* L.). Mol. Gen. Genet. 204: 204-207.

Van Deynze, A.E., Sorrells, M.E., Park, W.D., Ayres, N.M., Fu, H., Cartinhour, S.W., Paul, E. and McCouch, S.R. (1998) Anchor probes for comparative mapping of grass genera. Theor. Appl. Genet. 97: 356-369.

Wilson, W.A., Harrington, S.E., Woodman, W., Lee, M., Sorrells, M.E. and McCouch, S.R. (1998) Inferring the genome structure of progenitor maize through comparative analysis of rice, maize and the domesticated panicoids. Genetics (submitted).

Wu, K.S., Glaszamnn, J.C. and Khush, G.S. (1988) Chromosome location of ten isozyme loci in rice (*Oryza sativa* L.) through trisomic analysis. Biochem. Genet. 26: 311-328.

Xiong L, Liu, K.D., Dai, X.K., Wang, S.W., Xu, C.G., Zhang, D.P., Maroof, M.A.S., Sasaki, T. and Zhang, Q. (1997) A high density RFLP map based on the F_2 population of a cross between *Oryza sativa* and *O. rufipogon* using Cornell and RGP markers. Rice Genet. Newsletter 14: 110-116.

Yoshimura, S., Yamanouchi, U., Katayose, Y., Toki, S., Wang, Z.X., Kono, I., Kurata, N., Yano, M., Iwata, N. and Sasaki, T. (1998) Expression of *Xa1*, a novel bacterial blight resistance gene in rice, is induced by bacterial inoculation. Proc. Natl. Acad. Sci. 95: 1663-1668.

Zhang, H.B. and Wing, R.A. (1997) Physical mapping of the rice genome with BACs. Pl. Mol. Biol. 35: 115-127.

20. A framework genetic map of sorghum containing RFLP, SSR and morphological markers

JEFFREY L. BENNETZEN[1], VAIDYANATHAN SUBRAMANIAN[1],
JICHEN XU[1], SHANMUKHASWAMI S. SALIMATH[1],
SUJATHA SUBRAMANIAN[1],
DINAKAR BHATTRAMAKKI[2], *and* GARY E. HART[3]

[1]*Department of Biological Sciences, Purdue University, West Lafayette, Indiana 47907 U.S.A.*
<maize@bilbo.bio.purdue.edu>
[2]*Dupont Agricultural Products, Deleware Tech. Park, Suite 200, P.O. Box 6104,*
Newark, Delaware, 19714 U.S.A.
[3]*Department of Soil and Crop Sciences, Texas A&M University, College Station, Texas 77843 U.S.A.*

Contents

1. Introduction

Sorghum (*Sorghum bicolor* L. Moench) is a member of the tribe Andropogoneae, which also contains maize (*Zea mays* L.) and sugarcane (*Saccharum* L.). Despite having over 200 identified and named loci, only three linkage groups with greater than two mapped morphological loci had been identified by the late 1980's, and the largest of these contained only five genes (Schertz and Stephens, 1966).

Sorghum ranks fifth worldwide in grain production, and is the most important food crop for over 300 million people in Africa and other parts of the semi-arid tropics. Sorghum's exceptional tolerance to biotic and abiotic stresses, particularly drought and heat, partly account for its extensive use as a forage or food crop throughout the tropics and in dryland agriculture in temperate regions. Moreover, sorghum's relatively small genome (about 750 Mbp, compared to about 2500 Mbp for maize) (Michaelson et al. 1991) suggests that sorghum might be used as a surrogate for maize in some genomic studies (Bennetzen and Freeling 1993). Hence, efforts to generate a detailed genetic map of sorghum, employing DNA markers, were initiated in the late 1980's.

R.L. Phillips and I.K. Vasil (eds.), DNA-Based Markers in Plants, 347 - 355.

1.1. Sorghum maps based on DNA markers

The first DNA marker-based genetic map of sorghum was generated using restriction fragment length polymorphism (RFLP) markers from maize (Hulbert et al. 1990). This provided an initial sorghum map that could be directly compared to the maize genetic map, and such comparisons have continued to figure prominently in sorghum genetic mapping. From an initial crude map with just 37 markers covering less than 300 cM in several small linkage groups (Hulbert et al. 1990), many recent sorghum genetic maps have included over 200 markers each. These advanced maps cover at least 1200 cM in each case, and identify the ten linkage groups predicted by the ten chromosome pairs of sorghum (Chittenden al. 1994; Pereira et al. 1994; Peng et al. 1999).

1.2. Better marker systems

Most of the early genetic maps in sorghum primarily contained RFLPs as DNA markers. These robust markers proved to be directly comparable between different mapping populations, both within the genus *Sorghum* (Chittenden et al. 1994) and between sorghum and other grass species (Hulbert et al. 1990; Binelli et al. 1992; Whitkus et al. 1992; Berhan et al. 1993; Chittenden et al. 1994; Ragab et al. 1994; Pereira et al. 1994; Dufour et al. 1997; Peng et al. 1999). However, RFLP markers can be expensive and slow to utilize, particularly in cases where polymorphisms are infrequent within a segregating population. Some mapping projects utilized random amplified polymorphic DNA (RAPD) markers to speed up the mapping process (Weerasuriya 1995; Tuinstra et al. 1998), but the irreproducibility of these markers between populations even within the species made this largely a dead end approach.

Recently, a few laboratories have begun to use simple sequence repeat (SSR) markers in mapping sorghum (Taramino et al. 1997; Kong et al. 1999). Although these markers are not usually of value outside the species of their origin, they do provide a relatively inexpensive and dependable marker system for assembly of a framework map that can be compared to any other sorghum map containing the same SSRs. We hope that all sorghum mapping groups will routinely employ these SSRs to provide maps that can be easily compared and combined.

1.3. A framework map for sorghum

We have chosen to assemble a framework genetic map for sorghum that can be used by all subsequent sorghum mapping projects. This map (Figure 1) primarily comes from comparisons of the maps of Peng et al. (1999), Kong et al. (1999), Berhan et al. (1993), and Pereira et al. (1994). Markers were chosen by three primary criteria: (1) a spacing of about one marker for every 5 cM to 10 cM, (2) markers with low copy-numbers (preferably a copy number of one), and (3) RFLP markers that were from an anchor set that can be used for mapping any grass genome (Van Deynze et al. 1998). The map includes 194 DNA markers (covering about 1450 cM), of which 40 are SSRs (Kong et al. 1999; D. Bhattramakki and G. Hart, unpub. data). The designations of the linkage groups (A-J) are identical to those in Peng et al. (1999), and their relationships to other published linkage maps are shown in Table 1.

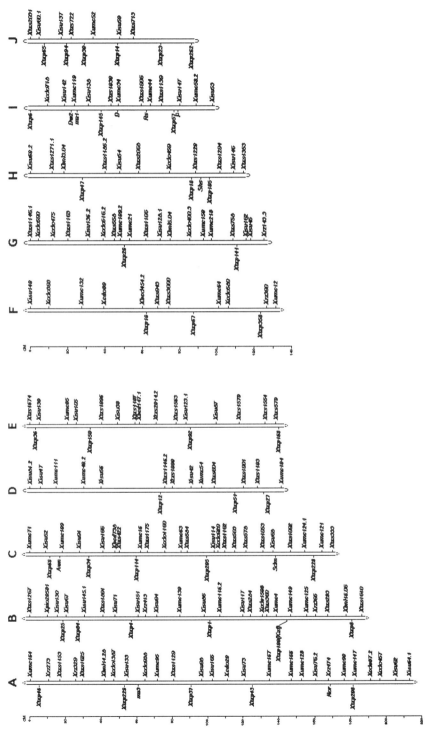

Figure 1. Framework linkage map of *Sorghum bicolor* composed of RFLPs, SSRs, and morphological markers. Symbols for RFLP loci are shown to the right and those for SSRs and morphological markers to the left of the linkage groups. See text for further explanation.

Table 1.   Relationships among the linkage groups that comprise several RFLP linkage maps of sorghum[1]

Peng et al., 1999	Whitkus et al., 1992	Berhan et al., 1993	Ragab et al., 1994	Chittenden et al., 1994[2]	Pereira et al., 1994	Dufour et al., 1997	Tao et al., 1998
A	B, C	E, J, K	B	C	C	C, K	3
B	D	B	A	B	F	F	5, 6
C	F, (M)	A	F	A	G	G	4
D	H	G	—	F	D	D, L	2
E	(A, G)	—	—	(J)	A	A	9
F	I	L	I	G	E	E	12
G	J	C	D	I	H	H	1
H	(K, L)	—	—	E	I	I	—
I	(E)	D	E	D	B	B	—
J	—	—	—	(H)	J	J	(7)
323	98	96	71	276	201	198	163

[1] Adapted from Peng et al. (1999). Parentheses surrounding a linkage-group designation indicate it is probably related to the other linkage groups listed on the same row but definitive evidence for this could not be deduced from the common markers in the linkage groups. Not shown are linkage groups whose probable relationship to other linkage groups could not be determined. Also not shown are the eight linkage groups composed of 37 markers of Hulbert et al. (1990) and the five linkage groups composed of 21 markers of Binelli et al. (1992). Extended versions of the former comprise the Berhan et al. (1993) linkage groups and each of the latter is composed of from two to six markers only. The last row in the table lists the number of markers on each map.

[2] See also Patterson et al. (1995).

1.4. Morphological and isozyme markers

Very few morphological markers have been placed on the sorghum map, although many such mapping projects are in progress. We placed a few morphological markers on the framework map, although their positions are only approximate because the mapping was performed in different populations than the ones used to assemble the framework map. These morphological markers, along with the SSRs, are shown to the left of the linkage groups.

The Mullet laboratory (Childs et al. 1997) located the photoperiod sensitivity gene *Ma3*, which encodes a phytochrome B, to linkage group A between *Xumc95* and *Xbnl14.28*. The Bennetzen lab mapped the nucleolar organizer (*Nor*) of sorghum as a DNA polymorphism by a polymerase chain reaction experiment across the ribosomal DNA spacer (S. Subramanian and J. Bennetzen, unpub. data). The *Nor* maps to linkage group A, between markers *Xisu76* and *Xumc90*. Interestingly, this location is not syntenic with the position of the *Nor* on the short arm of chromosome 6 in maize. This lack of synteny between nucleolar organizers also has been reported within the Triticeae, indicating that *Nor* position is one of the least conserved components of genome structure (Dubcovsky and Dvorak 1995).

Linkage group C is the location of a gene that determines the presence or absence of awns. The gene that determines this trait maps between *Xisu52* and *Xumc109* (G. Hart, Y. Peng, and K. Schertz, unpub. obs.).

The Fredericksen and Magill laboratories mapped a gene (*Sdm*) for resistance to sorghum downy mildew (*Peronosclerospora sorghi*) between markers *Xtxs1053* and *Xtxs1092* on linkage group C (Bhavanishankara Gowda et al. 1995). A second disease resistance gene (*Shs*), to head smut (*Sporisorium reilianum*), also was mapped by this group, between *Xtxs1220* and *Xtxs1294* in linkage group H (Oh et al. 1996).

Several morphological markers have been mapped to linkage group I, in a number of independent studies. The Bennetzen lab mapped the linkage group containing markers *D* (juicy stalk), *Rs* (red seedling) and *P* (adult plant color) (V. Subramanian, S. Salimath, J. Bennetzen, unpub. data) in this group. By inference, the *Lg* (liguleless) locus also should be in this region, because Webster (1965) mapped it about 7 cM from *P* and 20 cM from *Rs*. In addition, several labs have identified a dwarfing gene on this linkage group that is tightly linked to a maturity locus (Lin et al. 1995; Paterson et al. 1995; Pereira and Lee 1995). Most probably, these are the loci *Dw2* and *ma1*. *Mr* (presence or absence of midrib) also has been located in this linkage group (G. Hart, Y. Peng, and K. Schertz, unpub. data).

Many additional morphological and QTL markers need to be added to the sorghum genetic map. Sorghum also has a substantial number of identified isozyme polymorphisms that could be used to enrich a genetic map (Aldrich et al. 1992).

1.5. Comparative maps with other species: Is sorghum an ancient tetraploid?

Partly because sorghum was first mapped using maize DNA probes (Hulbert et al. 1990), most sorghum maps were generated using at least some markers that can be compared between species. These comparative maps indicate that sorghum has similar gene content and extensive regions of colinear gene order with the other grasses (Moore et al. 1995; Paterson et al. 1995; Gale and Devos 1998; Peng et al. 1999). However, many large rearrangements differentiate sorghum from the other grasses, including approximate whole arm translocations and inversions (Hulbert et al. 1990; Moore et al. 1995; Gale and Devos 1998). Not surprisingly, sorghum shows the greatest degree of colinearity with its closest mapped relatives, maize and sugarcane. In fact, sorghum has been proposed as the source of a foundation genetic map that can be used to help assemble maps for the complex polyploid and aneuploid genomes of various *Saccharum* species (daSilva et al. 1993; Glaszmann et al. 1997; Guimaraes et al. 1997). Recent DNA map comparisons suggest that sorghum and other grasses also have retained extensive genetic map colinearity with the genome of the dicotyledenous model plant, *Arabidopsis thaliana* (Paterson et al. 1996). However, genomic sequencing and clone map data indicate that there is surprisingly little colinearity, even at the level of adjacent genes, between Arabidopsis and any of the investigated grasses (Bennetzen et al. 1998; Tikhonov et al. 1999).

When the DNA sequences of long sorghum genomic segments (containing several genes) are compared with those of other grass species, substantial retention of both common gene content and gene order/orientation is observed (Chen et al. 1998; Tikhonov et al. 1999). Once again, sorghum exhibits more colinearity with its close relatives (maize) than it does with more distant relatives (rice, Arabidopsis). However, even in this short segment of less than 1 cM, maize was found to contain two short deletions (including three genes) relative to sorghum, and the DNA between the genes was totally unconserved (Tikhonov et al. 1999).

Because some parasorghums have only five chromosome pairs, and because the maize genome is an ancient tetraploid with ten chromosome pairs (Gaut and Doebley 1997), it has long been assumed that sorghum also is an ancient tetraploid. The map positions of duplicated DNA markers within sorghum populations indicate that many short segments of the sorghum genome have been duplicated at two unlinked locations (Chittenden et al. 1994; Peng et al. 1999). However, these mapped duplications currently fall far short of indicating a completely duplicated genome. When the (albeit incomplete) sorghum genetic maps are compared to the maps of other grasses, sorghum routinely appears to be a true diploid (Moore et al. 1994; Gale and Devos 1998). Obviously, additional research is needed to determine to what degree the sorghum genome is the product of segmental and/or chromosomal duplication.

1.6.   Future improvements in the sorghum genetic map

Although sorghum genetic maps have improved enormously over the last ten years, they are still in need of additional improvement. In most cases, it is still very difficult to determine to even a crude approximation where a gene or QTL that has been located on one sorghum map would be sited on any other sorghum map (for instance, see Table 1). This problem could be rectified if all future maps contain some of the SSR and RFLP anchor markers present on our framework map (Figure 1). The anchor RFLPs also will allow the sorghum maps to be oriented relative to all other grass genetic maps (Van Deynze et al. 1998). The mapping of centromeric (Jiang et al. 1996) and telomeric

DNA markers also would help determine the full scale of individual chromosomes, and how that scale relates to the genetic map. Similarly, it would be a major stride forward if *in situ* hybridization was utilized to correlate each of the linkage groups A-J with a particular sorghum chromosome 1-10. Finally, more of the 200 plus morphological markers (Schertz and Stephens 1966) and the many isozyme loci identified in sorghum (Aldrich et al. 1992) need to be located on the DNA marker-based genetic maps. Once mapped, these morphological markers can be compared to similar characteristics that have been positioned in other species, thereby providing an indication of the intraspecies commonalities and contrasts that can be used as the foundation for future crop study and improvement (Bennetzen and Freeling 1997). The economic importance and fascinating biology of sorghum justify major commitments to its study, and high-quality genetic maps can provide the foundation for those studies.

2. References

Aldrich, P.R., Doebley, J., Schertz, K.F., and Stec, A. (1992) Patterns of allozyme variation in cultivated and wild *Sorghum bicolor*. Theor. Appl. Genet. 85: 451-460.

Bennetzen, J.L., and Freeling, M. (1993) Grasses as a single genetic system: genome composition, collinearity and compatibility. Trends Genet. 9: 259-261.

Bennetzen, J.L., and Freeling, M. (1997) The unified grass genome: synergy in synteny. Genome Res. 7: 301-306.

Bennetzen, J.L., SanMiguel, P., Chen, M., Tikhonov, A., Francki, M., and Avramova, Z. (1998) Grass genomes. Proc. Natl. Acad. Sci. USA 95: 1975-1978.

Berhan, A. M., S. H. Hulbert, L. G. Butler, and J. L. Bennetzen (1993) Structure and evolution of the genomes of *Sorghum bicolor* and *Zea mays*. Theor. Appl. Genet. 86: 598-604.

Bhavanishankara Gowda, P.S., Xu, G.-W., Fredericksen, R.A., and Magill, C.W. (1995) DNA markers for downy mildew resistance in sorghum. Genome 38: 823-826.

Binelli, G., Gianfranceschi, L., Pe, M.E., Taramino, G., Busso, C., Stenhouse, J., and Ottaviano, E. (1992) Similarity of maize and sorghum genomes as revealed by maize RFLP probes. Theor. Appl. Genet. 84: 10-16.

Chen, M., SanMiguel, P., and Bennetzen, J.L. (1998) Sequence organization and conservation in *sh2/a1*-homologous regions of sorghum and rice. Genetics 118: 435-443.

Childs, K.L., Miller, F.R., Cordonnier-Pratt, M.-M., Pratt, L.H., Morgan, P.W., and Mullet, J.E. (1997) The sorghum photoperiod sensitivity gene, *Ma3*, encodes a phytochrome B. Plant Physiol. 113: 611-619.

Chittenden, L.M., Schertz, K.F., Lin, Y.-R., Wing, R.A., and Paterson, A.H. (1994) A detailed RFLP map of *Sorghum bicolor* X *S. propinquum*, suitable for high-density mapping, suggests ancestral duplication of sorghum chromosomes or chromosomal segments. Theor. Appl. Genet. 87: 925-933.

Dubcovsky, J., and Dvorak, J. (1995) Ribosomal RNA multigene loci: nomads of the Triticeae genomes. Genetics 140: 1367-1377.

Dufour, P., M. Deu, L. Grivet, A. D'Hont, F. Paulet, A. Bouet, D. Lanaud J. C. Glaszmann, and P. Hamon (1997) Construction of a composite sorghum genome map and comparison with sugarcane, a related complex polyploid. Theor. Appl. Genet. 94: 409-418.

Gale, M.D., and Devos, K.M. (1998) Plant comparative genetics after 10 years. Science 282: 656-659.

Gaut, B.S., and Doebley, J.F. (1997) DNA sequence evidence for the segmental allotetraploid origin of maize. Proc. Natl. Acad. Sci. USA 94: 6809-6814.

Glaszmann, J.C., Dufour, P., Grivet, L., D'Hont, A., Deu, M., Paulet, F., and Hamon, P. (1997) Comparative genome analysis between several tropical grasses. Euphytica 96: 13-21.

Guimaraes, C.T., Sills, G.R., and Sobral, B.W. (1997) Comparative mapping of Andropogoneae: *Saccharum* L. (sugarcane) and its relation to sorghum and maize. Proc. Natl. Acad. Sci. USA 94:14261-14266.

Jiang, J., Nasuda, S., Dong, F., Scherrer, C.W., Woo, S.-S., Wing, R.A., Gill, B.S., and Ward, D.C. (1996) A conserved repetitive DNA element located in the centromeres of cereal chromosomes. Proc. Natl. Acad. Sci. USA 93: 14210-14213.

Kong, L., Dong, J., and Hart, G.E. (2000) Characteristics, linkage-map positions, and allelic differentiation of Sorghum bicolor (L.) Moench DNA simple sequence repeats (SSRs). Theor. Appl. Genet., in press.

Lin, Y.-R., Schertz, K.F., and Paterson, A.H. (1995) Comparative analysis of QTLs affecting plant height and maturity across the Poaceae, in reference to an interspecific sorghum population. Genetics 141: 391-411.

Michaelson, M.J., Price, H.J., Ellison, J.R., and Johnston, J.S. (1991) Comparison of plant DNA contents determined by Feulgen microspectrophotometry and laser flow cytometry. Am. J. Bot. 78:183-188.

Moore, G., Devos, K.M., Wang, Z., and Gale, M.D. (1995) Grasses, line up and form a circle. Curr. Biol. 5: 737-739.

Oh, B.-J., Fredericksen, R.A., and Magill, C.W. (1996) Identification of RFLP markers linked to a gene for downy mildew resistance (*Sdm*) in sorghum. Can. J. Bot. 74: 315-317.

Paterson, A.H., Lan, T.-H., Reischmann, K.P., Chang, C., Lin, Y.-R., Liu, S.-C., Burow, M.D., Kowalski, S.P., Katsar, C.S., DelMonte, T.A., Feldmann, K.A., Schertz, K.F., and Wendel, J.F. (1996) Toward a unified genetic map of higher plants, transcending the monocot-dicot divergence. Nature Genet. 14: 380-382.

Paterson, A.H., Lin, Y.-R., Li, Z., Schertz, K.F., Doebley, J.F., Pinson, S.R.M., Liu, S.-C., Stansel, J.W., and Irvine, J.E. (1995) Convergent domestication of cereal crops by independent mutations at corresponding genetic loci. Science 269: 1714-1718.

Peng, Y., Schertz, K.F., Cartinhour, S., and Hart, G.E. (1999) Comparative genome mapping of *Sorghum bicolor* (L.) Moench using a RFLP map constructed in a population of recombinant inbred lines. Plant Breeding 118: 225-235.

Pereira, M.G., and Lee, M. (1995) Identification of genomic regions affecting plant height in sorghum and maize. Theor. Appl. Genet. 90: 380-388.

Pereira, M.G., Lee, M., Bramel-Cox, P., Woodman, W., Doebley, J., and Whitkus, R. (1994) Construction of an RFLP map in sorghum and comparative mapping in maize. Genome 37: 236-243.

Ragab, R.A., Dronavalli, S., Saghai-Maroof, M.A., and Yu, Y.G. (1994) Construction of a sorghum RFLP map using sorghum and maize DNA probes. Genome 37: 590-594.

Schertz, K.F., and Stephens, J.C. (1966) Compilation of gene symbols, recommended revision and summary of linkages for inherited characteristics of *Sorghum vulgare* Pers. Texas A&M University Technical Monograph, College Station, TX.

daSilva, J.A.G., Sorrells, M.E., Burnquist, W.L., and Tanksley, S.D. (1993) RFLP linkage map and genome analysis of *Saccharum spontaneum*. Genome 36: 782-791.

Tao, Y.Z., Jordan, D.R., Henzel, R.G., and McIntyre, C.L. (1998) Construction of a genetic map in a sorghum RIL population using probes from different sources and its comparison with other sorghum maps. Aust. J. Agric. Res. 49: 729-736.

Taramino, G., Tarchini, R., Ferrario, S., Lee, M., and Pe, M.E. (1997) Characterization and mapping of simple sequence repeats (SSRs) in *Sorghum bicolor*. Theor. Appl. Genet. 95: 66-72.

Tikhonov, A.P., SanMiguel, P.J., Nakajima, Y., Gorenstein, N.D., Bennetzen, J.L., and Avramova, Z. (1999) Colinearity and its exceptions in orthologous *adh* regions of maize and sorghum. Proc. Natl. Acad. Sci. USA 96: 7409-7414.

Tuinstra, M.R., Ejeta, G., and Goldsbrough, P. (1998) Evaluation of near-isogenic sorghum lines contrasting for QTL markers associated with drought tolerance. Crop Sci. 38: 835-842.

Van Deynze, A.E., Sorrells, M.E., Park, W.D., Ayres, N.M., Fu, H., Cartinhour, S.W., Paul, E., and McCouch, S.R. (1998) Anchor probes for comparative mapping of grass genera. Theor. Appl. Genet. 97: 356-369.

Webster, O.J. (1965) Genetic studies in *Sorghum vulgare* (Pers.). Crop Sci. 5: 207-210.

Weerasuriya, Y. (1995) The construction of a molecular linkage map, mapping of quantitative trait loci, characterization of polyphenols, and screening of genotypes for *Striga* resistance in sorghum. Ph.D. Dissertation, Purdue University.

Whitkus, R., Doebley, J., and Lee, M. (1992) Comparative mapping of sorghum and maize. Genetics 132: 1119-1130.

21. RFLP map of soybean

RANDY C. SHOEMAKER[1,2],
DAVID GRANT [1,2], *and* MARCIA IMSANDE [2]
[1]*USDA-ARS-CICGR*
[2]*Department of Agronomy, Iowa State University, Ames, IA 50011, USA.*
<resshoe@iastate.edu>

Contents

1. Historical perspective

The soybean is one of the oldest cultivated crops. It first emerged as a domesticated plant around the 11th century B.C. (Hymowitz 1970), although references to soybean appear in books written over 4500 years ago (Smith and Huyser 1987). The soybean was first planted in the United States in Thunderbolt, GA, in 1765 (Hymowitz and Harlan 1983). It was originally planted as a forage crop but by the early 1900's became of interest as an oilseed crop (Smith and Huyser 1987), an interest that continues today.

Soybean production was initiated in this country through the use of cultivars introduced from other countries. These plant introductions were evaluated and the most desirable ones were released for commercial use (Fehr 1987). During the 1940's cultivars were selected from crosses made between plant introductions. Often, the superior cultivars selected from this initial round of hybridizations were used as parents in the next cycle of hybridizations. This breeding scheme has allowed the map-based pedigree analysis of nearly all of the economically important cultivars released between 1939 and 1979 (Lorenzen et al. 1995). This breeding scheme continues today (Fehr 1987).

A limited number of accessions were used to derive the cultivars commonly grown in the U.S. today. An analysis of the pedigrees of cultivars in the Northern germplasm collection indicates that 88% of their collective genome was derived from just 10 accessions (Delannay et al. 1983; Specht and Williams 1984). Within the Southern germplasm the situation is even more extreme. Seventy percent of the genome of commonly grown Southern cultivars was shown to be contributed by as few as seven

R.L. Phillips and I.K. Vasil (eds.), DNA-Based Markers in Plants, 357 - 378.
© 2001 *Kluwer Academic Publishers. Printed in the Netherlands.*

accessions (Delannay et al. 1983). Other authors also report that the genetic foundation of modern soybean cultivars is limited and that just 12 ancestors can account for 80% of the soybean gene pool (Gizlice et al. 1994). This history of breeding and cultivar development in soybean tends to limit the amount of genetic diversity found among elite lines.

2.   Probe and map development

The soybean genome contains an estimated 1.1×10^9 (Arumuganthan and Earle 1991) to 1.29×10^9 bp (Gurley et al. 1979) to $1\ 81 \times 10^9$ bp (Goldberg 1978) for 1n DNA content. The genome is approximately 40-60% repetitive sequences (Gurley et al. l979; Goldberg 1978). The majority (65-70%) of single-copy sequences have a short period interspersion with single-copy sequences of 1.1-1.4 kb alternating with repetitive sequence elements of 0.3-0.4 kb (Gurley et al. 1979).

Analysis of pachytene chromosomes has shown that over 35% of the soybean genome is made up of heterochromatin - the short arms of six of the 20 bivalents are completely heterochromatic (Singh and Hymowitz 1988).

Low- or single-copy clones are required in establishing an RFLP map. An initial effort to create a soybean RFLP map involved the development of a random genomic library digested with Sau3AI and cloned into M13 digested with BamHI (Apuya et al. 1988). To differentiate single-copy clones from clones carrying repetitive DNA, the clones were probed with radioactive total soybean genomic DNA. Single- and low-copy clones were identified by their lack of hybridization (Apuya et al.1988). This approach is effective but labor intensive.

A high percentage of plant DNA is modified by methylation of the cytosine base. Many restriction endonucleases will not cleave DNA containing methylated cytosines within certain sequences. Keim and Shoemaker (1988) used the restriction endonuclease PstI, a methylation-sensitive enzyme to construct a genomic library. Sequences in and adjacent to transcribed regions generally are unmethylated and therefore will be cleaved by this enzyme (Burr et al. 1988). Using this technique, a library of recombinant DNA was developed that is primarily single-copy DNA sequence (Keim and Shoemaker 1988).

Total DNA was isolated from leaves of soybean seedlings (Keim et al. 1988). This DNA was digested with a 20 fold excess of PstI. DNA fragments were electrophoretically separated in low-melting agarose. Fragments in the estimated size range of 0.5-3.0 kb were extracted from the gel, and phenol extracted and cloned into the plasmid vector, pBS+. Recombinant molecules were transformed into the E. cold strain DH5 (a). Approximately 80-85% of the clones represent single- or low-copy DNA sequence (Keim and Shoemaker 1988; Keim et al. 1990a).

When tested against DNA from the *G. max* breeding line A81-356022 and *G. soja* accession PI 468.916 digested with restriction enzymes DraI, Taql, HindIII, EcoRI, or EcoRV, approximately 40% of the random genomic probes detect polymorphisms. About half of the polymorphic probes detect polymorphisms with two or more enzymes, suggesting that DNA rearrangements are the cause of the variability (Keim et al. 1990a). This was substantiated through RFLP mapping of polymorphic regions (Apuya et al. l988). Approximately 10% of the probes detect 'dominant' markers, i.e., the heterozygote classes are indistinguishable from dominant classes. These 'dominant' markers could be the result of additions or deletions within one of the genotypes.

The map represented in this chapter was generated by analysis of marker segregation among 60 F_2 individuals. The computer program 'MAPMAKER' (Lander et al. 1987) was used to create the best loci order. A minimum LOD score of 3.0 was used for the pairwise linkage analysis in all instances. More than 525 qualitative markers have been placed on this map. These markers include random genomic clones, clones of known genes, isoenzyme loci, and a variety of morphological and developmental classical markers. This map includes 20 linkage groups and encompasses approximately ~2550 cM (Fig. 1). Other genetic maps have been developed and an integration of several of these maps through the use of more than 500 SSRs is in the later stages of development (Cregan et al. in review).

3. Integrating classical and molecular markers

Historically, the construction of linkage maps comprised of phenotypic genetic markers has constituted the traditional means of genome analysis. However, conventional map development in soybean has proceeded slowly, due primarily to the difficulty in performing crosses and generating large numbers of hybrid seed, the lack of detailed cytogenetic markers, and a paucity of genetic variation in the germplasm. Thus, compared to other economically important species, the resolution of the classical map is poor. Currently, the classical soybean genetic linkage map contains 61 linked markers covering approximately 750 map units (Palmer and Hedges 1993). Many other qualitative traits remain unmapped.

Integrating classical phenotypic markers into a genetic map is a critical step to developing any organism as a genetic system. By screening near isogenic lines (NIL), many molecular markers that are linked to the conventional markers can be efficiently identified in a single experiment in donor and recurrent parents (Young et al. 1988).

Integrating classical markers into the molecular map is potentially an efficient process for soybeans, since the germplasm collection contains an extensive number of NILs (Bernard 1976). The potential of NILs in integrating soybean conventional and molecular linkage maps was discussed in depth by Muehlbauer et al. (1988). They calculated that if a BC_5S_1 NIL was screened with 100 randomly chosen loci, assuming polymorphisms existed between the recurrent parent and the donor parent at all loci, four loci should detect donor DNA and two or three of them could be expected to be genetically linked to the introgressed gene. Evaluating 63 NILs, each possessing an introgressed conventional gene with 12 isozyme loci, 5 presumptive linkages were observed by Muehlbauer et al. (1989). Muehlbauer et al. (1991) screened 116 NILs using RFLPs. Fifteen polymorphisms were observed where the NIL possessed the donor parent allele. Segregation analysis confirmed linkages of several genes with these markers. This approach now has been used many times with similar success. However, in many studies using this approach, even though linkages between genes and markers has been established, the position of some of these markers relative to other mapped molecular markers remains ambiguous (Fig. 1).

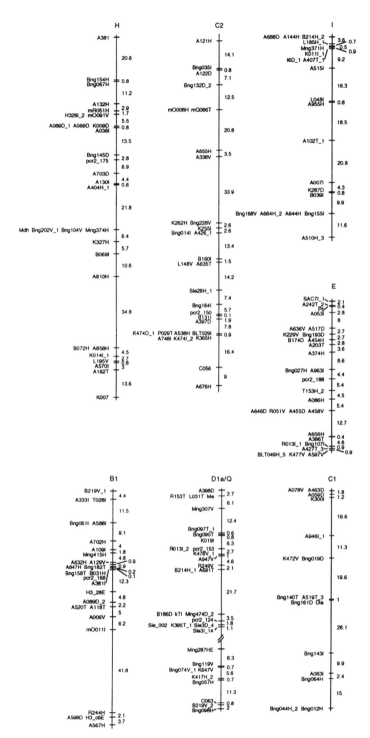

Figure 1.   Restriction fragment length polynorphism map of soybean. Lingkage groups have

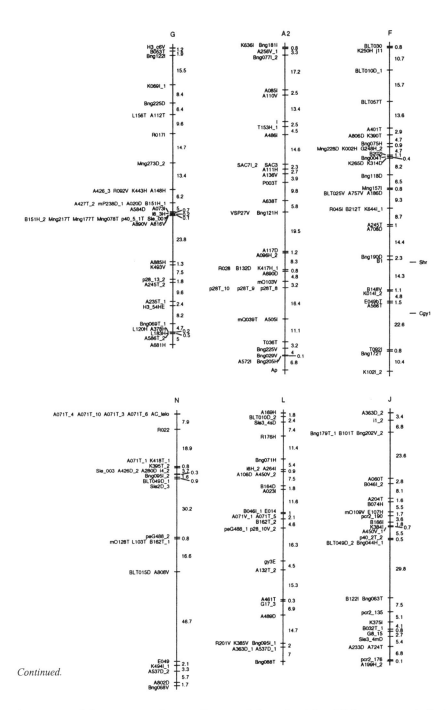

been arranged for economy of space. Gene symbols to the right of linkage groups indicate that the gene has been mapped to the linkage group, but that the exact position of the gene relative to other markers is ambiguous.

To fully exploit the potential of a molecular genetic map, it is necessary to unambiguously integrate molecular and conventional markers into a unified linkage map. This can be accomplished through painstaking segregation and linkage analysis in entire populations using both conventional and molecular markers; a technique used successfully by Shoemaker and Specht (1995) to integrate approximately half of the classical linkage groups into the molecular map in a single experiment. Many genes have been now been placed on linkage groups through mapping of cDNAs and genomic clones, isoenzymes, or mapping of phenotypic conventional markers (Table 1).

Table 1.   Probes detecting genetic linkages to qualitatively inherited genes/traits. Because of the possibility of duplicate loci the precise map position of these genes/traits remains ambiguous in some cases. For qualitatively inherited disease resistance genes see Table 3.

Gene/trait	Description	LG	Reference
SOYNOD26A	Nodulin 26	A1	Cregan et al. (in review)
I	Seed coat color	A2	Shoemaker and Specht, 1995
SAc7-2	Actin gene	A2	Diers et al. 1992a
SAc3-1	Actin gene	A2	Diers et al. 1992a
VSP27	Vegetative storage protein	A2	Shoemaker and Olson, 1993
Ap	Acid phosphatase	A2	Shoemaker and Specht, 1995
ENOD2	Early nodulin	A2	Ghassemi and Gresshoff, 1998
Lfl	Five-foliate leaf	A2	Shoemaker and Specht, 1995
D2	Yellow seed embryo	B1	Cregan et al. (in review)
Aco4	Aconitase	B1	Shoemaker and Specht, 1995
K1	Non-saddle seed coat	B1	Shoemaker and Specht, 1995
H3-hs28-1E	H3 histone gene	B1	Kanazin et al. 1996a
H3-c6E(S3)	H3 histone gene	B1	Kanazin et al. 1996a
Dia	Diaphorase	C1	Diers et al. 1992a
N	Normal hilum abscission	C1	Cregan et al. (in review)
T	Tawny pubescence	C2	Shoemaker and Specht, 1995
Pgi	Phosphoglucose isomerase	C2	Cregan et al. (in review)
Me	Malic enzyme	D1a/Q	Shoemaker and Specht, 1995
Sle_002	Late embryogenesis	D1a/Q	Calvo et al. 1997
Sle3D_4	Late embryogenesis	D1a/Q	Calvo et al. 1997
Sle3I_14	Late embryogenesis	D1a/Q	Calvo et al. 1997
G	Green seed coat	D1a/Q	Shoemaker and Specht, 1995
D1	Yellow seed embryo	D1a/Q	Shoemaker and Specht, 1995
Pd1	Dense pubescence	D1a/Q	Shoemaker and Specht, 1995
Fap2	Palmitate	D?	Nickell et al. 1994
Fr2	Fluorescent root	D2	Mansur et al. 1996
Mdh	Malate dehydrogenase	D2	Cregan et al. (in review)
SAc7-1	Actin gene	E	Diers et al. 1992a
Pb	Pubescent tip	E	Keim et al. 1990b

Table 1. Continued.

Gene/trait	Description	LG	Reference
Y9	Leaf color	E	Shoemaker and Specht, 1995
Bl	Seed coat luster	F	Shoemaker and Specht, 1995
Gy5	Glycinin subunit	F	Diers et al. 1994
Wl	Flower pigmentation	F	Mansur et al. 1996
Shr	Shrivelled seed	F	Chen and Shoemaker, 1997
Cgyl	a' subunit B-conglycinin	F	Chen and Shoemaker, 1997
Sle1	Late embryogenesis	G	Calvo et al. 1997
Mpi	Mannose-6-P isomerase	G	Shoemaker and Specht, 1995
H3-c6V(S4)	H3 histone gene	G	Kanazin et al. 1996a
H3-hs54	H3 histone gene	G	Kanazin et al. 1996a
H3-hs28-2E	H3 histone gene	H	Kanazin et al. 1996a
nts-1	Supernodulation	H	Landau-Ellis et al. 1991
Mdh	Malate dehydrogenase	H	Cregan et al. (in review)
Ps	Sparce pubescence	H	Cregan et al. (in review)
Enp	Endopeptidase	I	Shoemaker and Specht, 1995
Ln	Leaflet shape	I	Shoemaker and Specht, 1995
Rj2	Ineffective nodulation	J	Polzin et al. 1994
Sle4	Late embryogenesis	J	Calvo et al. 1997
R	Black seed coat	K	Mansur et al. 1996
Ep	Peroxidase	K	Mansur et al. 1996
Frl	Fluorescent root	K	Mansur et al. 1996
Pl	Pubescence	K	Shoemaker and Specht, 1995
Dtl	Indeterminate stem	L	Lee et al. 1996b
Ll	Pod color	L	Shoemaker and Specht, 1995
Sle3	Late embryogenesis	N	Calvo et al. 1997
L2	Pod color	N	Lee et al. 1996a
H3-hs55	H3 histone gene	N	Kanazin et al. 1996a
Pgm1	Phosophoglucomutase	O	Shoemaker and Specht, 1995
Gy4	Glycinin subunit	O	Diers et al. 1994
E2	Flowering and maturity	O	Shoemaker and Specht, 1995
H3-hs60H	H3 histone gene	P/B2	Kanazin et al. 1996a
Ep	Peroxidase	P/B2	Mansur et al. 1996
Frl	Fluorescent root	P/B2	Mansur et al. 1996
Fan	Linolenic acid	P/B2	Brummer et al. 1995
Idhl	Isocitrate dehydrogenase	W/D1b	Cregan et al. (in review)

4.   Quantitative trait loci

A number of studies have been conducted towards identifying quantitative trait loci (QTLs) in the soybean genome. These studies include morphological and reproductive traits, seed composition traits, yield components, nutrient efficiency and chemical sensitivity (Table 2).

An analysis of the distribution of QTL suggest a non-random dispersion of quantitative trait loci throughout the soybean genome. Some linkage groups contain more QTL than could be expected by random chance. Some linkage groups contain QTL for 'families' of traits. For example, related traits, e.g. various seed traits, are clustered on LG-C2 suggesting either that genes for similar traits are indeed clustered, or that single genes have pleitropic effects on related traits. These possible genetic relationships have been often discussed in relation to the inverse relationship between seed oil and seed protein content (Burton 1985).

On the other hand, seemingly unrelated QTL for agronomically important traits are also coincident on a single linkage group. For example, QTL for 18 different traits can be found on LG-L. These include not only seed related traits, but such traits as iron efficiency, flowering, lodging, leaf morphology, plant height, pod maturity, and stem diameter (Table 1). These observations suggest that some linkage groups are more important to breeding programs than others.

Table 2.   Markers and linkage groups that have been associated with significant variation for quantitative traits. All QTL relationships identified are significant at probabilities less than 0.01.

Trait	Loci	LG	Reference
Chlorimuron ethyl sensit.	cr168_1	E	Mian et al. 1997
	cr274_1	E	"
	A454_2	E	"
	EV6_1	E	"
Canopy height	R013_2	D1a/Q	Keim et al. 1990a
	K390	F	"
Drought tolerance	B031_1	G	Mian et al. 1996a
	A089_1	H	"
	A381	H	"
	cr497_1	J	"
	K375	J	"
	A233	J	"
Fe efficiency	K258_4−A256_2	A1	Lin et al. 1997
	I	A2	Diers et al. 1992b
	C063	D1a/Q	"
	K417_2	D1a/Q	"
	T036_2−K227	G	Lin et al. 1997
	A404_1−B069	H	"

Table 2. Continued.

Trait	Loci	LG	Reference
Fe efficiency	A515—K644_2	I	Lin et al. 1997
	A023—A132_2	L	"
	Satt9—K418_1	N	"
	Sat33—BLT015_3	N	"
	Satt70—A593	P/B2	"
	Satt63—A519_4	P/B2	"
First flower (R1)	K365	C2	Keim et al. 1990a
	K474_1	C2	"
	K474_2	C2	"
	A397—BLT029	C2	Mansur et al. 1993a
	A109_1	C2	Mansur et al. 1996
	A385—G17_3	L	Mansur et al. 1993a
	Satt6	L	Mansur et al. 1996
	A584—R079	M	Mansur et al. 1993a
	C9b	P/B2	Mansur et al. 1996
Hard seededness	*I*	A2	Keim et al. 1990b
	T153_1	A2	"
	G17_3	L	"
Leaf ash	K258	D2	Mian et al. 1996a
	A112	G	"
	A458_2	G	"
	cr509_1	G	"
	A638_2	G	"
	K443	G	"
	B031_1	G	"
	EV1_2	G	"
	K375	J	"
Lodging	A117	A2	Lee et al. 1996a
	gac197	C1	"
	A397	C2	Mansur et al. 1993b
	A109_1	C2	Mansur et al. 1996
	L183	G	Lee et al. 1996a
	cr517	J	Lee et al. 1996b
	A060	J	Mansur et al. 1996
	A385—G17_3	L	Mansur et al. 1993a
	Satt6	L	Mansur et al. 1996
	BLT7	L	"
	Dt1	L	Lee et al. 1996b
	A169	L	Lee et al. 1996a
	EV2	L	"

Table 2.   Continued.

Trait	Loci	LG	Reference
Leaf area (mm2)	A397—BLT029	C2	Mansur et al. 1993a
	BLT53NT3	H	Mansur et al. 1996
	A584—R079	M	"
	A085	A2	Mian et al. 1998
	BLT049	E	"
	K644_1	F	"
Leaf weight (dw/lf area)	BLT043	B1	Mian et al. 1998
	A122D	C2	"
	A381	H	"
	A089D_1	H	"
	EV2E_1	L	"
	A489	L	"
	A169	L	"
Leaf length (mm)	R013_2	D1a/Q	Keim et al. 1990a
	K478_1	D1a/Q	"
	A690	G	Mansur et al. 1996
	Satt6	L	"
	BLT4	N	"
Leaf width (mm)	A111	A2	Keim et al. 1990a
	BLT049_5	E	Mansur et al. 1996
	K390	F	Keim et al. 1990a
	BLT53NT3	H	Mansur et al. 1996
	R079	M	"
	K411	W/D1b	Keim et al. 1990a
Linoleate (% of oil)	A082	A1	Diers and Shoemaker, 1992
	A104	A1	"
	A170	A1	"
	A118	B1	"
	A242_2	E	"
	Pb	E	"
Linolenate (% of oil)	SAC7_1	E	Diers and Shoemaker, 1992
	K229	E	"
	A203	E	"
	A242_2	E	"
	A454	E	"
	A065	K	"
	A023	L	"
Oil (oil % of dw of seed)	T155	A1	Mansur et al. 1996
	A329	A1	"
	A975	A1	Brummer et al. 1997

Table 2.   Continued.

Trait	Loci	LG	Reference
Oil (oil % of dw of seed)	A104	A1	"
	T153_1—A111	A2	Mansur et al. 1993a
	A109	B1	Brummer et al. 1997
	Pb	E	Diers et al. 1992a
	A454	E	"
	SAC7_1	E	"
	K229	E	"
	A203	E	"
	A584	G	Brummer et al. 1997
	A816	G	"
	A890	G	"
	A069	H	"
	K011_1	I	Diers et al. 1992a
	A407_1	I	"
	BCl—A315	K	Mansur et al. 1993a
	K387	K	Brummer et al. 1997
	A023	L	Diers et al. 1992a
	Satt6	L	Mansur et al. 1996
	A242_2	P/B2	Diers et al. 1992a
Oleate (oleate % of oil)	A082	A1	Diers and Shoemaker, 1992
	A104	A1	"
	A170	A1	"
	A242_2	E	"
	Pb	E	"
	A619	P/B2	"
Palmitic (palmitic % of oil)	K375	J	Diers and Shoemaker, 1992
	A343_1	P/B2	"
	A018	P/B2	"
Plant height (cm)	BLT043	B1	Lee et al. 1996a
	A063	C1	Lee et al. 1996b
	A397	C2	Mansur et al. 1993b
	Satt79	C2	Mansur et al. 1996
	A295	D1a/Q	Lark et al. 1995
	K007	H	Lee et al. 1996b
	A060	J	Lark et al. 1995
	B166	J	Lee et al. 1996a
	A385	L	Mansur et al. 1993a
	G17_3	L	"
	Satt6	L	Mansur et al. 1996
	BLT7	L	"

Table 2.   Continued.

Trait	Loci	LG	Reference
Plant height (cm)	BLT7	L	"
	Dt1	L	Lee et al. 1996b
	A363_2	L	Lark et al. 1995
	L50MT3	L	Lark et al. 1995
	R079	M	Mansur et al. 1993b
	A135	W/D1b	Lee et al. 1996a
Pod dehiscence	cr274_1	E	Bailey et al. 1997
	B122	J	"
Pod maturity (R8)	R183	A1	Lark et al. 1994
	BLT043	B1	Lee et al. 1996a
	cr122	B1	"
	K472	C1	Keim et al. 1990a
	A063	C1	Lee et al. 1996a
	K365	C2	Keim et al. 1990a
	K474_1	C2	"
	K474_2	C2	"
	A397	C2	Mansur et al. 1993b
	Satt79	C2	Mansur et al. 1996
	R013_2	D1a/Q	Keim et al. 1990a
	B032_2	K	Lee et al. 1996b
	R051	K	"
	R051−N100	K	"
	Satt6	L	Mansur et al. 1996
	O109n	L	Lee et al. 1996a
	R079	M	Mansur et al. 1993b
	A584−R079	M	Mansur et al. 1993a
Protein (prot % dw of sd)	T155	A1	Mansur et al. 1996
	A329	A1	"
	A505	A2	Brummer et al. 1997
	A109	B1	"
	A063	C1	"
	A398	D1a/Q	"
	SAC7_1	E	Diers et al. 1992a
	A242_2	E	"
	A816	G	Brummer et al. 1997
	A890	G	"
	A144	I	"
	K011_1	I	Diers et al. 1992a
	A407_1	I	"
	A144	I	"

Table 2.   Continued.

Trait	Loci	LG	Reference
Protein (prot % dw of sd)	A688	I	"
	A023	L	"
	Satt6	L	Mansur et al. 1996
Beginning seed (R5)	A397—BLT029	C2	Mansur et al. 1993a
Beginning seed (R5)	A385—G17_3	L	Mansur et al. 1993a
Reproductive period	G17_3	L	Mansur et al. 1996
	R079	M	"
Seed filling (R1-R8)	A397—BLT029	C2	Mansur et al. 1993a
	G8_15	J	Keim et al. 1990a
	A584—R079	M	Mansur et al. 1993a
Seed number	A109_1	C2	Mansur, 1996a
	R079	M	Mansur, 1996a
Seed weight	T155	A1	Mansur, 1996a
	K443	A2	"
	A118	B1	Maughan et al. 1996
	A059	C1	Mian et al. 1996b
	A635	C2	"
	A262d	C2	Mansur et al. 1996
	A257	D2	Mian et al. 1996b
	BLT049_2	E	"
	B031_1	G	"
	A235_1	G	"
	A816	G	Maughan et al. 1996
	K384	J	"
	B166	J	Mian et al. 1996b
	Dt1	L	"
Seed yield	A109_1—A397	C2	Mansur et al. 1993a
	Satt79	C2	Mansur et al. 1996
	A584—R079	M	Mansur et al. 1993a
	OC01_650	N	Hnetkovsky et al. 1996
Stearate (stearate % of oil)	A233	J	Diers and Shoemaker, 1992
Stem diameter	G17_3	L	Keim et al. 1990a
	K385	L	"
	R201	L	"

Another group of QTL 'gene discovery' studies has included disease resistance traits (Table 3). These quantitatively inherited disease resistance genes are in addition to those mapped as qualitative traits (Table 2) or to the Resistance Gene Analogs (RGAs) reported by Kanazin et al. (1996b) and Yu et al. (1996b).

Table 3.   Disease resistance genes placed on the soybean molecular genetic map. Where gene symbols are used, the resistance gene was mapped as a qualitative trait. Where the locus is identified by molecular markers, the resistance gene was mapped as a quantitative trait.

Disease	Locus	LG	Reference
Soybean cyst nematode	A487	A1	Vierling et al. 1996
	A085	A2	Concibido et al. 1994
	OW15_400	A2	Chang et al. 1997
	BLT65	A2	"
	A136	A2	Mahalingam and Skorupska 1995
	S07a	A2	"
	I	A2	Webb et al. 1995
	Rhg4	A2	"
	A006	B1	Vierling et al. 1996
	A567	B1	"
	A112	G	"
	K069-1	G	Concibido et al. 1994
	05354a—05219a	G	Webb et al. 1995
	C006V-Bng122D	G	Concibido et al. 1996
	A378H	G	"
	OG13_490	G	Chang et al. 1997
	Bng112D	G	"
	OI03_450	G	"
	Satt038	G	Mudge et al. 1997
	Satt130	G	"
	B032-1	J	Concibido et al. 1994
	01175a—02301a	M	Webb et al. 1995
	A280Hae-1	N	Concibido et al. 1997
Sudden death syndrome	OO05_250	C2	Hnetkovsky et al. 1996
	K455D	C2	"
	OO05_250	C2	Chang et al. 1996
	OG13_490	G	"
	OI03_450	G	"
	OE04_450	G	"
	OE02_1000	G	"
	OF04_1600	N	"
	OC01_650	N	"
Javanese root-knot nem.	A806-1	F	Tamulonis et al. 1997a
	GmHSP	F	"
	A186-1	F	"
	A757-2	F	"
	B212-1	F	"
	R045-1	F	"

Table 3.   Continued.

Disease	Locus	LG	Reference
Javanese root-knot nem.	B174-2	F	Tamulonis et al. 1997a
Peanut root-knot nemat.	B212D-2—A111H-2	E	Tamulonis et al. 1997b
	B212V-1	F	"
S. root-knot nematode	K493H-1—Cs008D-1	G	Tamulonis et al. 1997c
	G248A-1	O	"
Brown stem rot	K375	J	Lewers et al. (in review)
Soybean mosaic virus	A186-1	F	Yu et al. 1994
	K644a	F	Yu et al. 1996a
	Rsv1	F	"
Phytophthora sojae	*Rps1*	N	Diers et al. 1992c
	Rps2	J	Polzin et al. 1994
	Rps3	F	Diers et al. 1992c
	Rps4	G	"
	Rps5	G	"
	Rps7	N	Lohnes and Schmitthenner, 1997
Powdery mildew	*Rmd*	J	Polzin et al. 1994
Bacterial blight	*Rpg4*	N	Mansur et al. 1996

5.   Genome duplication

Most genera of the Phaseoleae have a genome complement of 2n = 22. Lackey (1980) suggested that Glycine was probably derived from a diploid ancestor (n = 11) which underwent an aneuploid loss to n = 10 and subsequent polyploidation to yield the present 2n = 2x = 40 genome size (Palmer and Kilen 1987). Because soybean behaves like a diploid we regard 2n = 40 as the diploid chromosome number.

Genetic evidence of gene duplication suggests that soybean is a polyploid (Buttery and Buzzell 1976; Palmer and Kilen 1987). Hymowitz and Singh (1987) have suggested that the soybean be regarded as a stable tetraploid with diploidized genomes. The soybean genome possesses many examples of qualitative traits controlled by two loci (Zobel 1983; Palmer and Kilen 1987). Strong molecular evidence also suggests a polyploid origin for soybean. A random selection of 280 PstI probes digested with each of five restriction enzymes found that only 7.5% detected a single fragment, while 33.6% detected two fragments and 25.7% detected three fragments. Approximately one-third of the probes detected four or more fragments (Shoemaker et al. 1996).

There is a strong tendency for polyploids to evolve into a diploid state through sequence divergence and chromosome rearrangement (Leipold and Schmidtke 1982).

This type of genome rearrangement was observed during the analysis and comparison of the leghemoglobin genes of *Phaseolus vulgaris* and *Glycine max* (Lee and Verma 1984). Also, studies of ribosomal RNA (rRNA) gene sequences in soybean have indicated the rRNA gene sites have been eliminated during diploidization (Skorupska et al. 1989).

In a study in which data from nine soybean mapping populations was 'joined' it was found that as many as 33 markers from a single linkage group were duplicated on other linkage groups (Shoemaker et al. 1996). Homoeologous groups of markers (duplicated segments) were detected in all but two of the major linkage groups. The average size of the homoeologous regions ranged from 1.5 cM (three markers in common) to 106.4 cM (seven markers in common), with an average segmental duplication size of 45.3 cM (Shoemaker et al. 1996). Some regions of the genome existed in as many as six copies while on average, 2.55 duplications per segment were observed. It is difficult to reconcile the arrangement of homoeologous segments with the arrangement expected from a simple tetraploid (Helentjaris et al. 1988). However, the independent translocation events necessary to account for these arrangements could be expected during the 'diploidization' of the ancient polyploid soybean (Zobel 1983). These data strongly suggest that soybean is an ancient paleopolyploid.

Interestingly, the positions of QTLs for seed traits (protein and oil) showed correspondence across homoelogous regions (Shoemaker et al. 1996). This suggests that the genes or gene families contributing to seed composition in soybean have retained similar functions throughout the evolution of the genome. It also suggests that genome duplication may have played a role in the quantitative expression of many soybean traits.

Because soybean probes detect multiple fragments in Southern hybridization experiments, and because each fragment could represent a different locus on a chromosome, it is very important to map each polymorphism or to compare probe/enzyme/polymorphisms with a reference population ('anchoring'). Without this information, the map location of polymorphic fragments must remain ambiguous.

6.   Validity of an interspecific genetic map

The use of interspecific crosses in the construction of genetic maps creates the possibility of map aberrations resulting from cytogenetic variations such as inversions or translocations. However, pachytene chromosome analysis suggests that *G. max* and *G. soja* carry similar genomes. This genome is designated 'GG' (Singh and Hymowitz 1988).

Chromosome aberrations have been found in *G. max* at a low frequency, but occur at a higher frequency in *G. soja*. Forty-six of fifty-six *G. soja* accessions from China and the Soviet Union were determined to contain chromosome interchanges (Palmer et al. 1987). PI 468.916, the *G. soja* accession used in the development of our mapping population, did not contain a chromosome aberration relative to *G. max* (Palmer et al. 1987). Therefore, except for possible minor recombinational differences, the genetic map constructed from a cross between A81-356022 and PI 468916 represents a reasonably accurate linear representation of the soybean genome.

7. References

Apuya, N.R., Frazier, B.L., Keim, P., Roth, E.J. and Lark, K.G. (1988) Restriction fragment length polymorphisms as genetic markers in soybean, Glycine max (L.) Merrill. Theor. Appl. Genet. 75 889-901.

Arumuganthan, K. and Earle, E. (1991) Nuclear DNA content of some important plant species. Plant Mol. Biol. Rep. 9:208-218.

Bailey, M.A., Mian, M.A.R., Carter, T.E., Jr., Ashley, D.A., and Boerma, H.R. (1997) Pod dehiscence of soybean: identification of quantitative trait loci. J. Hered. 88:152-154.

Bernard, R.L. (1976) United States national germplasm collections. In: L.D. Hill (ed.), World Soybean Res., pp. 286-289. Interstate Printers and Publ., Danville, IL.

Brummer, E. C., Nickell, C., Wilcox, J. and Shoemaker, R. (1995) Mapping the *Fan* locus controlling linolenic acid content in soybean oil. J. Hered. 86:245-247.

Brummer, E.C., Graef, G.L., Orf, J., Wilcox, J.R., and Shoemaker, R.C. (1997) Mapping QTL for seed protein and oil content in eight soybean populations. Crop Sci. 37:370-378.

Burr, B., Burr, F.A., Thompson, K.H., Albettsen, M.C. and Stuber, C.W. (1988) Gene mapping with recombinant inbreds in maize. Genetics 118: 519-526.

Burton, J.W. (1985) Breeding soybeans for improved protein quantity and quality. In: R. Shibles (ed.), Proc. 3rd World Soybean Res. Conf., pp. 361-367. Westview Press, Boulder, CO.

Buttery, B.R. and Buzzell, R.I. (1976) Flavonol glycoside genes and photosynthesis in soybeans. Crop Sci. 16: 547-550.

Calvo, E., Wurtele, E. and Shoemaker, R. (1997) Cloning, mapping, and analyses of expression of the Em-like gene family in soybean [*Glycine max* (L.) Merr.]. Theor. Appl. Genet. 94:957-967.

Chang, S.J.C., Doubler, T.W., Kilo, V.Y., Abu-Tredeih, J., Prabhu, R., Freire, V., Suttner, R., Klein, J., Schmidt, M.E., Gibson, P.T., and Lightfoot, D.A. (1996) Two additional loci underlying durable field resistance to soybean sudden death syndrome (SDS). Crop Sci. 36:1684-1688.

Chang, S.J.C., Doubler, T.W., Kilo, V.Y., Suttner, R., Klein, J., Schmidt, M.E., Gibson, P.T., and Lightfoot, D.A. (1997) Association of loci underlying field resistance to soybean sudden death syndrome (SDS) and cyst nematode (SCN) race 3. Crop Sci. 37:965-971.

Chen, Z. and Shoemaker, R. (1997) Molecular mapping of the *Shr* and *Cgy1* genes in soybean. Soybean Genet. Newsl. 24:183-185.

Concibido, V.C., Denny, R.L., Boutin, S.R., Hautea, R., Orf, J.H., and Young, N.D. (1994) DNA marker analysis of loci underlying resistance to soybean cyst nematode (*Heterodera glycines* Ichinohe). Crop Sci. 34:240-246.

Concibido, V.C., Young, N.D., Lange, D.A., Denny, R.L., Danesh, D., and Orf, J.H. (1996) Targeted comparative genome analysis and qualitative mapping of a major partial-resistance gene to the soybean cyst nematode. Theor. Appl. Genet. 93:234-241.

Concibido, V.C., Lange, D.A., Denny, R.L., Orf, J.H., and Young, N.D. (1997) Genome mapping of soybean cyst nematode resistance genes in Peking, PI90763, and PI88788 using DNA markers. Crop Sci. 37:258-264.

Cregan, P., Jarvik, T., Bush, A., Shoemaker, R., Lark, K., Kahler, A., Van Toai, T., Lohnes, D., Chung, J. and Specht, J. (1998) An integrated genetic map of the soybean. Crop Sci. (in review).

Delannay, X., Rodgers, D.M. and Palmer, R.G. (1983) Relative genetic contributions among ancestral lines to North American soybean cultivars. Crop Sci. 23: 944-949.

Diers, B.W., Keim, P., Fehr, W.R. and Shoemaker, R.C. (1992a) RFLP analysis of soybean seed protein and oil content. Theor. Appl. Genet. 83: 608-612.

Diers, B.W., Cianzio, S.R., and Shoemaker, R.C. (1992b) Possible identification of quantitative trait loci affecting iron efficiency in soybean. J. Plant Nutr. 15:2127-2136.

Diers, B.W., Mansur, L, Imsande, J., and Shoemaker, R.C. (1992c) Mapping phytophthora loci in soybean with restriction fragment length polymorphism markers. Crop Sci. 32:377-383.

Diers, B.W. and R.C. Shoemaker. (1992) Restriction fragment length polymorphism analysis of soybean fatty acid content. JAOCS 69:1242-1244.

Diers, B., Beilinson, V., Nielsen, N. and Shoemaker, R. (1994) Genetic mapping of the *Gy*4 and *Gy*5 glycinin genes in soybean and the analysis of a variant of *Gy*4. Theor. Appl. Genet. 89:297-304.

Fehr, W.R. (1987) Breeding methods for cultivar development. In: J.R. Wilcox (ed.), Soybeans: Improvement, Production, and Uses, second ed., No. 16, pp. 249-293. American Society of Agronomy, Inc.,Crop Science Society of America, Inc., and Soil Science Society of America, Inc., Madison, Wl.

Ghassemi, F. and Gresshoff, P. (1998) The early *enod*2 and the leghemoglobin (LBC3) genes segregate independently from other known soybean symbiotic genes. Molecular Plant-Microbe Interactions 11:6-13.

Gizlice, Z., T. E. Carter, Jr. and Burton, J. (1994) Genetic base for North American public soybean cultivars released between 1947 and 1988. Crop Sci. 34:1143-1151.

Goldberg, R.B. (1978) DNA sequence organization in the soybean plant. Biochem. Genet. 16: 45-68.

Gurley, W.B., Hepburn, A.G. and Key, J.L. (1979) Sequence organization of the soybean genome. Biochem. Biophys. Acta 561: 167-183.

Helentjaris, T., King, G., Slocum, M., Siedenstrang, C. and Wegman, S. (1985) Restriction fragment polymorphisms as probes for plant diversity and their development as tools for applied plant breeding. Plant Mol. Biol. 5: 109-118.

Helentjaris, T., Weber, D. and Wright, S. (1988) Identification of the genomic locations of duplicate nucleotide sequences in maize by analysis of restriction fragment length polymorphisms. Genetics 118: 353-363.

Hnetkovsky, N., Chang, S.J.C., Doubler, Gibson, P.T., and Lightfoot, D.A. (1996) Genetic mapping of loci underlying field resistance to soybean sudden death syndrome (SDS). Crop Sci. 36:393-400.

Hymowitz, T. (1970) On the domestication of the soybean. Econ. Bot. 24: 408-421.

Hymowitz, T. and Harlan, J.R. (1983) Introduction of soybean to North America by Samuel Bowen in 1765. Econ. Bot. 37: 371-379.

Hymowitz, T. and Singh, R.J. (1987) Taxonomy and speciation. In: J.R. Wilcox (ed.), Soybeans: Improvement, Production, and Uses, second ed., No. 16, pp: 23-48. American Society of Agronomy, Inc., Crop Science Society of America, Inc. and Soil Science Society of America, Inc., Madison, Wl.

Hymowitz, T., Palmer, R.G. and Singh, R.J. (1991) Cytogenetics of the genus *Glycine*. In: T. Tsuchiya and P.K. Gupta (eds.), Chromosome Engineering in Plants: Genetics, Breeding, Evolution, Part B, pp. 53-63. Elsevier Science Publishers B.V., Amsterdam.

Kanazin, V., Blake, T. and Shoemaker, R. (1996a) Organization of the histone H3 genes in soybean, barley and wheat. Mol. Gen. Genet. 250:137-147.

Kanazin, V., Marek, L.F. and Shoemaker, R.C. (1996b) Resistance gene analogs are conserved and clustered in soybean. Proc. Natl. Acad. Sci. USA 93:11746-11750.

Keim, P. and Shoemaker, R.C. (1988) Construction of a random recombinant DNA library that is primarily single copy sequences. Soy. Genet. News 15: 147-148.

Keim, P., Olson, T.C. and Shoemaker, R.C. (1988) A rapid protocol for isolating soybean DNA. Soy. Genet. News 15: 150-152.

Keim, P., Shoemaker, R.C. and Palmer, R.G. (1989) Restriction fragment length polymorphism diversity in soybean. Theor. Appl. Genet. 77: 786-792.

Keim, P., Diers, B.W., Olson, T.C. and Shoemaker, R.C. (1990a) RFLP mapping in soybean: association between marker loci and variation in quantitative traits. Genetics 126: 735-742.

Keim, P., Diers, B.W. and Shoemaker, R.C. (1990b) Genetic analysis of soybean hard seededncss with molecular markers. Theor. Appl. Genet. 79: 465-469.

Lacky, J.A. (1980) Chromosome numbers in the *Phaseoleae* (Fabaceae: Faboideae) and their relation to taxonomy. Am. J. Bot. 67: 595-602.

Landau-Ellis, D., Angermuller, S., Shoemaker, R.C. and Gresshoff, P.M. (1991) The genetic locus controlling supernodulation in soybean (*Glycine max* L.) co-segregates tightly with a cloned molecular marker. Mol. Gen. Genet. 228: 221-226.

Lander, E.S., Green, P., Abrahamson, J., Barlow, A., Daly, M., Lincoln, S.E. and Newburg, L. (1987) MAPMAKER: An interactive computer package for constructing primary genetic linkage maps of experimental and natural populations. Genetics 1: 174- 181.

Lark, K.G., Orf, J., and Mansur, L.M. (1994) Epistatic expression of quantitative trait loci (QTL) in soybean [*Glycine max* (L) Merr] determined by QTL association with RFLP alleles. Theor. Appl. Genet. 88:486-489.

Lark, K.G., Chase, K., Adler, F., Mansur, L.M. and Orf, J. (1995) Interactions between quantitative trait loci in soybean in which trait variation at one locus is conditional upon a specific allele at another. Proc. Natl. Acad. Sci. USA 92:4656-4660.

Lee, J.S. and Verrna, D.S. (1984) Structure and chromosomal arrangement of leghemoglobin genes in kidney bean suggest divergence in soybean leghemoglobin gene loci following tetraploidization. EMBO J. 12: 2745-2752.

Lee, S.H., Bailey, M.A., Mian, M.A.R., Carter, T.E., Jr., Ashley, D.A., Hussey, R.S., Parrott, W.A. and Boerma, H.R. (1996a) Molecular markers associated with soybean plant height, lodging, and maturity across locations. Crop Sci. 36:728-735.

Lee, S.H., Bailey, M., Mian, M., Shipe, E., Ashley, D., Parrott, W., Hussey, R. and Boerma, R. (1996b) Identification of quantitative trait loci for plant height, lodging, and maturity in a soybean population segregating for growth habit. Theor. Appl. Genet. 92:516-523.

Leipold, M. and Schmidtke, J. (1982) Gene expression in phylogenetically polyploid organisms. In: G. Dover and R. Flavell (eds.), Genome Evolution, pp. 219-236. Academic Press, New York.

Lewers, K., Crane, E., Bronson, C., Schupp, J., Keim, P. and Shoemaker, R. Detection of linked QTL for soybean brown stem rot resistance in 'BSR 101' as expressed in a growth chamber environment. Molec. Breed. (in review).

Lin, S., Cianzio, S. and Shoemaker, R. (1997) Mapping genetic loci for iron deficiency chlorosis in soybean. Molec. Breed. 3:219-229.

Lohnes, D. and Schmitthenner, A. (1997) Position of the Phytophthora resistance gene *Rps*7 on the soybean molecular map. Crop Sci. 37:555-556.

Lorenzen, L., Boutin, S., Young, N., Specht, J. and Shoemaker, R. (1995) Soybean pedigree analysis using map-based molecular markers: I. Tracking RFLP markers in cultivars. Crop Sci. 35:1326-1336.

Mahalingam, R. and H.T. Skorupska. (1995) DNA markers for resistance to *Heterodera glycines* I. race 3 in soybean cultivar Peking. Breed. Sci. 45:435-443.

Maughan, P.J., Saghai Maroof, M. A. and Buss, G. (1996) Molecular marker analysis of seed weight: genomic locations, gene action, and evidence for orthologous evolution among three legume species. Theor. Appl. Genet. 93:574-579.

Mansur, L.M., Lark, K. G., Kross, H. and Oliveira, A. (1993a) Interval mapping of quantitative trait loci for reproductive, morphological, and seed traits of soybean (*Glycine max* L.). Theor. Appl. Genet. 86:907-913.

Mansur, L.M., Orf, J. and Lark, K.G. (1993b) Determining the linkage of quantitative trait loci to RFLP markers using extreme phenotypes of recombinant inbreds of soybean. Theor. Appl. Genet. 86:914-918.

Mansur, L.M. Orf, J., Chase, K., Jarvik, T., Cregan, P. and Lark, K.G. (1996) Genetic mapping of agronomic traits using recombinant inbred lines of soybean. Crop Sci. 36:1327-1336.

Mian, M.A.R., Bailey, M., Tamulonis, J., Shipe, E., Carter, T., Parrott, W., Ashley, D., Hussey, R. and Boerma, H. R. (1996a) Molecular markers associated with water use efficiency and leaf ash in soybean. Crop Sci. 36:1252-1257.

Mian, M.A.R., Bailey, M., Ashley, D., Wells, R., Carter, T., Parrott, W. and Boerma, H. R. (1996b) Molecular markers associated with seed weight in two soybean populations. Theor. Appl. Genet. 93:1011-1016.

Mian, M.A.R., Shipe, E., Alvernaz, J., Mueller, J., Ashley, D. and Boerma, H. R. (1997) RFLP analysis of chlorimuron ethyl sensitivity in soybean J. Hered. 88:38-41.

Mian, M.A.R., Wells, R., Carter, T., Ashley, D. and Boerma, H.R. (1998) RFLP tagging of QTLs conditioning specific leaf weight and leaf size in soybean. Theor. Appl. Genet. 96:354-360.

Mudge, J., Cregan, P., Kenworthy, J., Kenworthy, W., Orf, J. and Young, N. (1997) Two microsatellite markers that flank the major soybean cyst-nematode resistance locus. Crop Sci. 37:1611-1615.

Muehlbauer, G.J., Specht, J.E., Thomas-Compton, M.A., Staswick, P.E. and Bernard, R.L. (1988) Near isogenic lines - a potential resource in the integration of conventional and molecular marker linkage maps. Crop Sci. 28: 729-735.

Muehlbauer, G.J., Specht, J.E., Staswick, P.E., Graef, G.L. and Thomas-Compton, M.A. (1989) Application of the near-isogenic line gene mapping technique to isozyme markers. Crop Sci. 29: 1548-1553.

Muehlbauer, G.J., Staswick, P.E., Specht, J.E., Graef, G.L., Shoemaker, R.C. and Keim, P. (1991) RFLP mapping using near-isogenic lines in the soybean [*Glycine max* (L.) Merr.]. Theor. Appl. Genet. 81: 189-198.

Nickell, A., Wilcox, J., Lorenzen, L., Cavins, J., Guffy, R. and Shoemaker, R. (1994) The *Fap*2 locus in soybean maps to linkage group D.J. Hered. 85:160-162.

Palmer, R.G. and Hedges, B. (1993) Linkage map of soybean (*Glycine max* L. Merr.) In: S.J. O'Brien (ed.), Genetic Maps: Locus Maps of Complex Genomes, sixth ed., pp. 6.139-6.148.

Cold Spring Harbor Laboratory Press, Cold Spring Harbor, NY.

Palmer, R.G. and Kilen, T.C. (1987) Qualitative Genetics and Cytogenetics. In: J.R. Wilcox (ed.), Soybeans: Improvement, Production, and Uses, second ed., No. 16, pp. 135-209. American Society of Agronomy, Inc., Crop Science Society of America, Inc., and Soil Science Society of America, Inc., Madison, WI.

Palmer, R.G., Newhouse, K.E., Graybosch, R.A. and Delannay, X. (1987) Chromosome structure of the wild soybean: Accessions from China and the Soviet Union of *Glycine soja* Sieb. and Zucc. J. Hered. 78: 243-247.

Polzin, K.M., Lohnes, D., Nickell, C. and Shoemaker, R. (1994) Integration of *Rps*2, *Rmd*, and *Rj*2 into linkage group J of the soybean molecular map. J. Hered. 85:300-303.

Shoemaker, R.C., Polzin, K., Labate, J., Specht, J., Brummer, J., Olson, T., Young, N., Concibido, V., Wilcox, J., Tamulonis, J., Kochert, G. and Boerma, H.R. (1996) Genome duplication in soybean (*Glycine* subgenus *soja*). Genetics 144:329-338.

Shoemaker, R.C. and Specht, J.E. (1995) Integration of the soybean molecular and classical genetic linkage groups. Crop Sci. 35:436-446.

Singh, R.J. and Hymowitz, T. (1988) The genomic relationship between *Glycine max* (L.) Merr. and *G. soja* Sieb. and Zucc. as revealed by pachytene chromosomal analysis. Theor. Appl. Genet. 76: 705-711.

Skorupska, H., Albertsen, M.C., Langholz, K.D. and Palmer, R.G. (1989) Detection of ribosomal RNA genes in soybean, *Glycine max* (L.) Merr. by *in situ* hybridization. Genome 32: 1091-1095.

Smith, K.J. and Huyser, W (1987) World distribution and significance of soybean. In: J.R. Wilcox (ed.), Soybeans: Improvement, Production, and Uses, second ed., No. 16, pp. 1-22. American Society of Agronomy, Inc., Crop Science Society of America, Inc., and Soil Science Society of America, Inc., Madison, WI.

Specht, J.E. and Williams, J.H (1984) Contribution of genetic technology to soybean productivity retrospect and prospect. In: W.F.Fehr (ed.), Genetic Contributions to Yield Gains of Five Major Crop Plants, pp. 49-74. American Society of Agronomy, Madison, WI.

Tamulonis, J.P., Luzzi, B., Hussey, R., Parrott, W. and Boerma, H.R. (1997a) DNA markers associated with resistance to Javanese root-knot nematode in soybean. Crop Sci. 37: 783-788.

Tamulonis, J.P., Luzzi, B., Hussey, R., Parrott, W. and Boerma, H.R. (1997b) DNA marker analysis of loci conferring resistance to peanut root-knot nematode in soybean. Theor. Appl. Genet. 95:664-670.

Tamulonis, J.P., Luzzi, B., Hussey, R., Parrott, W. and Boerma, H.R. (1997c) RFLP mapping of resistance to southern root-knot nematode in soybean. Crop Sci. 37:1903-1909.

Vierling, R.A., Faghihi, J., Ferris, V. and Ferris, J. (1996) Association of RFLP markers with loci conferring broad-based resistance to the soybean cyst nematode (*Heterodera glycines*). Theor. Appl. Genet. 92:83-86.

Webb, D.M., Baltazar, B., Rao-Arelli, A., Schupp, J., Clayton, K., Keim, P. and Beavis, W. (1995) Genetic mapping of soybean cyst nematode race-3 resistance loci in the soybean PI437654. Theor. Appl. Genet. 91:574-581.

Wilcox, J.R. (1985) Breeding soybeans for improved oil quantity and quality. In: R. Shibles (ed.) Proc. 3rd World Soybean Res. Conf, pp. 380-386. Westview Press, Boulder, CO.

Young, N.D., Zamir, D., Ganal, M.W. and Tanksley, S.D. (1988) Use of isogenic lines and simultaneous probing to identify DNA markers tightly linked to the *Tm*-2a gene in tomato. Genetics 120: 579-585.

Yu, Y.G., Saghai Maroof, M., Buss, G., Maughan, P. and Tolin, S. (1994) RFLP and microsatellite mapping of gene for soybean mosaic virus resistance. Phytopath. 84:60-64.

Yu, Y.G., Saghai Maroof, M. and Buss, G. (1996a) Divergence and allelomorphic relationship of a soybean virus resistance gene based on tightly linked DNA microsatellite and RFLP markers. Theor. Appl. Genet. 92:64-69.

Yu, Y.G., Buss, G. and Saghai Maroof, M.A. (1996b) Isolation of a superfamily of candidate resistance genes in soybean based upon a conserved nucleotide-binding site. Proc. Natl. Acad. Sci. USA 93:11751-11756.

Zobel, R.W. (1983) Genic duplication: a significant constraint to molecular and cellular genetic manipulation in plants. Comments Mol. Cell Biophys. 1: 355-364.

22.   Genetic mapping in sunflowers

STEVEN J. KNAPP[1], SIMON T. BERRY[2], *and* LOREN H. RIESEBERG[3]

[1] *Department of Crop and Soil Science, Oregon State University,*
Corvallis, OR 97331 U.S.A. <Steven.J.Knapp@orst.edu>
[2] *Advanta Biotechnology Department, SES-Europe NV/SA,*
Industriepark, Soldatenplein Z2 nr. 15, B-3300, Tienen, Belgium
[3] *Department of Biology, Indiana University, Bloomington, IN 47405 U.S.A.*

Contents

1.   Introduction

Cultivated sunflower (*Helianthus annuus* L.) is a member of the subtribe Helianthinae of the Compositae (Asteraceae) family (Seiler and Rieseberg 1997). The genus is a polyploid complex, with diploid, tetraploid and hexaploid species, and a basic chromosome number of 17 (Heiser and Smith 1955). There are 12 annual diploid species and 37 perennial species. North America is the center of diversity for sunflowers. Sunflowers were first cultivated by Native Americans in 1000 BC, were introduced into Europe in the sixteenth century, and were first grown as a source of edible oil in nineteenth century Russia. Russian plant breeders increased sunflower seed oil concentrations from 330g/kg up to as much as 550g/kg between 1940 and 1965 and developed the first high oil cultivars (Putt 1997). These cultivars dramatically transformed sunflower as an oilseed crop and they have been widely used in the development of modern-day cultivars and hybrids — a significant fraction of

R.L. Phillips and I.K. Vasil (eds.), DNA-Based Markers in Plants, 379 - 403.

the diversity in elite inbred lines traces to high-oil germplasm developed in Russia (Korell et al. 1992; Cheres and Knapp 1998).

Selection for high oil, self-compatibility, fertility restoration, and branching created bottlenecks that narrowed the germplasm base (reduced the genetic diversity) of cultivated sunflower. Therefore, exotic germplasm and wild species are important genetic resources for enhancing the performance of sunflower hybrids (Jan 1997).

Apart from the development of high oil cultivars, the other factor in the rapid development of the sunflower industry was the discovery and development of cytoplasmic male sterility for producing hybrid seed (Leclerq 1969; Kinman 1970). The first commercial hybrids were grown in the US in 1972 (Fick and Zimmer 1976). Since then, the hybrid seed industry has grown from virtual obscurity to become a significant economic force today. Hybrids are principally developed and marketed by private seed companies.

Sunflower is primarily grown in the warm temperate regions of the five continents. The top five sunflower producing countries in rank order are Russia (the former republics of the Soviet Union), Argentina, France, the US, and China. Twenty million hectares of sunflower were grown and 24.3 million metric tons of sunflower oil were produced worldwide in 1997 (USDA Statistical Reporting Service).

Miller and Fick (1997) have published comprehensive lists of the known morphological and phenotypic markers for cultivated sunflower. These encompass the usual classes of morphological mutations (e.g. mutations affecting leaf, flower, and trichome morphology and flower color), in addition to known disease resistance genes, nuclear and cytoplasmic-genetic male sterility and fertility restorer genes, and mutations affecting fatty acid and tocopherol levels in seed oils. To date, very few of these phenotypic markers have been mapped (Table 1). In this chapter, we review the recent and rapid progress in the development of genetic markers and maps in cultivated and wild sunflowers.

2.   Biochemical and molecular markers

2.1. Biochemical markers

Isozymes have not been used extensively as markers in either domesticated or wild sunflower species primarily because very few are polymorphic (Torres 1983; Kahler and Lay 1985; Rieseberg and Seiler 1990; Quillet et al. 1992; Cronn et al. 1997). For example, a survey of over 700 achenes, representing 114 domesticated accessions, revealed only 53 alleles at 20 loci, with an average of 1.39 alleles per locus (Cronn et al. 1997). In cultivated germplasm, 8-9 polymorphic isozyme loci can discriminate between distantly related inbred lines/populations (Quillet et al. 1992), which, at best, allows them to be grouped according to their geographic origin (Tersac et al. 1994). However, Carrera and Poverene (1995) have demonstrated that wild *Helianthus* species have different alleles from cultivated sunflower and therefore may be useful in introgression studies. There have been several reports on the mapping of isozyme loci in various sunflower species (Kahler and Lay 1985; Torres 1983,

Table 1.   Major gene loci, which have been mapped using molecular markers in cultivated sunflower.

Gene	Original Reference	Phenotype	Map Reference
Pl1	Vranceanu (1970)	downy mildew resistance (race 1)	Mouzeyar et al. (1995) Gedil et al. (1999b)
Pl2	Zimmerman and Kinman (1972)	downy mildew resistance (race 2)	Vear et al. (1997)
Pl6	Miller (1992)	downy mildew resistance (race 6)	Roeckel-Drevet et al. (1996)
R_l	Putt and Sackston (1963)	rust resistance (race 0)	Lawson et al. (1996)
R_{ADV}	Kong and Kochman (1996)	rust resistance (race *Adv*)	Lawson et al. (1998)
Rf_l	Kinman (1970)	CMS fertility restorer	Gentzbittel et al. (1995) Quillet et al. (1995) Berry et al. (1997) Jan et al. (1998)
b_l and b_2	Kovacik and Skaloud (1990)	branching	Gentzbittel et al. (1995)
ms_l and *T*	Leclerq (1966)	genetic male sterility and anthocyanin pigmentation	Berry (1995)
Hyp		white pigmentation of seed hypodermis	Leon et al. (1996)

Rieseberg et al. 1993 and 1995b; Quillet et al. 1995; Mestries et al. 1998; Carrera et al., unpublished data) (Table 2).

Genetic markers also have been developed for several seed storage proteins in sunflower. Anisimova et al. (1995) used 1-D and 2-D SDS-PAGE electrophoresis to assay 2S albumins in sunflower and detected polymorphisms between several lines. Anisimova and Gavrlyuck (1989) and Anisimova et al. (1993) performed similar analyses on the 11S globulins and helianthinins. More recently, Serre et al. (1998) reported the mapping of an albumin locus on the public linkage map (Fig. 1). cDNAs have now been isolated for some members of the 2S albumin, the 11S globulin, and the helianthinin gene families (see GenBank for sequences and references). Their utility as RFLP markers is not known.

Table 2.   Isozyme loci mapped in intra- or inter-specific *Helianthus* populations.

Reference	Isozyme loci mapped	Linked loci
Torres and Diedenhoffen (1976)	*Adh1, Adh2, Acp1*	
Kahler and Lay, (1985)	*Prx3, Mdh1, Idh2, Gpi2, Pgm4*	*Prx3* and *Pgm4*
Rieseberg et al. (1993)	*Acp1*	
Quillet et al. (1995)	*Mdh1, Mdh2, Pgm1, Sdh1, Acp1, Me1*	*Mdh1* and *Sdh1*
Carrera et al. (unpublished)	*Acp1, Est1, Gdh2, Pgd3, Pgi2*	*Acp1* and *Pgd3*
Mestries et al. (1998)	*Idh1, Pgi3, Got1, Pgd2, Sdh*	*Idh1* and *Pgi3*

2.2.   Restriction Fragment Length Polymorphisms (RFLPs)

The history of genetic marker and map development in sunflower has in many ways been difficult to comprehend. This species has tremendous economic and biological significance, yet RFLP markers were not described until Gentzbittel et al. (1992, 1994) and Berry et al. (1994) published the first reports of RFLP fingerprinting in sunflowers. This work emerged nearly a decade after the advent of RFLP mapping in plants (Bernatzky and Tanksley 1986). The slow start in sunflower was partly caused by two factors. First, several laboratories had difficulty isolating large quantities of genomic DNA from sunflower that could be completely digested with restriction enzymes. Second, RFLP markers based on genomic *Pst*I clones (Berry et al. 1994), the traditional source of RFLP probes, had low polymorphism rates. Unlike the majority of crops, most of the published sunflower RFLP markers were developed from cDNA clones. On average, 40-45% of clones taken from various cDNA libraries reveal low copy, polymorphic restriction fragments (Berry et al. 1994; Gentzbittel et al. 1994; Jan et al. 1993). Sunflower RFLP markers have typically been screened for polymorphism using *Eco*RI, *Eco*RV, *Hind*III, *Dra*I, and *Bgl*II. *Dra*I generally produces the smallest molecular weight fragments, and these tend to be less polymorphic than higher molecular weight fragments (Miller and Tanksley 1990; Jan et al. 1993). However, Berry (1995) has shown that, to a large extent, the ability of a probe to detect polymorphism is independent of the restriction enzyme used. This finding suggests that most RFLPs in sunflower are caused by insertion-deletion events, as is the case in rice (McCouch et al. 1988).

The first RFLP fingerprinting study in sunflowers was carried out by Gentzbittel et al. (1992) on 44 *Helianthus* species. Using ten RFLP markers, they produced a molecular phylogeny that agreed with the existing classical taxonomy of the sub-tribe.

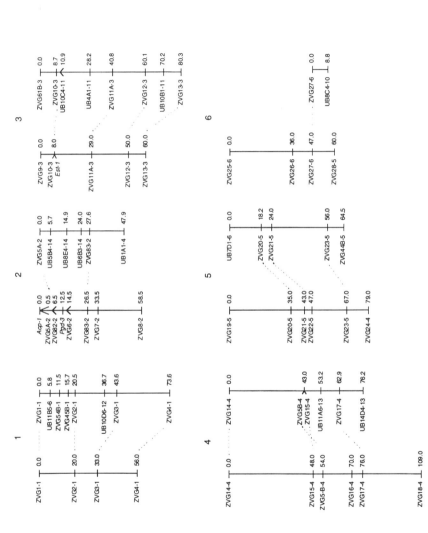

Figure 1. The public RFLP map of cultivated sunflower (*Helianthus annuus*). Two maps are shown for each linkage group. The left-hand map was developed by Berry et al. (1997) and is comprised of the 81 RFLP markers released to the public sector. The right-hand map was developed by Gedil et al. (2000a) by integrating 108 RFLP markers from the Berry et al. (1998) and Jan et al. (1997) maps with 10 RFLP markers for candidate disease resistance genes (Gedil et al. 2000b). The prefix for Berry et al. (1997) RFLP markers is ZVG, for Jan et al. (1997) RFLP markers is UB, and for RGC RFLP markers (Gedil et al. 1999b) is HR.

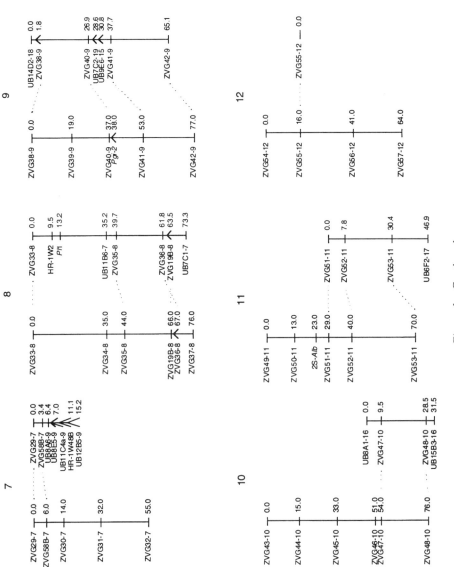

Figure 1. Continued.

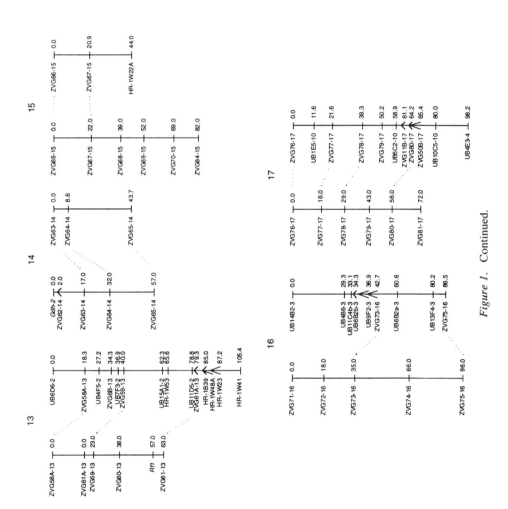

Figure 1. Continued.

Berry et al. (1994) and Gentzbittel et al. (1994) published the first RFLP fingerprinting studies in cultivated sunflower. They reached apparently different conclusions on the rate of nuclear DNA polymorphism in elite germplasm; however, the difference can be traced to how the markers were sampled for calculating the mean heterozygosity or polymorphic information content (PIC) (Marshall and Allard 1970). Gentzbittel et al. (1994) used random cDNAs to generate their data set (65% of which were monomorphic), whilst Berry et al. (1994) and Zhang et al. (1995) used a selected subset of markers that were known to be polymorphic.

2.3. Random Amplified Polymorphic DNAs (RAPDs)

High levels of RAPD (Williams et al. 1990) variation have been reported in sunflower species (Lawson et al. 1994; Tuelat et al. 1994; Arias and Rieseberg 1994), with the proportion of polymorphic loci averaging greater than 50% for most domesticated lines. Typically, 2-5 loci can be scored per primer in intra-specific mapping populations (Rieseberg et al. 1995b). Moreover, methods such as bulked segregant analysis (BSA; Michelmore et al. 1991) have allowed sunflower researchers to rapidly identify markers associated with agronomically important traits (e.g., Lawson et al. 1996 and 1998; Ji et al. 1996), and have been used for fingerprinting (Linder et al. 1998). RAPDs share the same limitations (dominance and bi-allelism) as AFLPs for some applications, and the reproducibility of RAPD fingerprints between laboratories can be poor (Jones et al. 1997). This can be overcome by converting RAPD bands into SCAR markers (Kesseli et al. 1994, Lawson et al. 1998).

RAPDs are not particularly good for comparative mapping studies because the homology between RAPD products tends to decrease as genetic distance increases (Williams et al. 1993; Thormann et al. 1994; Rieseberg 1996). Therefore, homology testing is essential for species-level comparisons (Rieseberg et al. 1995b; Rieseberg 1996). In sunflower, for example, approximately 9% of co-migrating RAPD fragments generated by the same primers in mapping populations, among closely related species, were found to be heterologous when tested (Rieseberg 1996). In addition, 13% of homologous loci mapped to genomic locations that were incongruent with the majority of loci. This finding suggests paralogous, rather than orthologous, relationships between loci.

2.4. Simple Sequence Repeats (SSRs) or microsatellites

Early studies by Mosges and Friedt (1994) and Berry (1995) used end-labeled oligo-nucleotides as RFLP probes to detect multiple polymorphic loci (oligo-fingerprinting), containing arrays of $(GACA)_n$, $(GGAT)_n$ and $(GATA)_n$ repeat motifs. Both studies concluded that the oligo-probe $(GATA)_4$ revealed the most variability between inbred lines. Berry (1995) demonstrated that at least five GATA-containing loci were dispersed across the sunflower genome. More recently, Dehmer and Friedt (1997) showed that, in addition to these three tetra-nucleotides, $(ACA)_6$, $(CAT)_6$, $(CATA)_4$ could also be used for oligo-fingerprinting. They reported that these motifs were among the most abundant tri- and tetra-nucleotide repeats in the sunflower genome. However, the most abundant simple sequence repeats in sunflower DNA were: $(A)_n$, $(GA)_n$, and $(CA)_n$ (Dehmer and Friedt 1997).

The presence of SSRs in the sunflower genome has led to the development of locus-specific PCR tests. Thus far fewer than 30 SSR markers have been described for cultivated sunflower (Brunel 1994; Whitton et al. 1996; unpublished). These have been developed from both cDNA sequences (Hongtrakul et al. 1998b; unpublished) and random genomic DNA libraries (Whitton et al. 1996). SSR markers were developed for 9 repeats found in 64 genes (unpublished). Eight were screened for polymorphisms among 28 elite inbred lines and 36 wild populations and land races (unpublished). The number of alleles ranged from one to seven for di-nucleotide and seven to 14 for tri-nucleotide repeats. PICs ranged from 0.54 to 0.90 and 0.00 to 0.58 for di- and tri-nucleotide repeats, respectively. The exotic germplasm used in this study was only slightly more polymorphic than elite germplasm. Whitton et al. (1996) assayed 13 SSR markers for polymorphism across several taxa in the Asteraceae. Nine amplified SSR loci from *H. giganteus*, *H. divaricatus*, and *H. annuus*.

SSR markers have not been widely used in sunflower research because very few have been described and the number of highly polymorphic SSRs is presently limited. SSR marker development projects are underway in a few laboratories and should create a supply of SSRs for genetic mapping and molecular breeding applications in sunflower.

2.5. Amplified Fragment Length Polymorphisms (AFLPs)

AFLPs (Vos et al. 1995) were rapidly adopted in sunflower, as in many other species, because they have a high multiplex ratio, are highly reproducible, require no prior sequence information, and can be assayed with very small amounts of DNA. The high multiplex ratio is a particularly powerful feature of this technique. For example, one AFLP reaction, using three selective nucleotides, typically yields 60 or more DNA fragments in sunflower (Hongtrakul et al. 1997; Gedil et al. 1999a).

Hongtrakul et al. (1997) used AFLPs to fingerprint a set of 23 elite inbred lines and reported polymorphism rates between 7 and 28% for the 253 possible crosses. Nearly half of the AFLP markers assayed were polymorphic in at least one cross (172 out of 359 markers) and the number of polymorphic bands per assay ranged from 4.2 to 14.3, with a mean heterozygosity of 0.11 for elite public inbred lines. This is significantly lower than the mean heteroygosities of 0.49 and 0.59 reported for RFLP markers by Berry et al. 1994 and Zhang et al. (1995) respectively. This difference can be partly attributed to the bi-allelic nature of AFLPs. The maximum PIC for bi-allelic markers is 0.5 compared to 1.0 for multi-allelic markers.

AFLPs should have a significant role in sunflower genome analysis and molecular breeding. The utility of AFLP analysis is greatest for projects where dominance is not disadvantageous, e.g., mapping in backcross, doubled haploid, or recombinant inbred populations and marker-assisted backcross breeding. Even though a significant percentage of AFLP markers can be scored co-dominantly using densitometry and specially developed software, most laboratories lack hardware and software for co-dominant scoring and are forced to score AFLPs dominantly. This reduces the utility of AFLPs for developing F_2 linkage maps (Knapp et al. 1996) and mapping QTLs in F_2 populations (Jiang and Zeng 1997).

2.6. Miscellaneous sequence-based markers

Hongtrakul et al. (1998a) developed single strand conformational polymorphism (SSCP) and intron fragment length polymorphism (IFLP) markers for two stearoyl-ACP desaturase genes. These polymorphisms were found to be caused by insertions-deletions, variation in the number of poly-T repeats, and single nucleotide polymorphisme. The heterozygosities for these markers ranged from 0.44 to 0.63 among several elite inbred lines. Because the IFLP primers were designed from exon sequences flanking introns, they may work across the different sunflower species and related genera (Hongtrakul et al. 1998a). The utility of intron markers in sunflower needs to be tested further; however, introns seem to be a promising source of genetic markers for mapping, candidate gene analysis, and marker-assisted selection.

Lawson et al. (1998) developed sequence characterized amplified region (SCAR) markers from the DNA sequence of RAPD markers linked to rust resistance genes. A dominant SCAR was found to be diagnostic for lines carrying the R_1 resistance gene. Similarly, another dominant SCAR was diagnostic for lines carrying the R_{Adv} resistance gene. Gedil et al. (2000b) developed SSCP and cleaved amplified polymorphic sequence (CAPS) (Konieczny and Ausubel 1993) DNA markers for candidate disease resistance genes linked to the downy mildew resistance locus *Pl1*. The presence of a *Tsp* 509 I fragment in the CAPS marker analysis was diagnostic for the resistance gene. SCAR, CAPS, and IFLP markers can be rapidly screened from small DNA samples using manual or automated screening methods.

3.   Genetic linkage maps

3.1. Proprietary RFLP maps

Several proprietary RFLP linkage maps have been published for cultivated sunflower. Berry et al. (1995) and Jan et al. (1998) published maps based on individual F_2 populations, whereas Gentzbittel et al. (1995) and Berry et al. (1996) published composite maps derived from data on several different mapping populations (Table 3.). The integration of different maps requires at least two common markers per linkage group between any two populations. These markers then act as anchor points allowing the merging of linkage groups in the correct orientation from the different crosses. Both Berry et al. (1996) and Gentzbittel et al. (1995) used the JoinMap program (Stam 1993) to produce a consensus or composite map. The prominent features of all these maps are highlighted below:
- The total genome lengths are close to the estimated length of the sunflower genome of 1650 cM (Gentzbittel et al. 1995).
- There are persistent gaps in several linkage groups (Berry et al. 1996). These gaps probably span regions of high recombination.
- Clusters of loci are found in most linkage groups (Berry et al. 1996), and these may correspond to centromeric regions, where recombination is suppressed (Tanksley et al. 1992; Kleinhofs et al. 1993).

- Distorted segregation is a frequent occurrence in the sunflower genome (Berry et al. 1995; Gentzbittel et al. 1995, unpublished data). This may be due to the presence of self-incompatibility loci or selection for or against individual genotypes during sporogenesis, gamtetogenesis, seed development, and plant growth.
- RFLP mapping has revealed that numerous genes are duplicated within the sunflower genome (Berry et al. 1996). This finding lends support to the hypothesis that *H. annuus* is an ancient polyploid (Heiser and Smith 1955; Jackson and Murray 1983). Berry et al. (1996) found that gene orders were not conserved between homoeologous linkage groups, which suggests that the genome has undergone a variety of rearrangements.

Table 3.　Summary of published, proprietary RFLP linkage maps of cultivated sunflower.

References	Mapping populations	No. of linkage groups	No. of mapped loci	Total genome length (cM)	Mean marker spacing (cM)
Berry et al. (1995)	One F_2	17	234	1380	5.9
Gentzbittel et al. (1995)	Three F_2, Two BC_1	23	237	1150	4.9
Jan et al. (1998)	One F_2	20	271	1164	4.6
Berry et al. (1996)	Nine F_2	17	635	1472	2.3

The relationships between the linkage groups of the various proprietary maps are unknown and until recently none of the probes were available for public research. This has stifled mapping and molecular breeding research in the public sector. Recently, Berry et al. (1997) released 81 re-coded probes to form the backbone of a public linkage map for cultivated sunflower. On average, these probes detect loci every 15 cM throughout the genome (Fig 1) and were used by Gedil et al. (2000a) to construct a public sunflower map (see Section 3.4). The release of additional RFLP probes and the development of public DNA markers, such as those released by Gedil et al. (2000b) and Hongtrakul et al. (1998a), should stimulate more collaborative work and facilitate cross-referencing maps and mapped genes.

3.2. AFLP maps

AFLP markers (Vos et al. 1995) are powerful tools for rapidly constructing genetic maps, increasing the density of genetic maps, and filling gaps in maps. Peerbolte and Peleman (1996) added 291 AFLP loci to two of the F_2 populations used by Gentzbittel et al. (1995). These markers pulled two linkage groups (LG 4 and LG 15) together and permitted several previously unlinked RFLP marker loci to be mapped (J. Peleman, pers. comm.). Gedil et al. (2000a) added 296 AFLP loci to the RFLP backbone of the public linkage map.

The lack of homology between some AFLP fragments of equal length can complicate the transfer of data between mapping populations. Peleman (1998) produced a dense AFLP map and data on the frequency of homolgous AFLP fragments in sunflower by integrating 13 proprietary AFLP maps. He reported that 9% of AFLP markers with the same molecular weight mapped to different loci, but one third of these could be attributed to errors in determining the size of AFLP bands (J. Peleman, pers. comm.). This problem can be partly alleviated by anchoring individual maps with locus-specific markers.

3.3. RAPD maps

RAPDs (Williams et al. 1990) have been extensively used for mapping in sunflower, particularly in wild species. The first map that provided broad coverage of the sunflower genome was for *H. anomalus* (a diploid species descended from *H. annuus* x *H. petiolaris*) and was based on 161 RAPD markers and one isozyme locus (Rieseberg et al. 1993). This map has been expanded over the past five years and now includes 549 RAPD, 151 AFLP, and one isozyme locus (Rieseberg et al. 1995b; Ungerer et al. 1998) covering 17 linkage groups and 1,983 cM. RAPD maps have been published for two other wild species, *H. annuus* and *H. petiolaris*, based on 212 and 400 markers, respectively (Rieseberg et al. 1995b). They reported 17 linkage groups for both species covering 1,084 and 1,761 cM, respectively. Comparison of map distances between conserved markers across the three genomes indicates that the recombination rates in *H. anomalus* and *H. petiolaris* are significantly higher than in *H. annuus*.

In addition to these dense maps, Quillet et al. (1995) generated a linkage map for an interspecific cross between *H. annuus* and *H. argophyllus* based on 48 RAPD loci and one isozyme locus. Linder et al. (1998) published a *H. annuus* a map with 80 RAPD markers as part of a study to assess gene flow between wild *H. annuus* and cultivated sunflower.

3.4. The public map of cultivated sunflower

Gedil et al. (2000a) developed an integrated RFLP-AFLP map using an F_2 population derived from the cross HA370 x HA372. This map was comprised of four hundred and four loci (108 RFLP and 296 AFLP loci) arranged in 17 linkage groups and covered 1,309 cM of the sunflower genome. Gedil et al. (2000a) mapped 64 RFLP loci with 56 public ZVG RFLP probes from the map of Berry et al. (1997) and 44 RFLP loci with 40 proprietary USDA-BGS RFLP probes from the map of Jan et al. (1997). The AFLPs were mapped using 42 AFLP primer combinations. Gedil et al. (2000b) mapped 10 RFLP loci using probes for resistance gene analogs (RGAs) homologous to the NBS-LRR families of resistance genes (Staskawicz et al. 1995). These loci were distributed among linkage groups 7, 8, 13, and 15 (Fig. 1).

The 17 linkage groups in the Berry et al. (1997) and Gedil et al. (1999a) maps presumably correspond to the 17 chromosomes of sunflower. The two maps are aligned side by side in Fig. 1. Two linkage groups (6 and 12) in the integrated map had only one ZVG RFLP locus and thus could not be aligned (oriented) with the Berry et al. (1997) map. RFLP markers from the Jan et al. (1997) map are lacking in linkage groups 12, 14, and 15 of the integrated map.

4. Applications of DNA markers to sunflower breeding

4.1. Germplasm fingerprinting

Statistical analysis of molecular fingerprinting data has consistently separated inbred lines of sunflower into sterility maintainer (B-line) and fertility restorer (R-line) groups (Berry et al. 1994; Gentzbittel et al. 1994; Zhang et al. 1995, and Hongtrakul et al. 1997). These studies reflect the fact that breeders tend to restrict inbred line development from either B x B or R x R crosses. Genetic distances have been widely studied as predictors of heterosis and hybrid performance in crop plants. Tersac et al. (1994) estimated genetic distance from isozyme data and found that genetic distance and heterosis were not correlated in sunflower. Cheres et al. (1999) estimated the correlation between genetic distance, heterosis, and hybrid performance using AFLPs and coancestries. They found that genetic distance alone was a weak predictor of hybrid performance in sunflower. As has been shown in a variety of empirical and theoretical studies, genetic distances and F_1 heterozygosities estimated from randomly selected markers are typically not correlated with heterosis or hybrid performance (Charcosset et al. 1991; Dudley et al. 1991; Bernardo 1992; Moser and Lee 1994).

Zhang et al. (1995) used RFLPs to screen inbred lines for intra-line polymorphisms. Although they found RFLPs within lines, there were no obvious phenotypic differences between the four lines they screened. They concluded that the polymorphisms stemmed from residual heterozygosity or outcrossing. Zhang (1995) proposed the use of RFLPs for distinctness, uniformity, and stability (DUS) testing in sunflower due to the changing effect of environment on the genotype from year to year. Molecular markers, such as SSRs, could also play an important role in protecting Plant Breeders' Rights in sunflower and assessing whether or not an inbred was essentially derived (UPOV Convention 1991).

4.2. Disease resistance

Cultivated sunflower is attacked by several pathogens and lacks strong resistance genes for several important diseases (Sackston 1992). Resistance genes have been found in several wild sunflower species and have been introgressed into cultivated sunflower using conventional methods (Seiler 1988; Skoric 1993). The process of introgressing genes from wild to cultivated sunflower could be sped up, and the associated linkage drag reduced, by using marker-assisted selection (MAS). Also, resistance genes could be pyramided using MAS, thereby increasing the durability of resistance.

Several disease resistance genes have been mapped in cultivated sunflower (Table 3). Downy mildew (*Plasmopara halstedii* (Farl.) Berl de Toni) resistance genes seem to comprise a tight cluster (Vear et al. 1997). The interaction between sunflower and *P. halstedii* genes follows the classic gene-for-gene model (Flor 1956). SCAR markers have been developed for two rust (*Puccinia helianthi* Schwein) resistance loci in sunflower (Lawson et al., 1998); however, the linkage arrangement of these loci has not been published.

Gentzbittel et al. (1998a) and Gedil et al. (2000b) cloned DNA fragments of candidate resistance genes or RGAs from genomic sunflower DNA using degenerate PCR primers. Gentzbittel et al. (1998a) mapped an RGA to a chromosome segment near *Pl1* and *Pl6* in other mapping populations. They suggested that the RGA might co-segregate with *Pl1* or *Pl6* and that the RGA was a candidate for a downy mildew resistance gene. Gedil et al. (2000b) independently cloned the same RGA (HR-4W2) and showed that HR-4W2 was a tandemly repeated gene family. HR-4W2 mapped close to but did not co-segregate with *Pl1* (~3 cM) (Gedil et al. 2000b). HR-4W2 should still be considered as a candidate for *Pl1* because recombinants could have arisen from phenotyping or genotyping errors. At the very least, HR-4W2 is tightly linked to *Pl1*. Gedil et al. (2000b) have developed cleaved amplified polymorphic sequence (CAPS) and SSCP markers for HR-4W2 for MAS.

The genetics of resistance to several economically important diseases of sunflower, e.g., *Phomopsis* stem canker (*Diaporthe helianthi*) and stalk and head rots caused by *Sclerotinium sclerotiorum*), seems to be complex. Mestries et al. (1998) mapped genes for resistance to *Sclerotinia sclerotiorum* and found four QTL for leaf resistance and two QTL for capitulum resistance. One QTL affected leaf and capitulum resistance. Gentzbittel et al. (1998) described a candidate gene (a serine-threonine protein kinase) for a major *Sclerotinia* resistance QTL. The candidate gene was homologous to the *Pto* gene in tomato (Martin et al. 1993) and the *Xa21* resistance gene in rice (Song et al. 1995). Besnard et al. (1997) used RAPD markers to search for chromosome segments introgressed from *H. argophyllus* into a *Phomopsis* resistant inbred line. The introgressed loci were associated with only a fraction of the genetic differences between the donor and recipient lines for *Phomopsis* resistance. This probably resulted from incomplete genome coverage and perhaps from non-allelic interactions. Additional studies are needed to identify QTL that confer *Phomopsis* resistance in the field. The lack of major genes for *Sclerotinia* and *Phomopsis* resistance has stimulated research on genetic engineering solutions to these problems in sunflower.

4.3. Agronomic traits

Apart from major disease resistance gene loci, a number of other single gene traits have been mapped in sunflower, including branching, fertility restoration, and genetic male sterility (Table 1). Until recently, very little had been published on quantitatively inherited traits in sunflower. For example, seed oil concentration, a trait with moderate heritability, is one of the most important traits in sunflower. Most of the genetic variance for this trait is additive. Leon et al. (1995) mapped six QTLs underlying 57% of the total genetic variance for seed oil content by partitioning the trait into total seed oil concentration, the hull and kernel oil concentration, and the hull and kernel percentage. Two QTLs affected kernel oil concentration, two QTLs affected kernel percentage, and two affected both.

The QTL with the largest effect on seed oil concentration on linkage group G was also found to be linked in repulsion to dominant gene for white hypodermis color (*Hyp*) (Leon et al. 1996). In a later study conducted in several environments using F2, F3 , and F4 families from the ZENB8 x HA89 cross, Leon et al. (2000c) reported eight QTLs on seven linkage groups that accounted for 88% of the genetic variation for seed oil percentage. Gene action was additive for four QTLs, whilst the remainder showed either dominance or over-dominance. QTL on linkage group G was detected in all environments and in all generations. The parental effects and relative magnitudes of the genetic effects were consistent across generations and environments.

Gentzbittel et al. (1998b) used F2, F3 , and F4 progeny to map QTLs affecting seed oil content, flowering date, and seed weight. They found two to three QTLs affecting seed oil content. These QTLs were associated with 19 to 54% of the phenotypic variance across generations. They also found two QTLs for seed weight and two QTLs for flowering date. Leon et al. (2000a) reported five QTLs associated with days to flowering on linkage groups A, B, H, I and L, which accounted for 73% and 89% of the phenotypic and genotypic variation across environments. QTLs on linkage groups A and B had the highest LOD scores in each environment and across environments (LOD 38.4 and 10.8). These two QTLs alone accounted for 84% of the genetic variation associated with the RFLP loci. In a later study, in additional environments covering a wider range of photoperiods (12.1 to 16.4 hrs), Leon et al. (2000b) identified six independent QTLs that were associated with growing degree days (GDD) to flowering on linkage groups A, B, F, I, J, and L. The six QTLs explained 67% of the total phenotypic variation and 76% of the total genetic variation. The QTLs on linkage groups A and B had the highest LOD scores (21.23 and 15.5, respectively) and explained 72% of the total genetic variation associated with RFLP loci.

Four of the six QTL for GDD to flowering (linkage groups A, B, F, and J) had significant QTL x E interactions (P < 0.01). The LOD scores for QTL in linkage groups A and B were highly dependent on PP. The LOD scores of QTL in linkage group A decreased while LOD values of QTL of linkage group B increased as PP increased from 12.1 to 16.4 h. Moreover, the LOD scores for QTL in linkage group B were not significant at PP of 12.1 and 13.1. Also, QTL mapping of the ratio of the GDD required by a progeny to flower at PP of 12.1 and 15.0 hours, defined as photo-period response (PPR), suggested that alleles at QTL in linkage groups A and B were responsive to PP. This finding suggests that these two QTLs are probably photoperiod genes. Gene action was additive at four of the six QTLs associated with GDD.

In a later study, Leon et al. (2000d) reported seven QTLs identified as been associated with plant height on linkage groups A, B, E, G, K, L, and M. These QTLs accounted for 76% of the genetic variation. There of these QTL on linkage groups A, B and L were also reported associated with GDD to flowering. The QTLs on groups A and B were also shown to affect leaf number, a finding consistent with the effects of these QTLs on photoperiod response.

5.   Marker studies in wild sunflower species

Most of the mapping work involving wild species has a strong evolutionary focus and primarily addresses questions concerning genomic evolution and the genetic architecture of reproductive barriers. In this section, we will briefly describe the results from these studies and their implications regarding speciation in *Helianthus*.

5.1. Genome evolution

Chromosomal evolution in *Helianthus* has been studied using a variety of techniques, including analyses of mitotic and meiotic chromosome configurations by light microscopy, and more recently, by genetic mapping. Unfortunately, *Helianthus* chromosomes are relatively small and numerous, making it difficult to differentiate chromosomes in mitotic preparations (Jan 1997). Several chromosome banding techniques have been applied to *H. annuus* chromosomes, and three chromosome pairs can be identified due to heterochromatin associated with secondary constrictions (Cuellar et al. 1996). *In-situ* hybridization with an 18S-26S ribosomal DNA (rDNA) probe indicates that these are rDNA clusters (Cuellar et al. 1996). A minor rDNA cluster has recently been identified on a fourth *H. annuus* chromosome pair as well (Schrader et al. 1997).

In contrast, meiotic studies abound for members of the genus, particularly for the 11 annual sunflowers belonging to sect. *Helianthus* (reviewed in Jan 1997). In the most comprehensive study, Chandler et al. (1986) examined meiotic abnormalities in the hybrids from 40 interspecific crosses. Based on multivalent configurations, chromosomal end arrangements for nine of the 11 species in sect. *Helianthus* were inferred. The chromosomes of these species were found to be highly differentiated, differing by numerous translocations and inversions.

Comparative genetic mapping studies of three species, *H.annuus, H. petiolaris*, and *H. anomalus*, largely confirm results from the meiotic studies of these taxa, but add considerable resolution (Fig. 2; Rieseberg et al. 1995b). For example, mapping data indicate that three rather than two inversions differentiate *H. annuus* and *H. petiolaris*, and that some linkage groups contain blocks homologous to three different linkages in other species. In addition, because *H. anomalus* appears to have been derived by hybridization between *H. annuus* and *H. petiolaris*, the mapping data provide a means for studying the genomic processes that accompany or facilitate diploid hybrid speciation.

For example, genetic models of diploid hybrid speciation suggest that it will be facilitated by rapid karyotypic evolution (Grant 1981; Templeton 1981). The rationale for this expectation is that a new hybrid lineage is unlikely to be evolutionarily stable unless it somehow becomes reproductively isolated from its parental species. Otherwise, it will be swamped by gene flow from parental populations. Chromosomal reorganization following hybridization provides a means for creating a chromosomal sterility barrier between the hybrid neospecies and its parents.

Rapid karyotypic evolution does appear to have facilitated the development of reproductive isolation between *H. anomalus* and its parents (Rieseberg et al. 1995b). The species has a distinctive set of chromosomal rearrangements (Fig. 2), that appear to

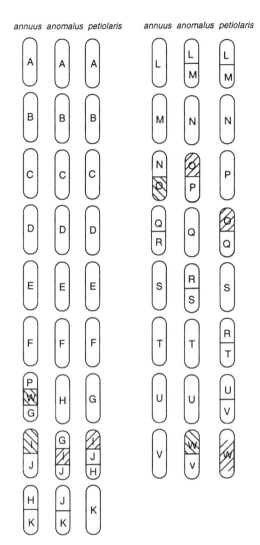

Figure 2. Linkage relationships between *Helianthus annuus, H. petiolaris*, and their putative hybrid derivative, *H. anomalus*, as inferred from comparative linkage mapping (Rieseberg et al. 1995b). Lines of shading within linkage groups indicate inversions.

result from both the merger of pre-existing structural differences between the parents, as well as chromosomal rearrangements apparently induced by recombination between the parental genomes.

In addition, to investigating karyotypic evolution, the genetic maps allowed Rieseberg et al. (1995b; 1996b) to estimate the parental contributions of *H. annuus* and *H. petiolaris* to the *H. anomalus* genome, and to compare the genomic composition of *H. anomalus* to three synthetic hybrid lineages that were created in the greenhouse.

These comparisons revealed that, although generated independently, all three synthetic hybrid lineages had converged to nearly identical gene combinations, and this set of gene combinations was statistically concordant with that of *H. anomalus* (P < 0.0001 for all comparisons). Concordance in genomic composition between the synthetic and ancient hybrids suggests that deterministic forces such as selection and genetic constraint, rather than stochastic forces, largely govern hybrid species formation. Because the synthetic hybrid lineages were generated in the greenhouse rather than under natural conditions, fertility selection probably played a greater role than ecological selection in shaping hybrid genomic composition. This conclusion is supported by the rapid increase in fertility observed in the three hybrid lineages; average pollen fertility increased from 5.6% in F_1 crosses to over 98% in fifth generation hybrids. Congruence in genomic composition also implies that the genomic structure and composition of hybrid species may be essentially fixed after a small number of generations of hybridization and probably remain relatively static thereafter.

Maps are currently being generated for two additional wild species, *H. deserticola* and *H. paradoxus* (Kim and Rieseberg, unpublished), which also appear to be derived by hybridization between *H. annuus* and *H. petiolaris* (Rieseberg 1991). Comparison of these maps, with the three already complete, will enable the identification of components of the speciation process that are in common to all three hybrid species. Documented congruence in the formation of these three hybrid taxa will provide a powerful means for deriving a more predictive theory of hybrid speciation.

5.2   Genetic architecture of reproductive barriers

Students of speciation have long been interested in the genetic architecture of reproductive barriers (i.e., the number, location, effects, and interactions of genetic factors that contribute to reproductive isolation) because it provides one of the few clues into how new species are formed. Knowledge of genetic architecture has practical value as well, because the ease with which agronomically important genes can be introgressed across species barriers depends on details of genetic architecture (Rieseberg et al. 1996a).

Genetic mapping data provides a powerful means for the dissection of genetic factors contributing to reproductive isolation. Two general approaches have been used in *Helianthus*. One approach involves genetic mapping of quantitative trait loci (QTLs) for sterility and is exemplified by studies of the genetic basis of reduced pollen viability in hybrids between *H. annuus* and *H. argophyllus* (Quillet et al. 1995). *Helianthus argophyllus* is the sister species of *H. annuus* (Rieseberg 1991), and cytogenetic analyses indicate that the two species differ by two reciprocal translocations (Chandler et al. 1986). To provide a quantitative estimate of the influence of chromosomal rearrangements on pollen viability, Quillet et al. (1995) analyzed the segregation of 48 genetic markers in a BC_1 progeny of an interspecific hybrid between *H. argophyllus* and the common sunflower, *H. annuus*. As predicted by cytogenetic studies, a wide range of variability in pollen viability was observed in the mapping family (27% to 93%). Over 80% of this variation was explained by three genetic intervals located on linkage groups 1, 2, and 3, respectively. Analyses of meiosis in the backcross hybrids revealed that meiotic abnormalities also were tightly correlated with the markers, indicating that chromosomal rearrangements are largely responsible for reducing fertility in hybrids between these species.

A second approach involves the analysis of patterns of differential introgression across species boundaries. Introgression of genes (and linked markers) contributing to isolation will be retarded, whereas neutral or positively selected chromosomal segments (and linked markers) will introgress at higher frequencies. If the markers have been genetically mapped, the observed patterns of introgression should also make it possible to locate the genomic intervals contributing to isolation.

This approach has been employed in controlled introgression experiments between *H. annuus* and *H. petiolaris* (Rieseberg et al. 1995a; 1996a). Comparative mapping experiments indicate that the two species differ by a minimum of seven interchromosomal translocations and three inversions (Fig. 2). Analyses of patterns of marker introgression, in 170 individuals from three introgression lines, revealed that it was reduced in all ten linkage groups that were structurally divergent. In addition, fourteen genomic intervals within the seven collinear linkages also resisted introgression. Presumably, these regions contain one or more genes that contribute to reduced hybrid fitness. Thus, reproductive isolation between these species appears to be complex and involves a large number chromosomal and genic sterility factors, as would be predicted by classical speciation theory (Coyne and Orr 1998).

6.   References

Anisimova, I.N. and Gavriluck, I.P. (1989) Heterogeneity and polymorphism of the 11S globulin of sunflower seeds. Genetics (in Russian) 25: 1248-1255.

Anisimova, I.N., Georgieva-Todorova, F. and Vassileva, R. (1993) Variability of helianthinin, the major seed storage protein in genus *Helianthus*. Helia 16: 49-58.

Anisimova, I.N., Fido, R.J., Tatham, A.S. and Shrewry, P.R. (1995) Genotypic variation and polymorphism of 2S albumins of sunflower. Euphytica 83: 15-23.

Arias, D.M. and Rieseberg L.H. (1994) Gene flow between cultivated and wild sunflowers. Theor. Appl. Genet. 89: 655-660.

Bernardo, R. (1992) Relationship between single-cross performance and molecular marker heterozygosity. Theor. Appl. Genet. 83:628-634.

Bernatzky, R. and Tanksley, S.D. Towards a saturated map in tomato based on isozyme and random cDNA sequences. Genetics 112:887-898

Berry, S.T. (1995) Molecular marker analysis of cultivated sunflower (*Helianthus annuus* L.). Ph.D. Thesis, Reading University, U.K.

Berry, S.T., Allen, R.J., Barnes S.R. and Caligari, P.D.S. (1994) Molecular analysis of *Helianthus annuus* L. 1. Restriction fragment length polymorphism between inbred lines of cultivated sunflower. Theor. Appl. Genet. 89: 435-441

Berry, S.T., Leon, A.J., Hanfrey, C.C., Challis, P., Burkholz, A., Barnes, S.R., Rufener, G.K., Lee, M. and Caligari, P.D.S. (1995) Molecular marker analysis of *Helianthus annuus* L. 2. Construction of an RFLP linkage map for cultivated sunflower. Theor. Appl. Genet. 91: 195-199.

Berry, S.T., Leon, A.J., Challis, P., Livini, C., Jones, R., Hanfrey, C.C., Griffiths, S. and Roberts, A. (1996) Construction of a high density, composite RFLP linkage map for cultivated sunflower

(*Helianthus annuus* L.). *In* "Proc. 14th Int. Sunflower Conf. Vol. 2", Beijing/Shenyang, China, 12-20 June, pp. 1155-1160.

Berry, S.T., Leon A.J., Peerbolte, R., Challis, C., Livini, C., Jones, R. and Feingold, S. (1997) Presentation of the Advanta sunflower RFLP linkage map for public research. *In* "Proc. 19th Sunflower Res. Workshop", Fargo, USA, January 9-10, pp. 113-118.

Besnard, G., Griveau, Y., Quillet, M.C., Serieys, H., Lambert, P., Vares, D. and Berville, A. (1997) Specifying the introgressed segments from *H. argophyllus* in cultivated sunflower (*Helianthus annuus* L.) to mark *Phomopsis* resistance genes. Theor. Appl. Genet. 94: 131-138.

Brunel, D. (1994) A microsatellite marker in *Helianthus annuus* L. Pl. Mol. Biol. 24: 397-400.

Carrera, A. and Poverene, M. (1995) Isozyme variation in *Helianthus petiolaris* and sunflower, *H. annuus*. Euphytica 81: 251-257.

Chandler, J.M., Jan, C.C. and Beard, B.H. (1986) Chromosomal differentiation among the annual *Helianthus* species. Syst. Bot. 11: 353-371.

Charcosset, A., Lefort-Buson, M., Gallais, A. (1991) Relationship between heterosis and heterozygosity at marker loci: A theoretical computation. Theor. Appl. Genet. 81:571-575.

Cheres, M.T. and Knapp, S.J. (1998) Ancestral origins and genetic diversity of cultivated sunflower: analysis of the pedigrees of public germplasm sources. Crop Sci. 38: 1476-1482.

Cheres, M.T., Miller, J.F., Crane, J.M. and Knapp, S.J. (1999a) Genetic distance as a predictor of heterosis and hybrid performance in sunflower. Theor. Appl. Genet. (in press).

Choumane, W. and Heizmann P. (1988) Structure and variability of nuclear ribosomal genes in the genus Helianthus. Theor. Appl. Genet. 76: 481-489.

Coyne, J.A. and H.A. Orr. (1998) The evolutionary genetics of speciation. Philosophical Transactions of the Royal Society of London Series B 353: 287-305.

Cronn, R., Brothers, M., Klier, K., Bretting, P.K. and Wendel, J.F. (1997) Allozyme variation in domesticated annual sunflower and its wild relatives. Theor. Appl. Genet. 95: 532-545.

Cuellar, T., Belhassen, E., Fern·ndez-Calv'n, B., Orellana, J. and Bella, J.L. (1996) Chromosomal differentiation in *Helianthus annuus* var. *macrocarpus*: Heterochromatin characterization and rDNA location. Heredity 76: 586-591.

Dehmer, K.J. and Friedt, W. (1998) Evaluation of different microsatellite motifs for analysing genetic relationships in cultivated sunflower (*Helianthus annuus* L.). Pl. Breed. 117: 45-48.

Dudley, J.W., Saghai Maroof, M.A., Rufener, G.K. (1991) Molecular markers and grouping of parents in maize breeding programs. Crop Sci. 31:660-668.

Fick, G.N. and Miller, J.F. (1997) Sunflower breeding. *In* Sunflower technology and production, Schneiter, A.A. (ed.). Amer. Soc. Agronomy, Madison, WI. pp. 395-439.

Fick, G.N. and Zimmer, D.E. (1976) Yield stability of sunflower hybrids and open pollinated varieties. In "Proc. 7th Int. Sunflower Conf., Krasnadar, Russia, June 27-July 3, Int. Sunflower Assoc., Paris, France, pp 253-258.

Flor, H.H. (1956) The complementary genic systems in flax and flax rust. Adv. Genet. 8:29-54

Gedil, M.A., Wye, C., Berry, S.K., Segers, B., Peleman, J., Jones, R., Leon, A., Slabaugh, M.B. and Knapp., S.J. 2000. An integrated RFLP-AFLP linkage map for cultivated sunflower. Genome (in review).

Gedil, M.A., Slabaugh, M.B., Berry, S.K., Jones, R., Michelmore, R., Miller, J.F., Gulya, T. and Knapp, S.J. 2000. Candidate disease resistance genes in sunflower cloned using conserved nucleotide binding site motifs: Genetic mapping and linkage to the *Pl1* gene for resistance to downy mildew. Genome (in review).

Gentzbittel, L., Perault, A. and Nicolas, P. (1992) Molecular phylogeny of the *Helianthus* genus, based on nuclear restriction-fragment-length polymorphism (RFLP). Mol. Biol. Evol. 9: 872-892.

Gentzbittel, L., Zhang, Y.-X, Vear, F., Griveau, B. and Nicolas, P. (1994) RFLP studies of genetic relationships among inbred lines of the cultivated sunflower, *Helianthus annuus* L.: evidence for distinct restorer and maintainer germplasm pools. Theor. Appl. Genet. 89:419-425.

Gentzbittel, L., Vear, F., Zhang, Y.-X. and Berville, A. (1995) Development of a consensus linkage map for cultivated sunflower (*Helianthus annuus* L.). Theor. Appl. Genet. 90: 1079-1086.

Gentzbittel, L., Mouzeyar, S., Badaoui, S., Mestries, E., Vear, F., Tourvielle De Labrouhe, D. and Nicolas, P. (1998a) Cloning of molecular markers for disease resistance in sunflower, *Helianthus annuus* L. Theor. Appl. Genet. 96: 519-525.

Grant, V. (1981) Plant speciation. Columbia University Press, New York.

Heiser, C.B. (1976) The sunflower. Univ. Oklahoma Press, Norman.

Heiser, C.B., Smith, D.M. (1955) New chromosome numbers in *Helianthus* and related genera. Proc. Ind. Acad. Sci. 64: 250-253

Hongtrakul, V., Husetis, G.M. and Kanpp S.J. (1997) Amplified fragment length polymorphisms as a tool for DNA fingerprinting sunflower germplasm: genetic diversity among oilseed inbred lines. Theor. Appl. Genet. 95: 400-407.

Hongtrakul, V., Slabaugh, M.B. and Knapp S.J. (1998a) DFLP, SSCP, and SSR marker polymorphisms for Δ9-stearoyl-acyl carrier protein desaturases strongly expressed in developing seeds of sunflower. Molec. Breed. 4: 195-203.

Hongtrakul, V., Slabaugh, M.B. and Knapp, S.J. (1998b) A seed specific Δ12 oleate desaturase gene is duplicated, rearranged, and weakly expressed in high oleic acid sunflower lines. Crop Sci. 38: 1245-1249.

Jackson, R.C. and Murray (1983) Colchicine induced quadrivalent formation in *Helianthus*: evidence for ancient polyploidy. Theor. Appl. Genet. 64: 219-222

Jan, C.C. (1997) Cytology and interspecific hybridization. *In* Sunflower Science and Technology (A.A. Schneiter, ed.), pp. 497-558. ASA, CSSA, and ASSA, Madison, WI.

Jan, C.C., Vick, B.A., Miller, J.F., Kahler, A.L. and Butler, E.T. (1993) Progress in the development of a genomic RFLP map of cultivated sunflower (*Helianthus annuus*). In "Proc. 15th Sunflower Res. Workshop", Fargo USA, pp 125-128.

Jan, C.C., Vick, B.A., Miller, J.F., Kahler, A.L. and Butler, E.T. (1998) Construction of an RFLP linkage map for cultivated sunflower. Theor. Appl. Genet. 96: 15-22.

Jiang, C. and Zeng, Z.-B. (1997) Mapping quantitative trait loci with dominant and missing markers in various crosses from two inbred lines. Genetica 101: 47-58.

Ji, J., Wang, G., Belhassen, E., Serieys, H. and Berville, A. (1996) Molecular markers of nuclear restoration gene *Rf1* in sunflower using bulked segregant analysis-RAPD. Sci. China Series C Life Sci. 39: 551-560.

Jones, C.J., Edwards, K.J., Castaglione, S., Winfield, M.O., Sala, F., Van De Wiel, C., Bredemeijer, G., Vosma, B., Matthes, M., Daly, A., Brettschnedider, R., Bettini, P., Buiatti, M., Maestri, E., Malcevschi, A., Marmiroli, N., Aert, R., Volckaert, G., Rueda, J., Linacero, R., Vazquez, A. and Karp, A. (1997) Reproducibility testing of RAPD, AFLP, and SSR markers in plants by a network of European laboratories. Molec. Breed. 3: 381-390.

Kahler, A.L. and Lay, C.L. (1985) Genetics of electorphoretic variants in the annual sunflower. J. Hered. 76: 335-340.

Kesseli, R.V., Paran I. and Michelmore, R.W. (1994) Analysis of a detailed genetic linkage map of *Lactuca sativa* (Lettuce) constructed from RFLP and RAPD markers. Genetics 136: 1435-1446.

Kinman, M.L. (1970) New developments in the USDA and state experimental station breeding programmes. In "Proc. 4th Int. Sunflower Conf.", Int. Sunflower Assoc., Toowoomba, Qld, Australia, pp181-183.

Kleinhofs, A., Kilian, A., Saghai-Maroof, M.A., Biyashev, R.M., Hayes, P., Chen, F.Q., Lapitan, N., Fenwick, A., Blake, T.K., Kanazin, V., Ananiev, E., Dahleen, L., Kudrna, D., Bollinger, J., Knapp, S.J., Liu, B., Sorrells, M., Heun, M., Franckowiak, J.D., Hoffman, D., Skadsen, R. and Steffenson, B.J. (1993) A molecular, isozyme and morphological map of the barley (*Hordeum vulgare*) genome. Theor. Appl. Genet. 86: 705-712.

Knapp, S.J., Holloway, J.L., Bridges, W.C. and Liu, B.-H. (1995) Mapping dominant markers using F_2 matings. Theor. Appl. Genet. 91:74-81.

Kong, G.A. and Kochman J.K. (1996) Understanding sunflower rust. *In* " Proc. 11th Int. Sunflower Conf.", Int. Sunflower Assoc., Toowoomba, pp 20-22

Korell, M., Mosges, G. and Friedt, W. (1992) Construction of a sunflower pedigree map. Helia 15:7-16

Konieczny, A. and Ausubel, F.M. (1993) A procedure for mapping *Arabidopsis* mutations using co-dominant ecotype-specific PCR-based markers. Plant J. 4: 403-410

Kovacik, M.V. and Skaloud, V. (1990) Results of inheritance evaluation of agronomically important traits in sunflower. Helia 13: 41-46.

Lawson, W.R., Goulter, K.C., Henry, R.J., Kong, G.A. and Kochman, J.K. (1994) Genetic diversity in sunflower (*Helianthus annuus* L.) as revealed by random amplified polymorphic DNA analysis. Aust. J. Agric. Res. 45: 1319-1327.

Lawson, W.R., Goulter, K.C., Henry, R.J., Kong, G.A. and Kochman, J.K. (1996) RAPD markers for a sunflower rust resistance gene. Aust. J. Agric. Res. 47: 395-401.

Lawson, W.R., Goulter, K.C., Henry, R.J., Kong, G.A. and Kochman, J.K. (1998) Marker-assisted selection for two rust resistance genes in sunflower. Molec. Breed. 4: 227-234.

Leclerq, P. (1966) Une sterilite male utilisable pour la production d'hybrid simples de tournesol. Ann. Amerlior Plantes 16: 135-144.

Leclerq, P. (1969) Une sterilite male cytoplasmique chez le tournesol. Ann. Amerlior Plantes 19: 99-106.

Leon, A.J., Lee, M., Rufener, G.K., Berry S.T. and Mowers, R.P. (1995) Use of RFLP markers for genetic linkage analysis of oil percentage in sunflower seed. Crop Sci. 35: 558-564.

Leon, A.J., Lee, M., Rufener, G.K., Berry S.T. and Mowers, R.P. (1996) Genetic mapping of a locus (*hyp*) affecting seed hypodermis color in sunflower. Crop Sci. 36: 1666-1668.

Leon, A.J., Andrade, F.H., and Lee, M. (2000a). Genetic mapping of factors affecting quantitative variation for flowering in sunflower (*Helianthus annuus* L.). Crop Sci. 40: 404- 407.

Leon, A.J., Lee, M., and Andrade, F.H. (2000b). Quantitative trait loci for growing degree days to flowering and photoperiod response in sunflower (*Helianthus annuus* L.). Theor. Appl. Genet. (in review).

Leon, A.J., Andrade, F.H. and Lee, M. (2000c). Genetic analysis of seed oil percentage across generations and environments in sunflower (*Helianthus annuus* L.). Crop Sci. (in review).

Leon, A.J., Andrade, F.H. and Lee, M. (2000d). Quantitative trait loci for plant height, leaf number and internode length in sunflower (*Helianthus annuus* L.). Theor. Appl. Genet. (in review).

Linder, C.R., Taha, I., Seiler, G.J., Snow, A.A. and Rieseberg, L.H. (1998) Long-term introgression of crop genes into wild sunflower populations. Theor. Appl. Genet. 96: 339-347.

McCouch, S.R., Kochert, G., Yu, Z.H., Wang, Z.Y., Khush, G.S., Coffman, W.R. and Tanksley, S.D. (1988) Molecular mapping of rice chromosomes. Theor. Appl. Genet. 76: 815-829.

Marshall, D.R. and Allard, R.W. (1970) Isozyme polymorphism in natural populations of *Avena fatus* and *A. barbata*. Heredity 25: 373-382.

Martin, G.B., Brommonschenkel, S.H., Chunwongse, J., Frary, A., Ganal, M.W., Spivey, R., Wu, T., Earle, E.D. and Tanksley, S.D. (1993) Map-based cloning of a protein conferring disease resistance in tomato. Science 262: 1432-1436.

Mestries, E., Gentzbittel, L., Tourvieille de Labrouhe, D., Nicolas, P. and Vear, F. (1998) Analyses of quantitative trait loci associated with resistance to *Sclerotinia sclerotiorum* in sunflower (*Helianthus annuus* L.) using molecular markers. Molec. Breed. 4: 215-226.

Miller, J.C. and Tanksley, S.D. (1990) Effect of different restriction enzymes, probe source, and probe length on detecting restriction fragment length polymorphisms in tomato. Theor. Appl. Genet. 80:385-389.

Miller, J.F. and Fick, G.N. (1997) The genetics of sunflower. *In* Sunflower technology and production, Schneiter, A.A. (ed.). Amer. Soc. Agronomy, Madison, WI. pp. 441-496.

Moser, H. and Lee, M. (1994) RFLP variation and genealogical distance, multivariate distance, heterosis and genetic variance in oats. Theor. Appl. Genet. 87:947-956.

Mosges, G. and Freidt, W. (1994) Genetic "fingerprinting" of sunflower lines and F₁ hybrids using isozymes, simple and repetitive sequences as hybridization probes, and random primers for PCR. Pl. Breed. 113: 114-124.

Mouzeyar, S., Roeckel-Drevet, P., Gentzbittel, L., Philippon, J., Tourvieille De Labrouhe, D., Vear, F. and Nicolas, P. (1995) RFLP and RAPD mapping of the sunflower *Pl1* locus for resistance to *Plasmopara halstedii* race 1. Theor. Appl. Genet. 91: 733-737.

Peerbolte, R.P. and Peleman, J. (1996) The Cartisol sunflower RFLP map (146 loci) extended with 291 AFLP™ markers. In "Proc. 18th Sunflower Res. Workshop", Fargo, USA, January 11-12, pp 174-178.

Peleman, J. (1998) Integration of genetic maps (AFLP) in lettuce and sunflower: observations and applications. Plant and Animal Genome VI Abstracts, San Diego, CA.

Pustovoit, V.S. (1964) Conclusions of work on the selection and seed production of sunflowers (in Russian). Agrobiology 5: 662-697.

Putt, E.D. (1997) Early history of sunflower. *In* Sunflower technology and production, Schneiter, A.A. (ed.). Amer. Soc. Agronomy, Madison, WI. pp 1-20.

Putt, E.D., Sackston, W.D. (1963) Studies on sunflower rust. IV. Two rust genes *R₁* and *R₂* for resistance in the host. Canadian J. Pl. Sci. 43: 490-496.

Quillet, M.C., Vear, F. and Branlard, G. (1992) The use of isozyme polymorphism for identification of sunflower (*Helianthus annuus*) inbred lines. J. Genet. Breed. 46: 295-304.

Quillet, M.C., Madjidian, N., Griveau, Y., Serieys, H., Tersac, M., Lorieux, M. and Berville, A. (1995) Mapping genetic factors controlling pollen viability in an interspecific cross in *Helianthus* sect. *Helianthus*. Theor. Appl. Genet. 91: 1195-1202.

Rieseberg, L.H. (1991) Homoploid reticulate evolution in *Helianthus*: Evidence from ribosomal genes. Am. J. Botany 78: 1218-1237.

Rieseberg, L.H. (1996) Homology among RAPD fragments in interspecific comparisons. Mol. Ecol. 5: 99-105.

Rieseberg, L.H. (1998) Genetic mapping as a tool for studying speciation. pp.459-487. *In* Molecular Systematics of Plants. 2nd edition (D.E. Soltis, P.S. Soltis, and J.J. Doyle, eds.). Chapman and Hall Inc., New York.

Rieseberg, L.H., D.M. Arias, M. Ungerer, C.R. Linder, and Sinervo, B. (1996a) The effects of mating design on introgression between chromosomally divergent sunflower species. Theor. Appl. Genet. 93: 633-644.

Rieseberg, L.H., Choi, H., Chan, R. and Spore, C. (1993) Genomic map of a diploid hybrid species. Heredity 70: 285-293.

Rieseberg, L.H., Linder, C.R. and Seiler, G. (1995a) Chromosomal and genic barriers to introgression in *Helianthus*. Genetics 141: 1163-1171.

Rieseberg, L.H. and Seiler, G. (1990) Molecular evidence and the origin and development of the domesticated sunflower (*Helianthus annuus*). Econ. Bot. 44S: 79-91.

Rieseberg, L.H., B. Sinervo, C.R. Linder, M. Ungerer and Arias, D.M. (1996b) Role of gene interactions in hybrid speciation: Evidence from ancient and experimental hybrids. Science 272: 741-745.

Rieseberg, L.H., Van Fossen, C. and Desrochers, A. (1995b) Hybrid speciation accompanied by genomic reorganization in wild sunflowers. Nature 375: 313-316.

Roeckel-Drevet, P., Gagne, G., Mouzeyar, S., Gentzbittel, L., Phillippon, J., Tourvieille de Labrouhe, D., Vear, F. and Nicolas, P. (1995) Co-location of downy mildew (*Plasmopara halstedii*) resistance genes in sunflower (*Helianthus annuus*). Euphytica 91: 225-228.

Sackston, W.E. (1992) On a treadmill: breeding sunflowers for resistance to disease. Ann. Rev. Phytopath. 30:529-551.

Schrader, O., Ahne, R., Fuchs, J. and Schubert, I. (1997) Karyotype analysis of *Helianthus annuus* using Giemsa banding and fluorescence *in situ* hybridization. Chromosome Res. 5: 451-456.

Seiler, G.J. (1988) The genus *Helianthus* as a source of genetic variability for cultivated sunflower. Proc. 12th Int. Sunfl. Conf., Novi Sad, Yugoslavia, pp. 17-58.

Seiler, G.J. and Rieseberg, L.H. (1997) Systematics, origin and germplasm resources of the wild and domesticated sunflower. *In* Sunflower technology and production, Schneiter, A.A. (ed.). Amer. Soc. Agronomy, Madison, WI. pp 21-66.

Serre, M., Feingold, S., Salaberry T., Leon, A. and Berry, S. (2000) The genetic map position of the locus encoding the 25 albumin seed storage proteins in cultivated sunflower (*Helianthus annuus* L.) Euphytica (in press).

Song, W-Y, Wang, G-L, Chen, L-L, Kim, H-S, Holsten, T., Wang, B., Zhai, W-X, Zhu, L-H., Fanquet, C. and Ronald, P. (1995) A receptor kinase-like protein encoded by the rice disease resistance gene, *Xa21*. Science 270:1804-1806.

Skoric, D. (1993) Wild species use in sunflower breeding — results and future directions. FAO/IBPR Plant Genetic Resources Newsletter 93: 17-23.

Stam, P. (1993) Construction of integrated genetic maps by means of a new computer package: JoinMap. Plant J 3:739-744.

Staskawicz, B.J., Ausubel, F.M., Baker, B.J., Ellis, J.G. and Jones, J.D.G. (1995) Moelcular genetics of plant disease resistance. Science 268:661-667.

Tanksley, S.D., Ganal, M.W., Prince, J.P., de Vincente, M.C., Bonierbale, M.W., Broun, P., Fulton, T.M., Giovannoni, J.J., Grandillo, S., Martin, G.B., Messeguer, R., Miller, J.C., Miller, L., Paterson, A.H., Pineda, O., Roder, M.S., Wing, R.A., Wu, W., Young, N.D.

(1992) High density molecular linkage maps of the potato and tomato genomes. Genetics 132: 1141-1160.

Templeton, A.R. (1981) Mechanisms of speciation — a population genetic approach. Ann. Rev. Ecology Syst. 12: 23-48.

Thormann C.E., Ferreira M.E., Camargo L.E.A., Tivang J.G., Osborn T.C. (1994) Comparison of RFLP and RAPD markers to estimating genetic relationships within and among cruciferous species. Theor. Appl. Genet. 88: 973-980.

Torres, A.M. (1983) Sunflowers (*Helianthus annuus* L.). In "Isozymes in Plant Genetics and Breeding" (S.D. Tanksley and T.J. Orton, eds.), pp. 329-338. Elsevier Publishers, Amsterdam.

Torres, A.M. and Diedenhofen, U. (1976) The genetic control of sunflower seed acid phosphatase. Canadian J. Genet. Cytol. 18: 709-716.

Tersac, M., Blanchard, P., Brunel, D. and Vincourt, P. (1994) Relationships between heterosis and enzymatic polymorphism in populations of cultivated sunflowers (*Helianthus annuus* L.). Theor. Appl. Genet. 88: 49-55.

Tuelat B., Zhang, Y.X. and Nicolas, N. (1994) Characteristics of random amplified DNA markers discriminating *Helianthus annuus* inbred lines. Agronomie 14:497-502

Ungerer, M.C., Baird, S., Pan, J. and Rieseberg, L.H. (1998) Rapid hybrid speciation in wild sunflowers. Proc. Natl. Acad. Sci. USA 95: 11757-11762.

Vear, F., Gentzbittel, L., Philippon, J., Mouzeyar, S., Mestries, E., Roeckel-Drevet, P., Tourvielle De Labrouhe, D. and Nicolas, P. (1997) The genetics of resistance to five races of downy mildew (*Plasmopara halstedii*) in sunflower (*Helianthus annuus* L.). Theor. Appl. Genet. 95: 584-589.

Vranceanu, V. (1970) Advances in sunflower breeding in Romania. In "Proc. 4th Int. Sunflower Conf.", Memphis, TN, pp 136-148.

Vos, P., Hogers, R., Bleeker, M., Reijans, M., van de Lee, T., Hornes, M., Frijters, A., Pot, J., Peleman, J., Kuiper, M. and Zabeau, M. (1995) AFLP: a new technique for DNA fingerprinting. Nucleic Acids Res. 23:4407-4414.

Whitton J., Rieseberg L.H. and Ungerer, M.C. (1997) Microsatellite loci are not conserved across the Asteraceae. Mol. Biol. Evol. 14: 204-209.

Williams J.G.K, Hanafey, M.K., Rafalski, J.A. and Tingey, S.V. (1993) Genetic analysis using random amplified polymorphic DNA markers. Meth. Enzym. 218: 704-740.

Williams, J.G.K., Kubelik, A.R., Livak, K.J., Rafalsky, J.A. and Tingey, S.V. (1990) DNA polymorphisms amplified by arbitrary primers are useful as genetic markers. Nucleic Acids Res 18: 6531-6535.

Zhang, Y.X. (1995) Evaluation of the potential of RFLPs for the study of distinctness, uniformity, and stability in sunflower. *In* "The working group on biochemical and molecular techniques and DNA profiling in particular," International Union for the Protection of New Varieties of Plants, Wageningen, Netherlands.

Zhang, Y.X., Gentzbittel, L., Vear, F. and Nicolas, N. (1995) Assessment of inter- and intra-inbred line variability in sunflower (*Helianthus annuus*) by RFLPs. Genome 38: 1040-1048.

Zimmerman, D.E. and Kinman M.L. (1972) Downy mildew resistance in cultivated sunflower and its inheritance. Crop Sci. 12: 749-751.

23.   The molecular map of tomato

ANNE FRARY *and* STEVEN D. TANKSLEY
Department of Plant Breeding and Biometry, Cornell University, ITHACA, NY 14853-1901, U.S.A.
<sdt4@cornell.edu>

Contents

1.   Introduction

Cultivated tomato (*Lycopersicon esculentum* L.) is a member of the nightshade family (*Solanaceae*) which includes several other crop plants including potato (*Solanum tuberosum*), pepper (*Capsicum annuum*), eggplant (*S. melongena*), and tobacco (*Nicotiana tabacum*). Of these solanaceous species, tomato is the most amenable to genetic analysis as it is diploid (2n = 2x = 24), autogamous and has a relatively small nuclear genome of approximately 950Mb (Arumuganathan and Earle 1991). The tomato genome contains nearly 80% single copy sequences as determined by high stringency hybridization conditions (Zamir and Tanksley 1988), a characteristic which also facilitates genetic analysis. The remainder of the genome is comprised of repetitive DNA. Repetitive elements that have been characterized and mapped in tomato include ribosomal DNA (Bernatzky and Tanksley 1986; Vallejos et al. 1986; Lapitan et al. 1991), the tomato genomic repeats TGRI, TGRII and TGRIII (Ganal et al. 1988; Lapitan et al. 1989), minisatellites (Broun and Tanksley 1993; Broun and Tanksley 1996), and microsatellites (Broun and Tanksley 1996; Vosman and Arens 1997).

In addition to cultivated tomato, the genus *Lycopersicon* includes eight wild species (*L. pimpinellifolium, L. pennellii, L. peruvianum, L. hirsutum, L. chilense, L. parviflorum, L. cheesmanii* and *L. chmielewskii*) which contain abundant genetic variability and which can be easily crossed to cultivated tomato, although embryo culture may be required to obtain hybrids for some combinations. Thus, these sources of genetic diversity can be exploited for the improvement of the tomato crop which was worth more than $600 million in 1996 (C. Garvey, personal communication).

R.L. Phillips and I.K. Vasil (eds.), DNA-Based Markers in Plants, 405 - 420.

2.   A brief history of mapping in tomato

Genetic mapping efforts in tomato began during the first part of the twentieth century (Jones 1917), and by 1975, the map contained more than 200 morphological markers assigned to the 12 linkage groups (Rick 1975). In the twenty five years since then, the map has expanded to include isozyme loci (Tanksley and Rick 1980) and restriction fragment length polymorphisms (RFLPs, Bernatzky and Tanksley 1986; Tanksley et al. 1992). The advent of DNA-based markers with neutral phenotypes and codominance provided greater genome coverage and also allowed the use of markers in gene tagging and quantitative trait loci (QTLs) experiments. In fact, tomato was the first species in which QTLs could be detected throughout an entire genome in a single segregating population (Paterson et al. 1988).

3.   Current status of the tomato molecular linkage map

The current tomato molecular linkage map was constructed using an F_2 population (67 individuals) derived from an interspecific cross between *L. esculentum* and *L. pennellii* and contains more than 1,000 markers covering a total of 1276 map units (cM) (Fig. 1; Pillen et al. 1996b). A majority of the markers are tomato cDNA and random genomic DNA clones (CT/CD and TG markers, respectively). However, genes of known function and/or phenotype have also been included (Table 1) which allows for comparison of this map with the classical linkage map of tomato. The map also includes major QTLs for seed and fruit weight and fruit shape.

4.   Utilization of the map

The availability of a tomato molecular linkage map with comprehensive coverage has facilitated the work of tomato breeders and geneticists and has allowed a greater understanding of various traits of agronomic importance and of the evolutionary relationships within the genus *Lycopersicon* and among tomato and its relatives such as potato and pepper. Examples of such research are described in the following sections.

4.1.   Gene tagging and map-based cloning

Molecular markers have been used to tag numerous plant resistance genes as well as other agronomically-important loci. This has accelerated and eased the process of introgressing favorable traits from unadapted germplasm into cultivated tomato. The advanced state of tomato molecular genetics has also enabled the map-based (positional) cloning of several tomato genes including loci for disease resistance and plant growth habit (Table 2).

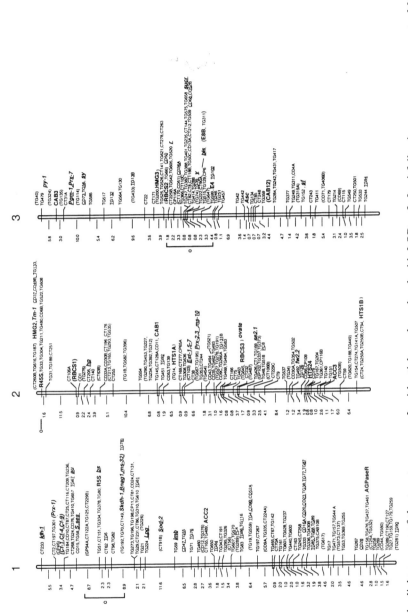

Figure 1. Molecular linkage map of the tomato genome. Loci by tick marks ordered with LOD > 3. Loci enclosed in parentheses have been located to corresponding intervals with LOD < 3. Position of underlined loci approximated from placement on previously published maps. Loci in bold correspond to known genes (see Tanle 1 for details). All other loci have been mapped directly on an F₂ population of 67 plants from *L. esculentum* x *L. pennellii* (Tanksley et al. 1992). Approximate centromere positions are indicated by brackets and bars with o's to the left of each chromosome (Grandillo and Tanksley 1996a). Black boxes on chromosomes 7 and 9 indicate the precisely mapped centromere postions (Frary et al. 1996).

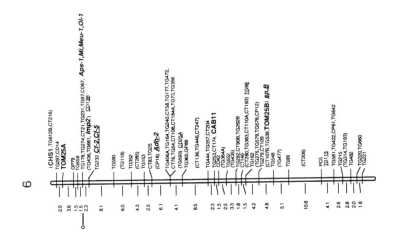

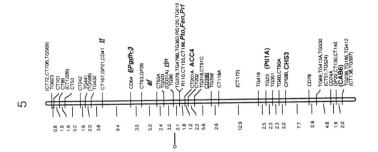

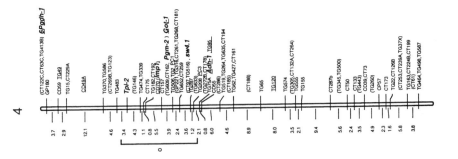

Figure 1. Continued.

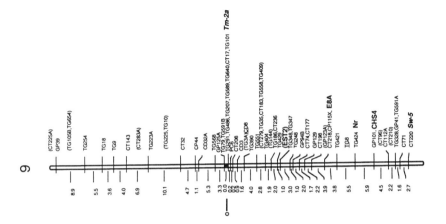

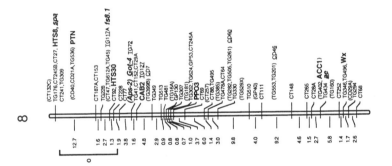

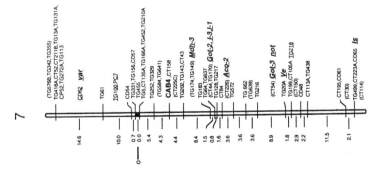

Figure 1. Continued.

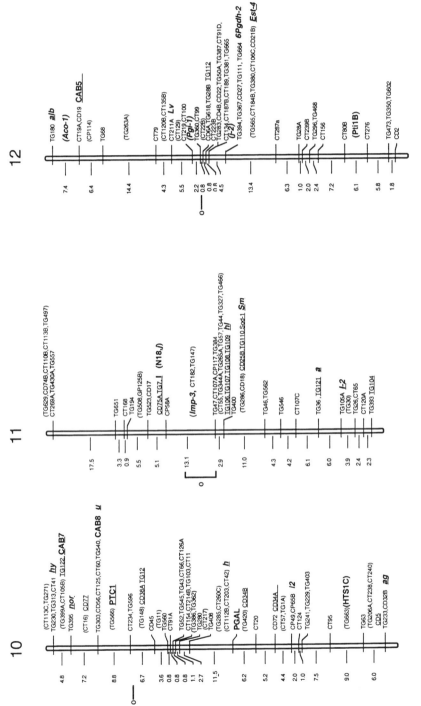

Figure 1. Continued.

Table 1.   Major gene loci, which have been mapped using molecular markers in cultivated sunflower.

Gene	Original Reference	Phenotype	Map Reference
Pl1	Vranceanu (1970)	downy mildew resistance (race 1)	Mouzeyar et al. (1995) *Gedil et al. (1999b)*
Pl2	*Zimmerman and Kinman (1972)*	*downy mildew resistance (race 2)*	*Vear et al. (1997)*
Pl6	*Miller (1992)*	*downy mildew resistance (race 6)*	*Roeckel-Drevet et al. (1996)*
R1	*Putt and Sackston (1963)*	*rust resistance (race 0). and Lawson et al. (1998)*	*Lawson et al. (1996)*
RADV	*Kong and Kochman (1996)*	*rust resistance (race Adv)*	*Lawson et al. (1998)*
Rf1	*Kinman (1970)*	*CMS fertility restorer*	*Gentzbittel et al. (1995) Quillet et al. (1995) Berry et al. (1997) Jan et al. (1998)*
b1 and b2	*Kovacik and Skaloud (1990)*	*Branching*	*Gentzbittel et al. (1995)*
ms1 and T	*Leclerq (1966)*	*genetic male sterility and anthocyanin pigment*	*Berry (1995)*
Hyp		*white pigment of seed hypodermis*	*Leon et al. (1996)*

Table 2.   Genes that have been positionally cloned in tomato.

Gene Name	Phenotype	Reference
Pto	*Pseudomonas syringae pv. tomato* resistance	Martin et al. 1993
Fen	fenthion sensitivity	Martin et al. 1994
Cf-2	*Cladosporium fulvum* resistance	Dixon et al. 1996
I2	*Fusarium oxysporum f.* sp. *lycopersici* resistance	Ori et al. 1997
Mi	root-knot nematode resistance	Milligan et al. 1998
Sp	self-pruning	Pnueli et al. 1998
Ls	lateral suppressor	Schumacher et al. 1999
Sw-5	tomato spotted wilt virus resistance	Brommonschenkel, pers. comm.

4.2. QTL analysis

The completeness of the map has also allowed tomato to be a pioneer system for the study of QTLs in plants. Quantitative traits have always presented a challenge to plant breeders, however, the use of molecular markers to identify and tag QTL can facilitate breeding for economically-important traits such as yield, fruit quality, color and stress tolerance. Research in this area has exploded since the earliest studies (Paterson et al. 1988; Paterson et al. 1991; de Vicente and Tanksley 1993) with projects focused on numerous characteristics. Traits of interest include yield and its components (*e.g.*, Goldman et al. 1995; Fulton et al. 1997; Bernacchi et al. 1998), horticultural and morphological characteristics such as plant height, mass and floral traits (*e.g.*, Grandillo and Tanksley 1996a; Tanksley et al. 1996; Bernacchi and Tanksley 1997; Paran et al. 1997), pest and disease resistance (*e.g.*, Mutschler et al. 1996; Thoquet et al. 1996; Chague et al. 1997), earliness (*e.g.*, Lindhout et al. 1994; Doganlar et al. submitted), and salt tolerance (*e.g.*, Foolad et al. 1997; Monforte et al. 1997). Some of this work has resulted in the identification of major QTLs for fruit weight (*fw2.2*, Alpert et al. 1995), fruit shape (*fs8.1*, Grandillo et al. 1996, and *ovate*, Ku et al. submitted), and seed weight (*sw4.1*, Doganlar et al. submitted). Currently, map-based cloning projects are underway to isolate *fw2.2* (Alpert and Tanksley 1996), *ovate* (Ku, personal communication), and *sw4.1* (Doganlar, personal communication). This research will provide further insight into the control of quantitative traits and fundamental processes of plant development.

4.3. Comparative mapping

The extensive work on mapping in tomato has facilitated mapping efforts in other solanaceous species. Many of the markers on the tomato map have also been mapped in potato (Bonierbale et al. 1988; Tanksley et al. 1992; Pillen et al. 1996b) and pepper (Prince et al. 1993; Pillen et al. 1996b; Livingstone et al. 1999). Thus, direct comparisions among the maps are possible. Such comparisons provide insight into the evolutionary processes that accompanied the divergence of these species from a common ancestor.

5.   Future prospects

The future of the tomato molecular linkage map promises to be as exciting and fruitful as its past. The recent emphasis on large-scale sequencing efforts and the study of genomics in plants will completely revolutionize the way in which tomato molecular genetics is done. The addition of expressed sequence tags (ESTs) to the map will provide a level of saturation that was unimaginable only twenty years ago. Moreover, the concomitant information about the expression patterns of these markers and their sequence similarity to known genes in other species will greatly expedite the process of gene cloning. The search for orthologous genes among widely divergent species such as tomato, rice and Arabidopsis will provide a new level of comparative mapping and an increased understanding of genome organization and evolution. In addition, the identification of the genes involved in various aspects of plant development such as fruit and seed production will allow the elucidation of complex developmental processes. Similar work will also reveal the biochemical pathways that produce important secondary metabolites like lycopene and betacarotene. Given recent advances in tomato genetics and the current revolution in biology, it is now possible to foresee that, eventually, the ultimate tomato molecular map may be available: the complete genomic sequence of *L. esculentum*.

6.   Acknowledgments

Thanks to Theresa Fulton and Dr. Hsin-Mei Ku for helpful comments on the manuscript.

7.   References

Alpert, K.B., Grandillo, S., Tanksley, S.D. (1995) *fw2.2*: A major QTL controlling fruit weight is common to both red and green-fruited tomato species. Theor. Appl. Genet. 91: 994-1000.

Alpert, K.B., Tanksley, S.D. (1996) High-resolution mapping and isolation of a yeast artificial chromosome contig containing *fw2.2*: a major fruit weight quantitative trait locus in tomato. Proc. Natl. Acad. Sci. USA 93: 15503-15507.

Arumuganathan, K., Earle, E.D. (1991) Nuclear DNA content of some important plant species. Plant Mol. Biol. Rep 9: 208-218.

Balint-Kurti, P.J., Dixon, M.S., Jones, D.A., Norcott, K.A., Jones, J.D.G. (1994) RFLP linkage analysis of the *Cf-4* and *Cf-9* genes for resistance to *Cladosporium fulvam* in tomato. Theor. Appl. Genet. 88: 691-700.

Balint-Kurti, P.J., Jones, D.A., Jones, J.D.G. (1995) Integration of the classical and RFLP linkage maps of the short arm of tomato chromosome 1. Theor. Appl. Genet. 90: 17-26.

Behare, J., Laterrot, H., Sarfatti, M., Zamir, D. (1991) Restriction fragment length polymorphism mapping of the *Stemphylium* resistance gene in tomato. Mol. Plant-Microbe Interact. 4: 489-492.

Bernacchi, D., Beck-Bunn, T., Eshed, Y., Lopez, Y., Petiard, V., Uhlig, J., Zamir, D., Tanksley, S;D. (1998) Advanced backcross QTL analysis in tomato. I. Identification of QTLs for traits of agronomic importance from *Lycopersicon hirsutum*. Theor. Appl. Genet. 97: 381-397.

Bernacchi, D., Tanksley, S.D. (1997) An interspecific backcross of Lycopersicon esculentum x L. hirsutum: linkage analysis and a QTL study of sexual compatibility factors and floral traits. Genetics 147: 861-877.

Bernatzky, R., Tanksley, S.D. (1986) Toward a saturated linkage map in tomato based on isozymes and random complementary DNA sequences. Genetics 112: 887898.

Bernatzky, R. (1993) Genetic mapping and protein product diversity of the selfincompatability locus in wild tomato *Lycopersicon peruvianum*. Biochem. Genet. 31: 173- 184.

Bonierbale, M.W., Plaisted, R.L., Tanksley, S.D. (1988) RFLP maps based on a common set of clones reveal modes of chromosomal evolution in potato and tomato. Genetics 120: 1095-1103.

Bournival, B.L., Scott J.W., Vallejos, C.E. (1989) An isozyme marker for resistance to race 3 of *Fusarium oxysporum* f. sp: *lycopersici* in tomato. Theor. Appl. Genet. 78: 489494.

Boynton, J.E. and Rick, C.M. (1965) Linkage tests with mutants of Stubbe's groups I, II, III and IV. Rep. Tomato Genet. Coop. 24-27.

Brommonschenkel, S.H. and Tanksley, S.D. (1997) Map-based cloning of the tomato genomic region that spans the *Sw-5* tospovirus resistance gene in tomato. Mol. Gen. Genet. 256:121-126.

Broun, P. and Tanksley, S.D. (1993) Characterization of tomato DNA clones with sequence similarity to human minisatellites 33.6 and 33.15. Plant Mol. Biol. 23: 231-242.

Broun, P. and Tanksley, S.D. (1996) Characterization and genetic mapping of simple repeat sequences in the tomato genome. Mol. Gen. Genet. 250:39-49.

Carland, F.M. and Staskawicz, B.J. (1993) Genetic characterization of the *Pto* locus of tomato. Semi-dominance and cosegregation of resistance to *Pseudomonas syringae* pathovar tomato and sensitivity to the insecticide Fenthion. Mol. Gen. Genet. 239: 17-27.

Chague, V., Mercier, J.C., Guenard, M., de Courcel, A. and Vedel, F. (1997) Identification of RAPD markers linked to a locus involved in quantitative resistance to TYLCV in tomato by bulked segregant analysis. Theor. Appl. Genet. 95:671-677.

Chetelat, R.T., DeVerna, J.W. and Bennett, A.B. (1995) Introgression into tomato *(Lycopersicon esculentum)* of the *L. chmielewskii* sucrose accumulator gene *(suer)* controlling fruit sugar composition. Theor. Appl. Genet. 91: 327-333.

Chunwongse, J., Bunn, T.B., Crossman, C., Jiang, J. and Tanksley, S.D. (1994) Chromosomal localization and molecular marker tagging of the powdery mildew resistance gene *(Lv)* in tomato. Theor. Appl. Genet. 89: 76-79.

DeVicente, M.C. and Tanksley, S.D. (1993) QTL analysis of transgressive segregation in an interspecific tomato cross. Genetics 134: 585-596.

Dickinson, M.J., Jones, D.A. and Jones, J.D.G. (1993) Close linkage between the *Cf-2, Cf-5* and *Mi* resistance loci in tomato. Mol. Plant-Microbe Interact. 6: 341-347.

Dixon, M.S, Jones, D.A., Hatzixanthis, K., Ganal, M.W, Tanksley, S.D. and Jones, J.D.G. (1995) High-resolution mapping of the physical location of the tomato *Cf-2* gene. Mol Plant-Mircobe Interact 8: 200-206.

Dixon, M.S., Jones, D.A., Keddie, J.S., Thomas, C.M., Harrison, K. and Jones, J.D.G. (1996) The tomato Cf-2 disease resistance locus comprises two functional genes encoding leucine-rich repeat proteins. Cell 84: 451-459.

Doganlar, S., Dodson, J., Gabor, B., Beck-Bunn, T., Crossman, C. and Tanskley, S.D. (1998) Molecular mapping of the *py-1* gene for resistance to corky root rot (*Pyrenochaeta lycopersici*) in tomato. Theor. Appl. Genet. 97:784-788.

Doganlar, S., Frary, A. and Tanksley, S.D. (submitted) The genetic basis of seed weight variation: tomato as a model system. Theor. Appl. Genet.

Doganlar, S., Tanksley, S.D. and Mutschler, M.A. (submitted) Identification and molecular mapping of loci controlling fruit ripening time in tomato. Theor. Appl. Genet.

Foolad, M.R., Stoltz, T., Dervinis, C., Rodriguez, R.L. and Jones, R.A. (1997) Mapping QTLs conferring salt tolerance during germination in tomato by selective mapping. Mol. Breed. 3: 269-277.

Frary, A., Presting, G.G. and Tanksley, S.D. (1996) Molecular mapping of the centromeres of tomato chromosomes 7 and 9. Mol. Gen. Genet. 250: 295-304.

Fulton, T.M., Beck-Bunn, T., Emmatty, D., Eshed, Y., Lopez, J., Petiard, V., Uhlig, J. Zamir, D. and Tanksley, S.D. (1997) QTL analysis of an advanced backcross of *Lycopersicon peruvianum* to the cultivated tomato and comparisons with QTLs found in other wild species. Theor. Appl. Genet. 95: 881-894.

Ganal, M.W., Bonierbale, M.W., Roeder, M.S., Park, W.D. and Tanksley, S.D. (1991) Genetic and physical mapping of the patatin genes in potato and tomato. Mol. Gen. Genet. 225: 501-509.

Ganal, M.W., Lapitan, N.L.V. and Tanksley, S.D. (1988) A molecular and cytogenetic survey of major repeated DNA sequences in tomato *Lycopersicon esculentum*. Mol. Gen. Genet. 213: 262-268.

Gebhardt, C., Ritter, E., Debener, T., Schachtsschabel, U., Walkemeier, B., Uhrig, H. and Salamini, F. (1989) RFLP analysis and linkage mapping in *Solanum tuberosum*. Theor. Appl. Genet. 78: 65-75.

Gillaspy, G.E., Keddie, J.S., Oda, K. and Gruissem, W. (1995) Plant inositol monophosphatase is a lithium-sensitive enzyme encoded by a multigene family. Plant Cell 7:2175-2185.

Giovannoni, J.J., Noensie, E.N., Ruezinsky, D.M., Lu, X., Tracy, S.L., Ganal, M.W., Martin, G.B., Pillen, K., Alpert, K. and Tanksley, S.D. (1995) Molecular genetic analysis of the *ripening-inhibitor* and *non-ripening* loci of tomato: A first step in genetic map-based cloning of fruit ripening genes. Mol. Gen. Genet. 248: 195-206.

Goldman, I.L., Paran and I., Zamir, D. (1995) Quantitative trait locus analysis of a recombinant inbred line population derived from a *Lycopersicon esculentum x L. cheesmanii*. Theor. Appl. Genet. 90: 925-932.

Grandillo, S. and Tanksley, S.D. (1996a) Genetic analysis of RFLPs, GATA microsatellites and RAPDs in a cross between *L. esculentum* and *L. pimpinellifolium*. Theor. Appl. Genet. 92: 957-965.

Grandillo, S. and Tanksley, S.D. (1996b) QTL analysis of horticultural traits differentiating the cultivated tomato from the closely related species *L. pimpinellifolium*. Theor. Appl. Genet. 92: 935-951.

Grandillo, S., Ku, H.M. and Tanksley, S.D. (1996) Characterization of fs8.1, a major QTL influencing fruit shape in tomato. Mol. Breed. 2: 251-260.

Iglesias, A.A., Barry, G.F., Meyer, C., Bloksberg, L., Nakata, P.A., Greene, T., Laughlin, M.J., Okita, T.W., Kishore, G.M. and Preiss, J. (1993) Expression of the potato tuber ADP-glucose pyrophosphorylase in *Escherichia coli*. J. Biol. Chem. 268: 1081-1086.

Jones, D.A., Dickinson, M.J., Balint-Kurti, P.J., Dixon, M.S. and Jones, J.D.G. (1993) Two complex resistance loci revealed in tomato by classical and RFLP mapping of the *Cf-2, Cf-4, Cf-5* and *Cf-9* genes for resistance to *Cladosporiumfulvum*. Mol. Plant-Microbe Interact. 6: 348-57.

Jones, D.F. (1917) Linkage in *Lycopersicon*. Am. Nat. 51: 608-621.

Kaloshian, I., Lange, W.H. and Williamson, V.M. (1995) An aphid resistance locus is tightly linked to the nematode resistance gene *Mi* in tomato. Proc. Natl. Acad. Sci. USA 92: 622-625.

Kawchuk, L.M., Lynch, D.R., Hachey, J., Bains, P.S. and Kulcsar, F. (1994) Identification of a codominant amplified polymorphic DNA marker linked to the *Verticillium* wilt resistance gene to tomato. Theor. Appl. Genet. 89: 661-664.

Kinzer, S.M., Schwager, S.J. and Mutschler, M.A. (1990) Mapping of ripening-related or -specific cDNA clones of tomato *(Lycopersicon esculentum)*. Theor. Appl. Genet. 79: 489-496.

Klein-Lankhorst, R., Rietveld, P., Machiels, B., Verkerk, R., Weide, R., Gebhardt, C., Koornneef, M. and Zabel, P. (1991) RFLP markers linked to the root knot nematode resistance gene *Mi* in tomato. Theor. Appl. Genet. 81: 661-667.

Ku, H.M., Doganlar, S., Chen, K. and Tanksley, S.D. (submitted) The genetic basis of pear-shaped tomato fruit. Theor. Appl. Genet.

Lapitan, N.L.V., Ganal, M.W. and Tanksley, S.D. (1989) Somatic chromosome karyotype of tomato based on *in situ* hybridization of the TGRI satellite repeat. Genome 32: 992998.

Lapitan, N.L.V., Ganal, M.W. and Tanksley, S.D. (1991) Organization of the 5S ribosomal RNA genes in the genome of tomato. Genome 34: 509-514.

Levesque, H., Vedel, F., Mathieu, C. and deCourcel, A.G.L. (1990) Identification of a short rDNA spacer sequence highly specific to a tomato line containing the *Tm-1* gene introgressed from *Lycopersicon hirsutum*. Theor. Appl. Genet. 80: 602-608.

Lincoln, J.E., Cordes, S., Read, E. and Fischer, R.L. (1987) Regulation of gene expression by ethylene during *Lycopersicon esculentum* (tomato) fruit development. Proc. Natl. Acad. Sci. USA 84: 2793-2797.

Lindhout, P., Van Heusden, S., Pet, G., van Ooijen, J.W., Sandbrink, H., Verkerk, R., Vrielink, R., and Zabel, P. (1994) Perspectives of molecular marker assisted breeding for earliness in tomato. Euphytica 79:279-286.

Lissemore, J.L., Colbert, J.T. and Quail, P.H. (1987) Cloning of cDNA for phytochrome from etiolated Cucurbita and coordinate photoregulation of the abundance of two distinct phytochrome transcripts. Plant Mol. Biol. 8: 485-496.

Livingstone, K.D., Lackney, V.K., Blauth, J.R., Van Wijk, R. and Jahn, M.K. (1999) Genome mapping in Capsicum and the evolution of genome structure in the Solanaceae. Genetics, in press.

Martin G.B., Williams J.G.K., Tanksley S.D. (1991) Rapid identification of markers linked to a *Pseudomonas* resistance gene in tomato by using random primers and near-isogenic lines. Proc. Natl. Acad. Sci. USA 88: 2336-2340.

Martin G.B., Brommonschenkel S.H., Chunwongse J., Frary A., Ganal M.W., Spivey R., Wu T., Earle E.D., Tanksley S.D. (1993) Map-based cloning of a protein kinase gene conferring disease resistance in tomato. Science 262:1432-1436.

Martin, G.B., Frary, A., Wu, T., Brommonschenkel, S., Chunwongse, J., Earle, E.D. and Tanksley, S.D. (1994) A member of the tomato *Pto* gene family confers sensitivity to Fenthion resulting in rapid cell death. Plant Cell 6: 1543-1552.

Messeguer, R., Ganal, M., DeVicente, M.C., Young, N.D., Bolkan, H.D. and Tanksley, S.D. (1991) High resolution RFLP map around the root knot nematode resistance gene *Mi* in tomato. Theor. Appl. Genet. 82: 529536.

Milligan, S.B., Bodeau, J., Yaghoobi, J., Kaloshian, I., Zabel, P. and Williamson, V.M. (1998) The root-knot nematode resistance gene Mi from tomato is a member of the leucine zipper, nucleotide binding, leucine-rich repeat family of plant genes. Plant Cell 10:1307-1319.

Monforte, A.J., Asins, M.J. and Carbonell, E.A. (1997) Salt tolerance in Lycopersicon species: VI. Genotype by salinity interaction in quantitative trait detection: constitutive and response QTLs. Theor. Appl. Genet. 95:706-713.

Mutschler, M.A., Doerge, R.W., Liu, S.C., Kuai, J.P., Liedl, B.E. and Shapiro, J.A. (1996) QTL analysis of pest resistance in the wild tomato *Lycopersicon pennellii*: QTLs controlling acylsugar level and composition. Theor. Appl. Genet. 92:709-718.

Newman, S.M., Eannetta, N.T., Yu, H., Prince, J.P., deVicente, M.C., Tanksley, S.D. and Steffens, J.C. (1993) Organization of the tomato polyphenol oxidase gene family. Plant Mol. Biol. 21: 1035-1051.

Ori, N., Paran I., Aviv, D., Eshed, Y., Tanksley, S.D., Zamir, D. and Fluhr, R. (1994) A genomic search for the gene conferring resistance to *Fusarium* wilt in tomato. Euphytica 79:201-204.

Ori, N., Eshed, Y., Paran, I., Presting, G., Aviv, D., Tanksley, S., Zamir, D. and Fluhr, R. (1997) The I2C family from the wilt disease resistance locus I2 belongs to the nucleotide binding, leucine-rich repeat superfamily of plant resistance genes. Plant Cell 9:521-532.

Paran, I., Goldman, I. and Zamir, D. (1997) QTL analysis of morphological traits in a tomato recombinant inbred line population. Genome 40:242-248.

Paterson, A.H., Damon, S., Hewitt, J.D., Zamir, D., Rabinowitch, H.D., Lincoln, S.E., Lander, E.S. and Tanksley, S.D. (1991) Mendelian factors underlying quantitative traits in tomato. Comparison across species, generations and environments. Genetics 127: 181-198.

Paterson, A.H., Lander, E.S., Hewitt, J.D., Peterson, S., Lincoln, S.E. and Tanksley, S.D. (1988) Resolution of quantitative traits into Mendelian factors, using a complete linkage map of restriction fragment length polymorphism. Nature 335: 721-726.

Pichersky, E., Brock, T.G., Nguyen, D., Hoffman, N.E., Piechulla, B., Tanksley, S.D. and Green, B.R. (1989) A new member of the CAB gene family: structure, expression and chromosomal location of Cab-8: the tomato gene encoding the type III chlorophyll *a/b*-binding polypeptide of photosystem I. Plant Mol. Biol. 12: 257-270.

Pichersky, E., Hoffman, N.E., Malik, V.S., Bernatzky, B. Tanksley, S.D., Szabo, L. and Cashmore, A.R. (1987a) The tomato *Cab-4 and Cab-S* genes encode a second type of CAB polypeptides localized in photosystem II. Plant Mol. Biol. 9: 109-120.

Pichersky, E., Hoffman, N.E., Bernatzky, R., Piechulla, B., Tanksley, S.D. and Cashmore, A.R. (1987b) Molecular characterization and genetic mapping of DNA sequences encoding the type I chlorophyll a/lo-binding polypeptide of photosystem I in *Lycopersicon esculentum* (tomato). Plant Mol. Biol. 9: 205216.

Pichersky, E., Tanksley, S.D., Piechulla, B.,Stayton, M.M. and Dunsmuir, P. (1988) Nucleotide sequence and chromosomal location of Cab-7: the tomato gene encoding the type II chlorophyll a/b-binding polypeptide of photosystem I. Plant Mol. Biol. 11: 69-71.

Pillen, K., Ganal, M.W. and Tanksley, S.D. (1996a) Construction of a high-resolution genetic map and YAC contigs in the tomato *Tm-2a* region. Theor. Appl. Genet. 93:228-233.

Pillen, K., Pineda, O., Lewis, C. and Tanksley, S.D. (1996b) Status of genome mapping tools in the taxon *Solanaceae*. In A. Paterson (ed.) Genome Mapping in Plants. R.G. Landes Co. pp. 281-308.

Pneuli, L., Carmel-Goren, L., Hareven, D., Gutfinger, T., Alvarez, J., Ganal, M., Zamir, D. and Lifschitz E. (1998) The Self-pruning gene of tomato regulates vegetative to reproductive switching of sympodial meristems and is the ortholog of CEN and TFL1. Development 125:1979-1989.

Prince, J.P., Pochard, E. and Tanksley, S.D. (1993) Construction of a molecular linkage map of pepper and a comparison of synteny with tomato. Genome 36: 404-417.

Reeves, A.F., Zobel, R.W. and Rick, C.M. (1968) Further tests with mutants of the Stubbe Series I, II, III and IV. Rep. Tomato Genet. Coop. 32-34.

Rick, C.M. (1975) The tomato. In: King R.C., ed. The handbook of genetics. Vol. 2, York:

Plenum Press, pp. 247-280.

Rick, C.M. (1980) Tomato linkage survey. Rep. Tomato Genet. Coop. 30: 2-17.

Rottmann, W.H., Peter, G.F., Oeller, P.W., Keller, J.A., Shen, N.F., Nagy, B.P., Taylor, L.P., Campbell, A.D. and Theologis, A. (1991) 1-Aminocylopropane-1-carboxylate synthase in tomato is encoded by a multigene family whose transcription is included during fruit and floral senescence. J. Mol Biol 222: 937-961.

Salmeron, J.M., Barker, S.J., Carland, F.M., Mehta, A., Staskawicz, B. (1994) Tomato mutants altered in bacterial disease resistance provide evidence for a new locus controlling pathogen recognition. Plant Cell 6: 511-520.

Sarfatti, M., Abu-Abied, M., Katan and J., Zamir, D. (1991) RFLP mapping of *I-1* a new locus in tomato conferring resistance against *Fusarium oxysporum* f. sp. *lycopersici* race 1. Theor. Appl. Genet. 82: 22-26.

Sarfatti, M., Katan, J., Fluhr, R. and Zamir, D. (1989) An RFLP marker in tomato linked to the *Fusarium oxysporum* resistance gene I-2. Theor. Appl. Genet. 78:755-9.

Scharf, K., Rose, S., Zott, W., Schoff, F. and Nover, L. (1990) Three tomato genes code for heat stress transcription factors with a region of remarkable homology to the DNA-binding domain of the yeast HSF. EMBO J 9: 4495-4501.

Schumacher, K., Ganal, M.G. and Theres, K. (1995) Genetic and physical mapping of the *lateral suppressor (ls)* locus in tomato. Mol. Gen. Genet. 246: 761-766.

Schumacher, K., Schmitt, T., Rossberg, M., Schmitz, G. and Theres, K. (1999) The Lateral suppressor (*Ls*) gene of tomato encodes a new member of the VHIID protein family. Proc. Natl. Acad. Sci. USA 96:290-295.

Schwartz, E., Shen, D., Aebersold, R., McGrath, J.M., Pichersky, E. and Green, B.R. (1991) Nucleotide sequence and chromosomal location of *Cabl 1* and *Cabl 2:* the genes for the fourth polypeptide of photosystem I light harvesting antenna (LHCI). FEBS Lett. 280: 229-234.

Segal, G., Sarfatti, M., Schaffer, M.A., Ori, N., Zamir, D. and Fluhr, R. (1992) Correlation of genetic and physical structure in the region surrounding the *1-2 Fusarium oxysporum* resistance locus in tomato. Mol. Gen. Genet. 231: 179-185.

Tanksley, S.D. and Costello, W. (1991) The size of the *L. pennellii* chromosome 7 segment containing the I-3 gene in tomato breeding lines measured by RFLP probing. Rep. Tomato Genet. Coop. 41: 60.

Tanksley, S.D., Ganal, M.W., Prince, J.P., de Vicente, M.C., Bonierbale, M.W., Broun, P., Fulton, T.M., Giovannoni, J.J., Grandillo, S., Martin, G.B., Messeguer, R., Miller, J.C., Miller, L., Paterson, A.H., Pineda, O., Roder, M., Wing, R.A., Wu, W. and Young, N.D. (1992) High density molecular linkage maps of the tomato and potato genomes. Genetics 132: 1141-1160.

Tanksley, S.D., Grandillo, S., Fulton, T.M., Zamir, D., Eshed, Y., Petiard, V., Lopez, J. and Beck-Bunn, T. (1996) Advanced backcross QTL analysis in a cross between an elite processing line of tomato and its wild relative *L. pimpinellifolium*. Theor. Appl. Genet. 92:213-224.

Tanksley, S.D. and Jones, R.A. (1981) Effect of stress on tomato ADH's: Description of a second ADH coding gene. Biochem. Genet. 19: 397-409.

Tanksley, S.D. and Kuehn, G. (1985) Genetics, subcellular localization and molecular characterization of 6-phosphogluconate dehydrogenase isozymes in tomato. Biochem. Genet. 23:

442-454.

Tanksley, S.D. and Loaiza-Figueroa, F. (1985) Gametophytic self-incompatibility is controlled by a single major locus on chromosome 1 in *Lycopersicon peruvianum*. Proc. Natl. Acad. Sci. USA 82: 5093-5096.

Tanksley, S.D. and Rick, C.M. (1980) Isozymic gene linkage map of the tomato: Applications in genetics and breeding. Theor. Appl. Genet. 57: 161-170.

Tanksley, S.D. and Zamir, D. (1988) Double tagging of a male-sterile gene in tomato using a morphological and enzymatic marker gene. Hort Science 23: 387-388.

Thoquet, P., Olivier, J., Sperisen, C., Rogowsky, P., Prior, P., Anais, G., Mangin, B., Bazin, B., Nazer, R. and Grimsley, N. (1996) Polygenic resistance of tomato plants to bacterial wilt in the French West Indies. Mol Plant-Micro Interact 9:837-842.

Vallejos, C.E., Tanksley, S.D. and Bernatzky, R. (1986) Localization in the tomato genome of DNA restriction fragments containing sequences homologous to the ribosomal RNA 45S, the major chlorophyll a-b binding polypeptide and the ribulose bisphosphate carboxylase genes. Genetics 112: 93-106.

Van der Beek, G., Pet, G. and Lindhout, P. (1994) Resistance of powdery mildew *(Oidium lycopersicum)* in *Lycopersicon hirsutum is* controlled by an incompletely dominant gene *Ol-1* on chromosome 6. Theor. Appl. Genet. 89: 467-473.

Van der Biezen, E.A., Glagotskaya, T., Overduin, B., Nijkamp, H.J.J. and Hille, J. (1995) Inheritance and genetic mapping of resistance to *Alternaria alternata* f.sp. *lycopersici* in *Lycopersicon pennellii*. Mol. Gen. Genet. 247: 453-461.

Van Eck, H.J. (1995) Localisation of morphological traits on the genetic map of potato using RFLP and isozyme markers. PhD. dissertation. Dept Plant Breeding, Wageningen: Agricultural University, 146p.

Vosman, B. and Arens, P. (1997) Molecular characterization of GATA/GACA microsatellite repeats in tomato. Genome 40: 25-33.

Whitham, S., Dinesh-Kumar, S.P., Choi, D., Hehl, R., Corr, C. and Baker, B. (1994) The product of the tobacco mosaic virus resistance gene N: Similarity to Toll and the interleukin-1 receptor. Cell 78: 1101-1115.

Wing, R.A., Zhang, H.B. and Tanksley, S.D. (1994) Map-based cloning in crop plants: Tomato as a model system: I. Genetic and physical mapping of jointless. Mol. Gen. Genet. 242: 681-688.

Witsenboer, H.M.A., Van de Griend, K.G., Tiersma, J.B., Nijkamp, H.J.J. and Hille, J. (1989) Tomato resistance to *Alternaria* stem canker localization in host genotypes and functional expression compared to non-host resistance. Theor. Appl. Genet. 78: 457-462.

Yen, H.C., Shelton, B.A., Howard, L.R., Lee, S., Vrebalov, J. and Giovannoni, J.J. (1997) The tomato high-pigment *(hp)* locus maps to chromosome 2 and influences plastome copy number and fruit quality. Theor. Appl. Genet. 95: 1069-1079.

Young, N.D., Zamir, D., Ganal, M.W. and Tanksley, S.D. (1988) Use of isogenic lines and simultaneous probing to identify DNA markers tightly linked to the *Tm-2a* gene in tomato. Genetics 120: 579-586.

Zamir, D., Bolkan, H., Juvik, J.A. et al. (1993) New evidence for placement of *Ve-* the gene for resistance to *Verticilium* race 1. Rep Tomato Genet Coop 43: 51-52.

Zamir, D. and Tanksley, S.D. (1988) Tomato genome is comprised largely of fast-evolving low copy number sequences. Mol. Gen. Genet. 213: 254-261.

Zhou, J., Loh, Y.T., Bressan and R.A., Martin, G.B. (1995) The tomato gene *Pti1* encodes a serine/threonine kinase that is phosphorylated by *Pto* and is involved in the hypersensitive response. Cell 83: 925-935.

24. Molecular-marker maps of the cultivated wheats and other *Triticum* species

GARY E. HART

Soil and Crop Sciences Department, Texas A&M University, College Station, TX 77843, U.S.A.
<g-hart@tamu.edu>

Contents

1. Introduction

The cultivated wheats, *Triticum aestivum* L. em Thell. ($2n = 6x = 42$, genomes A, B, and D) and *T. turgidum* L. var. *durum* ($2n = 4x = 28$, genomes A and B), provide approximately 20% of the food calories consumed by humankind, more than any other crop plant, including rice, maize, and potato (Reitz, 1967; Briggle and Curtis 1987). Wild relatives of the wheats also are important. They possess numerous agronomically important attributes that are superior to those present in the cultivated wheats (examples include disease resistance, drought tolerance, salt tolerance and high protein content) and a large number of tools have been developed by wheat geneticists for transferring alien genes into the cultivated wheats (Knott 1987; Maan 1987).

This chapter reviews and summarizes the development of molecular-markers maps for *T. aestivum*, *T. turgidum*, *T. monococcum* L. ($2n = 2x = 14$, genome A^m), *T. tauschii* (Coss.) Schmal. ($2n = 2x = 14$, genome D), and *T. umbellulatum* (Zhuk.) Bowden ($2n = 2x = 14$, genome U). Also described is the genetic nomenclature that is used for DNA-marker loci and alleles in the wheats and related species. It should be noted that a complete catalogue of Triticum genetic markers is published every five years in the Proceedings of the International Wheat Genetics Symposium (for the most recent edition, see McIntosh et al. 1998) and an annual supplement is published in both the Wheat Newsletter and the Wheat Information Service. The catalogue lists all Triticum genes, DNA restriction-fragment-length polymorphisms (RFLPs), simple-sequence repeats (SSRs), and sequence-tagged sites (STSs) that have been localized to a chromosome or chromosome arm, all known alleles of Triticum genes and a prototype

R.L. Phillips and I.K. Vasil (eds.), DNA-Based Markers in Plants, 421 - 441.

strain or strains for each allele, the chromosomal locations of all of the aforementioned genetic markers, the linkage positions of mapped genes, literature citations, and a large amount of other information.

2.   *Triticum* molecular markers

RFLPs comprise the vast majority of the molecular markers mapped in *Triticum* species to date; the most recent edition of the Catalogue of Gene Symbols for Wheat (McIntosh et al. 1998) lists approximately 1700 DNA clones that have been used for linkage and/or deletion and/or aneuploid mapping of one or more *Triticum* species. Also listed in the catalogue are approximately 220 SSR and STS loci, and the approximate map locations of 279 additional hexaploid wheat SSR loci were published recently (Röder et al. 1998). Only a small number of amplified-fragment-length polymorphisms (AFLPs) and random-amplified-polymorphic DNAs (RAPDs) have been mapped in wheat, usually for the purpose of tagging genes of interest (see, e.g., Parker et al. 1998 and Hu et al. 1997, respectively).

The vast majority of the DNA clones used to map RFLP loci in *Triticum* species are either anonymous cDNA clones or anonymous genomic DNA (gDNA) clones isolated from *T. aestivum, T. tauschii,* oat, barley, and a few other plant species. Devos et al. (1992) found gDNA clones to be almost twice as efficient as cDNA clones in detecting RFLPs and that some classes of gDNA clones, namely, chromosome-specific clones and clones that hybridize to sequences in non-homoeologous chromosomes, are the most polymorphic. Some 'known-function' probes (i.e., probes derived from sequenced genes) also have been used for linkage mapping of *Triticum* species, and sequencing of anonymous clones has revealed the possible function of a goodly number of other RFLP loci.

Most cDNA clones hybridize to fragments located in each of the three members of one or sometimes two homoeologous chromosome-arm groups of hexaploid wheat; very few clones hybridize to fragments from only one or two chromosomes in a group or to chromosomes in more than two groups (Sharp et al. 1989; Chao et al 1989a,b; Devos et al. 1992). Low-copy-number anonymous gDNA clones hybridize less frequently than cDNA clones to sequences located in each of the three members of a homoeologoous group. For example, Liu and Tsunewaki (1991) found that only 32% of 72 *Pst*I gDNA clones tested detected loci in all three groups and that 49% of the clones detected loci in the chromosomes of two or more groups. Ten of 15 homoeologous group-3 gDNA clones studied by Devos et al. (1992) hybridized to one or more fragments in each of the three arms of one of the group-3 arm groups and the remainder hybridized either to fragments located in non-homoeologous chromosomes or to one or more fragments located in one chromosome only. Anderson et al. (1992) determined the chromosome-arm locations of over 800 DNA fragments using 210 barley and oat cDNA clones and wheat gDNA clones (the number of cDNA versus gDNA clones studied was not reported). Seventy-seven per cent of the clones hybridized to fragments in one chromosome-arm group only and almost all of the clones hybridized to at least one fragment in each of the three arms. This high frequency is undoubtedly due in good part to the fact that preference in choosing clones and the restriction enzymes to be used with them was given in this study to clone-enzyme combinations that yielded only three fragments of approximately equal hybridization intensity.

3.   Mapping populations

F_2 populations, F_3 families, bulked F_4 families and recombinant inbred lines (RILs) are the types of populations that have been used most frequently for constructing molecular-marker linkage maps of *Triticum* species. Recombinant substitution lines (RSLs) and doubled haploids (DHs) also have been used, but in fewer instances. RILs, RSLs and DHs have the advantage of being permanent while F_2 populations, F_3 families, and bulked F_4 families are easier to produce.

Most *Triticum* molecular-marker maps have been constructed by analyzing derivatives of wide crosses. The difficulty in constructing RFLP linkage maps of wheat using varietal variation and the value of using wide crosses for this purpose were revealed in the studies of Chao et al. (1989b), who produced the first RFLP maps of wheat chromosomes. They used six mapping populations derived from ten varieties to construct maps of the homoeologous group-7 chromosomes of hexaploid wheat. They studied two F_2 populations obtained from varietal crosses, one population of single-seed descent lines derived from a varietal cross, one population of doubled haploids derived from a cross of two intervarietal chromosome-substitution lines and two populations of single-chromosome recombinant lines. Thirty-one RFLP loci were mapped, including five loci in 7A, 13 in 7B, and 13 in 7D. A test of the level of polymorphism among all pair-wise combinations of six of the 10 varieties using 18 group-7 cDNA clones and 13 restriction enzymes revealed polymorphism in an average of only 8.7% of the comparisons. However, a test for 7D polymorphism between Hobbit 'S' and *VPMI*, which contains a 7D that is mostly derived from *Aegilops ventricosa*, revealed 23.3% polymorphism. Similar findings were reported in a study of the homoeologous group-3 chromosomes in which only two F_2 mapping populations were analyzed (Devos et al. 1992).

In addition to linkage maps, physical maps of the 21 chromosomes of *T. aestivum* comprised of almost 1000 DNA markers have been produced by analyzing deletion lines derived in *T. aestivum* cv. Chinese Spring (see below).

4.   Genetic nomenclature

Genetic nomenclature guidelines for wheat and related species are presented in McIntosh et al. (1998). The guidelines specify that the basic symbol for genetic markers detected at the DNA level is a 'X'. For markers of unknown function, the X is followed by a laboratory designator, a number that identifies the probe or primer(s) used to detect the locus, a hyphen (-), and the symbol for the chromosome in which the locus is located. For example, *Xtam2-1A* designates a chromosome 1A RFLP locus detected with clone pTaTAM2 of *Texas A&M* University, and *Xtam2-1B* and *Xtam2-1D* designate loci located in 1B and 1D, respectively, that are detected with the same clone. Two or more loci detected with the same clone and located in the same chromosome are assigned the same designation except for the addition of a period and an Arabic numeral immediately after the chromosome designation, e.g., *Xcdo534-1B.1* and *Xcdo534-1B.2* designate two chromosome-1B loci detected with clone CDO534. Guidelines for choosing a laboratory designator are presented in McIntosh et al. (1998) and also are available electronically via the Internet Gopher from host "greengenes.cit.cornell.edu", port 70, menu "Grains files to browse" / "Nomenclature of Biochemical and Molecular Loci in Wheat".

Locus symbols for DNA markers of unknown function that are detected with 'known-function' probes may include, in parentheses following the probe designation, a symbol for the gene from which the probe was obtained. For example, the clone UMC207 was derived from a sucrose synthase gene and *Xumc207(Ss1)-7A* designates a chromosome 7A locus detected with the clone. Likewise, when the primers used to amplify a DNA marker of unknown function are of sufficient length and similarity to a known gene to amplify the gene, the DNA-marker symbol may include the gene symbol in parentheses following the number assigned to the primers. The set number and other numbers assigned by the Commission on Plant Gene Nomenclature also may be included inside the parentheses immediately after the symbol for genes for which the Commission has assigned mnemonic designations. A locus detected with a DNA probe or primer set and whose function is known or subsequently determined is designated with the same type of symbol that is used for markers identified by their morphological or physiological effects, i.e., a symbol that indicates the function of the locus. For example, *Nor-A1* and *Nor-B1* designate the A- and B-genome members, respectively, of the first-designated orthologous set of nucleolar organizer gene loci.

Alleles at wheat loci are designated with a lower-case letter following the locus symbol. For hexaploid wheat, the allele present in *T. aestivum* cv. Chinese Spring is designated '*a*' and, in the case of RFLP loci, the name of the restriction enzyme also is included in the allele designation. For example, *Xtam1-5A-EcoRIa* denotes the allele detected at the *Xtam1-5A* locus in Chinese Spring with *Eco*RI.

The chromosome designation is an integral part of the locus symbol for wheat DNA markers. Nevertheless, on chromosome maps and in a limited number of other contexts, the chromosome designation and the hyphen preceding it are often omitted. For example, on a map of chromosome 5A, *Xtam1* is an abbreviation for *Xtam1-5A*, and *Xbcd348-2A.1* and *Xbcd348-2A.2* may be abbreviated as *Xbcd348.1* and *Xbcd348.2* on a map of 2A.

5.   Molecular marker linkage maps of *Triticum* species

The 1998 Catalogue of Gene Symbols for Wheat (McIntosh et al. 1998) lists 3760 DNA markers that have been localized to a chromosome or a chromosome arm in one or both of the cultivated wheats and/or in *T. monococcum* and/or in *T. tauschii*. As Table 1 shows, the number of identified molecular markers per chromosome ranges from 140 (6D) to 225 (1A) and the number of identified molecular markers per homoeologous chromosome-group ranges from 415 (group 4) to 601 (group 5).

Table 1.   Chromosomal distribution of DNA markers that have been localized to a chromosome or chromosome arm in one or more of the species *T. aestivum, T. turgidum, T. monococcum* and *T. tauschii.* [1]

Chromosome Group	Genome A	Genome B	Genome D	Totals
1	225	181	166	572
2	216	218	202	636
3	172	171	158	501
4	153	141	121	415
5	241	184	176	601
6	179	163	140	482
7	202	170	181	553
Totals	1388	1228	1144	3760

[1] Reference: McIntosh et al. 1998.

Table 2 lists the principal RFLP linkage maps constructed for *T. aestivum, T. turgidum, T. monococcum, T. tauschii,* and *T. umbellulatum.* The two maps of *T. aestivum* that contain the largest number of markers were developed at the John Innes Centre in the United Kingdom, principally by analyzing F_2-, F_3-, and F_4-derivatives of *T. aestivum* cv. Chinese Spring x Sear's Synthetic and cv. Timgalen x RL4137, and by a group of U.S., Australian and French scientists, who analyzed F_{6-7} RILs derived from a cross of *T. tauschii* x *T. turgidum* var. *durum* cv. Altar 84 followed by a cross of the artificial hexaploid derived from the cross to *T. aestivum* cv. Opata. The former map is approximately 2600 cM in length and is comprised of more than 500 DNA markers. Included among the markers on the map are eight telomeres and 23 other telomeric-associated sequences (Mao et al. 1997). The latter map is approximately 3600 cM in length and is comprised of more than 1000 DNA markers. The vast majority of markers on both of the maps are RFLPs. A framework map of the (*T. tauschii* x *T. turgidum* var. *durum* cv. Altar 84) x *T. aestivum* cv. Opata RIL population consisting of RFLP loci mapped at a high LOD score and that includes numerous other markers is shown in Fig. 1. This map was kindly provided by Dr. Philippe Leroy, Station d'Amelioration des Plantes, Clermont-Ferrand, France.

The hexaploid wheat linkage maps of Liu and Tsunewaki (1991) and Cadelan et al. (1997) are comprised of 198 and 266 markers, respectively. Only 28 mapped D-genome markers are reported in the former publication. Three of the D-genome linkage groups reported in the latter are comprised of seven or fewer markers and linkage groups were not produced for three other D-genome chromosomes. As shown in Table 2, a recombined *T. aestivum* 1A and *T. monococcum* 1A^m was mapped by Dubcovsky et al. (1995), a recombined *T. aestivum* 4D and *T. turgidum* 4B was mapped by Dvorák et al. (1995), and 1D of *T. aestivum* was mapped by Dubcovsky et al. (1996).

Table 2.   RFLP linkage maps of Triticum species.

Species	Mapped genomes/ chromosome groups/ chromosomes	Reference(s)	Type(s) of population(s) analyzed	Parents of population(s) analyzed
T. aestivum	Genomes A, B, & D	Liu & Tsunewaki 1991	F$_3$ families	cv. Chinese Spring & *T. spelta* var. duhamelianum
T. aestivum	Group 2 Group 3 Groups 4, 5, & 7 Group 5 Group 6 Group 7 Genomes A, B, & D	Devos et al. 1993b Devos et al. 1992, Devos & Gale, 1993 Devos et al. 1995b Xie et al. 1993 Jia et al. 1996 Chao et al. 1989 Gale et al. 1995	F$_2$ families, F$_3$ families, and bulked F$_4$ families	cv. Chinese Spring & Sear's Synthetic/ cv. Timgalen & RL4137
T. aestivum	Group 1 Group 2 Group 3 Groups 4, 5, & 7 Group 6	Van Deynze et al. 1995 Nelson et al. 1995b Nelson et al. 1995c Nelson et al. 1995a Marino et al. 1996	F$_{6-7}$ recombinant inbred lines	(*T. tauschii* x *T. turgidum* var. *durum* cv. Altar 84) & cv. Opata
T. aestivum	Groups 1-7, except chromosomes 2D, 4D, & 5D	Cadelan et al. 1997	Doubled haploids	monosomic cv. Chinese Spring & cv. Courtot lines
T. aestivum	Chromosome 1D	Dubcovsky et al. 1996	Recombinant substitution lines	cv. Chinese Spring & cv. CS-var. *dicoccoides* 1D^{dic}(1D^{ss}) substitution line
T. aestivum & *T. monococcum*	*T. aestivum* 1A/ *T. monococcum* 1A^m	Dubcovsky et al. 1995	Lines segregating for 1A/ 1A^m recombinants	cv. Chinese Spring & cv. CS-*T. monococcum* 1A$^{m(cs)}$ substitution line
T. aestivum & *T. turgidum*	*T. aestivum* 4D/ *T. turgidum* 4B	Dvořák et al. 1995	Lines segregating for 4B/4D recombinants	var. Chinese Spring 4D-var. *durum* cv. Langdon 4B substitution line & *T. turgidum ph1c* mutant line

Table 2.   Continued

Species	Mapped genomes/ chromosome groups/ chromosomes	Reference(s)	Type(s) of population(s) analyzed	Parents of population(s) analyzed
T. turgidum	Genomes A & B	Blanco et al. 1998	F7 recombinant inbred lines	var. *durum* cv. Messapia & var. dicoccoides acc. MG4343
T. turgidum	Chromosomes 6A & 6B	Chen et al. 1994, Du & Hart 1998	Recombinant substitution lines & either F2 plants or F3 families	var. *durum* cv. Langdon & var. *durum* cv. Langdon-var. *dicoccoides* 6Adic(ldn) and 6Bdic(ldn) substitution lines
T. turgidum	Chromosome 1B	Dubcovsky et al. 1996	Recombinant substitution lines	var. *durum* cv. Langdon & var. durum cv. Langdon-var. dicoccoides 1Bdic(ldn) substitution line
T. monococcum	Chromosome 1Am	Dubcovsky et al. 1995	F3 families	spp. *monococcum* accs. G1777 & G2528
T. monococcum	Genome Am	Dubcovsky et al. 1996	F2 plants and F3 families	ssp. *monococcum* acc. DV92 & ssp. *aegilopoides* acc. G3116/ssp. monococcum accs. G1777 & G2528
T. tauschii	Genome D	Gill et al. 1991, Boyko et al 1999	F2 plants and F3 families	ssp. *eusquarrosa* var. *meyeri* acc. TA1691 & var. typica acc. TA1704
T. tauschii	Genome D	Lagudah 1991	F2 plants	ssp. *eusquarrosa* var. *typica* acc. AUS18902 & var. *meyeri* acc. AUS 18911/ ssp. *strangulata* acc. AUS 21929 & ssp. *eusquarrosa* var. *typica* acc. CPI 110730
T. umbellulatum	Genome U	Zhang et al. 1998	F3 families	acc. JIC2010001 & acc. JIC2010003

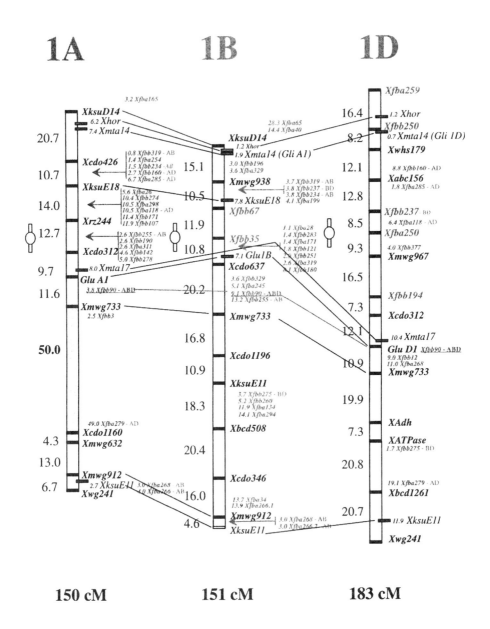

150 cM   **151 cM**   **183 cM**

Figure 1.   Framework molecular-marker map of the (*T. tauschii* x *T. turgidum* var. *durum* cv. Altar 84) x *T. aestivum* cv. Opata recombinant inbred population. The approximate positions of the centromeres are shown to the left of the linkage groups. Symbols for framework markers are located immediately to the right of the linkage groups and cM distances between the markers are shown on the left side of the linkage groups. Other markers are shown to the right of the interval in which they are located and the positions of these markers within the intervals are indicated by the cM values to the left of them. The genomes in which orthologous markers are located are listed to the right of the markers. Also, lines connect some orthologous markers.

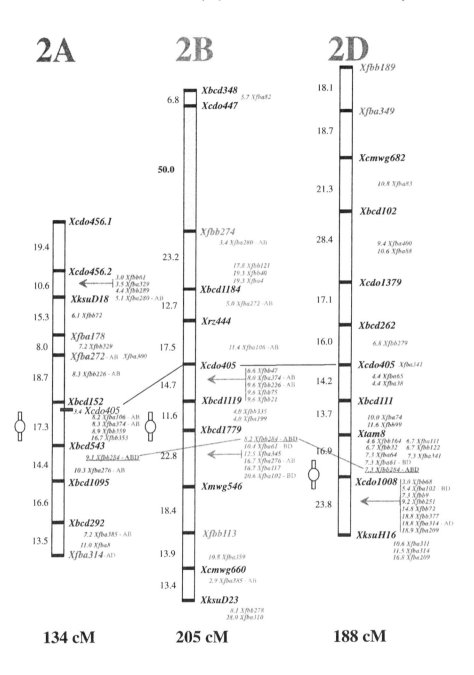

Figure 1. Continued.

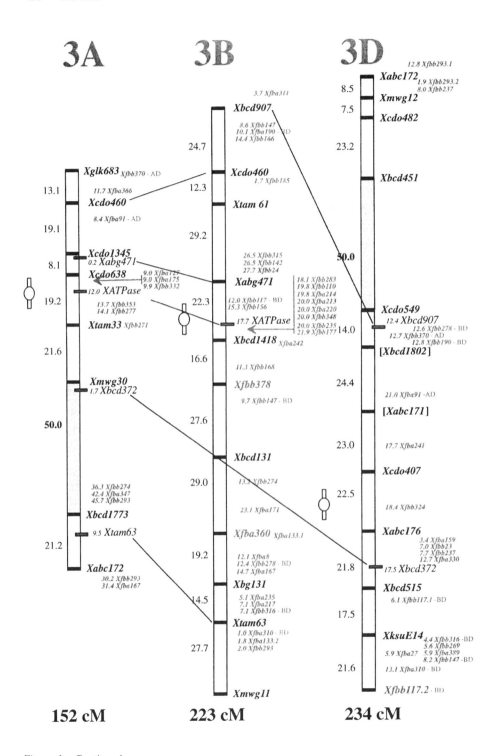

Figure 1. Continued.

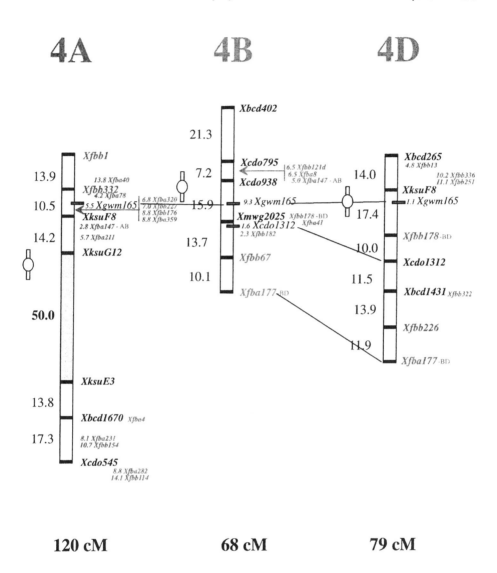

Figure 1.   Continued.

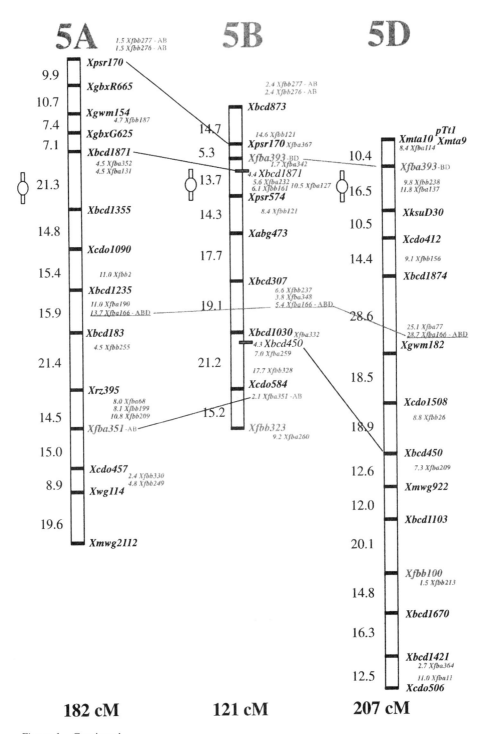

Figure 1. Continued.

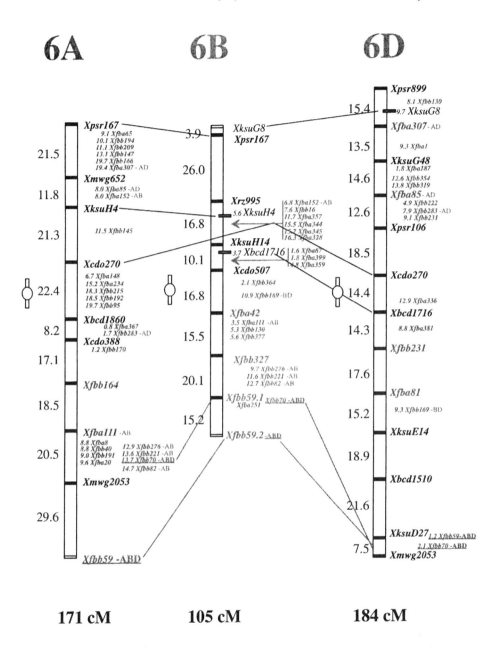

Figure 1.   Continued.

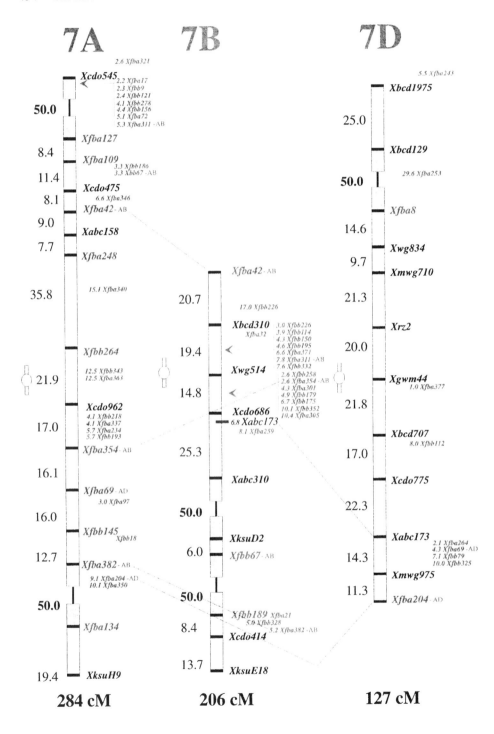

Figure 1.   Continued.

Blanco et al. (1998) reported a linkage map of *T. turgidum* that consists of 14 linkage groups and 213 markers (including 198 RFLPs) and is approximately 1350 cM in length. It contains sizable gaps in eight linkage groups. The linkage mapping of *T. turgidum* chromosomes 6A and 6B initiated by Chen et al. (1994) was extended by Du and Hart (1998), who developed LOD ≥ 3.0 6A and 6B linkage maps that are 111 cM and 123 cM in length, respectively, and are comprised of 73 and 56 markers, respectively. 1B of *T. turgidum* was mapped by Dubcovsky et al. (1996).

A linkage map of *T. monococcum* composed of 335 markers was developed by Dubcovsky et al. (1996). The map is 1079 cM in length. Most of the markers on the map are RFLPs but a sizable number of isozyme and seed-storage-protein genes, rRNA loci, and loci controlling morphological traits also were mapped. Mapping of $1A^m$ was reported earlier by Dubcovsky et al. (1995). The mapping of the D genome of *T. tauschii* that was performed by Gill et al. (1991) was extended by Boyko et al. (1999), who developed a linkage map consisting of approximately 550 markers. Also, a partial linkage map of the D genome of *T. tauschii* was produced by Lagudah et al. (1991). A linkage map of *T. umbellulatum*, produced principally to analyze the evolutionary relationships between the U genome chromosomes and those of hexaploid wheat, was produced by Zhang et al. (1998). It consists of 81 markers and is 776 cM in length.

Isolation, characterization, aneuploid mapping and linkage mapping of wheat SSRs has been reported by Devos et al. (1995), Röder et al. (1995a), Bryan et al. (1997), Röder et al. (1998), and Stephenson et al. (1998). These papers reported the sequences for 294 SSR primer sets and linkage mapping of approximately 330 SSR loci in hexaploid wheat.

6.   Physical maps of *Triticum aestivum* chromosomes

The discovery by Endo (1988) that a chromosome of *Aegilops cylindrica* L. causes breakage of wheat chromosomes when added to wheat led to the development of approximately 430 deletion lines in cv. Chinese Spring (Endo and Gill 1996). Most of the lines are homozygous for one terminally-deleted chromosome and a few of them contain two or more chromosomes with terminal deletions. These stocks were used systematically by Dr. B. S. Gill and colleagues to produce physical maps of all of the chromosomes of hexaploid wheat. The physical maps of the homoeologous group-6 chromosomes were recently extended by Weng et al. (2000). Table 3 lists the maps produced and the types of markers mapped. Table 4 shows the chromosomal distribution of the almost 1000 molecular markers, including about 50 SSRs, that comprise the physical maps.

A more elementary form of physical mapping, namely, the use of cv. Chinese Spring compensating nullisomic-tetrasomic lines and ditelosomic lines to determine the chromosomal locations and then the arm locations of RFLP loci before they are subjected to linkage mapping, has been an integral part of essentially all of the RFLP linkage mapping performed in *T. aestivum*. In addition to the papers cited in Table 2, the results of this form of physical mapping are reported in Anderson et al. (1992) and Devey and Hart (1993).

Table 3.   RFLP and SSR physical maps of *Triticum aestivum* chromosomes produced by analysis of cv. Chinese Spring lines containing terminally-deleted chromosomes.

Chromosome(s) mapped	Reference(s)	Marker type
1A, 1B, 1D	Gill et al. 1996	RFLPs
1B	Kota et al. 1993	RFLPs
2A, 2B, 2D	Delaney et al. 1995a	RFLPs
2A, 2B, 2D	Roder et al. 1998a	SSRs
3A, 3B, 3D	Delaney et al. 1995b	RFLPs
4A, 4B, 4D	Mickelson-Young et al. 1995	RFLPs
5A, 5B, 5D	Gill et al. 1996	RFLPs
6A, 6B, 6D	Gill et al. 1993	RFLPs
6A, 6B, 6D	Weng et al. 1999	RFLPs
7A, 7B, &D	Werner et al. 1992	RFLPs
7A, 7B, &D	Hohmann et al. 1994, 1995a, 1995b	RFLPs

Table 4.   Chromosomal distribution of DNA markers physically-mapped in *T. aestivum* by analysis of cv. Chinese Spring deletion lines. [1]

Chromosome group	Genome A	Genome B	Genome D	Totals
1	34	45	37	116
2	38	33	39	110
3	22	25	27	74
4	28	28	28	84
5	64	73	73	210
6	73	94	97	264
7	36	43	32	111
Totals	295	341	333	969

[1] For references, see Table 3

7.   Structural evolution of *Triticum* chromosomes

The molecular-marker maps of *Triticum* species indicate that there is a high degree of colinearity among chromosomes within homoeologous groups both within and between most species in the genus. In *T. aestivum*, with the possible exception of a short reciprocal 2BS-6BS translocation, the chromosomes in homoeologous groups 1, 2, 3 and 6 appear to be colinear. Likewise, 4B and 4D, 5B and 5D, and 7A and 7D appear to be colinear. Chromosomes 4A, 5A and 7B are extensively rearranged, however. 5AL contains a segment translocated to it from the original 4AL, the former long arm of 4A is now the short arm, the present-day 4AL includes pieces of 7BS, 5AL and the former 4AS, and 7BS contains a small segment of 5AL. Furthermore, 4A differs from its homoeologues by both a pericentric inversion and a paracentric inversion and a small segment of the former long arm of 4A is located between the 7BS and 5AL segments in the present-day 4AL (Naranjo et al. 1987; Liu et al. 1992; Devos et al. 1995b).

The chromosomes of *T. monococcum* and *H. vulgare* L. are highly conserved, the only rearrangements detected between them to date being a reciprocal translocation involving 4A^mL and 5HL and paracentric inversions in the long arms of chromosomes 1 and 4 (Dubcovsky et al. 1996). In contrast, the genomes of *Secale cereale* L. and *T. umbellulatum* are extensively rearranged relative to the genomes of hexaploid wheat. 1R is the only *S. cereale* chromosome that appears to be colinear with wheat homoeologues. 3R contains a short segment that is homoeologous to the wheat group-6 chromosomes and each one of the other five R-genome chromosomes is comprised of segments that are homoeologous to either three or four wheat chromosomes (Devos et al. 1993a). Chromosomes 4U, 5U, 7U and especially 6U of *T. umbellulatum* are rearranged relative to the chromosomes of wheat (Zhang et al. 1998). The centromeric region of 6U, an acrocentric chromosome, is homoeologous to the group-4 chromosomes of wheat, and the long arm consists of segments that, in order, are homoeologous to segments of the group-6, group-7, group-2, group-7, group-1 and possibly group-6 chromosomes of wheat.

8.   Concluding remarks

Due to polyploidy, autogamy, and large genomes, the cultivated wheats, in spite of their importance to humankind, were poorly genetically mapped compared to several other important crop plant species until recently. Most genes and other genetic markers in hexaploid and tetraploid wheat are triplicated and duplicated, respectively. Consequently, the effects of recessive or inactive alleles are usually masked by the effects of dominant or active alleles at orthologous loci. Autogamy causes a relatively low level of DNA and allozyme variation to be present among wheat varieties. The large amount of DNA in each wheat genome (about 18 pg per haploid nucleus) makes the techniques for study of DNA markers more difficult to perform than with most other crop species.

In recent years, however, the development of new and improved marker systems and the mapping of populations derived from wide crosses have resulted in the production of excellent genetic maps for all of the chromosomes of hexaploid wheat and for some or all of the chromosomes of several other *Triticum* species. These markers and marker systems are being widely used in wheat genetics and breeding programs. The most recent edition of the Catalogue of Gene Symbols for Wheat (McIntosh 1998) lists more than 75 wheat genes that have been tagged with molecular markers of one form or another, and the recent publication of the sequences for 230 wheat SSR primer sets (Röder et al. 1998), bringing the total number of wheat SSR primer sets available to almost 300, should greatly facilitate mapping of wheat genes in varietal crosses. It is clear that many of the impediments to genetic analysis of the cultivated wheat due to polyploidy, autogamy, and large genomes have now been surmounted.

9.   References

Anderson, J.A., Ogihara, Y., Sorrells, M.E., and Tanksley, S.D. (1992) Development of a chromosomal arm map for wheat based on RFLP markers. Theor. Appl. Genet. 83: 1035-1043.

Blanco, A., Bellomo, M.P., Cenci, A., De Giovanni, C., D'Ovidio, R., Iacono, E., Laddomada, B., Pagnotta, M.A., Porceddu, E., Sciancalepore, A., Simeone, R., and Tanzarella, O.A. (1998) A genetic linkage map of durum wheat. Theor. Appl. Genet. 97: 721-728.

Boyko, E.V., Gill, K.S., Mickelson-Young, L., Nasuda, S., Raupp, W.J., Ziegle, J.N., Singh, S., Hassawi, D.S., Fritz, A.K., Namuth, D., Lapitan, N.L.V., and Gill, B.S. (1999) A high-density genetic linkage map of *Aegilops tauschii*, the D-genome progenitor of bread wheat. Theor. Appl. Genet. 99: 16-26.

Briggle, L.W., and Curtis, B.C. (1987) Wheat worldwide. In: E.G. Heyne (ed.), Wheat and Wheat Improvement, 2nd Ed., pp. 1-32. American Society of Agronomy, Madison, WI.

Bryan, G.J., Collins, A.J., Stephenson, P., Orry, A., Smith, J.B., and Gale, M.D. (1997) Isolation and characterisation of microsatellites from hexaploid bread wheat. Theor. Appl. Genet. 94: 557-563.

Cadalen, T., Boeuf, C., Bernard, S., and Bernard, M. (1997) An intervarietal molecular marker map in *Triticum aestivum* L. em. Thell. and comparison with a map from a wide cross. Theor. Appl. Genet. 94: 367-377.

Chao, S., Raines, C.A., Longstaff, M., Sharp, P.J. Gale, M.D., and Dyer, T.A. (1989a) Chromosomal location and copy number in wheat and some of its close relatives of genes for enzymes involved in photosynthesis. Mol. Gen. Genet. 218: 423-430.

Chao, S., Sharp, P.J., Worland, A.J., Koebner, R.M.D., and Gale, M.D. (1989b) RFLP-based genetic maps of wheat homoeologous group 7 chromosomes. Theor. Appl. Genet. 78: 495-504.

Chen, Z., Devey, M., Tuleen, N.A., and Hart, G.E. (1994) Use of recombinant substitution lines in the construction of RFLP-based genetic maps of chromosomes 6A and 6B of tetraploid wheat (*Triticum turgidum* L.). Theor. Appl. Genet. 89:703-712.

Delaney, D.E., Nasuda, S., Endo, T.R., Gill,B.S., and Hulbert, S.H. (1995a) Cytologically based physical maps of the group-2 chromosomes of wheat. Theor. Appl. Genet. 91: 568-573.

Delaney, D.E., Nasuda, S., Endo, T.R., Gill, B.S., and Hulbert, S.H. (1995b) Cytologically based physical maps of the group 3 chromosomes of wheat. Theor. Appl. Genet. 91: 780-782.

Devey, M.E., and Hart, G.E. (1993) Chromosomal localization of intergenomic RFLP loci in hexaploid wheat. Genome 36: 913-918.

Devos, K.M., Atkinson, M.D., Chinoy, C.N., Liu, C., and Gale, M.D. (1992) RFLP-based genetic map of the homoeologous group 3 chromosomes of wheat and rye. Theor. Appl. Genet. 83: 931-939.

Devos, K.M., Atkinson, M.D., Chinoy, C.N., Harcourt, R.L., Koebner, R.M.D., Liu, C.J., Masojc, P., Xie, D.X., and Gale, M.D. (1993a) Chromosome rearrangements in the rye genome relative to that of wheat. Theor. Appl. Genet. 85: 673-680.

Devos, K.M., Millan, T., and Gale, M.D. (1993b) Comparative RFLP maps of the homoeologous group 2 chromosomes of wheat, rye and barley. Theor. Appl. Genet. 85: 784-792.

Devos, K.M., and Gale, M.D. (1993) Extended genetic maps of the homoeologous group 3 chromosomes of wheat, rye and barley. Theor. Appl. Genet. 85: 649-652.

Devos, K.M., Bryan, G.J., Collins, A.J., Stephenson, P., and Gale, M.D. (1995a) Application of two microsatellite sequences in wheat storage proteins as molecular markers. Theor. Appl. Genet. 90: 247-252.

Devos, K.M., Dubcovsky, J., Dvorák, J., Chinoy, C.N., and Gale, M.D. (1995b) Structural evolution of wheat chromosomes 4A, 5A and 7B and its impact on recombination. Theor. Appl. Genet. 91: 282-288.

Du, C., and G. E. Hart. 1997. *Triticum turgidum* L. 6A and 6B recombinant substitution lines: extended linkage maps and characterization of residual background alien genetic variation. Theor. Appl. Genet. 96: 645-653.

Dubcovsky, J., Luo M-C., and Dvorák, J. (1995) Differentiation between homoeologous chromosomes 1A of wheat and 1A^m of *Triticum monococcum* and its recognition by the wheat *Ph1* locus. Proc Nat Acad Sci, USA 92: 6645-6649.

Dubcovsky, J., Luo, M-C., Zhong, G-Y., Bransteitter, R., Desai, A., Kilian, A., Kleinhofs, A., and Dvorák, J. (1996) Genetic map of diploid wheat, *Triticum monococcum* L., and its comparison with maps of *Hordeum vulgare* L. Genetics 143: 983-999.

Dvorák, J., Dubcovsky, J., Luo, M.C., Devos, K.M., nd Gale, M.D. (1995) Differentiation between wheat chromosomes 4B and 4D. Genome 38: 1139-1147.

Endo, T.R. (1988) Induction of chromosome structural changes by a chromosome of *Aegilops cylindrica* L. in common wheat. J. Hered 79: 366-370.

Endo, T.R., and Gill, B.S. (1996) The deletion stocks of common wheat. J. Hered 87: 295-307.

Gale, M.D., Atkinson, M.D., Chinoy, C.N., Harcourt, R.L., Jia, J., Li, Q.Y., and Devos, K.M. (1995) Genetic maps of hexaploid wheat. In: Li, Z.S., and Xin, Z.Y. (eds.), Proc. 8th Inter. Wheat Genet. Symp., pp. 29-40. Beijing, China.

Gill, K.S., Lubbers, E.L., Gill, B.S., Raupp, W.J. and Cox, T.S. (1991) A genetic linkage map of *Triticum tauschii* (DD) and its relationship to the D genome of bread wheat (AABBDD). Genome 34: 362-374.

Gill, K.S., Gill, B.S., and Endo, T.R. (1993) A chromosome region-specific mapping strategy reveals gene-rich telomeric ends in wheat. Chromosoma 102: 374-381.

Gill, K.S., Gill. B.S., Endo. T.R., and Boyko E.V. (1996) Identification and high-density mapping of gene-rich regions in chromosome group 5 of wheat. Genetics 143: 1001-1012.

Gill, K.S., Gill B.S., Endo, T., and Taylor, T. (1996) Identification and high-density mapping of gene-rich regions in chromosome group 1 of wheat. Genetics 144: 1883-1891.

Hohmann, U., Endo, T.R., Gill, K.S., and Gill, B.S. (1994) Comparison of genetic and physical maps of group 7 chromosomes from *Triticum aestivum* L. Mol. Gen. Genet. 245: 644-653.

Hohmann, U., Endo, T.R., Herrmann, R.G., and Gill, B. S. (1995a) Characterization of deletions in common wheat induced by an *Aegilops cylindrica* chromosome: detection of multiple chromosome arrangements. Theor. Appl. Genet. 91: 611-617.

Hohmann, U., Graner, A., Endo, T.R., Gill, B.S., and Herrmann, R.G. (1995a) Comparison of wheat physical maps with barley linkage maps for group 7 chromosomes. Theor. Appl. Genet. 91: 618-626.

Hu, X.Y., Ohm, H.W., and Dweikat, I. (1997) Identification of RAPD markers linked to the gene *PM1* for resistance to powdery mildew in wheat. Theor. Appl. Genet. 94: 832-840.

Jia, J., Devos, K.M., Chao, S., Miller, T.E., Reader, S.M., and Gale, M.D. (1996) RFLP-based maps of homoeologous group-6 chromosomes of wheat and their application in the tagging of *Pm12*, a powdery mildew resistance gene transferred from *Aegilops speltoides* to wheat. Theor. Appl. Genet. 92: 559-565.

Knott, D.R. (1987) Transferring alien genes to wheat. In: Heyne, E.G. (ed.). Wheat and Wheat Improvement, 2nd Ed., pp. 462-471. American Society of Agronomy, Madison, WI.

Kota, R.S., Gill, K.S., Gill, B.S., and Endo, T.R. (1993) A cytogenetically based physical map of chromosome 1B in common wheat. Genome 36: 548-554.

Lagudah, E.S., Appels, R., Brown, A.H.D., and McNeil, D. (1991) The molecular genetic analysis of *Triticum tauschii*, the D genome donor to hexaploid wheat. Genome 34: 375-386.

Liu, C.J., Atkinson, M.D., Chinoy, C.N., Devos, K.M., and Gale, M.D. (1992) Nonhomoeologous translocations between group 4, 5 and 7 chromosomes within wheat and rye. Theor. Appl. Genet. 83: 305-312.

Liu, Y-G., and Tsunewaki, K. (1991) Restriction fragment length polymorphism (RFLP) analysis in wheat. II. Linkage maps of the RFLP sites in common wheat. Jap. J. Genet. 66: 617-633.

Maan, S.S. (1987) Interspecific and intergeneric hybridization in wheat. In: Heyne, E.G. (ed.). Wheat and Wheat Improvement, 2nd Ed., pp. 453-461. American Society of Agronomy, Madison, WI.

Mao, L., Devos, K.M., Zhu, L., and Gale, M.D. (1997) Cloning and genetic mapping of wheat telomere-associated sequences. Mol. Gen. Genet. 254: 584-591.

Marino, C.L., Nelson, J.C., Lu, Y.H., Sorrells, M.E., Leroy, P., Tuleen, N.A., Lopes, C.R., and Hart, G.E. (1996) Molecular genetic maps of the group 6 chromosomes of hexaploid wheat (*Triticum aestivum* L. em. Thell.). Genome 39: 359-366.

McIntosh, R.A., Hart, G.E., Devos, K.M., Gale, M.D., and Rogers, W.J. (1998) Catalogue of gene symbols for wheat. In: Slinkard, A.E. (ed.). Proc. 9th Inter. Wheat Genet. Symp., Vol. 5. University Extension Press, University of Saskatchewan, Canada.

Mickelson-Young, L., Endo, T.R., and Gill, B.S. (1995) A cytogenetic ladder map of the wheat homoeologous group-4 chromosomes. Theor. Appl. Genet. 90: 1007-1011.

Naranjo, T., Roca, A., Goicoecha, P. G., and Giraldz, R. (1987) Arm homoeology of wheat and rye chromosomes. Genome 29: 873-882.

Nelson, J.C., Sorrells, M.E., Van Deynze, A.E., Lu, Y.H., Atkinson, M.D., Bernard, M., Leroy, P., Faris, J.D., and Anderson, J.A. (1995a) Molecular mapping of wheat: Major genes and rearrangements in homoeologous groups 4, 5 and 7. Genetics 141: 721-731.

Nelson, J.C., Van Deynze, A.E., Autrique, E., Sorrells, M.E., Lu, Y.H., Merlino, M., Atkinson, M., and Leroy, P. (1995b) Molecular mapping of wheat. Homoeologous group 2. Genome 38: 516-524.

Nelson, J.C., Van Deynze, A.E., Autrique, E., Sorrells, M.E., Lu, Y.H., Negre, S., Bernard, M., and Leroy, P. (1995c) Molecular mapping of wheat. Homoeologous group 3. Genome 38: 525-533.

Parker, G.D., Chalmers, K.J., Rathjen, A.J., and Langridge, P. (1998) Mapping loci associated with flour colour in wheat (*Triticum aestivum* L.). Theor. Appl. Genet. 97: 238-245.

Reitz, L.P. (1967) World distribution and importance of wheat. In: Quisenberry, K.S., and Reitz, L.P. (eds.), Wheat and Wheat Improvement, pp. 1-18. American Society of Agronomy, Madison, WI.

Röder, M.S., Plaschke, J., König, S.U., Börner, A., Sorrells, M.E., Tanksley, S.D., and Ganal, M.W. (1995) Abundance, variability and chromosomal location of microsatellites in wheat. Mol. Gen. Genet. 246: 327-333.

Röder, M.S., Korzun, V., Gill, B.S., and Ganal, M.W. (1998a) The physical mapping of microsatellite markers in wheat. Genome 41: 278-283.

Röder, M.S., Korzun, V., Wendehake, K., Plaschke, J., Tixier, M-H., Leroy, P., and Ganal, M.W. (1998b) A microsatellite map of wheat. Genetics 149: 2007-2023.

Sharp, P.J., Chao, S., Desai, S., and Gale, M.D. (1989) The isolation, characterisation and application in the Triticeae of a set of RFLP probes identifying each homoeologous chromosome arm. Theor. Appl. Genet. 78: 342-348.

Stephenson, P., Bryan, G.J., Kirby, J., Collins, A.J., Devos, K.M., Busso, C.S., and Gale, M.D. (1998) Fifty new microsatellite loci for the wheat genetic map. Theor. Appl. Genet. 97: 946-949.

Van Deynze, A.E., Dubcovsky, J., Gill, K.S., Nelson, J.C., Sorrells, M.E., Dvořák, J., Gill, B.S., Lagudah, E.S., McCouch, S.R., and Appels, R. (1995) Molecular-genetic maps for group 1 chromosomes of Triticeae species and their relation to chromosomes in rice and oat. Genome 38: 45-59.

Weng, Y., Tuleen, N.A., and Hart, G.E. (2000) Extended physical maps and a consensus physical map of the homoeologous group-6 chromosomes of hexaploid wheat (*Triticum aestivum* L. em Thell.). Theor. Appl. Genet. 100: 519-527.

Werner, J.E., Endo, T.R., and Gill, B.S. (1992) Towards a cytogenetically based physical map of the wheat genome. Proc. Natl. Acad. Sci. USA 89: 11307-11311.

Xie, D.X., Devos, K.M., Moore, G., and Gale, M.D. (1993) RFLP-based genetic maps of the homoeologous group 5 chromosomes of bread wheat (*Triticum aestivum* L.). Theor. Appl. Genet. 87: 70-74.

Zhang, H., Jia, J., Gale, M.D., and Devos, K.M. (1998) Relationships between the chromosomes of *Aegilops umbellulata* and wheat. Theor. Appl. Genet. 96: 69-75.

25. Molecular marker linkage maps in diploid and hexaploid oat (*Avena* sp.)

S. F. KIANIAN[1], S. L. FOX[2], S. GROH[3], N. TINKER[4],
L. S. O'DONOUGHUE[5], P. J. RAYAPATI[6], R. P. WISE[7],
M. LEE[6], M. E. SORRELLS[8], G. FEDAK[4], S. J. MOLNAR[4],
H. W. RINES[3,9], *and* R. L. PHILLIPS[3]

[1] *Department of Plant Sciences, North Dakota State University, Fargo, ND 58105*
<kianian@badlands.nodak.edu>
[2] *Crop Development Centre, University of Saskatchewan, Saskatoon, Saskatchewan, S7N 5A8, Canada*
[3] *Department of Agronomy and Plant Genetics, and* [9]*USDA-ARS, University of Minnesota,*
St. Paul, MN 55108
[4] *Agriculture and Agri-food Canada, Eastern Cereals and Oilseeds Research Center,*
Ottawa, Ontario, K1A 0C6, Canada
[5] *DNA LandMarks Inc., P. O. Box 6, St. Jean Sur Richelieu, Quebec, J3B 6Z1, Canada*
[6] *Department of Agronomy, Iowa State University, Ames, IA 50011*
[7] *Corn Insects and Crop Genetics Research, USDA-ARS and Department of Plant Pathology,*
Iowa State University, Ames, IA 50011
[8] *Department of Plant Breeding and Biometry, 252 Emerson Hall, Cornell University,*
Ithaca, NY 14853

Contents

1. Introduction

The genus *Avena* is organized into 14 taxa (8 diploid, 5 tetraploid, and 1 hexaploid) classified on the basis of chromosome number, genome, diaspore (unit of dispersal), flower morphology, and cross fertility (Ladizinsky 1989). Based on chromosome pairing and structure, diploid, tetraploid, and hexaploid species were given the genomic designations A or C, AABB or AACC, and AACCDD, respectively (Rajhathy and

R.L. Phillips and I.K. Vasil (eds.), DNA-Based Markers in Plants, 443-462.

Thomas 1974). The primary cultivated oat species are hexaploid ($2n = 6x = 42$) *A. sativa* L. and *A. byzantina* C. Koch. Extensive cytological work has led to the development of karyotypes, as well as aneuploid stocks, (Rajhathy and Thomas 1974, Hacker and Riley 1965, Morikawa 1985, Linares et al. 1992, Jellen et al. 1993 and 1997). However, until recently identification of homoeologous groupings in hexaploid oat has been hindered by a lack of a complete aneuploid series, useful genetic markers, and easily identifiable chromosome morphology. The C-banding technique, and in certain cases the use of semi-automated digital image analysis/enhancement systems (Jellen et al. 1993), greatly facilitated the discrimination of individual chromosomes and/or genomes in diploid (Yen and Filion 1977, Fominaya et al. 1988a), tetraploid (Fominaya et al. 1988b), and hexaploid species (Linares et al. 1992, Jellen et al. 1993 and 1997). For instance, the C genome chromosomes are easily distinguished on the basis of their dark staining pattern from the A, B, or D genome chromosomes (Fominaya et al. 1988a, Jellen 1992). More recently, genomic *in situ* hybridization (GISH) and a demonstrated high hybridization specificity of DNA from C genome diploid species relative to A genome diploid species has allowed further delineation of C genome chromosomes from A and D genome chromosomes in AACC tetraploid and AACCDD hexaploids species. Several intergenomic translocations also have been identified within these allopolyploid species (Chen and Armstrong 1994, Jellen et al. 1994, Leggett and Markhand 1995).

Even though a significant number of physiological and metabolic studies have been conducted using oat, genetic studies have lagged behind those of other plant species (Guerin and Guerin 1993). However, recent advances in oat cytology (Jellen et al. 1992), transformation (Somers et al. 1992 and 1996, Torbert et al. 1998), isolation and charac-terization of important genes (Yun et al. 1993, Lin et al. 1996, Kianian et al. 1999), and use of maize pollen in the derivation of fertile haploids and maize alien chromosome addition lines (Rines and Dahleen 1990, Riera-Lizarazu et al. 1996) have enhanced the study and characterization of oat genomes.

Molecular markers have allowed the construction of extensive linkage maps in many crop species based on Restriction Fragment Length Polymorphism (RFLP) or Polymerase Chain Reaction (PCR) type markers. Due to the complexity of genetic analyses and segregation patterns in hexaploids (Sorrells 1992), a number of strategies have been followed to simplify the development of a cultivated oat genomic map. Diploid oat species corresponding to ancestors of hexaploid oat have been used to construct RFLP maps (O'Donoughue et al. 1992, Rayapati et al. 1994). The use of less complex diploid maps to predict linkage arrangements in hexaploids is based on the assumption that the genomes are relatively homosequential. Localization of RFLP markers to chromosomes using aneuploid stocks is another approach taken to reduce the complexities of mapping this large polyploid genome (Jellen 1992, Rooney 1992, Kianian et al. 1997). In this report we summarize the mapping efforts of a consortium of scientists at Agriculture Canada-Ottawa, Cornell University, Iowa State University, University of Minnesota, and the Agricultural Research Service-United States Department of Agriculture for the development of an RFLP map for cultivated hexaploid oat.

2.   Diploid maps

Two independent diploid maps have been constructed. In both cases 'A-genome' diploid species were used. The first map was based on 44 $F_{2:3}$ families from a cross of *A. atlantica* Baum et Fedak X *A. hirtula* Lag. (O'Donoughue et al. 1992, Van Deynze et al. 1995), and the other on an F_2 population (88 individuals) of *A. strigosa* Schreb. X *A wiestii* Steud. (Rayapati et al. 1994).

2.1.   *A. atlantica* X *A. hirtula* map

A total of 367 RFLP markers were used in the construction of the *A. atlantica* X *A. hirtula* map (O'Donoughue et al. 1992, Van Deynze et al. 1995). The main source of markers were oat (CDO) and barley (BCD) leaf cDNA libraries. Seven linkage groups, presumably corresponding to the seven chromosomes of the haploid A genome, and covering 737 cM were identified (Fig. 1). Specific areas of tightly clustered markers were noted on all linkage groups (e.g. group A region flanked by *Xbcd200* and *Xbcd454*) with several markers that were closely spaced or co-segregating (Fig. 1). These areas may reflect reduced recombination around the centromeres. Reduced recombination around centromeres is known to occur in several species such as maize (Rhoades 1955) and tomato (Tanksley et al. 1992) and may be due to the centromere itself or to peri-centromeric heterochromatin (Roberts 1965). C-banding of 'A-genome' *Avena* species shows no evidence of centromeric heterochromatin in mitotic chromosomes (Fominaya et al. 1988a, Jellen 1992), although large heterochromatic blocks are seen around the centromeres of all chromosomes at pachytene in hexaploid oat (Johnson et al. 1987).

Twenty-three of the mapped clones identified duplicated sequences within this diploid A genome; most of these involved either a locus on linkage group A and another linkage group or a locus on linkage group D and another one on either C, G, or E. In addition to single locus duplication detected by Southern analysis, segmental duplications were also evident. A duplicated segment is delineated by at least two loci and they account for about 25-35% of the total genome on average in the *Triticeae* (Anderson et al. 1992, Kleinhofs et al. 1993). Nineteen percent of the markers in this oat diploid map showed significant deviation from the expected Mendelian ratios. Many of these markers 18/37 (49%) were located on linkage group A, where 64% of all markers (11/18) gave skewed inheritance ratios. All of these markers except one were skewed toward the male parent *A. hirtula*. Similar trends for clustering and preferential transmission were observed in other parts of the genome. These distortions could be due to genes affecting gamete or hybrid viability and/or inadvertent selection due to small population size.

2.2.   *A. strigosa* X *A. wiestii* map

An F_2 population was produced by crossing accessions of the diploid species *A. strigosa* (CI 3815) and *A. wiestii* (CI 1994), which are resistant and susceptible, respectively, to 40 isolates of *Puccinia coronata*, the causal agent of oat crown rust. Eighty-eight F_2 individuals were used to construct an RFLP linkage map. Two hundred and eight

S.F. Kianian, S.L. Fox, S. Groh, N. Tinker, L.S. O'Donoughue, P.J. Rayapati, R.P. Wise, M. Lee, M.E. Sorrells, G. Fedak, S.J. Molnar, H.W. Rines, and R.L. Phillips

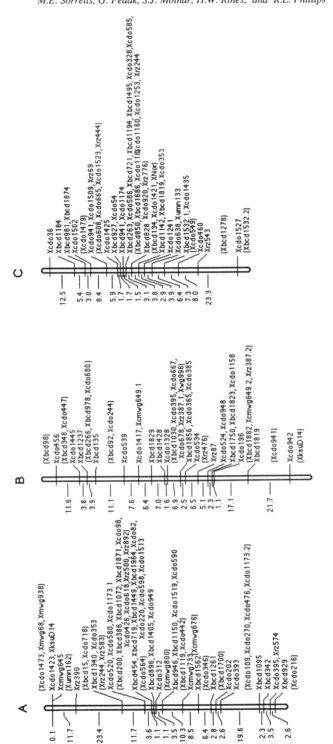

Figure 1. "A-genome" oat RFLP map based on an F₂ of the cross *A. atlantica* X *A. hirtula* (Van Deynze et al. 1995). Map distances are given in centi-Morgan (Kosambi function (Kosambi 1944)). Markers in parentheses have been assigned to intervals only.

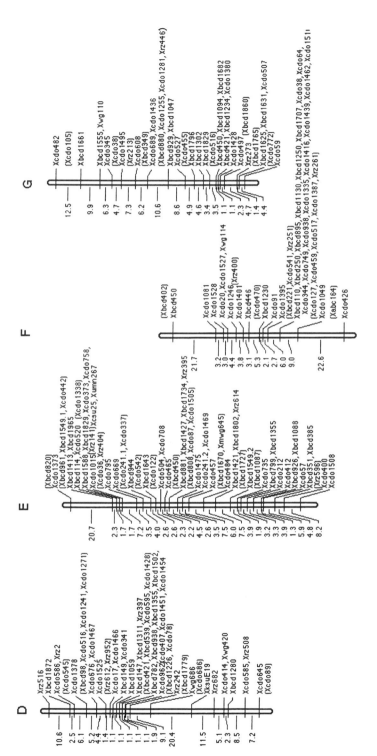

Figure 1. Continued.

RFLP loci have been placed in ten linkage groups (Fig. 2). One-hundred and seventy clones that detect 178 loci were obtained from an oat root cDNA library (ISU) and 30 additional clones detecting 30 loci were selected from other oat marker sets. This map covers 2416 cM, with an average of 12 cM between RFLP loci (Fig. 2).

Eighty-eight F_3 families, derived from the F_2 individuals used to construct the map, were screened for resistance to nine *P. coronata* isolates. The resistant phenotype in response to eight of these isolates is dominant and maps to a locus designated *Pca*. Genetic analysis indicated that resistance to a ninth isolate was regulated by two dominant genes, one of which maps to the *Pca* locus (Rayapati et al. 1994).

Five unique specificities within the *Pca* region were differentiated by recombination in a set of 100 RILs derived from the original F_2 *A. strigosa* X *A. wiestii* population. The observation of five specificities conferring disease resistance that are linked in coupling is unusual, because such genes are generally linked in repulsion and inherited from multiple donor parents (Wise et al. 1996). Seven RILs containing putative recombination breakpoints within *Pca* were backcrossed to the susceptible *A. wiestii* parent. Inoculation of the derived BC_1F_2 populations with six diagnostic crown rust isolates established that each of these specificities are encoded by unique genes within the *Pca* cluster (Yu and Wise, unpublished).

Bulked segregant analysis was used to identify AFLP markers closely linked and flanking the *Pca* region. Additionally, 450 AFLP markers have been positioned in reference to 45 previously mapped RFLP markers in the *A. strigosa* X *A. wiestii* RIL population. The increased density of markers has facilitated the re-estimation of framework linkage groups (Yu and Wise 2000). The DNA markers and genetic stocks generated by these analyses provide valuable tools for further investigations of the genetics of host-pathogen interactions in *Avena* and related grasses such as *Triticum*, *Hordeum*, and *Secale* (Yu et al. 1996).

2.3. Comparison of diploid oat maps

One of the major differences noted between the two diploid oat maps is map size. Greater parental divergence is often associated with reduced recombination rates and/or smaller map size. This phenomenon has been observed in rice (McCouch et al. 1988), potato (Gebhardt et al. 1991), tomato (Rick 1969), maize (Doebley and Stec 1991), and *Brassica* (Kianian and Quiros 1992). Chromosome inversions can explain reduced recombination frequencies in interspecific crosses (Kianian and Quiros 1992, Vallejos et al. 1992); however, reduced recombination was observed in interspecific crosses even when detailed cytogenetic analysis revealed normal pairing and chiasmata formation (Rick 1969). In interspecific hybrids of the genus *Triticum* it was suggested (Lukaszewski 1995) that positive chiasma interference reduces the amount of recombination. These observations suggest that the genomes of intraspecific crosses undergo greater recombination than that of interspecific genomes.

The frequency of RFLPs between *A. atlantica* and *A. hirtula* detected with single digests of two enzymes (*Eco*RI and *Eco*RV) was 74% (O'Donoughue et al. 1992), while only 39% of clones detected RFLPs in the *A. strigosa* and *A. wiestii* parents under the same conditions. Thus, the genomes of *A. strigosa* and *A. wiestii* are probably less divergent than

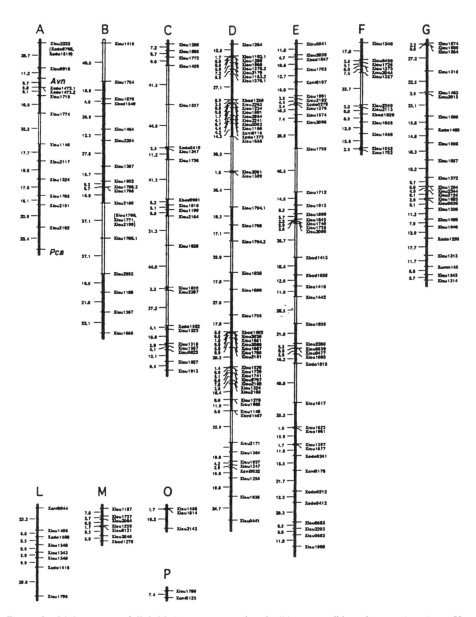

Figure 2. Linkage map of diploid *Avena* representing the "A-genome" based on an *A. strigosa* X *A. wiestii* cross (Rayapati et al. 1994). 208 RFLP markers and 1 locus (*Pca*) conferring resistance to *Puccinia coronata* were assigned to ten linkage groups (A-G, L, M, and O). Numbers on the left of a linkage group represent map distances in cM (Kosambi function (Kosambi 1944)). Designations on the right are marker or gene names. Some allele frequencies displayed segregation ratios skewed toward the *A. strigosa* (†), *A. wiestii* (#), and heterozygous (*) genotypes. Thin lines represent regions between groups of markers linked by a LOD score greater than or equal to 3.0, but with a distance greater than 30 cM. Thick lines represent regions between markers linked by less than 30 cM at LOD 5.0.

A. *atlantica* and A. *hirtula*, and likely represent the same species. This is consistent with Ladizinsky's (1989) estimate of the evolutionary divergence of these taxa describing A. *strigosa*, A. *wiestii*, and A. *hirtula* as morphological variants belonging to one biological species (A. *strigosa*), while A. *atlantica* is a member of a distinct biological species.

Map size, and therefore recombination, was considerably reduced in the A. *atlantica* X A. *hirtula* population as compared to the A. *strigosa* X A. *wiestii* population. To characterize the recombination frequencies in the two diploid oat crosses, 26 probes that detected RFLPs in the A. *atlantica/A. hirtula* cross were mapped in the A. *strigosa / A. wiestii* cross. Thirteen of these probes showed conserved synteny in both populations (Fig. 2, Table 1). The 3.9 fold reduction in recombination for regions detected by these probes in the A. *atlantica/A. hirtula* cross is consistent with the four-fold decrease (2416 vs. 737 cM) in total length for the A. *atlantica/A. hirtula* map (Table 1). The difference in map sizes produced by two different diploid oat crosses is consistent with previous comparisons of maps produced from intra- and interspecific crosses in maize and wheat. Thus, the likelihood that the A. *strigosa/A. wiestii* map is constructed from an intraspecific cross while the A. *atlantica/A. hirtula* map is constructed from a much more divergent interspecific cross is the best explanation for the greater recombination frequency and larger gaps in the A. *strigosa/A. wiestii* map.

Table 1.   Comparison of recombination frequencies among syntenic markers in two diploid oat maps.

Syntenic markers	Linkage group [a]	Map distances in cM [b]	
		A. strigosa/A. wiestii [c]	A. atlantica/A. hirtula [d]
CDO1473-CDO1519	A	70	56
CDO1502-BCD981	C	134	3
CDO1467-BCD1502	D	78	10
BCD1413-CDO1015	E	113	6
CDO1015-CDO241	E	97	32
CDO241-CDO212	E	37	33
CDO212-CDO412	E	12	3
CDO1495-CDO1255	G	117	24
Total		658	167
Total map length		2416	737

[a] When the same probes detected linked RFLP loci in two different mapping populations, the linkage group name designated by Van Deynze et al. 1995 was used.

[b] Map distances are given in centi-Morgan (cM) calculated by the Kosambi function (Kosambi 1944).

[c] Values are from Rayapati et al. 1994.

[d] Values are from Van Deynze et al. 1995.

3. Hexaploid maps

3.1. RFLP map in Kanota X Ogle

A recombinant inbred F_6-derived population of 71 oat lines from a cross of *A. byzantina* cv Kanota X *A. sativa* cv Ogle (KO) has been used to develop a linkage map in cultivated hexaploid oat (O'Donoughue et al. 1995). This is the most complete hexaploid oat map currently available; it forms an important point of reference for comparative mapping. Approximately 700 loci have been mapped in the KO cross. Most of these are RFLPs derived from approximately 480 different cDNA sequences originating from etiolated leaf tissue of oat (CDO) or barley (BCD), developing endosperm of oat (UMN), root tissue of oat (ISU), or inflorescence of oat (ACO). The current KO framework linkage map (Fig. 3) contains 221 markers on 32 linkage groups, spanning a total length of 1770 cM. Numerical linkage group names (Fig. 3) were initially based on homology with the linkage groups of the *A. atlantica* X *A. hirtula* diploid oat map (O'Donoughue et al. 1995). For continuity, this naming convention has been preserved, with names being concatenated as linkage groups are joined by aneuploid chromosome assignment and additional mapping (Table 2 and Fig. 3). Efforts to improve the KO map are focused on four primary areas: (1) further analysis of aneuploid stocks, (2) mapping of additional markers, (3) mapping in additional RILs, and (4) re-estimation of framework linkage groups based on the above three areas.

The KO mapping population has been extended to 137 RILs, and additional RFLP marker loci including genomic clones have been analyzed (Kianian et al. 1999). To date, over 110 RFLP loci initially mapped on the first 71 RILs have been placed on the extended population. The number of linkage groups has been reduced to a total of 32 from the original 36 (O'Donoughue et al. 1995), including four new linkage groups containing two or three markers each. Aneuploid analysis has confirmed five of these new associations (Table 2 and Fig. 3). Several associations revealed by aneuploid analysis remained genetically independent.

Despite additional mapping efforts, it has not yet been possible to collapse the hexaploid oat map into 21 clearly defined linkage groups to correspond to the 21 oat chromosomes. One contributing factor is the size of the mapping population relative to the large oat genome. Non-random spacing of cDNA markers can be expected to produce many recombination gaps in this population. Furthermore, as the number of markers increases, the stringency for assigning markers to linkage groups must also increase to prevent a high frequency of false linkages.

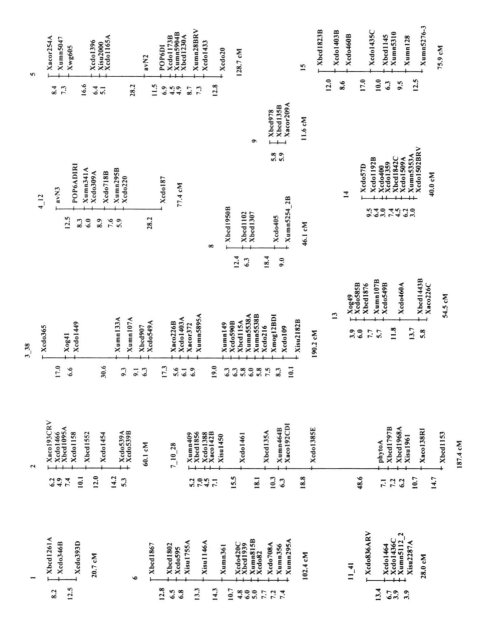

Figure 3.

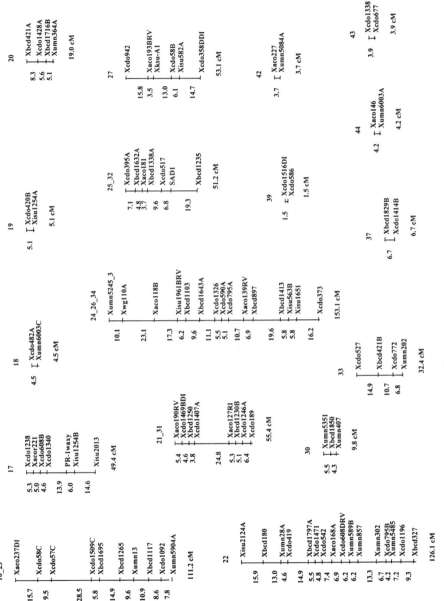

Figure 3. Continued.

Table 2.   Linkage groups associated with specific chromosomes by aneuploid analysis.

Monosomic chromosome [a]	Aneuploid [b]	Assignment by aneuploid analysis [c,d]
1C	K1, K2	21+31
2C	K3, K5, Sun II 8-1 1-1	15
3C		
4C		
5C	K8, SIV	5
6C		
7C	SVII, SXIV	3
8	SXVI, SXVII, SXVIII	23+ *Xumn107C*
9	Sn10-2 3-3	17
10		
11	SI, K19	13
12	SV	2
13	K16	6+9
14	K7, K13	16+23, 16+38
15	K10, K20, SXV	7+10+28
16	K18	20
17	K11, K17	3+24+34
18	K21	33
19	K12, K14, SXII	22
20		
21	SVIII, SIX, SX	4+12

[a] Chromosome designations are according to Jellen et al. 1993.
[b] K refers to aneuploids derived from the cultivar Kanota.
   S or Sn refer to aneuploids derived from the cultivar Sun II.
[c] Linkage group designations are based on O'Donoughue et al. 1995.
[d] Assignments by aneuploid analysis are from Kianian et al. 1997 and Fox et al. 2000 (in press).

3.2. Terra X Marion linkage map

Linkage mapping in additional populations, and comparative mapping among these populations will contribute to further knowledge of hexaploid oat genome structure. Additional populations will also provide opportunities to confirm quantitative trait loci. Linkage mapping is being performed using 102 recombinant inbred lines from an intraspecific *A. sativa* cross Terra x Marion (TM). To date, approximately 400 markers have been mapped in this population: 170 RFLP, 200 AFLP, and 25 RAPD markers (Tinker, unpublished).

The current TM map is considerably smaller than the KO map, spanning less than 1000 cM. This may reflect a higher degree of homogeneity in this cross, such that fewer chromosome regions are polymorphic for the markers tested. Many cDNA sequences represented on the KO map revealed few or no polymorphisms in TM (Tinker, unpublished).

Comparative RFLP mapping has allowed most of the TM map to be tentatively associated with linkage groups in the KO map. However, for many RFLP loci mapped in TM it was not possible to directly infer allelism with loci mapped in KO because different pairs of alleles appear to segregate in each cross. Partial mapping with additional enzymes and comparison with other oat maps may resolve some of these difficulties. Mapping of AFLP loci in KO will also facilitate comparisons (Tinker, unpublished). Eventually, locus-specific microsatellite markers could help sort out structural relationships.

3.3. Aneuploid analysis

Aneuploid stocks have been used to localize RFLP markers to specific chromosomes or chromosome arms in several crops including wheat (Devey and Hart 1988, Lagudah et al. 1991, Gill et al. 1991, Anderson et al. 1992), rice (McCouch et al. 1988) and maize (Helentjaris et al. 1986). In hexaploid cultivated oat, aneuploids including nullisomics (2n - 2 = 40), monosomics (2n - 1 = 41) and ditelosomics (2n - tt = 40 + tt) have been recovered. Although most of the aneuploid series are incomplete they have facilitated the assignment of markers to some chromosomes. Nullisomics have been effective for assigning RFLP markers to specific chromosomes, and ditelosomics allow for localization of such markers to specific chromosome arms (Kianian et al. 1997). Nullisomics and ditelosomics of cultivar Sun II were used to assign 134 DNA sequences to 10 syntenic groups and a subset of these to six chromosome arms (Kianian et al. 1997). Consequently, eight of the published oat RFLP linkage groups were assigned to five physical chromosomes (Table 2). Three of these linkage groups were assigned to their respective chromosome arms using ditelosomics (Table 2).

Monosomic lines derived from the cultivars Kanota (K) (Morikawa 1985) and Sun II (S) (Hacker and Riley 1963; Jellen et al. 1997) were used as female parents in crosses with euploid plants of the cultivars Ogle (O) and Kanota, producing a series of monosomic F_1s. Each of the 21 different oat chromosomes was thought to be represented at least once in this set of 97 KO, SO and SK crosses (Fox et al. 2000). The identity of the chromosomes in the K and S monosomic lines was determined using C-banding by Jellen et al. (1993 and 1997). From each RFLP linkage group, cDNA probes were selected based on expected polymorphism between the parents of each monosomic F_1 and the ability to observe the maternal allele of the polymorphic locus. Associations between a linkage group and a chromosome were made when the maternal form of a polymorphism was absent in the monosomic F_1s representing a specific chromosome deletion. Loss of a band indicated that the probe-binding site was located on the chromosome that was not transmitted to the monosomic F_1. Two or more loci from the same linkage group were used to make an association to a chromosome except when few or inadequate loci were present in a linkage group. Twenty-three linkage groups have been associated with 17 chromosomes (Table 2; Fox et al. 2000). Two or more genetically unlinked linkage groups have been physically associated to each of chromosomes 1C, 13, 14, 15, 17 and 21. These associations do not orient the linkage groups with respect to the chromosome or to each other.

Linkage groups 3, 6, 11, 17, 23 and 30 have areas of reduced recombination as indicated by clustering of loci (O'Donoughue et al. 1995) and may be due to proximity to centromeres or translocation breakpoints. Linkage group 3 appeared to involve a translocation between chromosomes 7C and 17. Six loci between *Xcdo395B* and *Xbcd342B* were associated with chromosome 17 (represented by K11 and K17); but eleven markers that had been mapped between *Xumn149* and *Xumn133A* were not associated with chromosome 17 (Fig. 4). The opposite situation was observed for chromosome 7C (represented by SVII and SXIV) where *Xbcd1735* was associated with chromosome 7C but *Xcdo346A* and *Xbcd342B* were not associated (Fox et al. 2000).

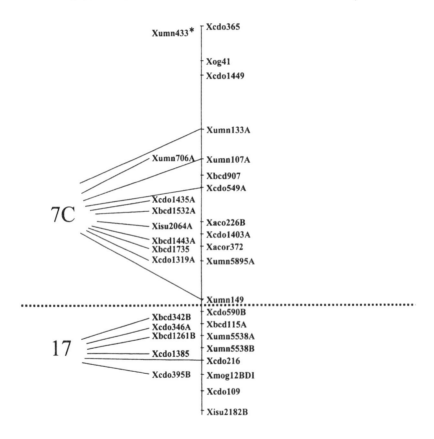

Figure 4. Linkage group 3 of the framework KO map (see Fig. 3) showing distances between selected loci that are associated with chromosome 7C or 17. Loci on the left of the linkage group axis are placed relative to the framework markers on the right. The dotted line is the suggested region where a translocation breakpoint would exist that separates the upper chromosome 7C portion from the lower chromosome 17 portion of this linkage group. The asterisk (*) indicates an anomalous association.

Linkage group 23 displayed clustered loci and was also shown to be involved in a translocation. This linkage group was found to be associated with both chromosome 14 (represented by K7) and chromosome 8 (represented by SXVI, SXVII, and SXVIII); however, linkage group 23 was clearly not associated with K13, also deficient for chromosome 14 (Fox et al. 2000). Supporting this observation, chromosome C-banding by Jellen et al. (1993) had shown that K7 represents the "normal" chromosome 14 and K13 represents a translocated chromosome 14. This suggests that the "normal" location of linkage group 23 is on chromosome 14.

The C genome chromosomes of oat are distinct, while the A and D genomes are indistinguishable from each other based on staining and genomic *in situ* hybridization techniques (Chen and Armstrong 1994, Jellen et al. 1994). Because chromosomal rearrangements such as translocations and inversions have played an important role in the evolution of hexaploid oat (Rajhathy and Thomas 1974), the distinctiveness of C genome chromosomes was used to determine remnant homoeologous segments that exist in the other two genomes (Kianian et al. 1997). Due to various evolutionary forces, segmental homoeology instead of whole chromosome homoeology appears to best describe the genome organization in hexaploid oat.

3.4.   Comparison of diploid and hexaploid oat maps

One hundred and two of the clones used on the hexaploid oat (KO) map had previously been mapped on the *A. atlantica* X *A. hirtula* diploid oat map (O'Donoughue et al. 1992). Comparisons of marker order between the two maps reveal that although large groups of markers remain syntenic, at least 5 major translocation differences exist between the A genome diploids and the hexaploid oat cultivars mapped (O'Donoughue et al. 1995). The hexaploid map linkage groups were initially ordered based on homology with the linkage groups of the *A. atlantica* X *A. hirtula* diploid oat map (O'Donoughue et al. 1995). However, it is apparent that within the genus *Avena* many structural rearrangements have taken place during the course of evolution. This fact may explain past difficulties in obtaining a complete aneuploid series in oat as well as difficulties in the clear characterization of individual genomes.

4.   Relationships between oat genomes and other grasses

Comparative linkage maps have provided researchers with a new tool for elucidating the genomic structural relationships among species. Probes hybridizing to loci in multiple species, known as anchor probes, are used to enhance species-specific maps. These maps are aligned according to common loci to identify regions with conserved gene order. Comparative maps can be used to transfer information on agronomically important traits and the underlying genes between species. Oat genome structural relationships with maize, rice, and *Triticeae* were examined using anchor probes mapped in each of the species (Van Deynze et al. 1995). The diploid oat map from *A. atlantica* X *A. hirtula* with 367 loci covering 737 cM had 192 markers in common with wheat, 138 with rice, and 129 with maize (Van Deynze et al. 1995). The highest degree of conservation was found between oat and wheat with 84% conserved regions, followed by rice (79% conserved regions), and maize (71% conserved regions). The diploid oat linkage group A showed almost complete conservation with wheat chromosomes 1, rice chromosome 5, and maize chromosome 8. The diploid oat linkage group D showed a high degree of conservation with wheat chromosome 7 and rice chromosome 6. The lowest degree of conservation was found for diploid oat linkage group G. Most regions of the diploid oat map were homoeologous to regions on two maize chromosomes except for portions of maize chromosomes 1, 2, 4, 8, and 9. Several rearrangements such as inversions and translocations were evident between oat and the other grasses.

The benefits from comparative mapping research will be greatest for species with limited genetic information such as oat, millet or other grasses which can be compared to species with high genetic information such as wheat, rice, and maize. Alignment of the conserved regions between species allows one to infer orthology for agronomically important genes. Evidence for the conservation of gene functions between oat and other grass species was found for several genes (Van Deynze et al. 1995), including a gene for the seed storage protein avenin (O'Donoughue et al. 1995), the stem rust gene *Pg9*, and several crown rust resistance genes (Chong et al. 1994, Yu et al. 1996, Bush and Wise 1998). Genomic regions controlling heading date in hexaploid oat (Siripoonwiwat et al. 1996) could be associated with genes or loci controlling vernalization sensitivity in wheat (Nelson et al. 1995), photoperiod response in barley (Pan et al. 1994), growing-degree units to silking/tasseling in maize (Phillips et al. 1992), and heading date in rice (Li et al. 1995). The merging of information from QTL mapping studies, mutant gene functions, mapped gene sequences and candidate genes will lead to rapid gene discovery and characterization across species.

5.   Conclusions

Construction of linkage maps in polyploids has lagged behind that of diploid species due to the complexities of genome size and genetic analysis (Sorrells 1992). The concurrent development of diploid oat maps, chromosome arm maps in hexaploids based on aneuploids, and hexaploid oat maps is providing a more comprehensive picture of genome organization in the genus *Avena*. This information is essential, as it is becoming increasingly clear that *Avena* is characterized by frequent structural rearrangements. However, these structural differences may offer oat geneticists certain advantages. Translocations identified cytologically are being mapped genetically with molecular markers and these markers can be used to identify duplication-deficiency (Dp-Df) lines. These lines can be employed in the generation of specific gene duplications such as those for crown rust resistance (Wilson and McMullen 1997). The difference in recombination frequencies (about 4 fold) between the two diploid populations raises interesting questions about factors affecting chromosome pairing, crossing over and recombination frequencies in specific crosses. The cultivated oat maps being developed will provide sets of recombinant inbred lines which will be well characterized both genetically and cytologically. This material will be invaluable for further genetic studies in oat such as the identification of chromosomal regions controlling traits of agronomic and biological significance.

6. Acknowledgments

The authors wish to thank the scientists who provided their unpublished results for inclusion in this chapter. We are also grateful to a number of technical staff at the four institutions for providing expertise throughout this project. The financial support of The Quaker Oats Company is especially appreciated.

7. References

Anderson, J.A., Ogihara, Y., Sorrells, M.E. and Tanksley, S.D. (1992) Development of a chromosomal arm map for wheat based on RFLP markers. Theor. Appl. Genet. 83: 1035-1043.

Bush, A.L. and Wise, R.P. (1998) High resolution mapping adjacent to the *Pc71* crown-rust resistance locus in hexaploid oat. Molecular Breeding 4:13-21.

Chen, Q. and Armstrong, K.C. (1994) Genomic in situ hybridization in *Avena sativa*. Genome 37: 607-612.

Chong, J., Howes, N.K., Brown, P.D. and Harder, D.E. (1994) Identification of the stem rust resistance gene *Pg9* and its association with crown rust resistance and endosperm proteins in 'Dumont' oat. Genome 37:440-447.

Devey, M. and Hart, G. (1988) Intergenomic restriction fragment length polymorphism in hexaploid wheat. Agron. Abstr. p.79.

Doebley, J. and Stec, A. (1991) Genetic analysis of the morphological differences between maize and teosinte. Genetics 129: 285-295.

Fominaya, A., Vega, C. and Ferrer, E. (1988a) Giemsa C-banded karyotypes of *Avena* species. Genome 30: 627-632.

Fominaya, A., Vega, C. and Ferrer, E. (1988b) C-banding and nucleolar activity of tetraploid *Avena* species. Genome 30: 633-638.

Fox, S.L., Jellen, E.N., Kianian, S.F., Rines, H.W., and Phillips, R.L. (2000) Assignment of RFLP linkage groups to chromosomes using monosomic F_1 analysis in hexaploid oat. Theor. Appl. Genet. (in press)

Gebhardt, C., Ritter, E., Barone, A., Debener, T., Walkemeier, B., Ganal, U.M.W., Tanksley, S.D. and Salamini, F. (1991) RFLP maps of potato and their alignment with the homoeologous tomato genome. Theor. Appl. Genet. 83: 49-57.

Gill, K.S., Lubbers, E.L., Gill, B.S., Raupp, W.J. and Cox, T.S. (1991) A genetic linkage map of *Triticum tauschii* (DD) and its relationship to the D genome of bread wheat (AABBDD). Genome 34: 362-374.

Guerin, T.F. and Guerin, P.M. (1993) Recent developments in oat molecular biology. Plant Molec. Biol. Rep. 11: 65-72.

Hacker, J.B. and Riley, R. (1963) Aneuploids in oat varietal populations. Nature 197:924-925.

Hacker, J.B. and Riley, R. (1965) Morphological and cytological effects of chromosome deficiency in *Avena sativa*. Can. J. Genet. Cytol. 7: 304-315.

Helentjaris, T., Weber, D. and Wright, S. (1986) Use of monosomics to map cloned DNA fragments in maize. Proc. Natl. Acad. Sci. (USA). 83: 6035-6039

Jellen, E.N. (1992) Characterization of image-enhanced, C-banded oat (*Avena* spp.) monosomics and identification of oat genomes and homoeologous chromosomes. Ph.D. Thesis, University of Minnesota, St. Paul.

Jellen, E.N., Phillips, R.L. and Rines, H.W. (1993) C-banded karyotypes and polymorphisms in hexaploid oat accessions (*Avena* spp.) using Wright's stain. Genome 36: 1129-1137.

Jellen, E.N., Gill, B.S. and Cox, T.S. (1994) Genomic in situ hybridization differentiates between A/D- and C-genome chromatin and detects intergenomic translocations in polyploid oat species (genus *Avena*). Genome 37: 613-618.

Jellen, E.N., Rines, H.W., Fox, S.L., Davis, D.W., Phillips, R.L. and Gill, B.S. (1997) Characterization of 'Sun II' oat monosomics through C-banding and identification of eight new 'Sun II' monosomics. Theor. Appl. Genet. 95: 1190-1195.

Johnson, S.S., Phillips, R.L. and Rines, H.W. (1987) Possible role of heterochromatin in chromosome breakage induced by tissue culture in oats (*Avena sativa* L.). Genome 29: 439-446.

Kianian, S.F. and Quiros, C.F. (1992) Generation of a *Brassica oleracea* composite RFLP map: linkage arrangements among various populations and evolutionary implications. Theor. Appl. Genet. 84:544-554.

Kianian, S.F., Wu, B.-C., Fox, S.L., Rines, H.W. and Phillips, R.L. (1997) Aneuploid marker assignment in hexaploid oat with the C genome as a reference for determining remnant homoeology. Genome 40: 386-396.

Kianian, S.F., Egli, M.A., Phillips, R.L., Rines, H.W., Somers, D.A., Gengenbach, B.G., Webster, F.H., Livingston, S.M., Groh, S., O'Donoughue, L.S., Sorrells, M.E., Wesenberg, D.M., Stuthman, D.D. and Fulcher, R.G. (1999) Association of a major groat oil content QTL and an acetyl-CoA carboxylase gene in oat. Theor. Appl. Genet. 98: 284-894.

Kleinhofs, A., Kilian, A., Saghai-Maroof, M.A., Biyashev, R.M., Hayes, P., Chen, F.Q., Lapitan, N., Fenwick, A., Blake, T.K., Kanazin, V., Ananiev, E., Dahleen, L., Kudrna, D., Bollinger, J., Knapp, S.J., Liu, B., Sorrells, M., Heun, M., Franckowiak, J.D., Hoffman, D., Skadsen, R. and Steffenson, B.J. (1993) A molecular, isozyme and morphological map of the barley (*Hordeum vulgare*) genome. Theor. Appl. Genet. 86: 705-712.

Kosambi, D.D. (1944) The estimation of map distance from recombination values. Ann. Eugen. 12: 172-175.

Ladizinsky, G. (1989) Biological species and wild genetic resources in *Avena*. IBPGR Report of a Working Group on *Avena*. p. 19-32.

Leggett, J.M. and Markhand, G.S. (1995) The genomic identification of some monosomics of *Avena sativa* L. cv. Sun II using genomic in situ hybridization. Genome 38: 747-751.

Lagudah, E.S., Appels, R., Brown, A.H.D. and McNell, D. (1991) The molecular genetic analysis of *Triticum tauschii*, the D-genome donor to hexaploid wheat. Genome 34: 375-386.

Li, Z., Pinson, S.R.M., Stansel, J.W. and Park, W.D. (1995) Identification of quantitative traits loci (QTLs) for heading date and plant height in cultivated rice (*Oryza sativa* L.). Theor. Appl. Genet. 91:374-381.

Lin, K.-C., Bushnell, W.R., Szabo, L.J. and Smith, A.G. (1996) Isolation and expression of a host response gene family encoding thaumatin-like proteins in incompatible oat stem rust fungus interactions. Mol. Plant-Microbe Interactions 9: 511-522.

Linares, C., Vega, C., Ferrer, E. and Fominaya, A. (1992) Identification of C-banded chromosomes in meiosis and the analysis of nucleolar activity in *Avena byzantina* C. Koch cv 'Kanota'. Theor. Appl. Genet. 83: 650-654.

Lukaszewski, A.J. (1995) Physical distribution of translocation breakpoints in homoeologous recombinants induced by the absence of *Ph1* gene in wheat and tritecale. Theor. Appl. Genet. 90: 714-719.

McCouch, S., Kochert, G., Yu, Z., Wang, Z., Khush, G., Coffman, W. and Tanksley, S. (1988) Molecular mapping of rice chromosomes. Theor. Appl. Genet. 76: 815-829.

Morikawa, T. (1985) Identification of the 21 monosomic lines in *Avena byzantina* C. Koch cv. 'Kanota'. Theor. Appl. Genet. 70: 271-278.

Nelson, J.C., Van Deynze, A.E., Autrique, E., Sorrells, M.E., Lu, Y.H., Merlino, M., Atkinson, M. and Leroy, P. (1995) Molecular mapping of wheat: Homoeologous group 2. Genome 38:516-524.

O'Donoughue, L.S., Wang, Z., Roder, M., Kneen, B., Leggett, M., Sorrells, M.E. and Tanksley, S.D. (1992) An RFLP-based linkage map of oats based on a cross between two diploid taxa (*Avena atlantica* X *A. hirtula*). Genome 35: 765-771.

O'Donoughue, L.S., Kianian, S.F., Rayapati, P.J., Penner, G.A., Sorrells, M.E., Tanksley, S.D., Phillips, R.L., Rines, H.W., Lee, M., Fedak, G., Molnar, S.J., Hoffman, D., Salas, C.A., Wu, B., Autrique, E. and Van Deynze, A. (1995) A molecular linkage map of cultivated oat. Genome 38:368-380.

Pan, A., Hayes, P.M., Chen, F., Chen, T.H.H., Blake, T., Wright, S., Karsai, I. and Bedö (1994) Genetic analysis of the components of winterhardiness in barley. Theor. Appl. Genet. 89: 900-910.

Phillips, R.L., Kim, T.S., Kaeppler, S.M., Parentoni, S.N., Shaver, L., Stucker, R.E. and Openshaw, S.J. (1992) Genetic dissection of maturity using RFLPs. p. 135-150. In Proceedings of the 47th Annual Corn and Sorghum Industry Research Conference.

Rajhathy, T. and Thomas, H. (1974) Cytogenetics of oats (*Avena* L.). A. Wilkes ed. Miscellaneous Publication of The Genetic Society of Canada-No. 2, Ottawa, Ontario, Canada.

Rayapati, P.J., Lee, M., Gregory, J.W. and Wise, R.P. (1994) An RFLP linkage map of diploid oat *Avena* based on RFLP loci and a locus conferring resistance to *Puccinia coronata* var. *avenae*. Theor. Appl. Genet. 89: 831-837.

Rhoades, M.M. (1955) The cytogenetics of maize. In: Corn and corn improvement. Edited by G.F. Sprague. Academic Press, N.Y. p. 123-219.

Rick, C.M. (1969) Controlled introgression of chromosomes of *Solanum pennellii* into *Lycopersicon esculentum*: segregation and recombination. Genetics 62: 753-768.

Riera-Lizarazu, O., Rines, H.W. and Phillips, R.L. (1996) Cytological and molecular characterization of oat x maize partial hybrids. Theor. Appl. Genet. 93: 123-135.

Rines, H.W. and Dahleen, L.S. (1990) Haploid oat plants produced by application of maize pollen to emasculated oat florets. Crop Sci. 30:1073-1078.

Roberts, P.A. (1965) Difference in behaviour of eu- and hetero-chromatin: crossing over. Nature 205: 725-726.

Rooney, W.L. (1992) Identification and characterization of RFLP markers linked to crown rust resistance in oat (*Avena* sp.) Ph.D. Thesis. University of Minnesota, St. Paul.

Siripoonwiwat, W., O'Donoughue, L.S., Wesenberg, D., Hoffman, D.L., Barbosa-Neto, J.F. and Sorrells, M.E. (1996) Chromosomal regions associated with quantitative traits in oat.

JQTL 2: Article 3.

Somers, D.A., Rines, H.W., Gu, W., Kaeppler, H.F. and Bushnell, W.R. (1992) Fertile, transgenic oat plants. Bio/Technology 10: 1589-1594.

Somers, D.A., Rines, H.W., Torbert, K.A., Pawlowski, W.P. and Milach, S.K.C. (1996) Genetic transformation in *Avena sativa* L. (oat). In Y.P.S. (ed.) Plant Protoplasts and Genetic Engineering VII, Biotechnology in Agriculture and Forestry, Vol. 38, Springer-Verlag, Berlin. p. 178-190.

Sorrells, M.E. (1992) Development and application of RFLPs in polyploids. Crop Sci. 32: 1086-1091.

Tanksley, S.D., Ganal, M.W., Prince, J.P., DeVincente, M.C., Bonierbale, M.W., Broun, P., Fulton, T.M., Giovanonni, J.J., Grandillo, S., Martin, G.B., Messeguer, R., Miller, J.C., Miller, L., Paterson, A.H., Pineda, O., Roder, M., Wing, R.A., Wu, W. and Young, N.D. (1992) High density molecular linkage maps of the tomato and potato genomes: biological inferences and practical applications. Genetics 132: 1141-1160.

Torbert, K.A., Rines, H.W. and Somers, D.A. (1998) Transformation of oat using mature embryo-derived tissue cultures. Crop Sci. 38: 226-231.

Vallejos, C.E., Sakiyama, N.S. and Chase, C.D. (1992) A molecular marker-based linkage map of *Phaseolus vulgaris* L. Genetics 131: 733-740.

Van Deynze, A.E., Nelson, J.C., O'Donoughue, L.S., Ahn, S.N., Siripoonwiwat, W., Harrington, S.E., Yglesias, E.S., Braga, D.P., McCouch, S.R. and Sorrells, M.E. (1995) Comparative mapping in grasses. Oat relationships. Mol. Gen. Genet. 249:349-356.

Wilson, W.A. and McMullen, M.S. (1997) Dosage dependent genetic suppression of oat crown rust resistance gene *Pc-62*. Crop Sci. 37: 1699-1705.

Wise, R.P., Lee, M. and Rayapati, P.J. (1996) Recombination within a 5-centimorgan region in diploid *Avena* reveals multiple specificities conferring resistance to *Puccinia coronata*. Phytopath. 86: 340-346.

Yen, S.-T. and Filion, W.G. (1977) Differential Giemsa staining in plants. V. Two types of constitutive heterochromatin in species of *Avena*. Can. J. Genet. Cytol. 19: 739-743.

Yu, G.X., Bush, A.L. and Wise, R.P. (1996) Comparative mapping of homoeologous group 1 regions and genes for resistance to obligate biotrophs in *Avena*, *Hordeum*, and *Zea mays*. Genome 39:155-164.

Yu, G-X., and Wise, R.P. (2000) An anchored AFLP and retrotransposon-based map of diploid Avena. Genome (in press).

Yun, S.J., Martin, D.J., Gengenbach, B.G., Rines, H.W. and Somers, D.A. (1993) Sequence of a (1-3, 1-4)-ß-glucanase cDNA from oat. Plant Physiol. 103: 295-296.

26. A compilation of molecular genetic maps of cultivated plants

OSCAR RIERA-LIZARAZU[1],
M. ISABEL VALES[1], and RONALD L. PHILLIPS
Department of Agronomy and Plant Genetics and Plant Molecular Genetics Institute,
University of Minnesota, St. Paul, MN 55108
[1] *Current Address: Department of Crop and Soil Science, Oregon State University,*
Corvallis, OR 97331 <Oscar.Riera@orst.edu>

Contents

1. Introduction?

In chapters of this book, detailed discussions of mapping efforts for various crop species are described. Similar treatments have also been summarized in other publications (see: Phillips and Vasil 1994; Jauhar 1996; Heslop-Harrison 1996; Paterson 1996) but no comprehensive summary of DNA-based marker maps of crops has been produced. The purpose of this chapter was to fill this void by succinctly presenting DNA-based marker mapping information of a larger number of plant species recognized as having economic importance.

2. Summary

Before 1980, the paucity of genetic markers, based on visible mutations and biochemicals, was the major obstacle for the construction of detailed genetic maps of plants. Botstein and co-workers (1980) showed that DNA restriction fragment length polymorphisms (RFLP's) were an abundant source of genetic markers in humans. This breakthrough marked the beginning of the age of DNA-based genetic markers. This revolution was also accelerated by the development of the polymerase chain reaction (PCR, Saiki et al. 1988).

DNA-based markers were adopted readily by crop scientists. In 1986, the construction of the first RFLP-based maps of maize and tomato (Helentjaris et al. 1986, Bernatzky and Tanksley 1986) marked the beginning of a storm of genetic map construction and use. The number of DNA-based marker maps published each year has

R.L. Phillips and I.K. Vasil (eds.), DNA-Based Markers in Plants, 463 - 497.

increased steadily in the past decade. In 1986, only three maps had been published but since 1993 an average of 30 maps are published each year (Fig. 1). DNA-based genetic maps of economically important plant species from about 66 different genera including monocots, dicots, and gymnosperms have been published. A list of these molecular genetic maps, the mapping populations, the number and types of markers, the genetic length of these maps, and references are presented in Table 1.

The majority of these molecular genetic maps have low to medium marker saturation. These maps have been primarily used to study and manipulate qualitative and quantitative trait loci of economic importance. Also, highly saturated linkage maps (of one marker per centiMorgan) of rice, tomato, maize, barley, and sugar beet have been constructed to facilitate map-based gene cloning and the construction of physical maps of these crop genomes.

Genome-mapping activities continue to move at a fast pace and the information that has been summarized here will inevitably become obsolete. Thus, a list of plant genome, germplasm, and reference resources is presented in Table 2. These internet-accessible databases, which were established by various research groups, should be consulted for current information.

We anticipate that the synergism between molecular biology, engineering and computer science will change the way DNA-based technologies will be utilized in the future. New cost-effective and high-throughput systems, which are under development, will undoubtedly become part of the crop scientist's repertoire of DNA-based tools. These new tools will advance our knowledge and understanding of agronomically important traits and will also create new opportunities for crop improvement.

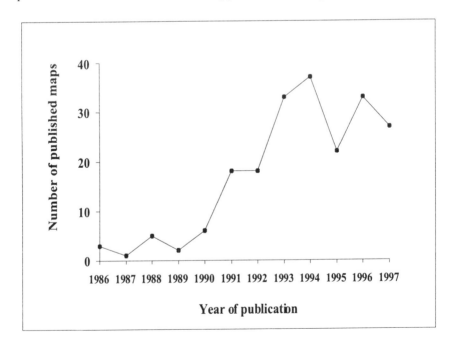

Figure 1. Number of published genetic maps from 1986 to 1997.

Table 1. List of published DNA-based marker maps of economically important plants.

Common name	Genus	Type of crop	Mapping population[a]	Number of markers	Types of markers[b]	Genetic map length (cM)	Reference
Alfalfa	*Medicago*	forage	138 F$_2$'s from 'MqK93' x 'Mcw2'	89	RFLP, RAPD, isozyme, morph.	659	Kiss et al. 1993; Kiss et al. 1997
			86 F$_2$'s from 'W2xiso' x 'PI440501'	108	RFLP	468	Brummer et al. 1993a; Brummer et al. 1993b
			87 BC$_1$'s from 'F2-16' x ('F2-16' x '74-25-20')	61	RFLP, RAPD	553	Echt et al. 1993; Kidwell et al. 1993
				86	"	603	"
			55 F$_1$'s from 'PG-F9' x 'W2x-1'	50	RFLP	234	Tavoletti et al. 1996
				55	"	261	"
Almond	*Prunus*	nut	60 F$_1$'s from 'Ferragnes' x 'Tuono'	127	RFLP, isozyme	393	Arus et al. 1994; Viruel et al. 1995
American chestnut	*Castanea*	timber and nut	102 F$_2$'s from 'R4T31' ('Roxbury East #1' x 'Mahogany') x 'R4T52' ('Roxbury #1' x 'Mahogany')	241	RAPD, RFLP, isozyme	530	Kubisiak et al. 1997
Apple	*Malus*	fruit	56 trees from 'Rome Beauty' x 'White Angel'	253	RFLP, isozyme	950	Weeden and Hemmat 1993; Hemmat et al. 1994
				156	"	950	"
			2 half-sib populations from 'Wijcik McIntosh' x NY75441-67 and 'Wijcik McIntosh' x NY75441-58	238	RAPD, isozyme, morph., dis. res.	1206	Conner et al. 1997
				110	"	692	"
				183	"	898	"
Asparagus	*Asparagus*	vegetable	40 to 80 BC progeny from 6 different families	43	RFLP, RAPD, morph	418	Restivo et al. 1995; Caporali et al. 1996
			63 F$_1$'s from 'A19' x 'MW25'	183	RFLP	1324	Lewis and Sink 1996; Jiang et al. 1997
				163	"	1281	"

Table 1. Continued.

Common name	Genus	Type of crop	Mapping population[a]	Number of markers	Types of markers[b]	Genetic map length (cM)	Reference
Aspen	*Populus*	wood	5 full-sib families	57	RFLP, allozyme	664	Liu and Furnier 1993
			Three-generation outbred pedigree	343	RFLP, RAPD, STS	1261	Bradshaw et al. 1994
Azuki bean	*Vigna*	Asian pulse	80 F_2's from *V. angularis* x *V. nashimae*	132	RAPD, RFLP, morph.	1250	Kaga et al. 1996
Bananas	*Musa*	fruit	92 F_2's from F1 hybrid 'SFB5' ('SF265' x 'Banksii')	90	RFLP, isozyme, RAPD	606	Faure et al. 1993
Barley	*Hordeum*	cereal and beer	91 DHs from 'Proctor' x 'Nudinka'	197	RFLP, RAMP, morph., dis. res.	1270	Heun et al. 1991; Becker and Heun 1995
			71 DHs from 'Igri' x 'Franka'	92	RFLP, SSR	870	Graner et al. 1991; Liu et al. 1996
			135 F_2's from 'Vada' x '1B-87'	160	RFLP	1408	Graner et al. 1991
			150 DHs from 'Steptoe' x 'Morex'	317	RFLP, RAPD, SAP, SSR, isozyme, morph., dis. res.	1250	Kleinhofs et al. 1993; von Wettstein-Knowles 1993; Liu et al. 1996
			150 DHs from 'Harrington' x 'TR306'	216	RFLP, SSR, isozyme, dis. res.	1327	Kasha et al. 1994; Liu et al. 1996
			58 F_2's from *H. spontaneum* x 'Shin Ebisu 16'	133	RFLP, morph.	1817	Sherman et al. 1995
			113 DHs from 'Proctor' x 'Nudinka'	282	RFLP, AFLP, SSR, protein	1873	Becker et al. 1995
			Integrated map of 4 populations: 'Proctor' x 'Nudinka', 'Igri' x 'Franka', 'Steptoe' x 'Morex', 'Harrington' x 'TR306'	898	RFLP, dis. res.	1060	Qi et al. 1996

Table 1.   Continued.

Common name	Genus	Type of crop	Mapping population[a]	Number of markers	Types of markers[b]	Genetic map length (cM)	Reference
			Consensus of 7 maps ('Igri' x 'Franka', 'Proctor' x 'Nudinka', 'Steptoe' x 'Morex', 'Clipper' x 'Sahara 377', 'Haruno Nijo' x 'Galleon', 'Chebec' x 'Harrington', 'Shannon' x 'Proctor')	587	RFLP, dis. res.	1087	Landridge et al. 1995
Black mustard	Brassica	condiment	83 F$_2$'s from 'B1164' x 'B1157'	124	RFLP, RAPD, isozyme	677	Truco and Quiros 1994
			88 BC's from 'Catania' x ('Catania' x 'CrGC-2')	288	RFLP	855	Lagercrantz and Lydiate 1995
Blueberry	Vaccinium	fruit	38 F$_1$'s from 'US388' x 'Knight'	72	RAPD	950	Rowland and Levi 1994
			F1's from tetraploid hybrid 'US75' x 'Bluetta' (4x)	140	RAPD	1288	Qu and Hancock 1997
Cabbage, broccoli, cauliflower, kale, kohlrabi, brussel sprouts	Brassica	vegetable	96 F$_2$'s from 'Packman' x 'Wisconsin Golden Acres'	258	RFLP	820	Slocum et al. 1990a; Slocum et al. 1990b; Slocum et al. 1993
			90 F$_2$'s from 'CrG-85' x '86-16-5'	310	RFLP, RAPD, STS, SCAR, isozyme, morph.	1606	Landry et al. 1992; Cheung et al. 1997a
			Composite map of 4 populations	108	RFLP, isozyme, morph.	747	Kianian and Quiros 1992
			180 F$_2$'s from 'EW' x 'CR7'	50	RFLP	576	Figdore et al. 1993
			169 DHs from 'GDDH33' x 'A12DHd'	303	RFLP	875	Bohuon et al. 1996

Table 1. Continued.

Common name	Genus	Type of crop	Mapping population[a]	Number of markers	Types of markers[b]	Genetic map length (cM)	Reference
Canola, mustard	*Brassica*	condiment and oil	124 F$_2$'s from 'BI-16' x 'OSU Cr-7'	159	RFLP, RAPD	921	Camargo et al. 1997
			119 DHs from 'J90-4317' x 'J90-2733'	343	RFLP	2073	Cheung et al. 1997
Cassava	*Manihot*	tuber	90 F$_1$'s from 'TMS30572' x 'CM2177-2'	427	RFLP, RAPD, SSR, isozyme	932	Fregene et al. 1997
Celery	*Apium*	vegetable	90 F$_2$'s 'A112' x 'A143'	135	RFLP, RAPD, isozyme, dis. res., morph.	803	Yang and Quiros 1995
Chickpea	*Cicer*	legume	3 interspecific populations of *C. arietinum* x *C. reticulatum*	91	RFLP, RAPD, isozyme, morph.	550	Simon and Muehlbauer 1997
Citrus	*Citrus*	fruit	60 BC$_1$'s of (*C. grandis* x *Poncirus trifoliata*) x *C. grandis*	160	RAPD, RFLP, isozyme	1192	Durham et al. 1992; Cai et al. 1994
			60 F$_1$'s from 'Sacaton' x 'Troyer'	53	RFLP, SSR, isozyme, protein	1700	Jarrell et al. 1992; Kijas et al. 1997
Cocoa	*Theobroma*	industrial	131 BC$_1$'s of ('Catongo' x 'Pound 12') x 'Catongo'	138	RAPD, RFLP, morph.	1068	Crouzillat et al. 1996
Common bean	*Phaseolus*	legume	BC population from 'XR-235-1-1' x ('Calima' x 'XR-235-1-1')	244	RFLP, isozyme, protein, morph.	960	Vallejos et al. 1992
			75 F$_2$'s from 'BAT93' x 'Jalo EEP558'	152	RFLP, isozyme, RAPD, dis. res., morph.	960	Nodari et al. 1993; Gepts 1993
			128 BC$_1$'s from ('Ms8EO2' x 'Corel') x 'Corel'	157	RFLP, RAPD, SCAR, morph.	568	Adam-Blondon et al. 1994
			70 RIL's from 'PC-50' x "XAN-159"	181	RAPD, morph.	426	Jung et al. 1997
Cotton	*Gossypium*	fiber and oil	57 F$_2$'s from 'Palmeri' x 'K101'	705	RFLP	4675	Reinisch et al. 1994
Cowpea	*Vigna*	legume	58 F$_2$'s from 'IT2246' x 'TVN1963'	97	RFLP	684	Menancio-Hautea et al.

Table 1. Continued.

Common name	Genus	Type of crop	Mapping population[a]	Number of markers	Types of markers[b]	Genetic map length (cM)	Reference
			94 RIL's from 'IT84S-2049' x '524B'	181	RAPD, RFLP, AFLP	972	1993a; Fatokun et al. 1993 Menendez et al. 1997
Cucumber	Cucumis	vegetable	101 F₂'s from 'Gy14' x '432860'	58	RFLP, RAPD, isozyme, morph., dis. res.	766	Kennard et al. 1994
				70	"	480	"
Cuphea	Cuphea	oil	140 F₂'s from 'LN43/1' x 'LN68/1'	37	RFLP, allozyme	288	Webb et al. 1992
Eastern Gammagrass	Tripsacum	pasture	F₂ population	61	RFLP	609	Blakey et al. 1994
Eastern white pine	Pinus	timber	72 megagametophytes from 'P-18'	91	RAPD, SSR, SSCP	745	Echt and Nelson 1997
Eucalyptus	Eucalyptus	wood and fiber	62 F₁'s from E. grandis x E. urophylla	240	RAPD	1552	Grattapaglia and Sederoff 1994
				251	"	1101	"
			93 F₁'s from E. urophylla x E. grandis	269	RAPD	1331	Verhaegen and Plomion 1996
				236	"	1415	"
Faba bean	Vicia	legume	20 F₂'s from 'Vf6' x 'Vf173' and 44 F₂'s from 'Vf6' x 'Vf35'	66	RFLP, RAPD, isozyme, allozyme	350	Torres et al. 1993
Foxtail millet	Setaria	cereal	127 F₂'s from 'B100' x 'A10'	160	RFLP	964	Wang et al. 1998; Devos et al. 1998
Grape	Vitis	fruit and wine	60 F₁'s from 'Cayuga White' x 'Aurore'	214	RAPD, RFLP, isozyme	1196	Lodhi et al. 1995
				225	"	1477	"
Lettuce	Lactuca	vegetable	66 F₂'s from 'Calmar' x 'Kordaat'	319	RFLP, RAPD,	1950	Landry et al. 1987;

Table 1.   Continued.

Common name	Genus	Type of crop	Mapping population[a]	Number of markers	Types of markers[b]	Genetic map length (cM)	Reference
					isozyme, dis. res., morph.		Kesseli et al. 1990; Kesseli et al. 1993; Kesseli et al. 1994
Lentil	*Lens*	legume	66 F_2's from *L. culinaris* x *L. orientalis*	34	RFLP, isozyme, morph.	333	Havey and Muehlbauer 1989; Simon et al. 1993
			40 F_2's from *L. ervoides* x *L. culinaris*	64	RFLP, isozyme, morph	560	Weeden et al. 1992
Loblolly pine	*Pinus*	wood	Three-generation of 177 individuals	75	RFLP	489	Groover et al. 1994
				87	"	521	"
			Three-generation pedigree of 101 individuals	75	RFLP, isozyme	632	Devey et al. 1994
Longleaf pine	*Pinus*	wood	72 megagametophytes from 'clone 3-356'	133	RAPD	1635	Nelson et al. 1994
Maize, Corn	*Zea*	cereal, vegetable, oil, forage	46 F_2's from 'H427' x '761'	113	RFLP	1000	Helentjaris et al. 1986
			47 F_2's from 'H427' x '761' and 196 F_2's from 'Tx303' x 'Co159'	284	RFLP	1393	Helentjaris et al. 1988
			48 RIL's from 'T232' x 'CM37', and 41 RIL's from 'Tx303' x 'Co159'	760	RFLP, SSR, isozyme	1460	Burr et al. 1988; Burr et al. 1993; Taramino and Tingey 1996; Senior and Heun 1993
			56 F_2's from 'Tx303' x 'Co159'	1736	RFLP, isozyme	1708	Coe et al. 1990; Coe and Neuffer 1993; Gardiner et al. 1993; Davis et al. 1998

Table 1. Continued.

Common name	Genus	Type of crop	Mapping population[a]	Number of markers	Types of markers[b]	Genetic map length (cM)	Reference
			260 F$_2$'s from 'Chapalote maize' x 'Chalco teosinte'	58	RFLP, morph.	850	Doebley and Stec 1991
			112 F$_2$'s from 'B73' x 'G35'	106	RFLP	2100	Beavis and Grant 1991
			112 F$_2$'s from 'B73' x 'Mo17'	148	"	2200	"
			144 F$_2$'s from 'K05' x 'W65'	78	"	1600	"
			144 F$_2$'s from 'J40' x 'V94'	78	"	1500	"
			88-99 F$_2$'s from 'A619' x 'Mangelsdorf's Multiple tester'	291	RFLP, isozyme, morph.	1868	Shoemaker et al. 1992
			93 F$_2$'s from 'B68 Ht' x 'B73 Ht rhm'	89	"	1413	"
			200 F$_2$'s from 'De811' x 'B73 Ht rhm'	120	"	1859	"
			290 F$_2$'s from 'Reventador maize' x 'Balsas teosinte'	82	RFLP, morph.	1037	Doebley and Stec 1993
			139 F$_2$'s form 'ADENT' x 'B73 rhm1'	87	RFLP	1066	Bubeck et al. 1993
			193 F$_2$'s from 'B73 rhm1' x 'NC250A'	87	"	1416	"
			144 F$_2$'s from 'B73' x 'NC250A'	87	"	1679	"
			300 F$_3$'s from 'B73' x 'B52'	86	RFLP	1716	Shon et al. 1993
			150 F$_2$'s from 'B89' x '33-16'	105	RFLP, RAPD	1539	Pe et al. 1993
			150 F$_2$'s from 'Mo17' x 'H99'	103	RFLP	1419	Veldboom et al. 1994, Veldboom and Lee 1994, Veldboom and Lee 1996
			380 F$_3$'s from 'KW1265' x 'D146', 40 TC's from ('KW1265' x 'D146')F3 x 'KW4115', and 40 TC's from ('KW1265' x 'D146')F3 x 'KW5361'	89	RFLP	1612	Schon et al. 1994
			85 F$_2$'s from 'Io' x 'Lc'	113	RFLP, protein,	1490, 1810	Damerval et al. 1994

Table 1.   Continued.

Common name	Genus	Type of crop	Mapping population[a]	Number of markers	Types of markers[b]	Genetic map length (cM)	Reference
			232 F₃'s from 'B73' x 'A7'	87	RFLP, RAPD	1600	Ajmone et al. 1994; Ajmone et al.1995
			~400 F₂'s from 'C7-351' x 'C7-369'	46	RFLP, isozyme	940	Ragot et al. 1995
			214 F₂:₃'s from 'W6786 su1Se1' x 'II731Asu1se1'	92	RFLP, morph.	1735	Tadmor et al. 1995, Azanza et al. 1996
			234 F₂'s from 'Ac7643' x 'Ac7729/TZSRWS5'	142	RFLP	1760	Ribaut et al. 1996
			171 F₂'s from 'CML131' x 'CML67'	100	RFLP	1536	Bohn et al. 1996, Bohn et al. 1997
			Composite map: 95 F2's from 'Io' x 'F₂', 145 F5:6's from 'Io' x 'F₂', 129 F₅,₆'s from 'F252' x 'F2', and 152 F₅,₆'s from 'Io' x 'F252'	275	RFLP, isozyme, EST	1765	Causse et al. 1996
			192 RIL's from 'B73' x 'Mo17'	129	RFLP, SSR, isozyme	1766	Senior et al. 1996
			185 RIL's from 'Mo17' x 'H99'	110	"	1598	"
			81 F₂'s from 'Polj17' x 'F-2'	84	RFLP, isozyme	1149	Lebreton et al. 1995
Maritime pine	*Pinus*	wood	62 megagametophytes	263	RAPD	1380	Plomion et al. 1995
Mungbean	*Vigna*	legume	58 F₂'s from 'VC3890' x 'TC 1966'	172	RFLP	1570	Menancio-Hautea et al. 1993a; Menancio-Hautea et al. 1993b
Norway spruce	*Picea*	timber	72 megagametophytes	185	RAPD	3584	Binelli and Bucci 1994
			8 megagametophytes from 48 trees	70	RAPD	475	Bucci et al. 1997
Oat	*Avena*	cereal	71 RILs from 'Kanota' x 'Ogle'	561	RFLP, RAPD, STS, isozyme, protein, morph.	1482	O'Donoughue et al. 1995

Table 1.   Continued.

Common name	Genus	Type of crop	Mapping population[a]	Number of markers	Types of markers[b]	Genetic map length (cM)	Reference
			44 F$_2$'s from 'M66/3' x 'Cc7090-CAV4490'	192	RFLP	614	O'Donoughue et al. 1992
Oil palm	*Elaeis*	oil	88 F$_2$'s from 'CI3815' x 'CI1994'	208	RFLP, dis. res.	2416	Rayapati et al. 1994
Onion	*Allium*	vegetable	98 F$_2$'s from AB7/30	97	RFLP	860	Mayes et al. 1997
Pacific yew	*Taxus*	timber and Taxol	60 F$_2$'s from 'BYG15-23' x 'AC43'	66	RFLP, RAPD	1000	Havey and King 1996
			39 seeds from a single tree	41	RAPD	306	Gocmen et al. 1996
Papaya	*Carica*	fruit	253 F$_2$'s from 'Sunrise' x 'UH 356'	62	RAPD	999	Sondur et al. 1996
Pea	*Pisum*	legume	Not stated	126	RFLP	779	Weeden and Wolko 1990; Weeden et al. 1993
			71 RIL's from 'JI281' x 'JI399'	151	RFLP, protein	1700	Ellis et al. 1992
			174 F$_2$'s from 'Erygel' x '661'	69	RFLP, RAPD, SSR, dis. res., morph.	550	Dirlewanger et al. 1994
			102 F$_2$'s from 'Primo' x 'OSU442-15'	207	RFLP, RAPD, STS, dis. Res.	1330	Gilpin et al. 1997
Peach	*Prunus*	fruit	71 F$_2$'s from 4 'New Jersey Pillar' x 'KV 77119' F$_1$'s	65	RFLP, RAPD, morph.	332	Belthoff et al. 1993; Rajapakse et al. 1995
			270 F$_2$'s from '(1161:12 x 2678:47)' x '1:55'	53	RAPD, morph.	350	Dirlewanger and Bodo 1994
			96 F$_2$'s from 'NC174RL' x 'Pillar'	83	RAPD, isozyme, morph.	396	Chaparro et al. 1994
			64 F$_2$'s from '54P455' x 'Padre'	118	RFLP, isozyme, morph., RAPD	800	Foolad et al. 1995; Warburton et al. 1996
			77 F$_1$'s from 'Summergrand' x '1908'	15	RAPD	83	Dirlewanger et al. 1996
				84	RAPD	536	"

Table 1.   Continued.

Common name	Genus	Type of crop	Mapping population[a]	Number of markers	Types of markers[b]	Genetic map length (cM)	Reference
Peanut	*Arachis*	legume	87 F$_2$'s from A. stenosperma x A. cardenasii	120	RFLP	1500	Halward et al. 1993a; Halward et al. 1993b
Pearl millet	*Pennisetum*	cereal	132 F$_2$'s from 'Tift 23DB' x 'IP18292'	181	RFLP	303	Liu et al. 1994
Pepper	*Capsicum*	vegetable	46 F$_2$'s from 'Doux des Landes' x 'CA4'	85	RFLP	603	Tanksley et al. 1988
			46 F$_2$'s from 'CA133' x 'CA4'	192	RFLP	720	Tanksley et al. 1993b; Prince et al. 1993
			Integrated map of 'Perennial' x 'Yolo Wonder', 'Vat' x 'CM334', and 'Yolo Wonder' x 'CM334' DH populations	85	RFLP, RAPD	820	Lefebvre et al. 1995
Potato	*Solanum*	tuber	65 F$_1$'s from '84510' x 'T704'	135	RFLP, isozyme	606	Bonierdale et al. 1988
			67 BC1's from 'H82.309/5' x ('H81.691/1' x 'H82.309/5')	304	RFLP, morph.	1034	Gebhardt et al. 1989; Gebhardt et al. 1991
			100 F$_1$'s from 'H82.337/49' x 'H80.696/4'	211	RFLP	636	Gebhardt et al. 1991
			155 BC's from ('USW2230' x 'PI473331') x 'PI473331'	261	RFLP, isozyme	684	Tanksley et al. 1992; Bonierdale et al. 1993
			56 AC-derived lines from PI458314	58	RFLP	308	Rivard et al. 1996
			31 AC-derived lines from PI230582	57	RFLP	463	"
			45 F$_2$'s from PI458314	50	RFLP	389	"
			23 F$_2$'s from PI230582	47	RFLP	358	"
			58 F$_1$'s from 'TBR' x 'SPG'	20	RFLP	472	Kreike and Stiekema 1997
				62	"	331	"

Table 1. Continued.

Common name	Genus	Type of crop	Mapping population[a]	Number of markers	Types of markers[b]	Genetic map length (cM)	Reference
			Combined maps from ('SH 76-128-1857' x 'Y66-13-628') x 'RH89-039-16'	179	AFLP	884	Rouppe van der Voort et al. 1997
				289	"	857	"
				235	"	735	"
Radiata pine	*Pinus*	wood	Three-generation pedigree of 102 individuals	208	RFLP, RAPD, SSR	1382	Devey et al. 1996
Rapeseed	*Brassica*	oil	90 F_2's from 'Westar' x 'Topas'	145	RFLP	1413	Landry et al. 1991; Cloutier et al. 1997
			105 DHs from 'Major' x 'Stellar'	199	RFLP, isozyme, dis. res.	1506	Ferreira et al. 1994; Thormann et al. 1996
			151 DHs from 'Manshot's Hamburger Raps' x 'Samourai'	207	RFLP, RAPD, morph.	1441	Uzunova et al. 1995
			50 DHs from 'SYN1' x 'N-0-9'	399	RFLP	1656	Parkin et al. 1995
			95 DHs from '90-DHW-1855-4' x '87-DHS-002'	342	RFLP, RAPD	2125	Cheung et al. 1997
Rice	*Oryza*	cereal	50 F_2's from 'IR34583' x 'Bulu Dalam'	135	RFLP	1389	McCouch et al. 1988
			144 F_2's from 'Kasalath' x 'Fl134'	347	RFLP	1836	Saito et al. 1991
			113 BC's from 'BS125' x ('BS125' x 'WLO2')	726	RFLP	1491	Tanksley et al. 1993a; Causse et al. 1994
			186 F_2's from 'Nipponbare' x 'Kasalath'	2275	RFLP, RAPD, STS, SSCP	1522	Kurata et al. 1994; Harushima et al. 1998
			135 DH's from 'IR64' x 'Azucena'	135	RFLP	1811	Huang et al. 1994
Rye	*Secale*	cereal	120F_2's from 'Ds 2' x 'RxL10'	156	RFLP, protein	1073	Devos et al. 1993
			25 to 48 plants from noninbreds of 'Danko' and 'Halo'	60	RFLP, RAPD, isozyme, morph.	350	Phillipp et al. 1994

Table 1.   Continued.

Common name	Genus	Type of crop	Mapping population[a]	Number of markers	Types of markers[b]	Genetic map length (cM)	Reference
			137 F$_2$'s from inbreds from a synthetic population	127	RFLP, RAPD	760	Senft and Wricke 1996
			54 F$_2$'s from 'E' x 'R'	89	RFLP, RAPD	340	Loarce et al. 1996
			110F$_2$'s from 'UC90' x 'E-line'	67	RFLP	1423	Wanous et al. 1997
			275 F$_2$'s from 'P87' x 'P105'	91	RFLP, morph., isozyme	670	Korzun et al. 1998
Slash pine	*Pinus*	wood	68 megagametophytes from 'clone 8-7'	68	RAPD	782	Nelson et al. 1993
Sorghum	*Sorghum*	cereal	55 F$_2$'s from 'Shanqui Red' x 'M91051'	37	RFLP	283	Hulbert et al. 1990
			149 F$_2$'s from 'IS18729' x 'IS24756'	35	RFLP	440	Binelli et al. 1992
			81 F$_2$'s from 'IS2482C' x 'IS18809'	98	RFLP, isozyme	949	Whitkus et al. 1992
			55 F$_2$'s from an unspecified cross	96	RFLP	709	Melake-Berhan et al. 1993
			78 F$_2$'s from 'CK60' x 'PI229828'	214	RFLP, SSR	1530	Pereira et al. 1994; Taramino et al. 1997
			93 F$_2$'s from 'BSC35' x 'BTX631'	71	RFLP	633	Ragab et al. 1994
			56 F$_2$'s from S. bicolor x S. propinquum	276	RFLP	1445	Chittenden et al. 1994
			50 F$_2$'s from 'IS3620C' x 'BTx623'	190	RFLP	1789	Xu et al. 1994
			110 RIL's from 'IS2807' x '379'	155	RFLP	977	Dufuor et al. 1997
			91 RIL's from 'IS2807' x '249'	129	RFLP	878	"
Soybean	*Glycine*	oil and protein	60 F$_2$'s from 'A81-356022' x 'PI468.916'	252	RFLP	2147	Keim et al. 1990; Diers et al. 1992; Shoemaker and Olson 1993
			69 F$_2$'s from 'Minsoy' x 'Noir 1'	132	RFLP, isozyme, morph., protein	1550	Lark et al. 1993
			68 F$_2$'s from 'Bonus' x 'PI81762'	600	RFLP, SSR	2679	Rafalski and Tingey 1993; Morgante et al. 1994

Table 1.   Continued.

Common name	Genus	Type of crop	Mapping population[a]	Number of markers	Types of markers[b]	Genetic map length (cM)	Reference
			60 F$_2$'s from 'Clark' x 'Harosoy'	530	RFLP. RAPD, SSR, morph.	1486	Shoemaker and Specht 1995; Akkaya et al. 1995
			223 RIL's from 'Minsoy' x 'Noir 1'	220	RFLP, SSR, morph.	2000	Mansur et al. 1996
			300 RIL's from 'BSR-101' x 'PI437654'	840	RFLP, RAPD, AFLP	3441	Keim et al. 1997
Strawberry	*Fragaria*	fruit	F$_2$'s from 'Baron Solemacher' x 'WC6'	80	RAPD, STS, isozyme, morph.	445	Davis and Yu 1997
Sugar beet	*Beta*	sugar	96 F$_2$'s from 'S227'	297	RFLP, AFLP, isozyme, morph.	1057	Pillen et al. 1992; Pillen et al. 1993; Schondelmaier et al. 1996
			49 plants from 'KWS'	298	RFLP, RAPD	815	Barzen et al. 1992; Barzen et al. 1995
			222 and 133 F$_2$'s from two populations	573	RFLP, RAPD	621	Hallden et al. 1996; Nilsson et al. 1997
Sugarcane	*Saccharum*	sugar	80 haploids from 'SES208'	216	RFLP	2107	da Silva et al. 1993
			88 F$_2$'s from 'ADP85-0068' x 'SES208'	527	RAPD, RFLP	1500	Al-Janabi et al. 1993; da Silva et al. 1995
			77 progeny from 'R570'	408	RFLP, isozyme	500	Grivet et al. 1996
			84 F$_1$'s from 'La Purple' x 'Molokai 5829'	161	RAPD, morph.	1152	Mudge et al. 1996
Sugi	*Cryptomeria*	timber	73 F$_2$'s from 'Kumotooshi' x 'Okinoyama-sugi'	164	RFLP, RAPD, STS, isozyme, morph.	887	Mukai et al. 1995; Tsumura et al. 1997
Sunflower	*Helianthus*	oil	56 individuals from *H. anomalus* x *H. annuus*	162	RAPD, isozyme	2338	Rieseberg et al. 1993

Table 1.   Continued.

Common name	Genus	Type of crop	Mapping population[a]	Number of markers	Types of markers[b]	Genetic map length (cM)	Reference
			Consensus map of 3 F$_2$ and 2 BC$_1$ populations	237	RFLP	1150	Gentzbittel et al. 1995
			289 F$_2$'s from 'HA89' x 'ZENB8'	234	RFLP	1380	Berry et al. 1995
			93 F$_2$'s from 'RHa271' x 'HA234'	271	RFLP	1164	Jan et al. 1998
Sweet cherry	Prunus	fruit	56 microspore-derived callus cultures	89	RAPD, allozyme	503	Stockinger et al. 1996
Tomato	Lycopersicon		F$_2$'s from L. hirsutum x L. esculentum	104	RFLP	506	Helentjaris et al. 1986
			46 BC and F$_2$'s from 'LA1500' x 'LA716'	112	RFLP, isozyme	760	Bernatzky and Tanksley 1986
			BC's from 'Vendor Tm2a' x ('Vendor Tm2a' x 'LA716')	85	RFLP	1097	de Vicente and Tanksley 1991
			BC's from ('Vendor Tm2a' x 'LA716') x 'LA716'	85	RFLP	1299	de Vicente and Tanksley 1991
			67 F$_2$'s from 'VF36 Tm2a' x 'LA716'	1030	RFLP, isozyme	1276	Tanksley and Matschler 1990; Tanksley et al. 1992; Tanksley 1993
			432 F$_2$'s from 'Vendor Tm2a' x 'LA716'	98	RFLP	1141	de Vicente and Tanksley 1993
			170 BC$_2$'s from ('M82' x LA1589')	135	RFLP	1277	Tanksley et al. 1996
			97 RIL's from 'UC204B' x 'LA483'	132	RFLP	1209	Paran et al. 1995
Turnip, pak-choi, 'Kwan Hoo Choi'	Brassica	vegetable	61 F$_2$'s from 'Yorii Spring' x 'Kwan Hoo Choi'	49	RFLP, isozyme	262	McGrath and Quiros 1991
			95 F$_2$'s from 'Michihili' x	280	RFLP	1850	Song et al. 1991;

Table 1. Continued.

Common name	Genus	Type of crop	Mapping population[a]	Number of markers	Types of markers[b]	Genetic map length (cM)	Reference
			'Spring broccoli' 104 F₂'s from 'R500' x 'Horizon'	360	RFLP	1876	Song et al. 1993
			91 F₂'s from 'R500' x 'Per'	139	RFLP	1785	Chyi et al. 1992 Teutonico and Osborn 1994
			F₂'s from ('Chinese cabbage' x 'Mizu-na'	71	RAPD, isozyme	624	Nozaki et al. 1997
Walnut	*Juglans*	nut	63 BC's from (*J. hindsii x J. regia*) x *J. regia*	48	RFLP	306	Fjellstrom and Parfitt 1994
Watermelon	*Citrullus*	fruit	78 individuals from 'H-7' x 'SA-1'	74	RAPD, RFLP, isozyme, morph.	524	Hashizume et al. 1996
Wheat	*Triticum*	cereal	66 F₂'s from 'Chinese Spring' x 'Spelta' (*T. spelta* var. *duhamelianum*)	204	RFLP	1800	Liu and Tsunewaki 1991; Hart et al. 1993
			60 F₂'s from 'TA1691' x 'TA1704'	152	RFLP, protein, dis. res.	1554	Gill et al. 1991
			39 F₃ from 'AUS18902' x 'AUS18911'	68	RFLP, isozyme	340	Lagudah et al. 1991
			120 F₂'s from 'Chinese Spring' x 'Synthetic' (*T. turgidum* ssp. *dicoccum* x *T. tauschii*)	1015	RFLP, protein, dis. res.	2828	Chao et al. 1989; Hart and Gale 1990; Devos et al. 1992a; Devos et al. 1992b; Devos et al. 1993; Xie et al. 1993; Hart et al. 1993; Chen et al. 1994; Devos et al. 1995; Jia et al. 1995; Gale et al. 1995
			114 RILs from 'Opata 85' x 'W7984' [*T. turgidum* cv. 'Altar 84' x *T. tauschii* CI 18 = WPI219(PR88-89')]	935	RFLP, protein, dis. res.	3551	Nelson et al. 1995a; Nelson et al. 1995b; Nelson et al. 1995c; Van

Table 1.   Continued.

Common name	Genus	Type of crop	Mapping population[a]	Number of markers	Types of markers[b]	Genetic map length (cM)	Reference
							Deynze et al. 1995; Marino et al. 1996
			74 F$_2$'s from 'DV92' x 'G3116'	335	RFLP, isozyme, protein, morph.	1079	Dubcovsky et al. 1996
			275 DHs from 'Chinese Spring' x 'Courtot'	266	RFLP, protein	1772	Cadalen et al. 1997

[a] Abbreviations: AC = anther culture, BC = backcross generation, DH = doubled haploid, F$_x$ = selfed generation, TC = testcross, RIL = recombinat inbred line.

[b] Abbreviations: AFLP = amplified fragment length polymorphism, dis. res. = disease resistance, morph. = morphological trait, RAMP = random amplified microsatellite polymorphism, RAPD = random amplified polymorphic DNA, RFLP = restriction fragment length polymorphism, SAP = specific amplicon polymorphism, SCAR = sequence characterized amplified region, SSCP = single strand conformation polymorphism, SSR = simple sequence repeat, STS = sequence tagged site.

Tabel 2.   List of plant genome, germplasm, and reference resources available through the World-Wide-Web.

Plant Genome, Germplasm and Reference Servers

Arabidopsis Biological Resource Center (ABRC), MSU
  (http://www.biosci.ohio-state.edu/~plantbio/facilities/abrc/ABRCHOME.HTM)
AtDB - Arabidopsis thaliana Database, Stanford
  (http://www.ba.cnr.it/Beanref)
BeanGenes - Phaseolus and Vigna Database, NDSU (http://beangenes.cws.ndsu.nodak.edu/)
BeanRef - Phaseolus and Vigna References, Germany
  (http://scaffold.biologie.uni-kl.de/Beanref/)
CottonDB Data Collection Site, TAMU (http://algodon.tamu.edu/htdocs-cotton/
  cottondb.html)
Dendrome - A Genome Database for Forest Trees, USDA (http://dendrome.ucdavis.edu/)
Germplasm Resources Information Network (GRIN), USDA (http://www.ars-grin.gov/)
GrainGenes - A Database for Small Grains and Sugarcane, USDA (http://wheat.pw.usda.gov/
  graingenes.html)
Lehle Seeds - Everything Arabidopsis (http://www.arabidopsis.com/)
Maize Genetic Database, Missouri (http://www.agron.missouri.edu/top.html)
Maize Genetic Cooperation - Stock Center, UIUC (http://www.uiuc.edu/ph/www/maize)
National Center for Genome Resources (http://www.ncgr.org/)
Nottingham Arabidopsis Stock Centre (NASC), UK (http://nasc.nott.ac.uk/)
Plant Genome Data and Information Center, NAL (http://www.nal.usda.gov/pgdic/)
Rice Genome Research Program (RGP), Japan (http://rgp.dna.affrc.go.jp/)
Soybase - Shoemaker Laboratory, ISU (http://macgrant.agron.iastate.edu/)
Angiosperm DNA C-Values Database, Kew Botanic Gardens (http://www.rbgkew.org.uk/
  cval/database1.html)

Plant Genome Databases hosted at the USDA-ARS Center for Bioinformatics and comparative
genomics, Cornell University
(http://ars-genome.cornell.edu)

AlfaGenes - alfalfa (*Medicago sativa*)
Arabidopsis Genome Resourse - Arabidopsis
BarleyDB - *Hordeum*
BeanGenes - *Phaseolus* and *Vigna*
BrassicaDB - *Brassica*
CassavaDB - *Manihot*
ChlamyDB - *Chlamydomonas reinhardtii*
CoolGenes - cool season food legumes
CottonDB - *Gossypium hirsutum*
FoggDB - Forage grasses
GrainGenes - wheat, barley, rye and relatives
MaizeDB - maize
MilletGenes - pearl millet
PathoGenes - fungal pathogens of small-grain cereals
RiceGenes - rice
RoseDB - Rosaceae
SolGenes - Solanaceae
SorghumDB - Sorghum bicolor
SoyBase - Soybeans
TreeGenes - Forest trees

---

3.   Acknowledgements

This work was partly supported by a postdoctoral fellowship from the "Comision Interministerial de Ciencia y Tecnologia" of Spain to M. Isabel Vales. We are grateful to Susan Wheeler and Peggy Mullett for their invaluable secretarial help.

4.   References

Adam-Blondon, A.-F., Sevignac, M., Dron, M. and Bannerot, H. (1994) A genetic map of common bean to localize specific resistance genes against anthracnose. Genome 37: 915-924.

Ajmone-Marsan, P., Monfredini, G., Ludwig, W.F., Melchinger, A.E., Franceschini, P., Pagnotto, G. and Motto., M. (1994) Identification of genomic regions affecting plant height and their relationship with grain yield in an elite maize cross. Maydica 39: 133-139.

Ajmone-Marsan, P., Monfredini, G., Ludwig, W.F., Melchinger, A.E., Franceschini, P., Pagnotto, G. and Motto, M. (1995) In an elite cross of maize a major quantitative trait locus controls one-fourth of the genetic variation for grain yield. Theor. Appl. Genet. 90: 415-424.

Akkaya, M.S., Shoemaker, R.C., Specht, J.E., Bhagwat, A.A. and Cregan, P.B. (1995) Integration of simple sequence repeat DNA markers into a soybean linkage map. Crop Sci. 35: 1439-1445.

Al-Janabi, S.M., Honeycutt, R.J., McClelland, M. and Sobral, B.W.S. (1993) A genetic linkage map of *Saccharum spontaneum* L. 'SES 208'. Genetics 134: 1249-1260.

Arus, P., Messeguer, R., Viruel, M., Tobutt, K., Dirlewanger, E., Santi, F., Quarta, R. and Ritter, E. (1994) The european *Prunus* mapping project. Euphytica 77: 97-100.

Azanza, F., Tadmor, Y., Klein, B.P., Rocheford, T.R. and Juvik, J.A. (1996) Quantitative trait loci influencing chemical and sensory characteristics of eating quality in sweet corn. Genome 39: 40-50.

Barzen, E., Mechelke, W., Ritter, E., Seitzer, J.F. and Salamini, F. (1992) RFLP markers for sugar beet breeding: chromosomal linkage maps and location of major genes for rhizomania resistance, monogermy and hypocotyl colour. Plant J. 2(4): 601-611.

Barzen, E., Mechelke, W., Ritter, E., Schulte-Kappert, E. and Salamini, F. (1995) An extended map of the sugar beet genome containing RFLP and RAPD loci. Theor. Appl. Genet. 90: 189-193.

Beavis, W.D. and Grant, D. (1991) A linkage map based on information from four F_2 populations of maize (*Zea mays* L.). Theor. Appl. Genet. 82: 636-644.

Becker, J. and Heun, M. (1995) Mapping of digested and undigested random amplified microsatellite polymorphisms in barley. Genome 38: 991-998.

Becker, J., Vos, P., Kuiper, M., Salamini, F. and Heun, M. (1995) Combined mapping of AFLP and RFLP markers in barley. Mol. Gen. Genet. 249: 65-73.

Belthoff, L.E., Ballard, R., Abbott, A., Baird, W.V., Morgens, P., Callahan, A., Scorza, R. and Monet, R. (1993) Development of a saturated linkage map of *Prunus persica* using molecular based marker systems. Acta Hort. 336: 51-56.

Bernatzky, R. and Tanksley, S.D. (1986) Toward a saturated linkage map in tomato based on isozymes and random cDNA sequences. Genetics 112: 887-898.

Berry, S.T., Leon, A.J., Hanfrey, C.C., Challis, P., Burkholz, A., Barnes, S.R., Rufener, G.K., Lee, M. and Caligari, P.D.S. (1995) Molecular marker analysis of *Helianthus annuus* L. 2. Construction of an RFLP linkage map for cultivated sunflower. Theor. Appl. Genet. 91: 195-199.

Binelli, G., Gianfranceschi, L., Pe, M., Taramino, G., Busso, C., Stenhouse, J. and Ottaviano, E. (1992) Similarity of maize and sorghum genomes as revealed by maize RFLP probes. Theor. Appl. Genet. 84: 10-16.

Binelli, G. and Bucci, G. (1994) A genetic linkage map of *Picea abies* Karst., based on RAPD markers, as a tool in population genetics. Theor. Appl. Genet. 88: 283-288.

Blakey, C.A., Coe, J., E.H. and Dewald, C.L. (1994) Current status of the *Tripsacum dacyloides* (Eastern gamagrass) RFLP molecular genetic map. Maize Genet. Newsl. 68: 35-38.

Bohn, M., Khairallah, M.M., Gonzalez-de-Leon, D., Hoisington, D.A., Utz, H.R., Deutsch, J.A., Jewell, D.C., Mihm, J.A. and Melchinger, A.E. (1996) QTL mapping in tropical maize: I. Genomic regions affecting leaf feeding resistance to sugarcane borer and other traits. Crop Sci. 36: 1352-1361.

Bohn, M., Khairallah, M.M., Gonzalez-de-Leon, D., Hoisington, D.A., Utz, H.R., Deutsch, J.A., Jewell, D.C., Mihm, J.A. and Melchinger, A.E. (1997) QTL mapping in tropical maize: II. Comparison of genomic regions for resistance to *Diatraea* spp. Crop Sci. 37: 1892-1902.

Bohuon, E.J.R., Keith, D.J., Parkin, I.A.P., Sharpe, A.G. and Lydiate, D.J. (1996) Alignment of the conserved C genomes of *Brassica oleracea* and *Brassica napus*. Theor. Appl. Genet. 93: 833-839.

Bonierbale, M.W., Plaisted, R.L. and Tanksley, S.D. (1988) RFLP maps based on a common set of clones reveal modes of chromosomal evolution in potato and tomato. Genetics 120: 1095-1103.

Bonierbale, M., Plaisted, R. and Tanksley, S. (1993) Molecular map of the potato (*Solanum tuberosum*) (2N = 48). In: O'Brien, S.J. (ed.), Genetic Maps: Locus Maps of Complex Genomes, pp. 6.88-6.90. Cold Spring Harbor Laboratory Press, Cold Spring Harbor, NY.

Bradshaw, J., H.D., Villar, M., Watson, B.D., Otto, K.G., Stewart, S. and Stettler, R.F. (1994) Molecular genetics of growth and development in *Populus*. III. A genetic linkage map of a hybrid poplar composed of RFLP, STS, and RAPD markers. Theor. Appl. Genet. 89: 167-178.

Brummer, E.C., Bouton, J.H. and Kochert, G. (1993a) RFLP linkage map of diploid alfalfa (*Medicago sativa* L.) (2n = 2x = 16). In: O'Brien, S.J. (ed.), Genetic Maps: Locus Maps of Complex Genomes, pp. 6.82-6.83. Cold Spring Harbor Laboratory Press, Cold Spring Harbor, NY.

Brummer, E.C., Bouton, J.H. and Kochert, G. (1993b) Development of an RFLP map in diploid alfalfa. Theor. Appl. Genet. 86: 329-332.

Bubeck, D.M., Goodman, M.M., Beavis, W.D. and Grant, D. (1993) Quantitative trait loci controlling resistance to gray leaf spot in maize. Crop Sci. 33: 838-847.

Bucci, G., Kubisiak, T.L., Nance, W.L. and Menozzi, P. (1997) A population 'consensus', partial linkage map of *Picea abies* Karst. based on RAPD markers. Theor. Appl. Genet. 95: 643-654.

Burr, B., Burr, F.A., Thompson, K.H., Albertson, M.C. and Stuber, C.W. (1988) Gene mapping with recombinant inbreds in maize. Genetics 118: 519-526.

Burr, B., Burr, F.A. and Matz, E.C. (1993) Maize molecular map (*Zea mays* L.) (2N = 20).

In: O'Brien, S.J. (ed.), Genetic Maps: Locus Maps of Complex Genomes, pp. 6.190-6.203. Cold Spring Harbor Laboratory Press, Cold Spring Harbor, NY.

Cadalen, T., Boeuf, C., Bernard, S. and Bernard, M. (1997) An intervarietal molecular marker map in *Triticum aestivum* L. Em. Thell. and comparison with a map from a wide cross. Theor. Appl. Genet. 94: 367-377.

Cai, Q., Guy, C.L. and Moore, G.A. (1994) Extension of the linkage map in *Citrus* using random amplified polymorphic DNA (RAPD) markers and RFLP mapping of cold-acclimation-responsive loci. Theor. Appl. Genet. 89: 606-614.

Camargo, L.E.A., Savides, L., Jung, G., Nienhuis, J. and Osborn, T.C. (1997) Location of the self-incompatibility locus in an RFLP and RAPD map of *Brassica oleracea.* J. Hered. 88: 57-60.

Caporali, E., Carboni, A., Spada, A. and Marziani Longo, G.P. (1996) Construction of a linkage map in *Asparagus officinalis* through RFLP and RAPD analysis. Acta Hort. 415: 435-440.

Causse, M.A., Fulton, T.M., Yong, G.C., Sang, N.A., Chunwongse, J., Wu, K., Xiao, J., Yu, Z., Ronald, P.C., Harrington, S.E., Second, G., McCouch, S.R. and Tanksley, S.D. (1994) Saturated molecular map of the rice genome based on an interspecific backcross population. Genetics 138: 12251-1274.

Causse, M., Santoni, S., Damerval, C., Maurice, A., Charcosset, A., Deatrick, J. and Vienne, D.de. (1996) A composite map of expressed sequences in maize. Genome 39: 418-432.

Chao, S., Sharp, P.J., Worland, A.J., Warham, E.J., Koebner, R.M.D. and Gale, M.D. (1989) RFLP-based genetic maps of wheat homoeologous group 7 chromosomes. Theor. Appl. Genet. 78: 495-504.

Chaparro, J.X., Werner, D.J., O'Malley, D. and Sederoff, R.R. (1994) Targeted mapping and linkage analysis of morphological isozyme, and RAPD markers in peach. Theor. Appl. Genet. 87: 805-815.

Chen, Z., Devey, M., Tuleen, N.A. and Hart, G.E. (1994) Use of recombinant substitution lines in the construction of RFLP-based genetic maps of chromosomes 6A and 6B of tetraploid wheat (*Triticum turgidum* L.). Theor. Appl. Genet. 89: 703-712.

Cheung, W.Y., Champagne, G., Hubert, N. and Landry, B.S. (1997a) Comparison of the genetic maps of *Brassica napus* and *Brassica oleracea.* Theor. Appl. Genet. 94: 569-582.

Cheung, W.Y., Friesen, L., Rakow, G.F.W., Seguin-Swartz, G. and Landry, B.S. (1997b) A RFLP-based linkage map of mustard [*Brassica juncea* (L.) Czern. and Coss.]. Theor. Appl. Genet. 94: 841-851.

Chittenden, L.M., Schertz, K.F., Lin, Y.-R., Wing, R.A. and Paterson, A.H. (1994) A detailed RFLP map of *Sorghum bicolor* x *S. propinquum*, suitable for high-density mapping, suggests ancestral duplication of Sorghum chromosomes or chromosomal segments. Theor. Appl. Genet. 87: 925-933.

Chyi, Y.-S., Hoenecke, M.E. and Sernyk, J.L. (1992) A genetic linkage map of restriction fragment length polymorphism loci for *Brassica rapa* (syn. *campestris*). Genome 35: 746-757.

Cloutier, S., Cappadocia, M. and Landry, B.S. (1997) Analysis of RFLP mapping inaccuracy in *Brassica napus* L. Theor. Appl. Genet. 95: 83-91.

Coe, J., E.H., Hoisington, D.A. and Neuffer, M.G. (1990) Linkage map of corn (maize) (*Zea mays* L.) (2N = 20). In: O'Brien, S.J. (ed.), Genetic Maps: Locus Maps of Complex Genomes, pp. 6.39-6.67. Cold Spring Harbor Laboratory Press, Cold Spring Harbor. NY.

Coe, E.H. and Neuffer, M.G. (1993) Gene loci and linkage map of corn (maize) (*Zea mays* L.)

(2N = 20). In: O'Brien, S.J. (ed.), Genetic Maps: Locus Maps of Complex Genomes, pp. 6.157-6.189. Cold Spring Harbor Laboratory, Cold Spring Harbor, NY.

Conner, P.J., Brown, S.K. and Weeden, N.F. (1997) Randomly amplified polymorphic DNA-based genetic linkage maps of three apple cultivars. J. Amer. Soc. Hort. Sci. 122: 350-359.

Crouzillat, D., Lerceteau, E., Petiard, V., Morera, J., Rodriguez, H., Walker, D., Phillips, W., Ronning, C., Schnell, R., Osei, J. and Fritz, P. (1996) *Theobroma cacao* L.: a genetic linkage map and quantitative trait loci analysis. Theor. Appl. Genet. 93: 205-214.

da Silva, J.A.G., Sorrells, M.E., Burnquist, W.L. and Tanksley, S.D. (1993) RFLP linkage map and genome analysis of *Saccharum spontaneum*. Genome 36: 782-791.

da Silva, J., Honeycutt, R.J., Burnquist, W., Al-Janabi, S.M., Sorrells, M.E., Tanksley, S.D. and Sobral, B.W.S. (1995) *Saccharum spontaneum* L. 'SES 208' genetic linkage map combining RFLP and PCR based markers. Molec. Breed. 1: 165-179.

Damerval, C., Maurice, A., Josse, J.M. and Vienne, D.d. (1994) Quantitative trait loci underlying gene product variation: a novel perspective for analyzing regulation of genome expression. Genetics 137: 289-301.

Davis, T.M. and Yu, H. (1997) A linkage map of the diploid strawberry, *Fragaria vesca*. J. Hered. 88: 215-221.

Davis, G., McMullen, M., Polacco, M., Grant, D., Musket, T., Baysdorfer, C., Staebell, M., Xu, G., Koster, L., Houchins, K., Melia-Hankcock, S. and Coe, E.H. (1998) UMC 1998 molecular map of maize. Maize Genet. Newsl. 72: 118-128.

de Vicente, M.C. and Tanksley, S.D. (1991) Genome-wide reduction in recombination of backcross progeny derived from male versus female gametes in an interspecific cross of tomato. Theor. Appl. Genet. 83: 173-178.

de Vicente, M.C. and Tanksley, S.D. (1993) QTL analysis of transgressive segregation in an interspecific tomato cross. Genetics 134: 585-596.

Devey, M.E., Fiddler, T.A., Liu, B.H., Knapp, S.J. and Neale, D.B. (1994) An RFLP linkage map for loblolly pine based on a three-generation outbred pedigree. Theor. Appl. Genet. 88: 273-278.

Devey, M.E., Bell, J.C., Smith, D.N., Neale, D.B. and Moran, G.F. (1996) A genetic linkage map for *Pinus radiata* based on RFLP, RAPD, and microsatellite markers. Theor. Appl. Genet. 92: 673-679.

Devos, K.M., Atkinson, M.D., Chinoy, C.N., Liu, C.J. and Gale, M.D. (1992a) RFLP-based genetic map of the homoeologous group 3 chromosomes of wheat and rye. Theor. Appl. Genet. 83: 931-939.

Devos, K.M., Millan, T. and Gale, M.D. (1992b) Comparative RFLP maps of the homoeologous group-2 chromosomes of wheat, rye and barley. Theor. Appl. Genet. 85: 784-792.

Devos, K.M. and Gale, M.D. (1993) Extended genetic maps of the homoeologous group 3 chromosomes of wheat, rye and barley. Theor. Appl. Genet. 85: 649-652.

Devos, K.M., Atkinson, M.D., Chinoy, C.N., Francis, H.A., Harcourt, R.L. and Koebner, R.M.D. (1993) Chromosomal rearrangements in the rye genome relative to that of wheat. Theor. Appl. Genet. 85: 673-6880.

Devos, K.M., Dubcovsky, J., Dvorak, J., Chinoy, C.N. and Gale, M.D. (1995) Structural evolution of wheat chromosomes 4A, 5A, and 7B and its impact on recombination. Theor. Appl. Genet. 91: 282-288.

Devos, K.M., Wang, Z.M., Beales, J., Sasaki, T. and Gale, M.D. (1998) Comparative genetic maps of foxtail millet (*Setaria italica*) and rice (*Oryza sativa*). Theor. Appl. Genet. 96: 63-68.

Diers, B.W., Keim, P., Fehr, W.R. and Shoemaker, R.C. (1992) RFLP analysis of soybean seed protein and oil content. Theor. Appl. Genet. 83: 608-612.

Dirlewanger, E., Isaac, P.G., Ranade, S., Belajouza, M., Cousin, R. and de Vienne, D. (1994) Restriction fragment length polymorphism analysis of loci associated with disease resistance genes and developmental traits in *Pisum sativum* L. Theor. Appl. Genet. 88: 17-27.

Dirlewanger, E. and Bodo, C. (1994) Molecular genetic mapping of peach. Euphytica 77: 101-103.

Dirlewanger, E., Pascal, T., Zuger, C. and Kervella, J. (1996) Analysis of molecular markers associated with powdery mildew resistance genes in peach (*Prunus persica* (L.) Batsch) x *Prunus davidiana* hybrids. Theor. Appl. Genet. 93: 909-919.

Doebley, J. and Stec, A. (1991) Genetic analysis of the morphological differences between maize and teosinte. Genetics 129: 285-295.

Doebley, J. and Stec, A. (1993) Inheritance of the morphological differences between maize and teosinte: Comparison of results for two F_2 populations. Genetics 134: 559-570.

Dubcovsky, J., Luo, M.-C., Zhong, G.-Y., Bransteitter, R., Desai, A., Killian, A., Kleinhofs, A. and Dvorak, J. (1996) Genetic map of diploid wheat, *Triticum monococcum* L., and its comparison with maps of *Hordeum vulgare* L. Genetics 143: 983-999.

Dufour, P., Deu, M., Grivet, L., D'Hont, A., Paulet, F., Bouet, A., Lanaud, C., Glaszmann, J.C. and Hamon, P. (1997) Construction of a composite sorghum genome map and comparison with sugarcane, a related complex polyploid. Theor. Appl. Genet. 94: 409-418.

Durham, R.E., Liou, P.C., Gmitter, F.G., Jr. and Moore, G.A. (1992) Linkage of restriction fragment length polymorphisms and isozymes in *Citrus*. Theor. Appl. Genet. 84: 39-48.

Echt, C.S., Kidwell, K.K., Knapp, S.J., Osborn, T.C. and Mc Coy, T.J. (1993) Linkage mapping in diploid alfalfa (*Medicago sativa*). Genome 37: 61-71.

Echt, C.S. and Nelson, C.D. (1997) Linkage mapping and genome length in eastern white pine (*Pinus strobus* L.). Theor. Appl. Genet. 94: 1031-1037.

Ellis, T.H.N., Turner, L., Hellens, R.P., Lee, D., Harker, C.L., Enard, C., Domoney, C. and Davies, D.R. (1992) Linkage maps in pea. Genetics 130: 649-663.

Fatokun, C.A., Danesh, D., Menancio-Hautea, D.I. and Young, N.D. (1993) A linkage map for cowpea (*Vigna unguiculata* L. Walp.) based on DNA markers (2N = 22). In: O'Brien, S.J. (ed.), Genetic Maps: Locus Maps of Complex Genomes, pp. 6.257-6.258. Cold Spring Harbor Laboratory Press, Cold Spring Harbor, NY.

Faure, S., Noyer, J.L., Horry, J.P., Bakry, F., Lanaud, C. and Gonzalez de Leon, D. (1993) A molecular marker-based linkage map of diploid bananas (*Musa acuminata*). Theor. Appl. Genet. 87: 517-526.

Ferreira, M.E., Williams, P.H. and Osborn, T.C. (1994) RFLP mapping of *Brassica napus* using doubled haploid lines. Theor. Appl. Genet. 89: 615-621.

Figdore, S.S., Ferreira, M.E., Slocum, M.K. and Williams, P.H. (1993) Association of RFLP markers with trait loci affecting clubroot resistance and morphological characters in *Brassica oleracea* L. Euphytica 69: 33-44.

Fjellstrom, R.G. and Parfitt, D.E. (1994) RFLP inheritance and linkage in walnut. Theor. Appl. Genet. 89: 665-670.

Foolad, M.R., Arulsekar, S., Becerra, V. and Bliss, F.A. (1995) A genetic map of *Prunus* based on an interspecific cross between peach and almond. Theor. Appl. Genet. 91: 262-269.

Fregene, M., Angel, F., Gomez, R., Rodriguez, F., Chavarriaga, P., Roca, W., Tohme, J. and

Bonierbale, M. (1997) A molecular genetic map of cassava (*Manihot esculenta* Crantz). Theor. Appl. Genet. 95: 431-441.

Gale, M.D., Atkinson, M.D., Chinoy, C.N., Harcourt, R.L., Jia, J., Li, Q.Y. and Devos, K.M. (1995) Genetic maps of hexaploid wheat. In: Li, Z.S. and Xin, Z.Y. (ed.), 8th International Wheat Genetics Symposium, pp. 29-40. Agricultural Scientech Press, Beijing, China.

Gardiner, J.M., Coe, E.H., Melia-Hancock, S., Hoisington, D.A. and Chao, S. (1993) Development of a core RFLP map in maize using and immortalized F_2 population. Genetics 134: 917-930.

Gebhardt, C., Ritter, E., Debener, T., Schachtschabel, U., Walkemeier, B., Uhrig, H. and Salamini, F. (1989) RFLP analysis and linkage mapping in *Solanum tuberosum*. Theor. Appl. Genet. 78: 65-75.

Gebhardt, C., Ritter, E., Barone, A., Debener, T., Walkemeier, B., Schachtschabel, U., Kaugmann, H., Thompson, R.D., Bonierbale, M.W., Ganal, M.W., Tanksley, S.D. and Salamini, F. (1991) RFLP maps of potato and their alignment with the homoeologous tomato genome. Theor. Appl. Genet. 83: 49-57.

Gentzbittel, L., Vear, F., Zhang, Y.X., Berville, A. and Nicolas, P. (1995) Development of a consensus linkage RFLP map of cultivated sunflower (*Helianthus annuus* L.). Theor. Appl. Genet. 90: 1079-1086.

Gepts, P. (1993) Linkage map of common bean (*Phaseolus vulgaris* L.) (2N = 22). In: O'Brien, S.J. (ed.), Genetic Maps: Locus Maps of Complex Genomes, pp. 6.101-6.109. Cold Spring Harbor Laboratory Press, Cold Spring Harbor, NY.

Gill, K.S., Lubbers, E.L., Gill, B.S., Raupp, W.J. and Cox, T.S. (1991) A genetic linkage map of *Triticum tauschii* (DD) and its relationship to the D genome of bread wheat (AABBDD). Genome 34: 362-374.

Gilpin, B.J., McCallum, J.A., Frew, T.J. and Timmerman-Vaughan, G.M. (1997) A linkage map of the pea (*Pisum sativum* L.) genome containing cloned sequences of known function and expressed sequence tags (ESTs). Theor. Appl. Genet. 95: 1289-1299.

Gocmen, B., Jermstad, K.D., Neale, D.B. and Kaya, Z. (1996) Development of random amplified polymorphic DNA markers for genetic mapping in Pacific yew (*Taxus brevifolia*). Can. J. For. Res. 26: 497-503.

Graner, A., Jahoor, A., Schondelmaier, J., Siedler, H., Pillen, K., Fishbeck, G., Wenzel, G. and Herrmann, R.G. (1991) Construction of an RFLP map of barley. Theor. Appl. Genet. 83: 250-256.

Grattapaglia, D. and Sederoff, R. (1994) Genetic linkage maps of *Eucalyptus grandis* and *Eucalyptus urophylla* using a pseudo-testcross: mapping strategy and RAPD markers. Genetics 137: 1121-1137.

Grivet, L., D'Hont, A., Roques, D., Feldmann, P., Lanaud, C. and Glaszmann, J.C. (1996) RFLP mapping in cultivated sugarcane (*Saccharum* spp.): genome organization in a highly polyploid and aneuploid interspecific hybrid. Genetics 142: 987-1000.

Groover, A., Devey, M., Fiddler, T., Lee, J., Megraw, R., Mitchel-Olds, T., Sherman, B., Vujcic, S., Williams, C. and Neale, D. (1994) Identification of quantitative trait loci influencing wood specific gravity in an outbred pedigree of loblolly pine. Genetics 138: 1293-1300.

Hallden, C., Hjerdin, A., Rading, I.M., Sall, T., Fridlundh, B., Johannisdottir, G., Tuvesson, S., Akesson, C. and Nilsson, N.O. (1996) A high density RFLP linkage map of sugar beet. Genome 39: 634-645.

Halward, T., Stalker, H.T. and Kochert, G. (1993a) RFLP Linkage map of diploid peanut (*Arachis* sp.) (2n = 2x = 20). In: O'Brien, S.J. (ed.), Genetic Maps: Locus Maps of Complex Genomes, pp. 6.86-6.87. Cold Spring Harbor Laboratory Press, Cold Spring Harbor, NY.

Halward, T., Stalker, H.T. and Kochert, G. (1993b) Development of an RFLP linkage map in diploid peanut species. Theor. Appl. Genet. 87: 379-384.

Hart, G.E. and Gale, M.D. (1990) Biochemical/molecular loci of hexaploid wheat (*Triticum aestivum*, 2n = 42, genomes AABBDD). In: O'Brien, S.J. (ed.), Genetic Maps: Locus Maps of Complex Genomes, pp. 6.28-6.38. Cold Spring Harbor Laboratory Press, Cold Spring Harbor, NY.

Hart, G.E., Gale, M.D. and McIntosh, R.A. (1993) Linkage maps of *Triticum aestivum* (Hexaploid wheat, 2N = 42, Genomes A,B, and D) and T. tauschii (2N = 14, Genome D). In: O'Brien, S.J. (ed.), Genetic Maps: Locus Maps of Complex Genomes, pp. 6.204-6.219. Cold Spring Harbor Laboratory Press, Cold Spring Harbor, NY.

Harushima, Y., Yano, M., Shomura, A., Sato, M., Shimano, T., Kuboki, Y., Yamamoto, T., Lin, S.Y., Antonio, B.A., Parco, A., Kajiya, H., Huang, N., Yamamoto, K., Nagamura, Y., Kurata, N., Khush, G.S. and Sasaki, T. (1998) A high-density rice genetic linkage map with 2275 markers using a single F-2 population. Genetics 148: 479-494.

Hashizume, T., Shimamoto, I., Harushima, Y., Yui, M., Sato, T., Imai, T. and Hirai, M. (1996) Construction of a linkage map for watermelon (*Citrullus lanatus* (Thunb.) Matsum and Nakai) using random amplified polymorphic DNA (RAPD). Euphytica 90: 265-273.

Havey, M.J. and Muehlbauer, F.J. (1989) Linkages between restriction fragment length, isozyme, and morphological markers in lentil. Theor. Appl. Genet. 77: 395-401.

Havey, M.J. and King, J.J. (1996) Molecular markers and mapping in bulb onion, a forgetten monocot. Hort Sci 31: 1116-1118.

Helentjaris, T., Slocum, M., Wright, S., Schaefer, A. and Nienhuis, J. (1986) Construction of genetic linkage maps in maize and tomato using restriction fragment length polymorphisms. Theor. Appl. Genet. 72: 761-769.

Helentjaris, T., Weber, D. and Wright., S. (1988) Identification of the genomic locations of duplicate nucleotide sequences in maize by analysis of restriction fragment length polymorphisms. Genetics 118: 355-363.

Hemmat, M., Weeden, N.F., Manganaris, A.G. and Lawson, D.M. (1994) Molecular marker linkage map for apple. J. Hered. 85: 4-11.

Heslop-Harrison, S.J., Ed. (1996) Unifying Plant Genomes. Symposia of the Society for Experimental Biology. The Company of Biologists Limited, Cambridge, U.K.

Heun, M., Kennedy, A.E., Anderson, J.A., Lapitan, N.L.V., Sorrells, M.E. and Tanksley, S.D. (1991) Construction of a restriction fragment length polymorphism map for barley (*Hordeum vulgare*). Genome 34: 437-447.

Huang, N., McCouch, S., Mew, T., Parco, A. and Guiderdoni, E. (1994) Development of an RFLP map from a doubled haploid population in rice. Rice Genet. Newsl. 11: 134-137.

Hulbert, S.H., Richter, T.E., Axtell, J.D. and Bennetzen, J.L. (1990) Genetic mapping and characterization of sorghum and related crops by means of maize DNA probes. Proc. Natl. Acad. Sci., USA 87: 4251-4255.

Jan, C.C., Vick, B.A., Miller, J.F., Kahler, A.L. and Butler, I., E.T. (1998) Construction of an RFLP linkage map for cultivated sunflower. Theor. Appl. Genet. 96: 15-22.

Jarrell, D.C., Roose, M.L., Traugh, S.N. and Kupper, R.S. (1992) A genetic map of citrus based on the segregation of isozymes and RFLPs in an intergeneric cross. Theor. Appl. Genet. 84: 49-56.

Jauhar, P.P., Ed. (1996) Methods of Genome Analysis in Plants. CRC Press, Boca Raton, FL.

Jia, J., Devos, K.M., Chao, S., Miller, T.E., Reader, S.M. and Gale, M.D. (1995) RFLP-based maps of the homoeologous group-6 chromosomes of wheat and their application in the tagging of Pm12, a powdery mildew resistance gene transferred from *Aegilops speltoides* to wheat. Theor. Appl. Genet. 92: 559-565.

Jiang, C., Lewis, M.E. and Sink, K.C. (1997) Combined RAPD and RFLP molecular linkage map of asparagus. Genome 40: 69-76.

Jung, G., Skroch, P.W., Coyne, D.P., Nienhuis, J., Arnaud-Santana, E., Aiyarathne, H.M., Kaeppler, S.M. and Bassett, M.J. (1997) Molecular-marker-based genetic analysis of tepary bean-derived common bacterial blight resistance in different developmental stages of common bean. J. Amer. Soc. Hort. Sci. 122: 329-337.

Kaga, A., Ohnishi, M., Ishii, T. and Kamijima, O. (1996) A genetic linkage map of azuki bean constructed with molecular and morphological markers using an interspecific population (*Vigna angularis* x *V. nakashimae*). Theor. Appl. Genet. 93: 658-663.

Kasha, K.J., Kleinhofs, A. and N.A.B.G.M.P. (1994) Mapping of the barley cross Harrington x TR306. Barley Genet. Newsl. 23: 65-69.

Keim, P., Diers, B.w., Olson, T.C. and Shoemaker, R.C. (1990) RFLP mapping in soybean: association between marker loci and variation in quantitative traits. Genetics 126: 735-742.

Keim, P., Schupp, J.M., Travis, S.E., Clayton, K., Zhu, T., Shi, L., Ferreira, A. and Webb, D.M. (1997) A high-density soybean genetic map based on AFLP markers. Crop Sci. 37: 537-543.

Kennard, W.C., Poetter, K., Dijkhuizen, A., Meglic, V., Staub, J.E. and Havey, M.J. (1994) Linkages among RFLP, RAPD, isozyme, disease-resistance, and morphological markers in narrow and wide crosses of cucumber. Theor. Appl. Genet. 89: 42-48.

Kesseli, R.V., Paran, I. and Michelmore, R.W. (1990) Genetic linkage map of lettuce (*Lactuca sativa*, 2N = 18). In: O'Brien, S.J. (ed.), Genetic Maps: Locus Maps of Complex Genomes, pp. 6.100-6.102. Cold Spring Harbor Laboratory Press, Cold Spring Harbor, NY.

Kesseli, R.V., Paran, I., Ochoa, O., Wang, W.-C. and Michelmore, R.W. (1993) Linkage map of lettuce (*Lactuca sativa*) (2N = 18). In: O'Brien, S.J. (ed.), Genetic Maps - Locus Maps of Complex Genomes, pp. 6.229-6.233. Cold Spring Harbor Laboratory Press, Cold Spring Harbor, NY.

Kesseli, R.V., Paran, I. and Michelmore, R.W. (1994) Analysis of a detailed genetic linkage map of *Lactuca sativa* (Lettuce) constructed from RFLP and RAPD markers. Genetics 136: 1435-1446.

Kianian, S.F. and Quiros, C.F. (1992) Generation of a *Brassica oleracea* composite RFLP map: linkage arrangements among various populations and evolutionary implications. Theor. Appl. Genet. 84: 544-554.

Kidwell, K.K., Echt, C.S., Osborn, T.C. and McCoy, T.J. (1993) RFLP and RAPD linkage map of alfalfa (*Medicago sativa* L.). In: O'Brien, S.J. (ed.), Genetic Maps: Locus Maps of Complex Genomes, pp. 6.84-6.85. Cold Spring Harbor Laboratory Press, Cold Spring Harbor, NY.

Kijas, J.M.H., Thomas, M.R., Fowler, J.C.S. and Roose, M.L. (1997) Integration of trinucleotide microsatellites into a linkage map of *Citrus*. Theor. Appl. Genet. 94: 701-706.

Kiss, G.B., Csanadi, G., Kalman, K., Kalo, P. and Okresz, L. (1993) Construction of a basic genetic map for alfalfa using RFLP, RAPD, isozyme and morphological markers. Mol. Gen. Genet. 238: 129-137.

Kiss, G.B., Kalo, P., Felfoldi, K., Kiss, P. and Endre, G. (1997) Genetic linkage map of alfalfa (*Medicago sativa*) and its use to map seed protein genes as well as genes involved in leaf morphogenesis and symbiotic nitrogen fixatio. In: Biological fixation of nitrogen for ecology and sustainable agriculture, pp. 279-282. Springer, Berlin.

Kleinhofs, A., Kilian, A., Saghai Maroof, M.A., Biyashev, R.M., Hayes, P., Chen, F.Q., Lapitan, N., Fenwick, A., Blake, T.K., Kanazin, V., Ananiev, E., Dahleen, L., Kudrna, D., Bollinger, J., Knapp, S.J., Liu, B., Sorrells, M., Heun, M., Franckowiak, J.D., Hoffman, D., Skadsen, R. and Steffenson, B.J. (1993) A molecular, isozyme and morphological map of barley (*Hordeum vulgare*) genome. Theor. Appl. Genet. 86: 705-712.

Korzun, V., Malyshev, S., Kartel, N., Westermann, T., Weber, W.E. and Borner, A. (1998) A genetic linkage map of rye (*Secale cereale* L.). Theor. Appl. Genet. 96: 203-208.

Kreike, C.M. and Stiekema, W.J. (1997) Reduced recombination and distorted segregation in a *Solanum tuberosum* (2x) x *S. spegazzinii* (2x) hybrid. Genome 40: 180-187.

Kubisiak, T.L., Hebard, F.V., Nelson, C.D., Zhang, J., Bernatzky, R., Huang, H., Anagnostakis, S.L. and Doudrick, R.L. (1997) Molecular mapping of resistance to blight in an interspecific cross in the genus *Castanea*. Phytopathology 87: 751-759.

Kurata, N., Nagamura, Y., Yamamoto, K., Harushima, Y., Sue, N., Wu, J., Antonio, B.A., Shomura, A., Shimizu, Lin, S.-Y., Inoue, T., Fukuda, A., Shimano, T., Kuboki, Y., Toyama, T., Miyamoto, Y., Kirihara, T., Hayasaka, K., Miyao, A., Monna, L., Zhong, H.S., Tamura, Y., Wang, Z.-X., Momma, T., Umehara, Y., Yano, M., Sasaki, T. and Minobe, Y. (1994) A 300 kilobase interval genetic map of rice including 883 expressed sequences. Nature Genetics 8: 365-372.

Lagercrantz, U. and Lydiate, D.J. (1995) RFLP mapping in *Brassica nigra* indicates differing recombination rates in male and female meioses. Genome 38: 255-264.

Lagudah, E.S., Appels, R., Brown, A.H.D. and McNeil, D. (1991) The molecular-genetic analysis of *Triticum tauschii*, the D-genome donor to hexaploid wheat. Genome 34: 375-386.

Landry, B.S., Kesseli, R.V., Farrara, B. and Michelmore, R.W. (1987) A genetic map of lettuce (*Lactuca sativa* L.) with restriction fragment length polymorphism, isozyme, disease resistance and morphological markers. Genetics 116: 331-337.

Landry, B.S., Hubert, N., Etoh, T., Harada, J.J. and Lincoln, S.E. (1991) A genetic map for *Brassica napus* based on restriction fragment length polymorphisms detected with expressed DNA sequences. Genome 34: 543-552.

Landry, B.S., Hubert, N., Crete, R., Chang, M.S., Lincoln, S.E. and Etoh, T. (1992) A genetic map for *Brassica oleracea* based on RFLP markers detected with expressed DNA sequences and mapping of resistance genes to race 2 of *Plasmodiophora brassicae* (Woronin). Genome 35: 409-420.

Langridge, P., Karakousis, A., Collins, N., Kretschmer, J. and Manning, S. (1995) A consensus linkage map of barley. Molec. Breed. 1: 389-395.

Lark, K.G., Weisemann, J.M., Matthews, B.F., Palmer, R., Chase, K. and Macalma, T. (1993) A genetic map of soybean (*Glycine max* L.) using an intraspecific cross of two cultivars: 'Minosy' and 'Noir 1'. Theor. Appl. Genet. 86: 901-906.

Lebreton, C., Lazic-Jancic, V., Steed, A., S, P. and Quarrie, S.A. (1995) Identification of QTL for drought responses in maize and their use in testing causal relationships between traits. J. Exp. Bot. 46: 853-865.

Lefebvre, V., Palloix, A., Caranta, C. and Pochard, E. (1995) Construction of an intraspecific integrated linkage map of pepper using molecular markers and doubled-haploid progenies. Genome 38: 112-121.

Lewis, M.E. and Sink, K.C. (1996) RFLP linkage map of asparagus. Genome 39: 622-627.

Liu, Y.-G. and Tsunewaki, K. (1991) Restriction fragment length polymorphism (RFLP) analysis in wheat II. Linkage maps of the RFLP sites in common wheat. Jpn. J. Genet. 66: 617-633.

Liu, Z. and Furnier, G.R. (1993) Inheritance and linkage of allozymes and restriction fragment length polymorphisms in trembling aspen. J. Hered. 84: 419-424.

Liu, C.J., Witcombe, J.R., Pittaway, T.S., Nash, M., Hash, C.T., Busso, C.S. and Gale, M.D. (1994) An RFLP-based genetic map of pearl millet (*Pennisetum glaucum*). Theor. Appl. Genet. 89: 481-487.

Liu, Z.-W., R.M., B. and Saghai Marrof, M.A. (1996) Development of simple sequence repeat DNA markers and their integration into a barley linkage map. Theor. Appl. Genet. 93: 869-876.

Loarce, Y., Hueros, G. and Ferrer, E. (1996) A molecular linkage map of rye. Theor. Appl. Genet. 93: 1112-1118.

Lodhi, M.A., Daly, M.J., Ye, G.-N., Weeden, N.F. and Reisch, B.I. (1995) A molecular marker based linkage map of *Vitis*. Genome 38: 786-794.

Mansur, L.M., Orf, J.H., Chase, K., Jarvik, T., Cregan, P.B. and Lark, K.G. (1996) Genetic mapping of agronomic traits using recombinant inbred lines of soybean. Crop Sci. 36: 1327-1336.

Marino, C.L., Nelson, J.C., Lu, Y.H., Sorrells, M.E., Leroy, P., Tuleen, N.A., Lopes, C.R. and Hart, G.E. (1996) Molecular genetic maps of the group 6 chromosomes of hexaploid wheat (*Triticum aestivum* L. em. Thell.). Genome 39: 359-366.

Mayes, S., Jack, P.L., Marshall, D.F. and Corley, R.H.V. (1997) Construction of a RFLP genetic linkage map for oil palm (*Elaeis guineensis* Jacq. Genome 40: 116-122.

McCouch, S.R., Kochert, G., Yu, Z.H., Wang, Z.Y., Khush, G.S., Coffman, W.R. and Tanksley, S.D. (1988) Molecular mapping of rice chromosomes. Theor. Appl. Genet. 76: 815-829.

McCouch, S.R., Chen, X., Panaud, O., Temnykh, S., Xu, Y., Gu Cho, Y., Huang, N., Ishii, T. and Blair, M. (1997) Microsatellite marker development, mapping and applications in rice genetics and breeding. Plant Mol. Biol. 35: 89--99.

McGrath, J.M. and Quiros, C.F. (1991) Inheritance of isozyme and RFLP markers in *Brassica campestris* and comparison with B. oleracea. Theor. Appl. Genet. 82: 668-673.

Melake Berhan, A., Hulbert, S.H., Butler, L.G. and Bennetzen, J.L. (1993) Structure and evolution of the genomes of *Sorghum bicolor* and *Zea mays*. Theor. Appl. Genet. 86: 598-604.

Menancio-Hautea, D., Fatokun, C.A., Kumar, L., Danesh, D. and Young, N.D. (1993a) Comparative genome analysis of mungbean (*Vigna radiata* L. Wilczek) and cowpea (*V. unguiculata* L. Walpers) using RFLP mapping data. Theor. Appl. Genet. 86: 797-810.

Menancio-Hautea, D., Kumar, L., Danesh, D. and Young, N.D. (1993b) A genome map for mungbean (*Vigna radiata* L. Wilczek) based on DNA genetic markers (2N = 2X = 22). In: O'Brien, S.J. (ed.), Genetic Maps: Locus Maps of Complex Genomes, pp. 6.259-6.261. Cold Spring Harbor Laboratory Press, Cold Spring Harbor, NY.

Menendez, C.M., Hall, A.E. and Gepts, P. (1997) A genetic linkage map of cowpea (*Vigna unguiculata*) developed from a cross between two inbred, domesticated lines. Theor. Appl. Genet. 95: 1210-1217.

Morgante, M., Rafalski, A., Biddle, P., Tingey, S. and Olivieri, A.M. (1994) Genetic mapping and variability of seven soybean simple sequence repeat loci. Genome 37: 763-769.

Mudge, J., Anderson, W.R., Kehrer, R.L. and Fairbanks, D.J. (1996) A RAPD genetic map of *Saccharum officinarum*. Crop Sci 36: 1362-1366.

Mukai, Y., Suyama, Y., Tsumura, Y., Kawahara, T., Yoshimaru, H., Kondo, T., Tomaru, N., Kuramoto, N. and Murai, M. (1995) A linkage map for sugi (*Cryptomeria japonica*) based on RFLP, RAPD, and isozyme loci. Theor. Appl. Genet. 90: 835-840.

Nelson, C.D., Nance, W.L. and Doudrick, R.L. (1993) A partial genetic linkage map of slash pine (*Pinus elliottii* Engelm. var. elliottii) based on random amplified polymorphic DNAs. Theor. Appl. Genet. 87: 145-151.

Nelson, C.D., Kubisiak, T.L., Stine, M. and Nance, W.L. (1994) A genetic linkage map of longleaf pine (*Pinus palustris* Mill.) based on random amplified polymorphic DNAs. J. Hered. 85: 433-439.

Nelson, J.C., Deynze, A.E.V., Autrique, E., Sorrells, M.E., Lu, Y.H., Merlino, M., Atkinson, M. and Leroy, P. (1995a) Molecular mapping of wheat. Homoeologous group 2. Genome 38: 516-524.

Nelson, J.C., Deynze, A.E.V., Autrique, E., Sorrells, M.E., Lu, Y.H., Negre, S., Bernard, M. and Leroy, P. (1995b) Molecular mapping of wheat. Homoeologous group 3. Genome 38: 525-533.

Nelson, J.C., Sorrells, M.E., Deynze, A.E.V., Lu, Y.H., Atkinson, M., Bernard, M., Leroy, P., Faris, J.D. and Anderson, J.A. (1995c) Molecular mapping of wheat: major genes and rearrangements in homoeologous groups 4, 5, and 7. Genetics 141: 721-731.

Nilsson, N.O., Hallden, C., Hansen, M., Hjerdin, A. and Sall, T. (1997) Comparing the distribution of RAPD and RFLP markers in a high density linkage map of sugar beet. Genome 40: 644-651.

Nodari, R.O., Tsai, S.M., Gilbertson, R.L. and Gepts, P. (1993) Towards an integrated linkage map of common bean 2. Developmetn of an RFLP-based linkage map. Theor. Appl. Genet. 85: 513-520.

Nozaki, T., Kumazaki, A., Koba, T., Ishikawa, K. and Ikehashi, H. (1997) Linkage analysis among loci for RAPDs, isozymes and some agronomic traits in *Brassica campestris* L. Euphytica 95: 115-123.

O'Donoughue, L.S., Wang, Z., Roder, M. and Kneen, B. (1992) An RFLP-based linkage map of oats based on a cross between two diploid taxa (*Avena atlantica* x *A. hirtula*). Genome 35: 765-771.

O'Donoughue, L.S., Kianian, S.F., Rayapati, P.J., Penner, G.A., Sorrells, M.E., Tanksley, S.D., Phillips, R.L., Rines, H.W., Lee, M., Fedak, G., Molnar, S.J., Hoffman, D., Salas, C.A., Wu, B., Autrique, E. and Van Deynze, A. (1995) A molecular linkage map of cultivated oat. Genome 38: 368-380.

Paran, I., Goldman, I., Tanksley, S.D. and Zamir, D. (1995) Recombinant inbred lines for genetic mapping in tomato. Theor. Appl. Genet. 90: 542-548.

Parkin, I.A.P., Sharpe, A.G., Keith, D.J. and Lydiate, D.J. (1995) Identification of the A and C genomes of amphidiploid *Brassica napus* (oilseed rape). Genome 38: 1122-1131.

Paterson, A.H. (1996) Genome Mapping in Plants. Austin, TX, R.G. Landes Company, Academic Press, Inc.

Pe, M.E., Gianfranceschi, L., Taramino, G., Tarchini, R., Angelini, P., Dani, M. and Binelli, G.

(1993) Mapping quantitative trait loci (QTLs) for resistance to *Gibberella zeae* infection in maize. Mol. Gen. Genet. 241: 11-16.

Pereira, M.G., Lee, M., Bramel-Cox, P., Woodman, W., Doebley, J. and Whitkus, R. (1994) Construction of an RFLP map in sorghum and comparative mapping in maize. Genome 37: 236-243.

Philipp, U., Wehling, P. and Wricke, G. (1994) A linkage map of rye. Theor. Appl. Genet. 88: 243-248.

Phillips, R.L. and Vasil, I.K., Eds. (1994) DNA-Based Markers in Plants. Advances in Cellular and Molecular Biology of Plants. Kluwer Academic Publishers, Dordrecht.

Pillen, K., Steinrucken, G., Wricke, G., Herrmann, R.G. and Jung, C. (1992) A linkage map of sugar beet (*Beta vulgaris* L.). Theor. Appl. Genet. 84: 129-135.

Pillen, K., Steinrucken, G., Herrmann, R.G. and Jung, C. (1993) An extended linkage map of sugar beet (*Beta vulgaris* L.) including nine putative lethal genes and the restorer gene x. Plant Breed. 111: 265-272.

Plomion, C., O'Malley, D.M. and Durel, C.E. (1995) Genomic analysis in maritime pine (*Pinus pinaster*).Comparison of two RAPD maps using selfed and open-pollinated seeds of the same individual. Theor. Appl. Genet. 90: 1028-1034.

Prince, J.P., Pochard, E. and Tanksley, S.D. (1993) Construction of a molecular linkage map of pepper and a comparison of synteny with tomato. Genome 36: 404-416.

Qi, X., Stam, P. and Lindhout, P. (1996) Comparison and integration of four barley genetic maps. Genome 39: 379-394.

Qu, L. and Hancock, J.F. (1997) Randomly amplified polymorphic DNA-(RAPD-) based genetic linkage map of blueberry derived from an interspecific cross between diploid *Vaccinium darrowi* and tetraploid *V. corymbosum*. J. Amer. Soc. Hort. Sci. 122: 69-73.

Rafalski, A. and Tingey, S. (1993) RFLP map of soybean (*Glycine max*) (2N = 40). In: O'Brien, S.J. (ed.), Genetic Maps: Locus Maps of Complex Genomes, pp. 6.149-6.156. Cold Spring Harbor Laboratory Press, Cold Spring Harbor, NY.

Ragab, R.A., Dronavalli, S., Maroof Saghai, M.A. and Yu, Y.G. (1994) Construction of a sorghum RFLP linkage map using sorghum and maize DNA probes. Genome 37: 590-594.

Ragot, M., Sisco, P.H., Hoisington, D.A. and Stuber, C.W. (1995) Molecular-marker-mediated characterization of favorable exotic alleles at quantitative trait loci in maize. Crop Sci. 35: 1306-1315.

Rajapakse, S., Belthoff, L.E., He, G., Estager, A.E., Scorza, R., Verde, I., Ballard, R.E., Baird, W.V., Callahan, A., Monet, R. and Abbott, A.G. (1995) Genetic linkage mapping in peach using morphological, RFLP and RAPD markers. Theor. Appl. Genet. 90: 503-510.

Rayapati, P.J., Gregory, J.W., Lee, M. and Wise, R.P. (1994) A linkage map of diploid *Avena* based on RFLP loci and a locus conferring resistance to nine isolates of *Puccinia coronata* var. 'avenae'. Theor. Appl. Genet. 89: 831-837.

Reinisch, A.J., Dong, J.-M., Brubaker, C.L., Stelly, D.M., Wendel, J.F. and Paterson, A.H. (1994) A detailed RFLP map of cotton, *Gossypium hirsutum* x *Gossypium barbadense*: chromosome organization and evolution in a disomic polyploid genome. Genetics 138: 829-847.

Restivo, F.M., Tassi, F., Biffi, R., Falavigna, A., Caporali, E., Carboni, A., Doldi, M.L., Spada, A. and Marziani, G.P. (1995) Linkage arrrangement of RFLP loci in progenies from crosses between doubled haploid *Asparagus officinalis* L. clones. Theor. Appl. Genet. 90: 124-128.

Ribaut, J.M., Hoisington, D.A., Deutsch, J.A., Jiang, C. and Gonzalez-de-Leon, D. (1996) Identification of quantitative trait loci under drought conditions in tropical maize. Flowering parameters and the anthesis-silking interval. Theor. Appl. Genet. 92: 905-914.

Rieseberg, L.H., Choi, H., Chan, R. and Spore, C. (1993) Genomic map of a diploid hybrid species. Heredity 70: 285-293.

Rivard, S.R., Cappadocia, M. and Landry, B.S. (1996) A comparison of RFLP maps based on anther culture derived, selfed, and hybrid progenies of *Solanum chacoense*. Genome 39: 611-621.

Roupe van der Voort, J.N.A.M., van Zandvoort, P., van Eck, H.J., Folkertsma, R.T., Hutten, R.C.B., Draaistra, J., Gommers, F.J., Jacobsen, E., Helder, J. and Bakker, J. (1997) Use of allele specificity of comigrating AFLP markers to align genetic maps from different potato genotypes. Mol. Gen. Genet. 255: 438-447.

Rowland, L.J. and Levi, A. (1994) RAPD-based genetic linkage map of blueberry derived from a cross between diploid species (*Vaccinium darrowi* and *V. elliottii*). Theor. Appl. Genet. 87: 863-868.

Saiki, R.K., Gelfand, D.H., Stoffels, S., Scharf, S., Higuchi, R.H., Horn, G.T., Mullis, K.B. and Erlich, H.A. (1988) Primer-directed enzymatic amplification of DNA with thermostable DNA polymerase. Science 239: 487-491.

Saito, A., Yano, M., Kishimoto, N., Nakagahra, M., Yoshimura, A., Saito, K., Kuhara, S., Ukai, Y., Kawase, M., Nogamine, T., Yoshimura, S., Ideta, O., Ohsawa, R., Hayano, Y., Iwata, N. and Sugiura, M. (1991) Linkage map of restriction fragment length polymorphism loci in rice. Jpn. J. Breed. 41: 665-670.

Schon, C.C., Lee, M., Melchinger, A.E., Guthrie, W.D. and Woodman, W.L. (1993) Mapping and characterization of quantitative trait loci affecting resistance against second generation European corn borer in maize with the aid of RFLPs. Heredity 70: 648-659.

Schon, C.C., Melchinger, A.E., Boppenmaier, J., Brunklaus-Jung, E., Herrmann, R.G. and Seitzer, J.F. (1994) RFLP mapping in maize: Quantitative trait loci affecting testcross performance of elite European flint lines. Crop Sci. 34: 378-389.

Schondelmaier, J., Steinrucken, G. and Jung, C. (1996) Integration of AFLP markers into a linkage map of sugar beet (*Beta vulgaris* L.). Plant Breed. 115: 231-237.

Senft, P. and Wricke, G. (1996) Short Communication-An extended genetic map of rye (*Secale cereale* L.). Plant Breed. 15: 508-510.

Senior, M.L. and Heun, M. (1993) Mapping maize microsatellites and polymerase chain reaction confirmation of the targeted repeats using a CT primer. Genome 36: 884-889.

Senior, M.L., Chin, E.C.L., Lee, M., Smith, J.S.C. and Stuber, C.W. (1996) Simple sequence repeat markers developed from maize sequences found in the GENBANK database: Map construction. Crop Sci. 36: 1676-1683.

Sherman, J.D., Fenwick, A.L., Namuth, D.M. and Lapitan, N.L.V. (1995) A barley RFLP map: alignment of three barley maps and comparisons to Gramineae species. Theor. Appl. Genet. 91: 681-690.

Shoemaker, J., Zaitlin, D., Horn, J., DeMars, S., Kirschman, J. and Pitas, J. (1992) A comparison of three Agrigenetics maize RFLP linkage maps. Maize Genet. Newsl. 66: 65-69.

Shoemaker, R.C. and Olson, T.C. (1993) Molecular linkage map of soybean (*Glycine max* L. Merr.) (2N = 40). In: O'Brien, S.J. (ed.), Genetic Maps: Locus Maps of Complex Genomes, pp. 6.131-6.138. Cold Spring Harbor Laboratory Press, Cold Spring Harbor, NY.

Shoemaker, R.C. and Specht, J.E. (1995) Integration of the soybean molecular and classical genetic linkage groups. Crop Sci. 35: 436-446.

Simon, C.J., Tahir, M. and Muehlbauer, F.J. (1993) Linkage map of lentil (*Lens culinaris*) (2N = 14). In: O'Brien, S.J. (ed.), Genetic Maps: Locus Maps of Complex Genomes, pp. 6.97-6.100. Cold Spring Harbor Laboratory Press, Cold Spring Harbor, NY.

Simon, C.J. and Muehlbauer, F.J. (1997) Construction of a chickpea linkage map and its comparison with maps of pea and lentil. J. Hered. 88: 115-119.

Slocum, M.K., Figdore, S.S., Kennard, W., Suzuki, J.Y. and Osborn, T.C. (1990a) RFLP linkage map of *Brassica oleracea* 2N = 18. In: O'Brien, S.J. (ed.), Genetic Maps: Locus Maps of Complex Genomes, pp. 6.103-6.105. Cold Spring Harbor Laboratory Press, Cold Spring Harbor, NY.

Slocum, M.K., Figdore, S.S., Kennard, W., Suzuki, J.Y. and Osborn, T.C. (1990b) Linkage arrangement of restriction fragment length polymorphism loci in *Brassica oleracea*. Theor. Appl. Genet. 80: 57-64.

Slocum, M.K., Figdore, S.S., Kennard, W.C., Suzuki, J.Y. and Osborn, T.C. (1993) RFLP Linkage map of *Brassica oleracea* (2N = 18). In: O'Brien, S.J. (ed.), Genetic Maps: Locus Maps of Complex Genomes, pp. 6.91-6.93. Cold Spring Harbor Laboratory Press, Cold Spring Harbor.

Sondur, S.N., Manshardt, R.M. and Stiles, J.I. (1996) A genetic linkage map of papaya based on randomly amplified polymorphic DNA markers. Theor. Appl. Genet. 93: 547-553.

Song, K.M., Suzuki, J.Y., Slocum, M.K., Williams, P.H. and Osborn, T.C. (1991) A linkage map of *Brassica rapa* (syn. *campestris*) based on restriction fragment length polymorphism loci. Theor. Appl. Genet. 82: 296-304.

Song, K.M., Suzuki, J.Y., Slocum, M.K., Williams, P.H. and Osborn, T.C. (1993) RFLP linkage map of *Brassica rapa* (syn. campestris) (2N = 20). In: O'Brien, S.J. (ed.), Genetic Maps: Locus Maps of Complex Genomes, pp. 6.94-6.96. Cold Spring Harbor Laboratory Press, Cold Spring Harbor, NY.

Stockinger, E.J., Mulinix, C.A., Long, C.M., Brettin, T.S. and Iezzoni, A.F. (1996) A linkage map of sweet cherry based on RAPD analysis of a microspore-derived callus culture population. J. Hered. 87: 214-218.

Tadmor, Y., Azanza, F., Han, T., Rocheford, T.R. and Juvik, J.A. (1995) RFLP mapping of the *sugary enhancer*1 gene in maize. Theor. Appl. Genet. 91: 489-494.

Tanksley, S.D., Bernatzky, R., Lapitan, N.L. and Prince, J.P. (1988) Conservation of gene repertoire but not gene order in pepper and tomato. Proc. Natl. Acad. Sci., USA 85: 6419-6423.

Tanksley, S.D. and Mutschler, M.A. (1990) Linkage map of tomato (*Lycopersicon esculentum*) (2N = 24). In: O'Brien, S.J. (ed.), Genetic Maps: Locus Maps of Complex Genomes, pp. 6.3-6.15. Cold Spring Harbor Laboratory Press, Cold Spring Harbor, NY.

Tanksley, S.D., Ganal, M.W., Prince, J.P., de Vicente, M.C., Bonierbale, M.W., Broun, P., Fulton, T.M., Giovannoni, J.J., Grandillo, S., Martin, G.B., Messeguer, R., Miller, J.C., Miller, L., Paterson, A.H., Pineda, O., Roder, M.S., Wing, R.A., Wu, W. and Young, N.D. (1992) High density molecular linkage maps of the tomato and potato genomes. Genetics 132: 1141-1160.

Tanksley, S.D. (1993) Linkage map of tomato (*Lycopersicon esculentum*) (2N = 24). In: O'Brien, S.J. (ed.), Genetic Maps: Locus Maps of Complex Genomes, pp. 6.39-6.60. Cold Spring

Harbor Laboratory Press, Cold Spring Harbor, NY.

Tanksley, S.D., Fulton, T.M. and McCouch, S.R. (1993a) Linkage map of rice (*Oryza sativa*). In: O'Brien, S.J. (ed.), Genetic Maps: Locus Maps of Complex Genomes, pp. 6.61-6.81. Cold Spring Harbor Laboratory Press, Cold Spring Harbor, NY.

Tanksley, S.D., Prince, J.P. and Kyle, M.M. (1993b) Linkage map of pepper (*Capsicum annuum*) (2N = 24). In: O'Brien, S.J. (ed.), Genetic Maps: Locus Maps of Complex Genomes, pp. 6.220-6.227. Cold Spring Harbor Laboratory Press, Cold Spring Harbor, NY.

Tanksley, S.D., Grandillo, S., Fulton, T.M., Zami, D., Eshed, Y., Petiard, V., Lopez, J. and Beck-Bunn, T. (1996) Advanced backcross QTL analysis in a cross between an elite processing line of tomato and its wild relative *L. pimpinellifolium*. Theor. Appl. Genet. 92: 213-224.

Taramino, G. and Tingey, S. (1996) Simple sequence repeats for germplasm analysis and mapping in maize. Genome 39: 277-287.

Taramino, G., Tarchini, R., Ferrario, S., Lee, M. and Pe, M.E. (1997) Characterization and mapping of simple sequence repeats (SSRs) in *Sorghum bicolor*. Theor. Appl. Genet. 95: 66-72.

Tavoletti, S., Veronesi, F. and Osborn, T.C. (1996) RFLP linkage map of an alfalfa meiotic mutant based on an F_1 population. J. Hered.: 167-170.

Teutonico, R.A. and Osborn, T.C. (1994) Mapping of RFLP and qualitative trait loci in *Brassica rapa* and comparison to the linkage maps of *B. napus*, *B. oleracea*, and *Arabidopsis thaliana*. Theor. Appl. Genet. 89: 885-894.

Thormann, C.E., Romero, J., Mantet, J. and Osborn, T.C. (1996) Mapping loci controlling the concentrations of erucic and linolenic acids in seed oil of *Brassica napus* L. Theor. Appl. Genet. 93: 282-286.

Torres, A.M., Weeden, N.F. and Martin, A. (1993) Linkage among isozyme, RFLP and RAPD markers in *Vicia faba*. Theor. Appl. Genet. 85: 937-945.

Truco, M.J. and Quiros, C.F. (1994) Structure and organization of the B genome based on a linkage map in *Brassia nigra*. Theor. Appl. Genet. 89: 590-598.

Tsumura, Y., Suyama, Y., Yoshimura, K., Shirato, N. and Mukai, Y. (1997) Sequence-tagged-sites (STSs) of cDNA clones in *Cryptomeria japonica* and their evaluation as molecular markers in conifers. Theor. Appl. Genet. 94: 764-772.

Uzunova, M., Ecke, W., Weissleder, K. and Robbelen, G. (1995) Mapping the genome of rapeseed (*Brassica napus* L.). I. Construction of an RFLP linkage map and localization of QTLs for seed glucosinolate content. Theor. Appl. Genet. 90: 194-204.

Vallejos, C.E., Sakiyama, N.S. and Chase, C.D. (1992) A molecular marker-based linkage map of *Phaseolus vulgaris* L. Genetics 131: 733-740.

Van Deynze, A.E., Dubcovsky, J., Gill, K.S., Nelson, J.C., Sorrells, M.E., Dvorak, J., Gill, B.S., Lagudah, E.S., McCouch, S.R. and Appels, R. (1995) Molecular-genetic maps for group 1 chromosomes of Triticeae species and their relation to chromosomes in rice and oat. Genome 38: 45-59.

Veldboom, L.R., Lee, M. and Woodman, W.L. (1994a) Molecular marker-facilitated studies in an elite maize population: I. Linkage analysis and determination of QTL for morphological traits. Theor. Appl. Genet. 88: 7-16.

Veldboom, L.R. and Lee, M. (1994b) Molecular-marker-facilitated studies of morphological traits in maize. II. Determination of QTLs for grain yield and yield components. Theor.

Appl. Genet. 89: 451-458.

Veldboom, L.R. and Lee, M. (1996) Genetic mapping of quantitative trait loci in maize in stress and nonstress environments: I. Grain yield components. Crop Sci. 36: 1310-1319.

Verhaegen, D. and Plomion, C. (1996) Genetic mapping in *Eucalyptus urophylla* and *Eucalyptus grandis* using RAPD markers. Genome 39: 1051-1061.

Viruel, M.A., Messeguer, R., deVicente, M.C., Garcia-Mas, J., Puigdomenech, P., Vargas, F. and Arus, P. (1995) A linkage map with RFLP and isozyme markers for almond. Theor. Appl. Genet. 91: 964-971.

von Wettstein-Knowles, P. (1993) Barley (Hordeum vulgare) (2N = 14). In: O'Brien, S.J. (ed.), Genetic Maps: Locus Maps of Complex Genomes, pp. 6.110-6.120. Cold Spring Harbor Laboratory Press, Cold Spring Harbor, NY.

Wang, Z.M., Devos, K.M., Liu, C.J., Wang, R.Q. and Gale, M.D. (1998) Construction of RFLP-based maps of foxtail millet, *Setaria italica* (L.) P. Beauv. Theor. Appl. Genet. 96: 31-36.

Wanous, M., Heredia-Diaz, O., Ma, X., Goicoechea, P., Wricke, G., Ferrer, E. and Gustafson, J.P. (1997) Progress of molecular and cytological mapping in rye (*Secale cereale* L.). In: McGuire, P.E. and Qualset, C.O. (eds.), Progress in Genome Mapping of Wheat and Related Species: Joint Proc. 5th and 6th Public Workshops of the International Triticeae Mapping Initiative, 1-3 September 1995, Norwich UK and 30-31 August 1996, Sydney Australia. Report No. 18, University of California Genetics Resource Conservation Program. Davis, CA.

Warburton, M.L., Becerra-Velasquez, V.L., Goffreda, J.C. and Bliss, F.A. (1996) Utility of RAPD markers in identifying genetic linkages to genes of economic interest in peach. Theor. Appl. Genet. 93: 920-925.

Webb, D.M., Knapp, S.J. and Tagliani, L.A. (1992) Restriction fragment length polymorphism and allozyme linkage map of *Cuphea lanceolata*. Theor. Appl. Genet. 83: 528-532.

Weeden, N.F. and Wolko, B. (1990) Linkage map for the garden pea (Pisum sativum) based on molecular markers. In: O'Brien, S.J. (ed.), Genetic Maps: Locus Maps of Complex Genomes, pp. 6.106-6.112. Cold Spring Harbor Laboratory Press, Cold Spring Harbor, NY.

Weeden, N.F., Muehlbauer, F.J. and Ladizinsky, G. (1992) Extensive conservation of linkage relationships between pea and lentil genetic maps. J. Hered. 83: 123-129.

Weeden, N.F. and Hemmat, M. (1993) *Malus domestica*, Apple (2N = 34). In: O'Brien, S.J. (ed.), Genetic Maps: Locus Maps of Complex Genomes, pp. 6.35-6.38. Cold Spring Harbor Laboratory Press, Cold Spring Harbor, NY.

Weeden, N.F., Ambrose, M. and Swiecicki, W. (1993) *Pisum sativum*, Pea (2N = 14). In: O'Brien, S. (ed.), Genetic Maps: Locus Maps of Complex Genomes, pp. 6.24-6.34. Cold Spring Harbor Laboratory Press, Cold Spring Harbor, NY.

Whitkus, R., Doebley, J. and Lee, M. (1992) Comparative genome mapping of sorghum and maize. Genetics 132: 1119-1130.

Xie, D.X., Devos, K.M., Moore, G. and Gale, M.D. (1993) RFLP-based genetic maps of the homoeologous group 5 chromosomes of bread wheat (*Triticum aestivum* L.). Theor. Appl. Genet. 87: 70-74.

Xu, G.-W., Magill, C.W., Schertz, K.F. and Hart, G.E. (1994) A RFLP linkage map of *Sorghum bicolor* (L.) Moench. Theor. Appl. Genet. 89: 139-145.

Yang, X. and Quiros, C.F. (1995) Construction of a genetic linkage map in celery using DNA-based markers. Genome 38: 36-44.

List of contributors

WILLIAM D. BEAVIS, Virginia Bioinformatics Institute (0477) 1750 Kraft Drive Corporate Research Center Bldg. 10, Suite 1400 Blacksburg, VA 24061, U.S.A. <bioinfo@vt.edu>

JEFFREY L. BENNETZEN, Department of Biological Sciences, Purdue University, West Lafayette, Indiana 47907, U.S.A. <maize@bilbo.bio.purdue.edu>

SIMON T. BERRY, Advanta Biotechnology Department, SES-Europe NV/SA, Industriepark, Soldatenplein Z2 nr. 15, B-3300, Tienen, Belgium.

DINAKAR BHATTRAMAKKI, Dupont Agricultural Products, Deleware Tech. Park, Suite 200, P.O. Box 6104, Newark, Delaware, 19714, U.S.A.

J. H. BOUTON, University of Georgia, Department of Crop and Soil Science, Athens, GA 30602, U.S.A.

E. C. BRUMMER, Iowa State University, Agronomy Department, Ames, IA 50011, U.S.A.

BENJAMIN BURR, Biology Department, Brookhaven National Laboratory, Upton, NY 11973, U.S.A. <burr@bnl.gov>

SAM CARTINHOUR, Department of Genetics, Harvard Medical School and Molecular Biology, Massachusetts General Hospital, Boston, MA 02114, U.S.A.

D.-H. CHEN, Department of Plant Pathology, University of California, Davis, 1 Shield Ave, Davis; CA 95616, U.S.A.

J. MICHAEL CHERRY, Department of Genetics, Harvard Medical School and Molecular Biology, Massachusetts General Hospital, Boston, MA 02114, U.S.A.

EDWARD H. COE, Plant Genetics Research Unit, ARS-USDA, and Department of Agronomy, University of Missouri, Columbia, MO 65211, U.S.A. <ed@teosinte.agron.missouri.edu>

GEORGIA DAVIS, Plant Genetics Research Unit, ARS-USDA, and Department of Agronomy, University of Missouri, Columbia, MO 65211, U.S.A.

G. FEDAK, Agriculture and Agri-food Canada, Eastern Cereals and Oilseeds Research Center, Ottawa, Ontario, K1A 0C6, Canada.

S. L. FOX, Crop Development Centre, University of Saskatchewan, Saskatoon, Saskatchewan, S7N 5A8, Canada.

ANNE FRARY, Department of Plant Breeding and Biometry, Cornell University, ITHACA, NY 14853-1901, U.S.A.

CHRISTIANE GEBHARDT, Max-Planck-Institut für Züchtungsforschung, Carl-von-Linné-Weg, D-50829 Köln, Germany.
<gebhardt@mpiz-koeln.mpg.de>

HOWARD M. GOODMAN, Department of Genetics, Harvard Medical School and Molecular Biology, Massachusetts General Hospital, Boston, MA 02114, U.S.A.
<Goodman@frodo.mgh.harvard.edu>

ANDREAS GRANER, Institute for Plant Genetics and Crop Plant Research (IPK), Corrensstr 3, D-06466 Gatersleben, Germany

DAVID GRANT, USDA-ARS-CICGR and Department of Agronomy, Iowa State University, Ames, IA 50011, U.S.A.

S. GROH, Department of Agronomy and Plant Genetics, University of Minnesota, St. Paul, MN 55108-6026, U.S.A.

TRACY HALWARD, Department of Biology, Colorado State University, Fort Collins, CO 80523, U.S.A.

SUSAN HANLEY, Department of Genetics, Harvard Medical School and Molecular Biology, Massachusetts General Hospital, Boston, MA 02114, U.S.A.

GARY E. HART, Soil and Crop Sciences Department, Texas A&M University, College Station, TX 77843, U.S.A.
<g-hart@tamu.edu>

BRIAN HAUGE, Department of Genetics, Harvard Medical School and Molecular Biology, Massachusetts General Hospital, Boston, MA 02114, U.S.A.

MARCIA IMSANDE, Department of Agronomy, Iowa State University, Ames, IA 50011, U.S.A.

SHERRY KEMPIN, Division of Biology 156-29, California Institute of Technology, Pasadena, CA 91124, U.S.A.

S. F. KIANIAN, Department of Plant Sciences, North Dakota State University, Fargo, ND 58105, U.S.A.
<kianian@badlands.nodak.edu>

ANDRIS KLEINHOFS, Departments of Crop and Soil Sciences and Genetics and Cell Biology, Washington State University, Pullman, Washington 99164-6420, U.S.A. <andyk@wsu.edu>

STEVEN J. KNAPP, Department of Crop and Soil Science, Oregon State University, Corvallis, OR 97331, U.S.A. <Steven.J.Knapp@orst.edu>

GARY KOCHERT, University of Georgia, Department of Crop and Soil Science, Athens, GA 30602 and University of Georgia, Department of Botany, Athens, GA 30602 7271, U.S.A. <kochert@dogwood.botany.uga.edu>

MAARTEN KOORNNEEF, Department of Genetics, Agricultural University of Wageningen, Dreijenlaan 2, 6703 HA Wageningen, The Netherlands

M. LEE, Department of Agronomy, Iowa State University, Ames, IA 50011, U.S.A.

D. J. MACKILL, USDA-ARS, Department of Agronomy and Range Science, University of California, 1 Shields Ave, Davis, CA 95616, U.S.A.

S. R. McCOUCH, Department of Plant Breeding, Cornell University, Ithaca, NY 14853-1901, U.S.A. <srm4@cornell.edu>

MICHAEL D. McMULLEN, Plant Genetics Research Unit, ARS-USDA, and Department of Agronomy, University of Missouri, Columbia, MO 65211, U.S.A.

LEONARD MEDRANO, Division of Biology 156-29, California Institute of Technology, Pasadena, CA 91124, U.S.A.

ELLIOT MEYEROWITZ, Division of Biology 156-29, California Institute of Technology, Pasadena, CA 91124, U.S.A.

S. J. MOLNAR, Agriculture and Agri-food Canada, Eastern Cereals and Oilseeds Research Center, Ottawa, Ontario, K1A 0C6, Canada.

R. J. NELSON, Centro Internacional de la Papa, Lima 12, Peru.

JAMES NIENHUIS, Department of Horticulture. 1575 Linden Drive. University of Wisconsin. Madison, WI 53706, U.S.A.

L. S. O'DONOUGHUE, DNA LandMarks Inc., P.O. Box 6, St. Jean Sur Richelieu, Quebec, J3B 6Z1, Canada.

502

ANDREW H. PATERSON, Department of Crop and Soil Science, University of Georgia, Athens GA 30602, U.S.A.
<paterson@uga.edu>

RONALD L. PHILLIPS, Department of Agronomy and Plant Genetics, University of Minnesota, St. Paul, MN 55108-6026, U.S.A.
<phill005@umn.edu>

MARY L. POLACCO, Plant Genetics Research Unit, ARS-USDA, and Department of Agronomy, University of Missouri, Columbia, MO 65211, U.S.A.

CARLOS F. QUIROS, Department of Vegetable Crops, University of California, Davis, CA 95616, U.S.A.
<cfquiros@ucdavis.edu>

P. J. RAYAPATI, Department of Agronomy, Iowa State University, Ames, IA 50011, U.S.A.

ROBERT REITER, Monsanto, Agricultural Technology, Molecular Breeding Group, Ankeny, IA 50021, U.S.A.
<robert.s.reiter@monsanto.com>

OSCAR RIERA-LIZARAZU, Department of Crop and Soil Science, Oregon State University, Corvallis, OR 97331, U.S.A.
<Oscar.Riera@orst.edu>

LOREN H. RIESEBERG,Department of Biology, Indiana University, Bloomington, IN 47405, U.S.A.

H. W. RINES, Department of Agronomy and Plant Genetics, and USDA-ARS, University of Minnesota, St. Paul, MN 55108-6026, U.S.A.

ENRIQUE RITTER, NEIKER, Instituto Vasco de Investigacion y Desarrollo Agraria, Apdo 46, E-01080 Victoria, Spain.

P. C. RONALD, Department of Plant Pathology, University of California, Davis, 1 Shield Ave, Davis; CA 95616, U.S.A.
<pcronald@ucdavis.edu>

FRANCESCO SALAMINI, Max-Planck-Institut für Züchtungsforschung, Carl-von-Linné-Weg, D-50829 Köln, Germany.

SHANMUKHASWAMI S. SALIMATH, Department of Biological Sciences, Purdue University, West Lafayette, Indiana 47907, U.S.A.

RANDY C. SHOEMAKER, USDA-ARS-CICGR and Department of Agronomy, Iowa State University, Ames, IA 50011, U.S.A.
<rcsshoe@iastate.edu>

PAUL W. SKROCH, Department of Horticulture. 1575 Linden Drive. University of Wisconsin. Madison, WI 53706, U.S.A.

M. K. SLEDGE, University of Georgia, Department of Crop and Soil Science, Athens, GA 30602, U.S.A.

BRUNO W. S. SOBRAL, Virginia Bioinformatics Institute (0477) 1750 Kraft Drive Corporate Research Center Bldg. 10, Suite 1400 Blacksburg, VA 24061, U.S.A.
<sobral@vt.edu>

M. E. SORRELLS, Department of Plant Breeding and Biometry, 252 Emerson Hall, Cornell University, Ithaca, NY 14853, U.S.A.

H. THOMAS STALKER, Department of Crop Science, North Carolina State University, Raleigh, NC 27695-7629, U.S.A.

PIET STAMM, Department of Genetics, Agricultural University of Wageningen, Dreijenlaan 2, 6703 HA Wageningen, The Netherlands

CHARLES W. STUBER, U.S. Department of Agriculture, Agricultural Research Service Department of Genetics, North Carolina State University Raleigh, North Carolina 27695-7614, U.S.A.
<cstuber@ncsu.edu>

SUJATHA SUBRAMANIAN, Department of Biological Sciences, Purdue University, West Lafayette, Indiana 47907, U.S.A.

VAIDYANATHAN SUBRAMANIAN, Department of Biological Sciences, Purdue University, West Lafayette, Indiana 47907, U.S.A.

STEVEN D. TANKSLEY, Department of Plant Breeding and Biometry, Cornell University, ITHACA, NY 14853-1901, U.S.A.
<sdt4@cornell.edu>

N. TINKER, Agriculture and Agri-food Canada, Eastern Cereals and Oilseeds Research Center, Ottawa, Ontario, K1A 0C6, Canada.

M. ISABEL VALES, Department of Crop and Soil Science, Oregon State University, Corvallis, OR 97331, U.S.A.
<Oscar.Riera@orst.edu>

C. EDUARDO VALLEJOS, Department of Horticultural Sciences, and Graduate Program in Plant Molecular and Cellular Biology. 1143 Fifield Hall. University of Florida. Gainesville, FL 32611, U.S.A.
<vallejos@ufl.edu>

INDRA K. VASIL, Laboratory of Plant Cell and Molecular Biology, University of Florida, Gainesville, FL 32611-0690, U.S.A.

G.-L WANG, The Institute of Molecular Agrobiology, The National University of Singapore, 1 Research Link, NUS, Singapore, 117604.

MARK E. WAUGH, Virginia Bioinformatics Institute (0477) 1750 Kraft Drive Corporate Research Center Bldg. 10, Suite 1400 Blacksburg, VA 24061, U.S.A.
<bioinfo@vt.edu>

R. P. WISE, Corn Insects and Crop Genetics Research, USDA-ARS and Department of Plant Pathology, Iowa State University, Ames, IA 50011, U.S.A.

JICHEN XU, Department of Biological Sciences, Purdue University, West Lafayette, Indiana 47907, U.S.A.

NEVIN DALE YOUNG, Department of Plant Pathology, 495 Borlaug Hall, University of Minnesota, St. Paul, Minnesota 55108, U.S.A.
<neviny@tc.umn.ed>

Subject Index

Advances in Cellular and Molecular Biology of Plants

1. R.L. Phillips and I.K. Vasil (eds.): *DNA-Based Markers in Plants.* 1994
    ISBN 0-7923-2714-4

2. E.G. Williams, A.E. Clarke and R.B. Knox (eds.): *Genetic Control of Self-Incompatibility and Reproductive Development in Flowering Plants.* 1994 ISBN 0-7923-2574-5

3. Ch.S. Levings III and I.K. Vasil (eds.): *The Molecular Biology of Plant Mitochondria.* 1995   ISBN 0-7923-3224-5

4. B.A. Larkins and I.K. Vasil (eds.): *Cellular and Molecular Biology of Plant Seed Development.* 1997   ISBN 0-7923-4645-9

5. I.K. Vasil (ed.): *Molecular Improvement of Cereal Crops.* 1999
    ISBN 0-7923-5471-0

6. R.L. Phillips and I.K. Vasil (eds.): *DNA-Based Markers in Plants.* 2nd edition. 2001
    ISBN 0-7923-6865-7

For further information about the series and how to order please visit our Website
http: //www.wkap.nl/series.htm/cmbp

KLUWER ACADEMIC PUBLISHERS – DORDRECHT / BOSTON / LONDON